MOLECULAR TECHNIQUES *in* BIOCHEMISTRY and BIOTECHNOLOGY

MOLECULAR TECHNIQUES *in* BIOCHEMISTRY and BIOTECHNOLOGY

Siddhartha Shrivastava MSc, PhD, PDCR, IPIPR
OPPI-CSIR Young Scientist, NSF (USA) Fellow, Research Fellow
Department of Biochemistry, Institute of Medical Sciences
Banaras Hindu University, Varanasi

New Central Book Agency (P) Ltd
LONDON
DELHI KOLKATA PUNE HYDERABAD ERNAKULAM

NCBA

REGD OFFICE

8/1 Chintamoni Das Lane, Kolkata 700 009, India
email: ncbapvtltd@eth.net

OVERSEAS

NCBA (UK) Ltd, 149 Park Avenue North, Northampton, NN3 2HY, UK

EXPORT

NCBA Exports Pvt. Ltd, 212 Shahpur Jat, New Delhi 110 049
email: ncbaexp@ncbapvtltd.com

BRANCHES

208B Peacock Lane, Shahpur Jat
New Delhi 110 049
email: ncbadel@ncbapvtltd.com

Shop Nos. 3 & 4, Vinayak Towers
681/B Budhwar Peth, Appa Balwant Chowk
Pune 411 002
email: ncbapune@ncbapvtltd.com

House No. 3-1-315, 1st Floor
Nimboliadda, Kachiguda, **Hyderabad** 500 027
email: ncbahydb@ncbapvtltd.com

GSS Shopping Complex, 1st Floor
Opposite College Post Office, Convent Road
Ernakulam 682 035
email: ncbaernk@ncbapvtltd.com

Shop No. 15, Bharati Towers, Ground Floor
Forest Park, **Bhubaneswar** 751 001
email: ncbabub@ncbapvtltd.com

Molecular Techniques in Biochemistry and Biotechnology

First Published: January 2010
Reprinted: 2011
Reprinted: 2012

PUBLISHER AND TYPESETTER
New Central Book Agency (P) Ltd
8/1 Chintamoni Das Lane, Kolkata 700 009

PRINTER
New Central Book Agency (P) Ltd
Web-Offset Division, Dhulagarh, Sankrail, Howrah

COVER DESIGNER
Soumen Paul

COVER PRINTER
Liba Graphics, Kolkata

PROJECT TEAM
Prabhat Jas, Prabir Ghosh

ISBN: 978-81-7381-601-7

Price: ₹ 835.00

Contents

Preface

Molecular techniques are indispensable part of any biochemistry and biotechnology research laboratories. Students pursuing their carrier in life-sciences come across these techniques right from their undergraduate levels to their doctoral and post-doctoral levels. The present book *"Molecular Techniques in Biochemistry and Biotechnology"* intends to provide the readers with basic principles, instrumental setup and its maintenance, modifications of basic techniques and common queries on the subject. The objective of the book is to provide readers with adequate in-depth knowledge of basic molecular techniques which renders them ample confidence to use the techniques on their own and also to take on with zeal all sorts of examinations based on molecular techniques.

The book covers a vast area of molecular techniques—from basic lab maintenance to classical gel electrophoresis and from gene cloning, PCR to modern cell culture, RNAi and Microarray. As the technical developments in the field are proceeding fast the content of this book will need regular updations. The author hereby promises to regularly revise the contents of the book to provide its readers with the latest and updated cutting-edge techniques and their principles.

Varanasi

Siddhartha Shrivastava

Preface

Acknowledgements

A journey is easier when you travel together. Interdependence is certainly more valuable than independence. This book is the result of work whereby I have been accompanied and supported by many a noble-hearted people. It is a pleasant aspect that I have now the opportunity to express my gratitude to all of them.

First of all I would like to thank Mr S Sinha Roy, CEO, New Central Book Agency, Kolkata, for realizing the importance and publishing this topic. I am also indebted to Mr Prabir Ghosh, Mr Dipan Roy, Mr P Halder and other members of NCBA who monitored my work and took painstaking effort in providing me with valuable comments during production and were also available when I needed their advices.

I had the pleasure to cooperate, guide and work with several students who are also pursuing their MBBS, MD, or PhD courses at the Department of Biochemistry, Institute of Medical Science, BHU. They made me realize the requirements of the topics which need to be covered in this type of book. I am also much grateful to my ideal; Prof Debabrata Dash, my PhD supervisor & Head, Department of Biochemistry, IMS, BHU, and an ultimate standard of perfection or excellence, who taught me the basic skills of all the molecular techniques and provided me his lab for the research work. I am also thankful to Miss Deeksha, my sister-in-law and research scholar, Department of Applied Chemistry, IT, BHU, for critical reading of the manuscript.

I would like to forward a special vote of thanks for my beloved wife Dr Divya Ranjan, Asst Prof, Computer Science, BHU, who taught me the basic art of computational analysis and kept an eye on the progress of my work and was always available when I needed her valuable suggestions and advice. I gratefully acknowledge the contribution of my one-year old son Yathartha, who has granted me invaluable time everyday from his quota to complete this manuscript. Without his contribution this manuscript would have never been completed!

Last but not the least, I thank my parents (Mr U C Srivastava and Mrs Manju Srivastava) who formed part of my vision in my formative years, and also taught me the noble things that really matter in life. I am grateful to my brother Purshartha for rendering me the sense and the value of brotherhood. It is my great pleasure to express my hearty and eternal gratitude to my in-laws (Late Mr Priyaranjan Sinha and Mrs Manjula Sinha) for their encouragements and blessing during all stages, and Abhijeet (my brother-in-law) for his kindness and affection. Also I can't forget Bruno and Mitthu who are always there to relax me whenever I feel tired and depressed. I feel pride to be a unit of this family, and this humble effort of mine is dedicated to the energising companionship, help, affection, and infinite love they have bestowed upon me.

Varanasi

Siddhartha Shrivastava

CHAPTER 1

Elucidation of Basics

1.1 LABORATORY PRECAUTIONS AND STANDARD OPERATION PROCEDURES

Scientific work is conducted under a legal framework, which governs not only the health and safety of workers but also the conducting and recording of experiments for protection of intellectual property rights (IPR). Industrial laboratories work to good laboratory practice (GLP) conditions, which specify the organizational process and conditions which studies are planned, performed, monitored, recorded and repeated. GLP seeks to ensure as far as possible, the quality and integrity of data for non-clinical laboratory studies. GLP regulations were first published by the USA's Food and Drug Administration in 1976. Essential elements of GLP include personnel, management, and the responsibility of study directors, quality assurance, laboratory operations, equipment testing and standard operating procedures. Although the academic institutions such as universities work outside the limits of GLP regulations, the model is followed as closely as practicable. Following are some standard operating procedures which should be always kept in mind during work in biochemical or biotechnological laboratory.

1.1.1 Lab Standard Operations Procedure (SOP)

Personal Safety

- Always wear a buttoned lab coat or suits whose ergonomic design is critically important for efficient work and management.
- Lab Coat Laundry instructions: Wash separately from other clothes with detergent and bleach. When taking lab coat home for washing, carry separately from all other personal effects, i.e., not in your back pack.
- Long hair must be tied back. Keep hands away from hair.

- Wash hands with antibacterial soap (SWISH contains sodium lauryl sulphate (SDS) a detergent, coco diethanolamide, coco amido betaine, and copolymer of acrylamide) before leaving the lab.
- Dispose of the used gloves.
- No eating or drinking in the lab.
- Never mouth pipette. Always use a pro-pipette.
- Cover any cuts with a bandage.
- It is advisable to wear closed toed shoes.

Lab Environment

- Wash bench area before and after work with proper disinfectant and cleaning agent like BDD (Backdown Detergent Disinfectant containing nonylphenoxy polyethoxy ethanol, alkyl-aryl ammonium chloride and ethyl benzyl ammonium chlorides).This is especially important since biochemistry students use strong acids and bases and other chemicals which may be toxic. Never assume that the lab bench has been cleaned by the previous students or colleagues.
- First aid kit should always be present in lab.
- All the members of lab must know the location of exits, fire extinguisher, eye wash, full body shower, and first aid kit in the lab.
- Every member in the lab should know how to operate equipment before use. DO NOT use equipment unless you know exactly how to operate the equipment. It is advisable to take assistance from senior or demonstrator.
- Before leaving always clean your bench area. All equipments and supplies should be returned to original location.

1.1.2 Standard Disposal Practices

Treat animal tissue as a biohazard.

- All biohazard disposable containers must be labelled with a biohazard label.
- All biohazards or biomaterials must be autoclaved. Biohazards include any surface that has come in contact with bacteria or toxic tissues or any other biological sample which is toxic and harmful. The autoclave should be monitored regularly to ensure all organisms are destroyed. After autoclaving the container it is removed, tied and placed in a plastic bag before disposing and then disposed.
- Discard all non-sharp biologically contaminated items in separate covered container, this includes micro tubes, API strips, antibiotic strips, micro titration plates, disposable gloves, blood stained towels, etc.

- Any 'pointy' item must be disposed in deep welled covered bucket (not in ordinary container), this includes pipetteman tips, sticks, toothpicks, slides, Pasteur pipettes, broken glassware, brittle plastic objects, metal objects (needles or blades), etc.

Biohazard sharps disposal

Dispose of all sharps (needles, syringe tops, razors, scalpel blades) in specified container. Dispose of syringe with needle attached – do not take apart. Do not replace the needle cap before disposing (high frequency of accidents occur when replacing cap). Sharp's containers must be autoclaved before disposing. Always dispose of the syringe top in the biohazard sharps container even if not used for biologicals as it is a perceived hazard by the general public.

- If broken and need disposal pipettes are autoclaved and boxed before disposing.

Chemical hazardous material

Organic solvents must be disposed of in organic solvent container. The lab-in-charge/senior/demonstrator should instruct proper disposal methods for labs that contain hazardous materials. Never pour solvents down the sink. Use extreme care with flammable solvents. Alcohol used to flame spread rod should never be positioned within 40 cm of flame. Never put a very hot spread rod into a beaker of alcohol. The alcohol may catch fire. Some supplied chemicals may be preserved with 0.1% Na azide, handle with gloved hands. Handle caustic (acids and bases) solutions with care. Never discard an acid or a base greater than one molar down the sink. Discard in labelled glass containers. Use lots of water when discarding caustic solutions (< 1M). When handling stains or reagents, wear disposable gloves as the majority of stains or reagents contain hazardous material.

General garbage disposal

Nothing 'pointy' should be disposed in the general waste basket, e.g., disposal pipettes, tips, can top, toothpicks, sticks, etc. Nothing that has come into contact with biological material should be disposed in general waste container.

Despite even the most stringent attention to good safety practice, accidents do happen and it is very important to have qualified first aiders to deal with emergencies. Key telephone numbers for emergency services should be clearly and conveniently posted. The disposal of unused chemicals, radioactive isotopes and infected material should be subject to clear procedural guidelines and detailed records be kept of such disposals. Corrosive chemicals such as acids and alkalis must be neutralized before being disposed. All organic solvents must be stored in metal drums, with chlorinated solvents stored separately, before chemical incineration at high temperature radioactive waste in particular is subject to stringent disposal regulations that are legally enforceable. All spent or contaminated tissue or cell cultures or microbial cultures must be sterilized by autoclaving before disposal.

1.2 EQUIPMENT SAFETY

Whenever lab equipment is purchased, preference should be given to equipment that

- limits contact between the operator and hazardous material, and mechanical and electrical energy
- is corrosion-resistant, easy to decontaminate and impermeable to liquids
- has no sharp edges or burrs.

Every effort should be made to prevent equipment from becoming contaminated. To reduce the likelihood of equipment malfunction that could result in leakage, spill or unnecessary generation of aerosolized pathogens:

- Review the manufacturer's documentation. Keep for future reference.
- Use and service equipment according to the manufacturer's instructions.
- Ensure that anyone who uses a specific instrument or piece of equipment is properly trained in set-up, use and cleaning of the item.
- Decontaminate equipment before it is sent out for repairs or discarded.

The following sections outline some of the precautions and procedures to be observed with some commonly used laboratory equipment.

1.2.1 Centrifuges

Improperly used or maintained centrifuges can present significant hazards to users. Failed mechanical parts can result in release of flying objects, hazardous chemicals and biohazardous aerosols. The high speed spins generated by centrifuges can create large amounts of aerosol if a spill, leak or tube breakage occurs. To avoid contaminating your centrifuge:

- Check glass and plastic centrifuge tubes for stresslines, hairline cracks and chipped rims before use. Use unbreakable tubes whenever possible.
- Avoid filling tubes to the rim.
- Use caps or stoppers on centrifuge tubes. Avoid using light weight materials such as aluminum foil as caps.
- Use sealed centrifuge buckets (safety cups) or rotors that can be loaded and unloaded in a biological safety cabinet. Decontaminate the outside of the cups or buckets before and after centrifugation. Inspect O-rings regularly and replace if cracked or dry.
- Ensure that the centrifuge is properly balanced.
- Do not open the lid during or immediately after operation, attempt to stop a spinning rotor by hand or with an object, or interfere with the interlock safety device.
- Decant supernatants carefully and avoid vigorous shaking when resuspending.

When using high-speed or ultra centrifuges, follow the additional practices:

- Connect the vacuum pump exhaust to a trap.
- Record each run in a logbook, keep a record of speed and run time for each rotor.
- Install a HEPA filter between the centrifuge and the vacuum pump when working with biohazardous material.
- Never exceed the specified speed limitations of the rotor.

1.2.2 Electrophoresis equipment

- Ensure that electrophoresis equipment is properly grounded and has electrical interlocks. Do not bypass safety interlocks.
- Inspect electrophoresis equipment regularly for damage and potential tank leaks.
- Locate equipment away from high traffic areas, and away from wet areas such as sinks or washing apparatus.
- Display warning signs.

1.2.3 Heating baths, water baths

Heating baths keep immersed materials immersed at a constant temperature. They may be filled with a variety of materials, depending on the bath temperature required; they may contain water, mineral oil, glycerin, paraffin or silicone oils, with bath temperatures ranging up to 300°C. The following precautions are appropriate for heating baths:

- set up on a stable surface, away from flammable and combustible materials including wood and paper
- relocate only after the liquid inside has cooled
- ensure baths are equipped with redundant heat controls or automatic cut-offs that will turn off the power if the temperature exceeds a preset limit
- use with the thermostat set well below the flash point of the heating liquid in use
- equip with a thermometer to allow a visual check of the bath temperature.

The most common heating bath used in laboratories is the water bath. When using a water bath:

- clean regularly; a disinfectant, such as a phenolic detergent, can be added to the water
- avoid using sodium azide to prevent growth of micro-organisms; sodium azide forms explosive compounds with some metals
- raise the temperature to 90°C or higher for 30 minutes once a week for decontamination purposes
- unplug the unit before filling or emptying, and have the continuity-to-ground checked regularly.

1.2.4 Shakers, Blenders and Sonicators

When used with infectious agents, mixing equipment such as shakers, blenders, sonicators, grinders and homogenizers can release significant amounts of hazardous aerosols, and should be operated inside a biological safety cabinet whenever possible. Equipment such as blenders and stirrers can also produce large amounts of flammable vapours. The hazards associated with this type of equipment can be minimized by:

- selecting and purchasing equipment with safety features that minimize leaking
- selecting and purchasing mixing apparatus with non-sparking motors.
- checking integrity of gaskets, caps and bottles before using. Discard damaged items.
- allowing aerosols to settle for at least one minute before opening containers
- covering tops of blenders with a disinfectant-soaked towel during operation, when using biohazardous material
- when using a sonicator, immersing the tip deeply enough into the solution to avoid creation of aerosols
- decontaminating exposed surfaces after use.

1.2.5 Ovens and Hot Plates

Laboratory ovens are useful for baking or curing material, off-gassing, dehydrating samples and drying glassware.

- Select and purchase an oven whose design prevents contact between flammable vapours and heating elements or spark-producing components.
- Discontinue use of any oven whose backup thermostat, pilot light or temperature controller have failed.
- Avoid heating toxic materials in an oven unless it is vented outdoors (via a canopy hood, for example).
- Never use laboratory ovens for preparation of food for human consumption.
- Glassware that has been rinsed with an organic solvent should be rinsed with distilled water before it is placed in a drying oven.

1.2.6 Analytical Equipment

The following instructions for safe use of analytical equipment are general guidelines; consult the user's manual for more detailed information on the specific hazards:

- Ensure that installation, modification and repairs of analytical equipment are carried out by authorized service personnel.
- Read and understand the manufacturer's instructions before using this equipment.

- Make sure that preventive maintenance procedures are performed as required.
- Do not attempt to defeat safety interlocks.
- Wear safety glasses and lab coats (and other appropriate personal protective equipments as specified) for all procedures.

1.2.6.1 Scintillation Counters

- Use sample vials that meet the manufacturer's specifications.
- Keep counters clean and free of foreign materials.
- To avoid contaminating the counter and its accessories with radioactivity, change gloves before loading racks in the counter or using the computer keyboard. Verify on a regular basis (by wipe testing) that the equipment has not become contaminated.

1.2.6.2 Atomic absorption (AA) spectrometers

Sample preparation for atomic absorption procedures often require handling of flammable, toxic and corrosive products. Familiarize yourself with the physical, chemical and toxicological properties of these materials and follow the recommended safety precautions. Atomic absorption equipment must be adequately vented, as toxic gases, fumes and vapours are emitted during operation. Other recommendations to follow when carrying out atomic absorption analysis are:

- Wear safety glasses for mechanical protection.
- Check the integrity of the burner, drain and gas systems before use.
- Inspect the drain system regularly; empty the drain bottle frequently when running organic solvents.
- Allow the burner head to cool to room temperature before handling.
- Never leave the flame unattended. A fire extinguisher should be located nearby.
- Avoid viewing the flame or furnace during atomization unless wearing protective eyewear.
- Hollow cathode lamps are under negative pressure and should be handled with care and disposed of properly to minimize implosion risks.

1.2.6.3 Mass spectrometers (MS)

Mass spectrometry requires the handling of compressed gases and flammable and toxic chemicals. Consult MSDSs for products before using them. Specific precautions for working with the mass spectrometer include:

- Avoid contact with heated parts while the mass spectrometer is in operation.
- Verify gas, pump, exhaust and drain system tubing and connections before each use.

- Ensure that pumps are vented outside the laboratory, as pump exhaust may contain traces of the samples being analyzed, solvents and reagent gas.
- Used pump oil may also contain traces of analytes and should be handled as hazardous waste.

1.2.6.4 Gas Chromatographs (GC)

Gas chromatography requires handling compressed gases (nitrogen, hydrogen, argon, helium), and flammable and toxic chemicals. Consult product MSDSs before using such hazardous products. Specific precautions for working with gas chromatographs include:

- Perform periodic visual inspections and pressure leak tests of the sampling system plumbing, fittings and valves.
- Follow the manufacturer's instructions when installing columns. Glass or fused capillary columns are fragile – handle them with care and wear safety glasses to protect eyes from flying particles while handling, cutting or installing capillary columns.
- Turn off and allow heated areas such as the oven, inlet and detector, as well as connected hardware, to cool down before touching them.
- To avoid electrical shock, turn off the instrument and disconnect the power cord at its receptacle whenever the access panel is removed.
- Turn off the hydrogen gas supply at its source when changing columns or servicing the instrument.
- When using hydrogen as fuel (flame ionization FID and nitrogen-phosphorus detectors NPD), ensure that a column or cap is connected to the inlet fitting whenever hydrogen is supplied to the instrument to avoid build-up of explosive hydrogen gas in the oven.
- Measure hydrogen gas and air separately when determining gas flow rates.
- Perform a radioactive leak test (wipe test) on electron capture detectors (ECDs) at least every 6 months for sources of 50 MBq (1.35 mCi) or greater.
- Ensure that the exhaust from (ECDs) is vented to the outside.
- When performing split sampling, connect the split vent to an exhaust ventilation system or appropriate chemical trap if toxic materials are analyzed or hydrogen is used as the carrier gas.
- Use only helium or nitrogen gas, never hydrogen, to condition a chemical trap.

1.2.6.5 Nuclear Magnetic Resonance (NMR) Equipment

The superconducting magnet of NMR equipment produces strong magnetic and electromagnetic fields that can interfere with the function of cardiac pacemakers. Users of pacemakers and other implanted ferromagnetic medical devices are advised to consult with their physicians, the pacemaker's manual and pacemaker manufacturer before entering facilities which house NMR equipment. Precautions for work with NMR include the following:

- Post clearly visible warning signs in areas with strong magnetic fields.
- Measure stray fields with a gaussmeter, and restrict public access to areas of 5 gausses or higher.
- The strong magnetic field can suddenly pull nearby unrestrained magnetic objects into the magnet with considerable force. Keep all tools, equipment and personal items containing ferromagnetic material (e.g., steel, iron) at least 2 metres away from the magnet.
- Though not a safety issue, advise users that the magnetic field can erase magnetic media such as tapes and floppy disks, disable credit and automated teller machine (ATM) cards, and damage analog watches.
- Avoid skin contact with cryogenic (liquid) helium and nitrogen; wear a protective face mask and loose-fitting thermal gloves during dewar servicing and when handling frozen samples. Refer to Section 11, "Compressed Gases and Cryogenics".
- Ensure that ventilation is sufficient to remove the helium or nitrogen gas exhausted by the instrument.
- Avoid positioning your head over the helium and nitrogen exit tubes.
- NMR tubes are thin-walled; handle them carefully and reserve them for NMR use only.

1.2.6.6 High-pressure Liquid Chromatography (HPLC) Equipment

HPLC procedures may require handling of compressed gas (helium) and flammable and toxic chemicals. Familiarize yourself with the hazardous properties of these products, as well as recommended precautionary measures by referring to MSDSs.

- Inspect the drain system regularly; empty the waste container frequently when using organic solvents.
- Ensure that waste collection vessels are vented.
- Never use solvents with autoignition temperatures below 110°C.
- Be sure to use a heavy walled flask if you plan to use vacuum to degas the solvent.
- Never clean a flowcell by forcing solvents through a syringe: syringes under pressure can leak or rupture, resulting in sudden release of syringe contents.
- High voltage and internal moving parts are present in the pump. Switch off the electrical power and disconnect the line cord when performing routine maintenance of the pump.
- Shut down and allow the system to return to atmospheric pressure before carrying out maintenance procedures.

1.2.6.7 Liquid Chromatography (LC/MS) Equipment

LC/MS requires the handling of compressed nitrogen and flammable and toxic chemicals. Consult product MSDSs before using them. Take the following specific precautions for working with LC/MS equipment:

- Verify gas, pump exhaust and drain system tubing and connections before each use.
- Test the pressure switch for the exhaust line before each use.
- Ensure that pumps are vented outside the laboratory.

1.3 UNITS OF MEASUREMENT

The international nature of science demands standardization of nomenclature and units of measurement to simplify communication. Although, formulae and labelling of biochemicals generally use letters form Greek alphabet, the accepted convention for units of measurement is French Système International d' Unité (the SI system). Following tables give some important SI units and physical constants expressed in SI terms. Since some published work also contains non-SI units so it is important to have some appreciation of interconversion of units for interpretation. Table 1.3.3 gives some common units and their SI equivalents.

Table 1.3.1: Commonly used physical quantities and units in SI

Physical quantity	Name of SI unit	Symbol
Length	metre	m
Area	square metre	m^2
Volume	cubic metre	m^3
Time	second	s
Velocity	metres per second	ms^{-1}
Acceleration	metres per square second	ms^{-2}
Concentration	moles per cubic metre	$mol\ m^{-3}$
Density	kilograms per cubic metre	$kg\ m^{-3}$
Temperature	kelvin	K
Pressure	pascal	Pa
Electric charge	coulomb	C
Electric current	ampere	A
Electric potential difference	volt	V
Electric field strength	volts per metre	$V\ m^{-1}$
Electric resistance	ohm	Ω
Electric capacitance	farad	F
Wavelength	metre	m

Contd.

Physical quantity	Name of SI unit	Symbol
Luminous intensity	candela	cd
Force	newton	N
Energy	joule	J
Power	watt	W
Frequency	hertz	Hz
Magnetic flux density	tesla	T
Magnetic field strength	amperes per meter	Am^{-1}
Dipole moment	columb meter	Cm

Table 1.3.2: Some physical constants in SI units

Avogadro's number (molecules per mole, N)	6.0225×10^{23}
Boltzmann constant (k)	1.38×10^{-23} JK^{-1}
Dalton (atomic mass unit, Da)	1.661×10^{24}
Elementary charge (of proton) (e)	1.602×10^{-19} C
Faraday constant (F)	9.648×10^{4} C mol^{-1}
Molar or universal gas constant	8.314 J mol^{-1} K^{-1}
Molar volume of an ideal gas at standard Temperature and pressure (S.T.P.)	22.41 dm^3 mol^{-1}
Planck constant	6.626×10^{-34} Js
Velocity of light in a vacuum (c)	2.998×10^{8} m s^{-1}

Table 1.3.3: Conversion factors for changing common units to their SI equivalents

Unit	Symbol	SI equivalent
angstrom	A	10^{-10} m
inch	in	0.0254 m
ounce	oz	28.3 g
pounds	lb	0.4536 kg
centigrade degree	°C	(t °C +273) K, [(5/9) (°F–32)] °
millimetres of mercury	mmHg (torr)	133.322 Pa
atmosphere	atm	101325 Pa
bar	bar	100000 Pa
calorie	cal	4.186 J
erg	erg	10^{-7} J
electron volt	eV	1.602×10^{-19} J
curie	ci	3.7×10^{10} Bq

Conversion of non-SI and SI units of volume

Volume of liquids used in biochemical experiments is frequently very small. They are expressed in units which are non-SI and are also accepted terms of most scientific journals. Since they are non-SI, it becomes necessary to know their conversion. Table 1.3.4 gives conversion of non-SI and SI units of volume. And table 1.3.5 lists interconversion of mol, mmol and μmol in different volume to give different concentrations.

Table 1.3.4: Gives conversion of non-SI and SI units of volume

1 litre (l)	$=10^3$ ml	$=1$ dm^3	$= 10^{-3}$ m^3
1 millilitre (ml)	$=1$ ml	$=1$ cm^3	$=10^{-6}$ m^3
1 microlitre (ml)	$=10^{-3}$ ml	$=1$ mm^3	$=10^{-9}$ m^3
1 nanolitre (nl)	$=10^{-6}$ ml	$=1$ nm^3	$=10^{-12}$ m

Table 1.3.5: Interconversion of mol, mmol and μ mol

M	mM	μM
1 mol dm^{-3}	1 m mol dm^{-3}	1 μ mol dm^{-3}
1 m mol cm^{-3}	1 m mol cm^{-3}	1 n mol cm^{-3}
1 μ mol mm^{-3}	1 n mol mm^{-3}	1 p mol mm^{-3}

1.4 THE ANALYTICAL BALANCE

Knowledge of the operation of the analytical balance is essential to every experimenter in the biotechnology or biochemistry laboratory.

Analytical balances vary in capacity and sensitivity, and in their method of weighing solids and liquids. The sensitivity of the balance is defined as the minimum load that will produce a noticeable change in the scale divisions of the read out. Single-pan balances with manual operation have a capacity of 200 mg with a precision of ± 0.05 mg. Electric single-pan balances with electromagnetic force compensation can have a capacity of 160 g with precision of ± 0.1mg

The balance must be handled carefully and kept clean always. Samples are seldom weighed directly on pan; a weighing container is generally used. Containers may be aluminum or plastic pans, paper or glass weighing bottles. Some balances have automatic taring which will compensate for the weight of container into which sample is placed.

Many types of balances are available for laboratory use. Electronic single-pan, top-loading balances with microprocessor control, printer and taring memories can have capacities of 1200 g with precision of 0.005 g. These offer fast operation, high capacity, and acceptable accuracy for many laboratory purposes.

Fig. 1.1: Electronic single pan, microprocessor control analytical balance (*Courtesy:* Dr D. Dash, Deptt. of Biochemistry, IMS, BHU)

1.5 EXPRESSION OF CONCENTRATION

The three most useful expressions of concentration, *molarity, molality* and *normality*, assume that molecular weight or equivalent weight of the solute is known.

Molarity is defined as the number of moles of solute per litre of solution and is denoted by **M**. This is the most common expression for the solutions of the solids with known molecular weight. Molarity usually describes solutions of accurate concentrations; solutes are weighed on a analytical balance and solution volumes are measured in volumetric flasks. A more common concentration range for solutions in biochemistry is milimolarity, denoted by **mM**.

Molality is defined as the number of moles of solute per 1000 grams of solvent and is denoted by m. Molality can be more precise than molarity, since both solute and solvent are weighed. (A 1.00 millimolal solution contains 1.00 millimole of solute dissolved in 1000 grams of solvent.)

Normality is defined as the number of gram-equivalent weights of solute per litre of solution, and is denoted by N. this designation is useful for solutions of acids and bases. One gram-equivalent weight of acid is the quantity of the acid that can donate one mole of hydrogen ions to base; one gram-equivalent of the base is that amount which can accept one mole of hydrogen ions. For a monoprotic acid such as acetic acid, which yields one mole of hydrogen ions upon complete dissociation, the gram-equivalent weight is equal to the molecular weight.

$CH_3\text{–}COOH \rightleftharpoons CH_3\text{–}COO^- + H^+$ 1 gram-mol = 60 g

1 gram-equiv = 60 g

Fumaric acid, a diprotic acid, yields two moles of protons when reacted completely with a base.

$$HOOC{-}(CH)_2{-}COOH \rightleftharpoons {}^{-}OOC{-}(CH)_2{-}COO^- + 2H^+ \qquad \begin{aligned} 1 \text{ gram-mol} &= 126 \text{ g} \\ 1 \text{ gram-equiv} &= 63 \text{ g} \end{aligned}$$

The gram-equivalent weight of Fumaric acid is half the molecular weight.

The gram-equivalent weight of an oxidizing or a reducing agent in an oxidation-reduction (redox) reaction is the amount of compound that will transfer one mole of electrons. In the reduction of nicotinamide adenine dinuceotide (NAD^+), two moles of electrons are transferred per mole of NAD^- . Therefore, gram-equivalent weight of NAD^+ is half its molecular weight.

$$NAD^+ + 2e^- + H^- \rightleftharpoons NADH$$

Expression of concentration as **percentage of solute** is frequently for liquids and solids of undetermined molecular weights. Three types of percentage of solute designations are used: volume per volume (v/v), weight per volume (w/v), and weight per weight (w/w). For example, a solution with 5.5 ml of liquid solute in 100 ml of solution would be 5.5% (v/v) solution, and 5.5 g of solid solute in 100 g of solution is 5.5% (w/w) solution. The most commonly used is weight per volume, in which 5.5 g of solid solute is dissolved in 100 ml of solvent, is a 5.5% (w/v) solution.

1.6 MAKING DILUTIONS

Experimental conditions frequently require the dilution of an extract or a standard solution. There are several conventions for indicating the method of dilution. For example, "1-to-5 dilution" can indicate two cases. First, one part of the original solution could be diluted with four part of solvent to give a final diluted volume of five parts. This dilution is 1/5, since the concentration of the diluted solution is 1/5 of the concentration of the original solution. However, "1-to-5 dilution" could also indicate that one part of the original solution is diluted with five parts of the solvent to give a solution with 1/6 of the original concentration. The former convention appers to be more suitable and is more frequently used because the dilution factor is immediately apparent. In this case it is best to read "dilute 1:5" as "dilute one part solution with solvent to give five parts of total volume".

Serial dilution involves the systemic dilution of an original solution in fixed steps, such as 1:2, 1:4, 1:8, 1:16 or 1:10, 1:100, 1:1000. This type of dilution is used in immunology and microbiology. Serial dilutions require careful attention and accuracy of measurement because errors made in an early part of the process will be carried out to all subsequent dilutions.

1.7 STANDARD SOLUTIONS

Many experiments require the use of a standard solution containing a precisely measured amount of a pure substance. For example, in the quantitative determination of amino acids in a protein hydrolyzate, a solution of each pure amino acid is reacted with ninhydrin and the amount of colour developed is measured and used for comparison. The standard solutions are prepared by accurately weighing each anhydrous amino acid and adding an accurately measured volume of solvent. Care must be taken to control all variables, such as temperature, rate of hydration of solute, and pH. The accuracy of the prepared standard should be 1%. Sometimes standard solutions of single amino acid or single carbohydrate are required at different concentrations. These can be prepared by making a stock solution from which dilutions are made.

1.8 MICROSCOPY

1.8.1 Principles

Microscopy is necessary to evaluate the integrity of samples and to correlate structure with function. Microscopy serves two independent functions of enlargement (magnification) and improved resolution (rendering of two objects as separate entities). Light microscopes employ optical lenses to sequentially focus the image of objects, whereas electron microscope uses electromagnetic lenses. Light and electron microscopes work either in transmission or scanning mode depending on whether the light or electron beam either passes through the specimen and is diffracted or deflected by specimen surface. Polarized light microscopes detect optically active substances in cells; for example, particles of silica or asbestos in lung tissue.

Phase contrast microscopes are often used to improve image contrast of unstained material which is caused either by diffraction by the specimen or even by differences in thickness of the specimen. At their point of focus, the converging light rays shows interference, resulting in either increase or decrease in the amplitude of the resultant wave (constructive or destructive interference, respectively), which the eye detects as differences in brightness.

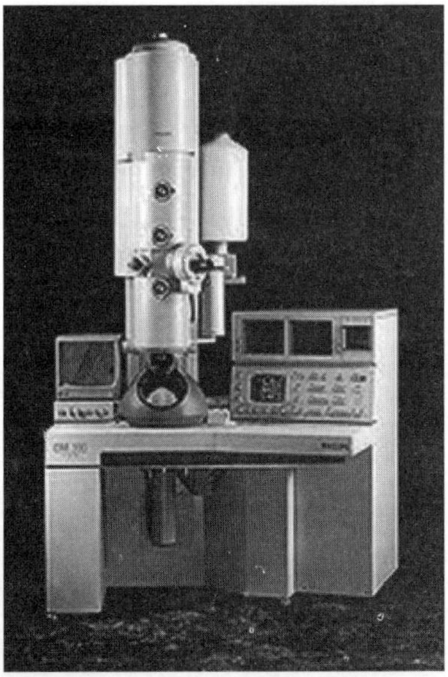

Fig. 1.2: Transmission electron microscope (*Courtesy:* Dr D. Dash, Deptt. of Biochemistry, IMS, BHU)

Confocal microscopy is an imaging technique used to increase micrograph contrast and/or to reconstruct three-dimensional images by using a spatial pinhole to eliminate out-of-focus light or flare in specimens that are thicker than the focal plane. This technique has been gaining popularity in the scientific and industrial communities. Typical applications include life sciences and semiconductor inspection.In a conventional (i.e., wide-field) fluorescence microscope, the entire specimen is flooded in light from a light source. Due to the conservation of light intensity transportation, all parts of specimen throughout the optical path will be excited and the fluorescence will be detected by a photodetector or a camera.

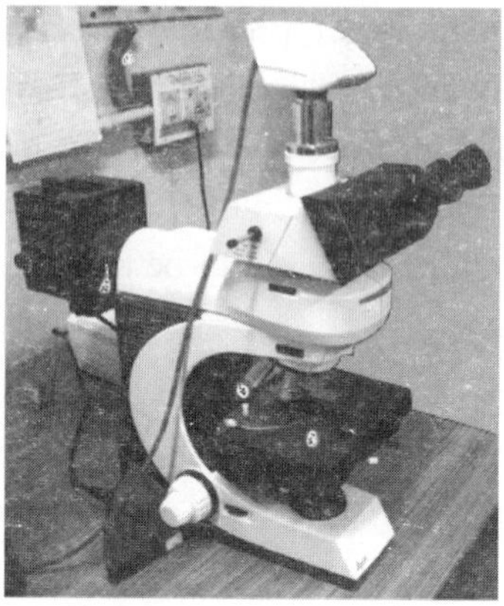

Fig. 1.3: Fluorescence microscope (*Courtesy:* Dr D. Dash, Deptt. of Biochemistry, IMS, BHU)

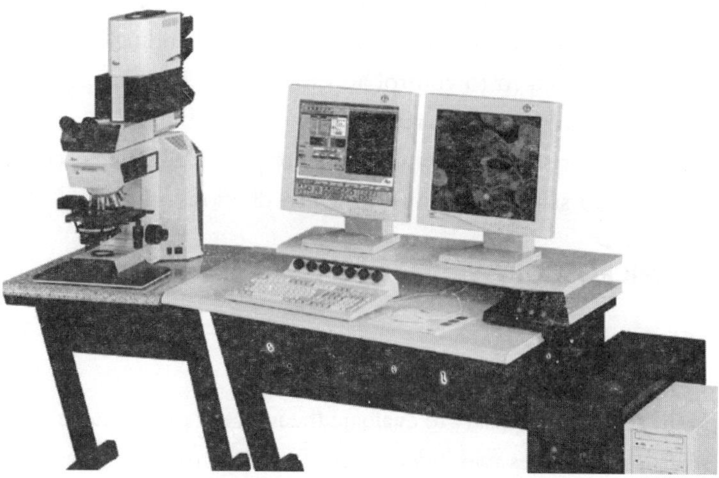

Fig. 1.4: Confocal microscope (*Courtesy:* Cytogenetics laboratory, Department of Zoology, BHU)

In contrast, a confocal microscope uses point illumination and a pinhole in an optically conjugate plane in front of the detector to eliminate out-of-focus information. Only the light within the focal plane can be detected, so the image quality is much better than that of wide-field images. As only one point is illuminated at a time in confocal microscopy, 2D or 3D imaging requires scanning over a regular raster (i.e., a rectangular pattern of parallel scanning lines) in the specimen. The thickness of the focal plane is defined mostly by the square of the numerical aperture of the objective lens, and also by the optical properties of the specimen and the ambient index of refraction

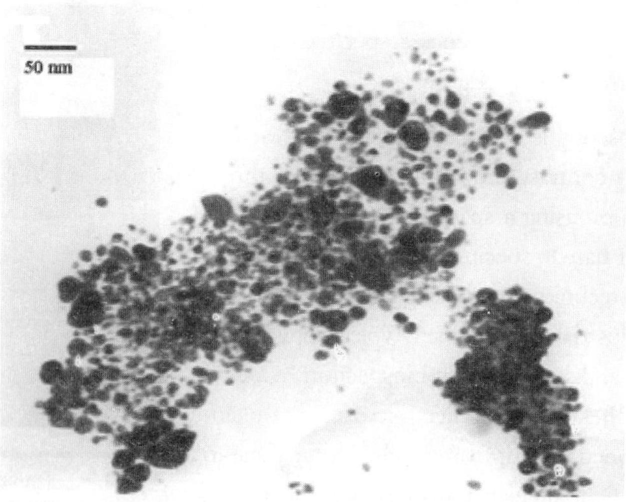

Fig. 1.5: Transmission electron micrograph of silver nanoparticles (*Courtesy:* S. Shrivastava et al. Nanotechnology-2007)

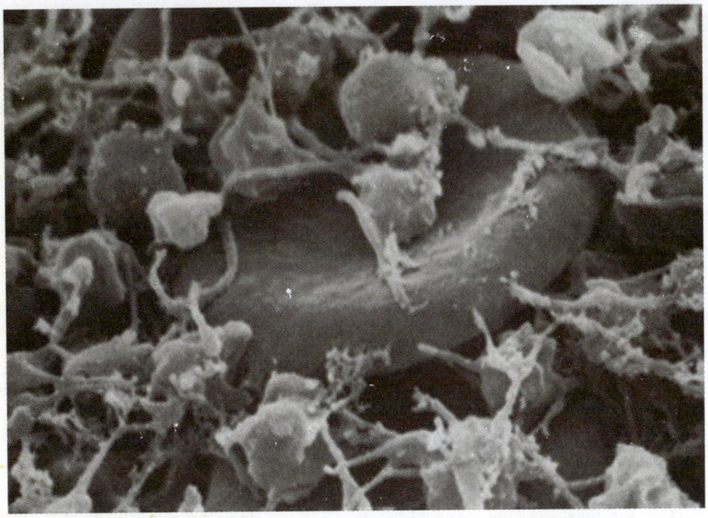

Fig. 1.6: Scanning electron micrograph of aggregated platelets and RBC

Microscopes using visible light will magnify approximately 1500 times and have a resolution limit of about 0.2 mm whereas a transmission electron microscope is capable of magnifying approximately 2,00,000 times and has a resolution limit for biological specimens of about 1 nm. This capability of TEM is largely a function of the very short wavelength of electrons accelerated under the influence of an applied electric field. (An accelerating voltage of 100 kV produces a wavelength of 4×10^{-3} nm.)

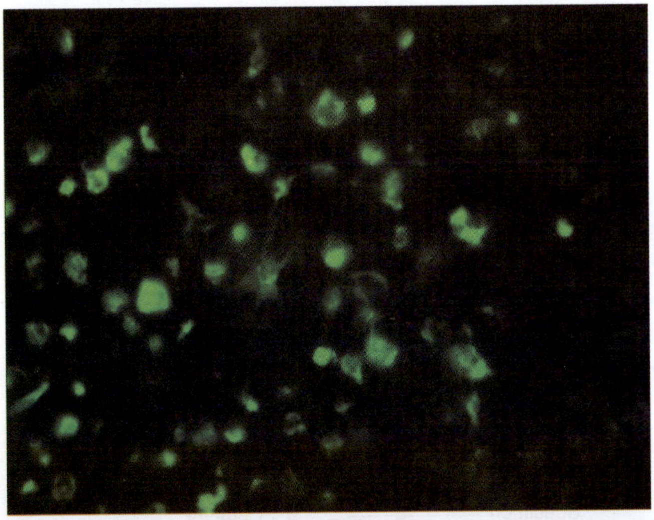

Fig. 1.7: Fluorescence microscopic image of activated platelets tagged with FITC conjugated Phalloidin (*Courtesy:* Dr D. Dash, Deptt. of Biochemistry, IMS, BHU)

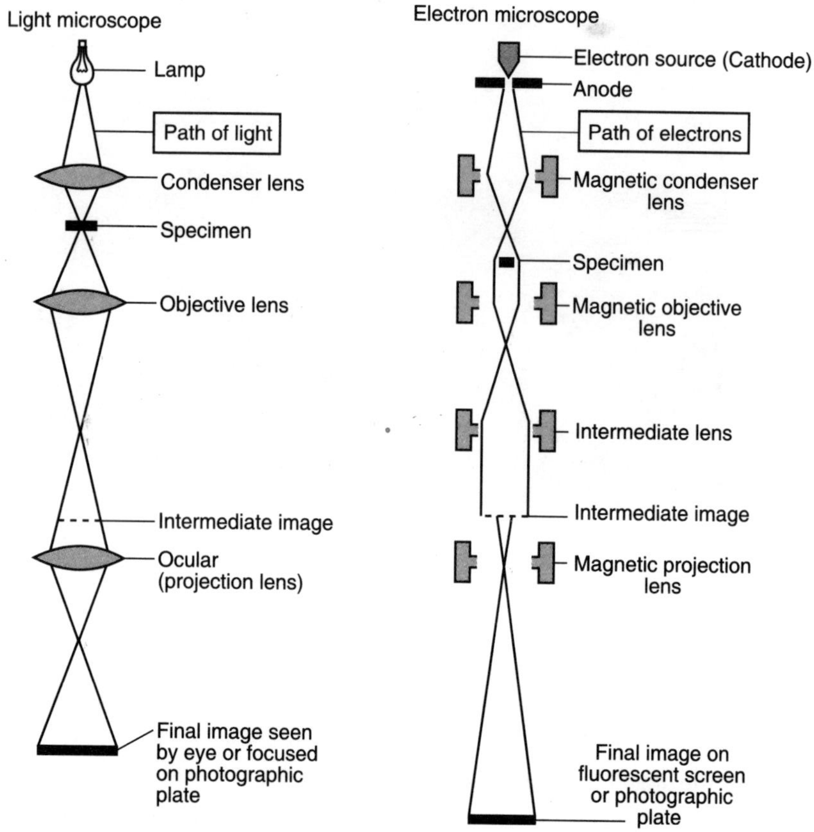

Fig. 1.8: Schematic diagrams of Light Microscope and Electron Microscope

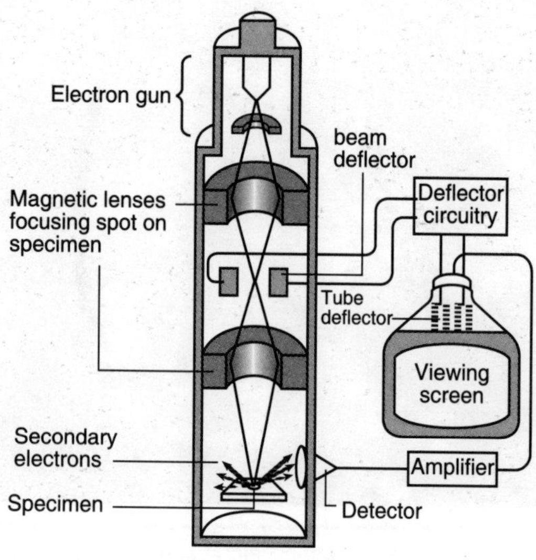

Fig. 1.9: Schematic diagram of Scanning Electron Microscope

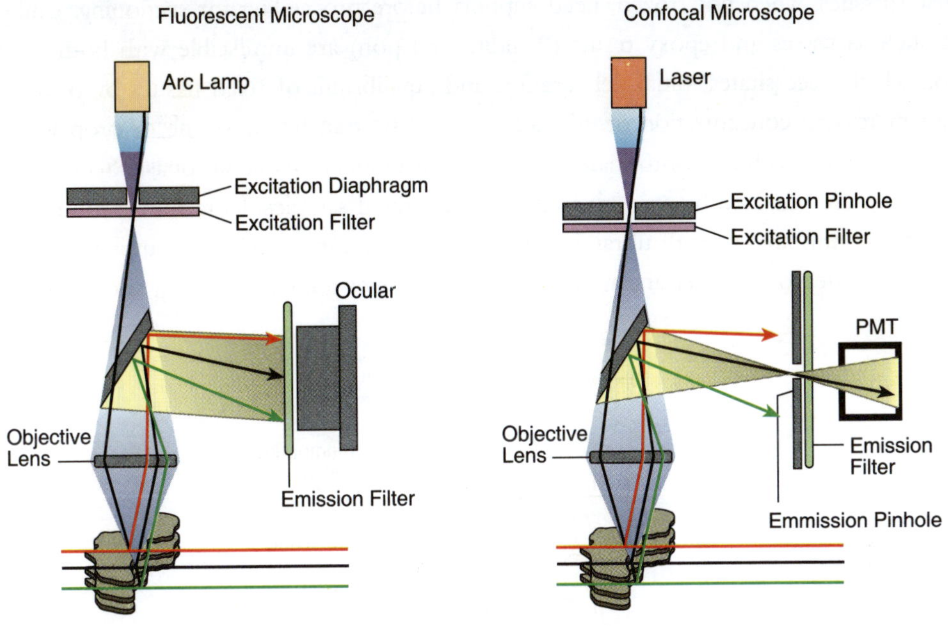

Fig. 1.10: Comparison of fluorescent and confocal microscope

Scanning electron microscopes (SEM) use a fine beam of electrons to scan back and forth across the metal coated specimen surface. The secondary electrons are generated from this surface collected by a scintillation crystal, which converts each electron impact into a flash of light. Each light flash inside the crystal is amplified by a photomultiplier and used to build up an image on a fluorescent screen. The principle application of SEM is the study of surfaces such as those of cells. The resolution limit of scanning electron microscope is about 6 nm

1.8.2 Specimen Preparation

This includes three steps:

1. Fixation
2. Sectioning, and
3. Staining.

TEM requires thin specimens, i.e., squashes, smear, hanging drops or very thin sections. For preservation of cellular integrity first sample is fixed either by rapid freezing or by chemical treatment. Generally used fixatives are formaldehyde (primarily used in light microscopy) and glutaraldehyde (for electron microscopy) is the result of formation of methylene bridges with side-chain amino groups of proteins, whilst osmium tetroxide, a common fixative in electron microscopy, cross links mainly with unsaturated fatty acid side chains. Fixed tissue is then subjected to sequential process of sectioning and staining.

Fixed tissues, when not frozen, need support before proceeding for sectioning. Embedding media such as waxes and epoxy resins (Araldite or Epon) are immiscible with both water and alcohol, which necessitates initial dehydration and equilibrium of fixed tissues by passing them through increasing concentration of ethanol followed by transfer to xylene or propylene oxide, before infiltration with an appropriate embedding medium in its liquid phase. Sections of 5-10 μm thickness are obtained by cryostat from 20 refrigerated samples. Ultra thin sections for TEM (less than 100 nm) are cut with ultramicrotomes using diamond or glass as cutting edge. Section ribbons are floated onto water and mounted on fine copper grids for staining and examination.

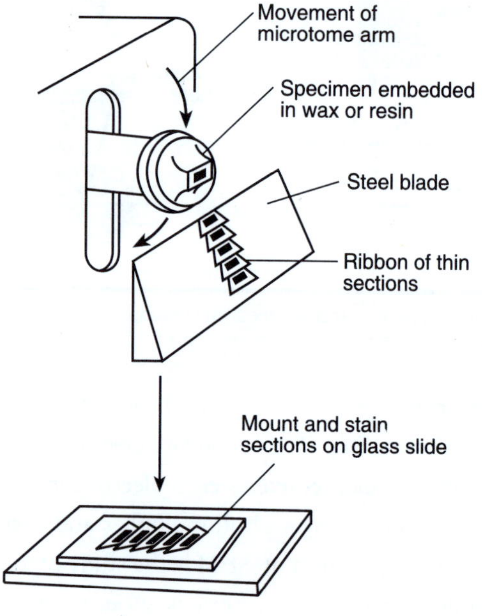

Fig. 1.11: Schematic representation of sample preparation for electron microscopy

In negative staining, a heavy metal stain, often phosphotungstic acid, is allowed to dry in a puddle around the surfaces of isolated cell particles supported on a thick carbon or plastic film. The stain molecules are deposited into surface crevices in the specimen during drying and produce ghost image in which the specimen appears light against dark background, often outlining details very clearly. The contrast can further improve by technique of shadowing.

Antibodies prepared against specific cellular proteins, which have absorbed colloidal gold are frequently used for staining in TEM as gold is electron dense.

Freeze-fracture techniques in which frozen samples are cleaved with a knife along facture planes in a membrane, utilize this shadowing techniques to produce replicas of broken cellular materials that show the membrane surface structure on and within the cell organelles.

Samples for ion probe analysis are prepared either by chemical fixation or, more successfully, by ultra rapid freezing methods, such as immersion in liquid propane or nitrogen slush , which serves to immobilize ions. Where frozen tissue is used, sections must be ultra thin and are obtained by ultracryomicrotome. Histological stains are used to produce contrast, which aids resolution. Many stains used in light microscopy depend upon the anionic or cationic behaviour of intracellular ampholytes and their efficiency to influence pH. Cytoplasmic contents are generally cationic in slightly acid pH range, implying preferential binding to anionic (acid) stains such as eosin. Chromatin and DNA at pH 6.0 are anionic and thus binds to cationic stains such as methylene blue. Contrast in material for TEM is improved by incorporating heavy metal salts into the specimens to induce a greater extent of electron absorption. This can be exploiting the binding of uranium to nucleic acids and proteins and of lead to lipids

1.9 TIPS TO REMEMBER

Light Microscopy Modifications

- Phase Contrast
- Differential Interference
- Contrast (Normarski)
- Confocal Scanning
- Fluorescence
- Dark Field (diffracted light)
- Image Enhancement.

Variations of Electron Microscopy

- Transmission (TEM)
- Scanning (SEM)
- Shadow-casting
- Freeze-fracture
- Freeze-etching
- CryoEM
- Negative Staining.

Image Enhancement

- Video cameras + computers used to enhance images
- Correct imperfections in optical systems
- Overcome limitations of human eye

- Seeing image in dim light
- Seeing small intensity differences
- Against bright background
- Does not increase actual resolution.

1.10 VIVA VOCE

Q.1. An enzyme reaction has been carried out in a volume of 10 cm^3 and calcium chloride has to be added to stop the reaction to give a final concentration of 10 mM calcium chloride. How many milligrams of calcium chloride needs to be added? Given that M_r of $CaCl_2.6H_2O = 219.08$.

Ans. 21.91 mg.

Q.2. Trypsin in powdered form has to be added to 10 cm^3 of a 1 mg cm^{-3} of bovine serum albumin in buffer (pH 8) to give an enzyme to substrate ratio of 1:500 (w/w). How much solid BSA is added?

Ans. 0.02 mg.

Q.3. List the main differences between SEM and TEM.

Q.4. What is the resolution limit for electron microscope?

Q.5. List the precautions necessary for working in biotechnological or biochemical lab.

Q.6. How are biological samples processed for electron microscopy?

Q.7. What are SI units and why are they necessary?

Q.8. How is serial dilution of any solution obtained?

Q.9. What are standard solutions? List their importance.

Q.10. How is fluorescent microscopy different from confocal microscopy?

Q.11. What is the principle of phase contrast microscopy?

Q.12. How can you check the accuracy of 1 ml micro pipette?

Q.13. Differentiate between electron microscope and confocal microscope.

Electrochemical Techniques

CHAPTER 2

All biochemical reactions take place at definite ion concentration and hence for proper proceeding and analysis of any biochemical reaction, pH of the reaction media needs to be stable. This resistance to pH fluctuations is provided by different buffers in biochemistry both *in vivo* and *in vitro*. Here, in this chapter we shall discuss various concepts of this basic requirement and different techniques employed in the process.

2.1 pH

The negative logarithm of the hydrogen ion concentration, the pH, is expressed as follows:

$$pH = -\log_{10} [H^+]$$

The pH scale is a measure of hydrogen ion concentration that eliminates dealing with large powers of 10 and compresses a large range of concentrations onto a more convenient scale, between 1 and 14. At a high concentration of H^- (10^{-1}M), the pH value is low, pH = 1. While at low concentration (10^{-12} M), the pH is high, pH = 12.

Hydrogen ions in the solution arise from the dissociation of acids.

$$HA \rightarrow H^+ + A$$

Strong acids are considered to be completely dissociated into ions in dilute solution. However, weak acids (or bases) are only partially dissociated in solution, and thus equilibrium is established between ions and the undissociated molecules.

$$HA \rightleftharpoons H^+ + A^-$$

The equlibrium constant K_a, is defined as follows:

$$K_a = [H^+] [A^-] / [HA]$$

Where HA = undissociated acid, H^+ = hydronium ion, and A^- = conjugate base. From this expression we can derive Henderson-Hasselbach equation, commonly called buffer equation, which relates pH of solution to the pK_a of the acid and the relative concentration of undissociated acid and the conjugate base forms. The Henderson-Hasselbach equation is

$$pH = pK_a + \log [\text{Conjugate base}] / [\text{acid}].$$

It is most useful in the preparation of buffers and in understanding how the concentration of the acid and conjugate base forms of a weak acid affect the pH.

2.2 BUFFERS

A buffer, by definition, resists changes in the pH of the solution. A buffer must contain the chemical species for "neutralizing" added amounts of acid or base. Generally, a buffer is a solution of a weak acid and its conjugate base (e.g., ammonia and ammonium chloride). A buffer is selected on the basis of its pK_a and its chemical nature. As shown by Henderson-Hasselbach equation, when concentrations of the conjugate base and the undissociated acid are equal, the pH of the solution equals pK_a of the buffer

$$pH = pK_a + \log 1$$

$\because \quad \log 1 = 0,$

$\therefore \quad pH = pK_a.$

When [conjugate base] = 10 × [acid], then

$$\log [\text{conjugate base}]/[\text{acid}] = \log 10 = 1,$$

$$pH = pK_a + 1$$

When [conjugate base] = 1/10 [acid], then

$$\log [\text{conjugate base}]/[\text{acid}] = \log 1/10 = -1$$

and

$$pH = pK_a - 1.$$

Thus, buffers are most effective in the range pH= $pK_a \pm 1$. Outside that range, the concentration of either the acid or the conjugate base is too small to effectively resist the effects of added hydrogen or hydroxide ion.

Once the desired pH range for an experiment has been decided, one can select a buffer on the basis of pK_a value. Since temperature affects the dissociation of some weak acids and bases and thus the pK_a value, buffer should always be pH adjusted for the temperature at which they will be used. Following table shows the effect of temperature on some important buffers.

For example, widely used buffer Tris [trid (hydroxymethy) aminomethane] shows a change in pK_a of – 0.031 for each degree change in temperature from reference temperature of 20°C(ΔpK_a/°C). Therefore, at 10°C the change in pK_a = (10°C – 20°C) (–0.031) = 0.31; the pK_a at 10°C is 8.3 + 0.31 = 8.6.

2.3 BUFFER CAPACITY AND IONIC STRENGTH

The molar strength of the two components of the buffer should be chosen to give adequate buffering capacity. The total ionic strength of the buffer may also be an important consideration; for example, in enzyme studies. The buffer strength is expressed in terms of the total concentration of conjugate base or conjugate acid. Again, let us use Tris as an example.

$$\underset{(I)}{(CH_2OH)_3\text{–}C\text{–}NH_3^+} \rightleftharpoons \underset{(II)}{(CH_2OH)_3\text{–}C\text{–}NH^+} \quad [pK_a = 8.3]$$

The pK_a for tris is 8.3 at 20°C; therefore, this substance can effectively serve as a buffer between pH 7.3 and 9.3. To make a 50 mM Tris buffer with pH = 7.8, start with either the free base (II) or the hydrochloride salt, Tris-HCol. A 50 mM solution contains 50 mmol of solute per litre of solution. Tris, the free base form, has a molecular weight of 121.1 g. the amount of solid Tris needed to make one litre of 50 mM solution is

$$50 \text{ mmol} \times 121.1 \text{ mg/mol} = 6055 \text{ mg Tris.}$$

However, pH of the solution is needed to be adjusted to 7.8 with acid. Hydrochloric acid would be the best choice because Tris-HCl is soluble and chloride ion is best compatible with most of the biological systems.

The chemical nature of the buffer is especially important in biological investigations – solubility, stability, defined interaction with other ions in the system, light absorption, and potential inhibition or stimulation of the function are all criteria to consider in buffer selection. The zwitterionic buffers do not pass through the biological membrane and do not complex cation such as Ca^{2+} and Mg^{2+}. All these buffer has tertiary amine function that dissociates in pH range 6-9.

Table 2.1: Zwitterionic Buffers

Name	pK_a	range
MES	6.15	5.8 - 6.5
ADA	6.62	6.2 - 7.2
PIPES	6.80	6.4 - 7.2
MOPS	7.20	6.5 - 7.9
HEPES	7.55	7.0 - 8.0
EPPS	8.00	7.6 - 8.6
Tricine	8.15	7.6 - 8.8
Bicine	8.35	7.8 - 8.0
CHES	9.55	9.0 - 10.1
CAPS	10.4	9.7 - 11.1

2.4 THE pH ELECTRODE

Principles: Perhaps the most convenient and accurate way of determining pH is by using a glass electrode.

The pH electrode depends on ion exchange in the hydrated layers formed on the glass electrode surface. Glass consists of a silicate network amongst which are metal ions coordinated to oxygen atom, and it is the metal ions that exchange with H^+. The glass electrode acts like a battery whose voltage depends on the H^+ activity of the solution in which it is immersed. The size of the potential (E) due to H^+ is given by the equation:

$$E = 2.303\ RT/F \log ([H^+]_i / [H^+]_o),$$

where $[H^+]_i$ and $[H^+]_o$ are the molar concentrations of H^+ inside and outside the glass electrode respectively.

In practice, $[H^+]_i$ is generally 10^{-1}, because the electrode contains 0.1 M HCl. Since pH= – log $[H^+]$, it follows that the developed potential is directly proportional to the pH of the solution outside the electrode. Glass electrodes are particularly useful because of lack of interference from the components of solution. On the whole these molecules are not readily contaminated by molecules in solution, and if other ions are present they do not cause any significant interference. However, at high pH they do respond to sodium. Inaccuracies also occur under very acid conditions.

A glass electrode consists of a thin, soft glass membrane that is situated at the end of a hard glass tube, or sometimes an epoxy body. Also present in the glass electrode is an internal reference electrode of the silver/silver chloride (Ag/AgCl) surrounded by electrolyte of 0.1 M HCl. This internal reference electrode gives rise to a steady potential. Thus the varying potential of the glass electrode can be compared with a steady potential produced by an external reference electrode such as the standard calomel electrode by joining internal and external reference electrode.

The external reference electrode can either be a separate probe or built around glass electrode giving a combination electrode. If a combination electrode is used, the level of the test solution must be high enough to cover the porous plug (liquid junction) but not as high as the level of salt bridge solution (KCl) in the external electrode because it is essential for KCl to diffuse out slowly into the test solution.

Whatever reference electrode is used the measured voltage is the result of the difference between that of the refrence and the glass electrodes. In practice, however, there are other potentials present in the system. These include so-called asymmetric potential, which is poorly understood but which is present across the glass membrane even when the H^+ concentration is the same on both sides. Also included are the potentials due to Ag/AgCl and to the liquid junction to the reference electrode, which gives the potential because the K^+ and Cl^- do not diffuse at exactly the same rate and, therefore, generate a small potential at the boundary between the sample and the KCl in the reference electrode. The measured potential for glass electrode should thus also include constants to account for the additional potential within the device. Therefore, the equation becomes:

$$E = E^* + 2.303\ RT/F \log ([H^+]_i/[H^+]_o),$$

where E* includes the standard electrode potential for glass electrode, and the constant junction potential present in the system.

At 25°C this equation becomes:

$$E = E^* + 0.059\ pH,$$

where E* now also includes a term to account for the internal H^+ concentration. As already known there is a 59 mV change for a 10-fold change in the activity of a monovalent ion; this means that a change of one pH unit produces a 59 mV change.

A pH electrode is used in conjunction with a pH meter. This records the potential due to H^+ concentration but is designed to take a little current from the circuit. A large current flow will cause changes in the ion concentration and hence changes in pH; this is prevented by having a high resistance present. The pH meter, glass electrode and reference calomel electrode are designed so that pH gives a zero potential.

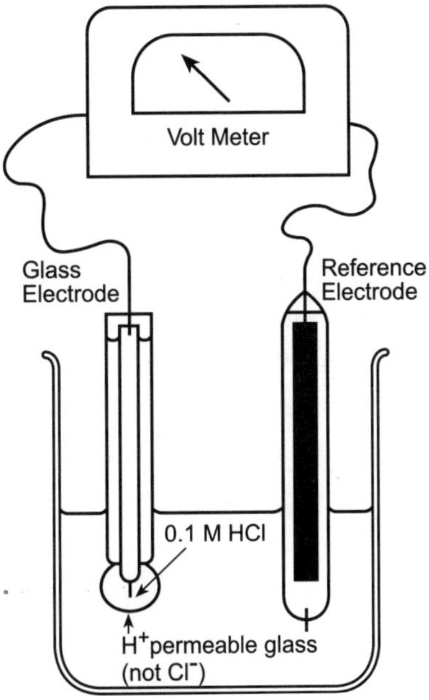

Fig. 2.1: Schematic diagram of pH meter

2.5 OPERATION OF pH ELECTRODE/METER

pH electrodes are available in variety of different shapes and sizes for many different applications. Intracellular pH can also be measured by using a miniature probes (micro electrodes). However, majority of them are based on same principle and operated in a similar fashion.

It is important that the outer layer of glass on glass electrode remains hydrated, and so it is normally immersed in a solution. Thus thin glass layer is fragile and thus care must be taken not to break it or scratch it, or to cause a build up of static electric charge by rubbing it. Gelatinous and protein containing solutions should not be allowed to dry out on the glass surface as they would inhibit response. As it is clear from above equations that potential produced is temperature dependent (each pH unity change represents 54.2 mV at 0°C and 61.5 mV at 37°C). This effect is predictable and can be compensated for. The pH meter will thus have a temperature compensation dial that must be correctly set before the meter is calibrated.

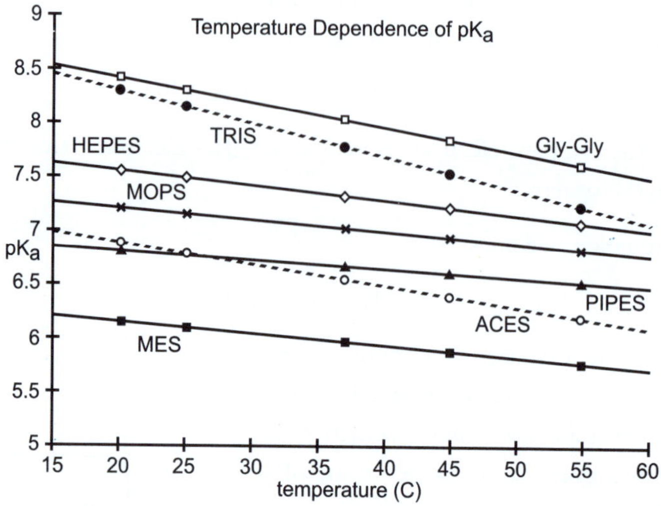

Fig. 2.2: Temperature dependence of pK_a

Calibration will necessitate the use of two solutions of widely differing pH. Usually calibration is first performed with buffer of pH 7, followed by a pH 4 buffer (if the sample is expected to be a acid) or a pH 9 buffer (if the sample is expected to be basic). Once the pH electrode is calibrated, it can simply be immersed in the solution to be measured and a rapid and accurate measurement estimate of pH can be made.

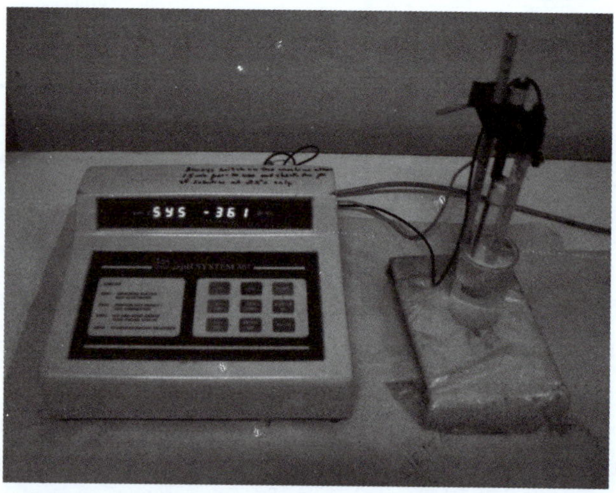

Fig. 2.3: A pH meter (*Courtesy:* Dr D. Dash, Deptt. of Biochemistry, IMS, BHU)

2.6 pH-STAT

It is a form of automatic titrator that can be used to maintain a constant pH during a reaction that involves either the production or removal of H^+ from solution. The rate of reaction can thus be determined because a recorder draws a curve representing the volume of the reagent added by titrator against time. A glass pH electrode, pH meter, recorder, controller, burette and magnetic stirrer are necessary; a controller is necessary to break the current to burette motor when the end point is reached. The best sorts are used so that less and less reagent is used as the end point is approached. This is to avoid the danger of overshoot. The kind of pH glass electrode used needs to be very accurate and stable for kinetic work.

The pH stat has certain limitations that solution should be stirred continuously which may cause the denaturation of the proteins and may introduce atmospheric components that may affect either the pH or the reaction proceeding in the vessel. Other problem is the existence of unknown junction potential and the tendency of liquid to leak. The latter effect can be counteracted in part by making the density of the burette liquid lower than that of the reaction mixture.

2.7 ION SELECTIVE AND GAS SENSING ELECTRODES

The glass pH electrode is really a kind of ion-selective electrode (ISE) that is sensitive to H^+. Similar potentiometric electrodes have been developed which are responsive to other ions, e.g., Na^+, NH_4^+, Cl^- and NO_3^-. The active material within these devices may be glass, an insoluble organic salt, or an ion exchange material. Glass is the active material within the pH electrode, but modified aluminium silicate glasses can also be used to produce a variety of monovalent cation responsive electrodes. Insoluble inorganic salts like silver sulphite can be used to produce electrodes responsive to Cu^{2+}, Pb^{2+} and Cd^{2+}, whereas lanthanum fluoride may be used to produce electrodes responsive to F^-.

Ion selective electrode responds to the activity of particular ion. However, if the instrument is calibrated with a standard of known concentration then, provided the ionic strength of solution are similar, the concentration of test solution will be recorded. If some of the ions are not free and exist in complex form or an insoluble precipitate, these electrodes will give a much lower reading, then with a method that detects all of the ions present. Generally used ion selective electrodes are Ca^{2+}, K^+, and NO_3^-.

An electrode may be ion selective but not ion specific. As with glass electrodes, these can be fouled by proteins forming a surface film. A reference electrode is also needed with these ISE so that the varying potential of these ISE can be compared with the steady potential produced by reference electrode.

Gas sensing electrodes: They are used generally to estimate the concentration of gas by its interaction in a thin layer surrounding an ion sensitive electrode, commonly a pH electrode. Carbon dioxide, sulphur dioxide, ammonia can all be measured by their dissolution in a thin layer surrounding the pH electrode, and measuring the resultant pH of the layer.

2.7.1 Miniaturisation and applications of ion sensitive electrodes

Miniaturisation of ion selective electrodes has been achieved by modification of field effect transistor to respond to specific ions. Such ion selective field effect transistors (ISFETs) are likely to have great clinical value. Multifunctional ISFETs are already available which are used to measure pH, Na^+, K^+, and Ca^{2+}.

2.8 THE CLARK OXYGEN ELECTRODE

It consists of a platinum cathode and silver anode, both immersed in same solution of saturated potassium chloride and separated from the test solution by a oxygen permeable membrane. When a potential difference of –0.6 V is applied across the electrodes such that platinum cathode is made negative with respect to silver anode, electrons are generated at anode and are then used to reduce oxygen at cathode. The oxygen tension at cathode drops and so to make this deficit more oxygen moves towards cathode. Since the rate of diffusion of oxygen from the membrane is the limiting step in the reduction process, the current produced by the electrode is proportional to the oxygen tension in the sample. These electrode reactions may be summarized as under:

At silver anode $4Ag + Cl^- \longrightarrow 4AgCl + 4e^-$

At platinum cathode $O_2 + 4H^+ + 4e^- \longrightarrow 2H_2O$

2.8.1 Operation of rank oxygen electrode (Clark electrode)

These allow the sample to be placed in upper reaction chamber by an oxygen permeable and ion impermeable membrane. Teflon is the usual choice, though cellophane, polythene, silicon rubber and cling film have been used with varying degree of success. Care must be taken that membrane must not become contaminated. Thinner membranes give more response but are more fragile. The membrane covers the electrodes and allows oxygen to diffuse towards them whilst preventing other reaction elements to reach electrode and poison them. The electrodes are maintained in electric continuity with potassium chloride solution.

Oxygen electrode is mounted above a stirring motor, which is able to rotate a magnetic follower (flea) when inserted into the reaction vessel which is important as the platinum cathode reduces oxygen to produce the electric current. A correct set-up will show reduction in current when stirrer is switched off due to depletion of oxygen in the potassium chloride filled electrode chamber. Resumption of stirring will result in a return of current (oxygen tension in potassium chloride) to its previous level prior to the stirrer being switched off. Since both solubility and rate of diffusion is affected by temperature, therefore some form of temperature control is necessary for better results which are done by circulating water bath. Calibration of the instrument should be carried out at the same temperature as that of experiment.

Many chemicals are adsorbed onto the surface of the membrane and reaction vessel; hence it is important that the apparatus is thoroughly cleaned after each experiment.

2.8.2 Applications of rank oxygen electrode

Due to their ability to give continuous trace, oxygen electrodes have largely replaced manometric techniques in the study of reactions involving oxygen uptake and evolution.

- **Mitochondrial studies**: The study of respiratory control and effect of inhibitors on mitochondrial respiration and the measurement of phosphorylation: oxidation (P : O) ratios are best done by oxygen electrodes.
- The sites of action of electron transport inhibitors can also be determined using an oxygen electrode.
- Micro-organism that uses oxygen as the terminal electron acceptor of respiratory electron transport can be studied using oxygen electrode and the effect of electron transport inhibitors determined.
- **Enzyme assays**: Enzymes are readily studied using Clark oxygen electrode, provided oxygen is involved in the reaction. Glucose oxidase, D-amino acid oxidase, and catalase are examples whose properties can be studied in this way.

2.8.3 Probe type Clark electrode

These rely on same principle of operation as the rank electrode. However, the cathode and retaining membrane are arranged at the end of the probe to enable insertion into a liquid phase. It has disadvantage that it does not have stirring arrangements. It has a variety of uses.

Measurement of oxygen in bulk liquids: Oxygen concentrations are routinely monitored in fermentation processes, sewage and industrial waste treatment and in inland, coastal and oceanic waters. This involves the variation of the Clark electrode called flush top sensor.

Clinical uses: Early clinical use of oxygen electrode is to measure heart-lung machines during open heart surgery. They are also used for testing patients who were treated with oxygen. Small samples of blood are taken from patient and oxygen content is measured in a small Clark type pO_2 electrode.

2.9 THE LEAF DISC ELECTRODE

Whilst the rank oxygen electrode is ideally suited to many applications requiring a measurement of oxygen in aqueous samples, a leaf disc electrode such as the Hanasatech LD2 is of more use if gaseous oxygen measurement is required. Since the measurement of oxygen evolution is one of the easiest ways of following photosynthetic process in leaves, this instrument has found much biological application.

This device measures oxygen amperometrically using the same principle as the rank electrode. However, instead of being a liquid -filled reaction vessel, the reaction chamber is designed to allow a leaf to be held in place and provided with saturating carbon dioxide (or bicarbonate as a source of carbon dioxide). Illumination is usually provided by an array of light-emitting diodes (which produce little heat) and the oxygen emitted by leaf during photosynthesis can be measured.

Calibration of this electrode is bit complex as compared to rank electrode. A zero oxygen signal can be produced by passing nitrogen through the reaction chamber. Once this is stopped and air is passed through the chamber, signal corresponding to 21% of the oxygen can be determined. However, in closed chamber system, the amount of oxygen is related to the oxygen concentration and to volume of the chamber. In practice, because the leaf disc itself may reduce the effective volume of the chamber., calibration involves injecting known volumes of air into the chamber and measuring the voltage response to obtain the effective volume of the chamber and hence a precise calibration of the electrode.

The leaf disc electrode has been used extensively for the study of the relationship between photosynthetic oxygen evolution under saturating carbon dioxide and the intensity of illumination, enabling calculation of quantum yield. the inclusion of probes to measure emitted fluorescence from the leaf disc at the same time as the oxygen evolutions measurement are made has resulted in a device that provides variety of information. Applications of these devices are diverse, ranging from studies of micropropagated plants to those plants suffering from atmospheric pollution. Though leaf disc electrode is clearly designed for whole leaf studies, photosynthetic rates of microalgae have also been studied using theses electrodes.

2.10 TIPS TO REMEMBER

$$\underset{\text{Acid}}{HA} \rightleftharpoons \underset{\text{Conjugate Base}}{H^+ + A^-}$$

$$\underset{\text{Base}}{B + H^+} \rightleftharpoons \underset{\text{Conjugate Acid}}{BH^+}$$

$$pH = -\log [H^+]$$

Henderson-Hasselbach

$pH = pK_a + \log [A^-]/[HA]$

A^- = conjugate base (or salt), base

HA = acid, conjugate acid

pK_a = – log of dissociation constant

pH measurement

$E = Eo + (2.3RT/F) \cdot pH$

E = electrical potential (or voltage)

Factors Affecting

- ionic concentration
- Na^+ error
- temperature

Choosing a Buffer

- literature or empirical
- ± 1 pH unit from pK_a
- buffering capacity ($\beta = \Delta B/\Delta pH$)
- compensate for known

Changes in H^+

- compatibility with experimental system

Good's Buffer Criteria

- pK_a between 6 and 8
- high solubility in aqueous solutions
- excluded by biological membranes
- minimal effects due to conc., temp. and ionic composition
- well-defined or non-existent interactions with cations
- chemical stability
- insignificant absorption between 240-700 nm
- readily available in pure form

2.11 VIVA VOCE

Q.1. What do you understand by pH? What are factors effecting pH?

Q.2. What do you understand by buffer strength. Discuss the criteria for good buffers.

Q.3. What is pK_a and how is it related to pH?

Q.4. Explain the principle of pH electrode.

Q.5. How will you measure the pH of HCl?

Q.6. Explain pH-stat.

Q.7. What is buffer equation?

Q.8. Give differences between ion selective and gas sensing electrodes.

Q.9. Give applications of rank oxygen electrodes.

Q.10. Explain leaf disc electrodes.

Q.11. What should be the pH of physiological buffers?

Q.12. How does temperature affects pH.

Q.13. Calculate the pH of a solution containing 0.001 mol $H^+ l^{-1}$.

Ans. 3.

Q.14. What is the H^+ concentration at pH 7.8?

Ans. 1.58×10^{-8} mol l^{-1}.

Manipulation of Nucleic Acids: Basic Tools and Techniques

CHAPTER 3

Nucleic acids are central point to molecular techniques. All molecular techniques ultimately have fate of extracting or modifying the information from/of nucleic acids. Here in this chapter we shall be considering different techniques which are employed for dealing with nucleic acid.

3.1 ISOLATION AND SEPARATION OF NUCLEIC ACIDS

3.1.1 Isolation of DNA

DNA is recovered from cells by the gentlest possible method of cell rupture to prevent DNA from fragmenting by mechanical shearing. This is usually in the presence of EDTA which chelate the Mg^{2+} needed for enzymes that degrade DNA, termed deoxyribonucleases (DNases). Ideally, cell walls, if present, should be digested enzymatically (e.g., lysozyme treatment of bacteria), and the cell membrane should be solubilised using detergent. If physical disruption is necessary, it should be kept to a minimum and should involve cutting and squashing of cells, rather than the use of shear forces. Cell disruption and most subsequent steps should be performed at 4°C using glassware and solutions that have been autoclaved to destroy DNase activity.

After release of nucleic acid from cell, RNA can be removed by treatment with ribonucleases (RNase) that have been heat treated to remove any DNase contaminants as RNase is stable to heat because of its disulphite bonds, which ensure rapid renaturation of molecule on cooling. The other major contaminant, protein, is removed by shaking the solution gently with water saturated phenol, or with a phenol/chloroform mixture, subsequently after proteinase K treatment, either of which denatures proteins, not nucleic acids. Centrifugation of the emulsion formed by this mixing, produces a lower organic phase and upper aqueous phase separated by interface of denatured proteins. The aqueous phase is recovered and deproteinised repeatedly until no more material is seen in interface. Finally the deproteinised DNA preparation is mixed with two volumes of absolute alcohol, and DNA pellet is redissolved in a buffer containing EDTA to inactivate any DNase present. The solution can be stored for at least a month at 4°C. DNA can also be stored frozen but repeated freezing and thawing tends to damage long DNA molecules by shearing.

If DNA from cell organelle or viral particle is needed, it is better to isolate the cell organelle or viral particle first as recovery of DNA from a mixture of DNA is difficult. Where high degree of purity is required, DNA may be subjected to density gradient ultra centrifugation through caesium chloride, which is particularly useful for the preparation of plasmid DNA. It is possible to check the integrity of DNA by agarose gel electrophoresis and determine the concentration of DNA by using the fact that 1 absorbance unit equates to 50 μg DNA cm^{-3} and so,

$$50 \times A_{260} = \text{concentration of DNA sample } (\mu g\ cm^{-3})$$

Contaminants may also be identified by scanning ultraviolet spectrophotometry from 200-300 nm. A ratio of absorbance at 260 nm to that at 280 nm of approximately 1.8 indicates that the sample is free from protein contamination which absorbs strongly at 280 nm.

Step 1.	Isolated or homogenized cells subject to lysis: 0.5% SDS + proteinase K (55°C for several hours)
Step 2.	Phenol Extraction: Gentle rocking for several hours
Step 3.	Ethanol Precipitation
Step 4.	RNase followed by proteinase K treatment
Step 5.	Repeat phenol extraction and EtOH ppt.

Procedure I – Typical procedure for isolation of genomic DNA

3.1.2 Isolation of RNA

RNA molecules are relatively short and thus are less prone to damage by shearing, so can be subjected to vigorous cell disruption. RNA is very vulnerable to digestion by RNase which are present endogenously as well as exogenously on finger tips. Gloves should, therefore, be worn and strong detergent should be included in the isolation medium to immediately denature any RNase. Subsequent deproteinisation is done vigorously as RNA is usually tightly associated with protein. DNase treatment is then followed to remove DNA contamination and at last RNA is precipitated with ethanol. Commonly used treatment in RNA extraction is guanadinium thiocyanate, which is both strong inhibitor of RNase and strong denaturant.

The integrity of RNA extracted can be checked by agarose gel electrophoresis. The most abundant RNA species are rRNA molecules 23s and 16s in prokaryotes and 18s and 28s in eukaryotes. The concentration of RNA may be estimated by using ultraviolet spectrophotometry. At 260 nm 1 absorbance unit equates 40 μg cm^{-3} of RNA and therefore:

$$40 \times A_{260\text{-}}\ \text{concentration of RNA sample } (\mu g\ cm^{-3})$$

Contaminants may also be identified in the same way as that for DNA by scanning ultra violet spectroscopy; however, in the case of RNA a 260 nm : 280 nm ratio of approximately 2 would be expected for a sample containing no protein.

In many cases it is desirable to isolate eukaryotic mRNA, which is only 2-5% of total RNA molecules. At high salt concentrations, the mRNA containing poly (A) tails binds to the complimentary oligo (dT) molecules of the affinity column so mRNA will be retained and all other RNA molecules will be washed off by further high salt solutions. Finally, bound mRNA can be eluted by using low concentrations of salt. Nucleic acid species may also be sub-fractionated by more physical means like electrophoresis and chromatographic separation based on the differences of nucleic acid fragment sizes or physicochemical characteristics.

Step 1.	Reagents and other utilities with RNase inhibitors
Step 2.	Isolated or homogenized cells subject to lysis – Detergents/lysozyme
Step 3.	Proteinase agents / RNA solvents treatment
Step 4.	Phenol Extraction phenol / chloroform
Step 5.	Alcohol precipitation 70% / 100% EtOH
Step 6.	Redissolve RNA.

Procedure II – Typical procedure for isolation of RNA

3.2 ELECTROPHORESIS OF NUCLEIC ACIDS

Electrophoresis in agarose or polyacrylamide gel is the usual way to separate DNA molecules according to the size. The technique can be used analytically or preparatively, and can be qualitative or quantitative. Large fragments of DNA such as chromosomes can be separated by a modification of electrophoresis termed as pulsed-field gel electrophoresis (PFGE). The easiest and common method is electrophoresis in horizontal agarose gels, followed by staining with ethidium bromide. This dye binds to DNA by insertion between stacked base pairs (intercalation), and exhibits a strong orange/red fluroscence when illuminated with ultraviolet light. Very oftenly electrophoresis is used to check the purity or intactness of a DNA preparation or to assess the enzymatic activity during cloning process.

Agarose gels are used to separate DNA molecules larger than 100 bps while shorter DNA fragments can be separated easily with polyacrylamide gels.

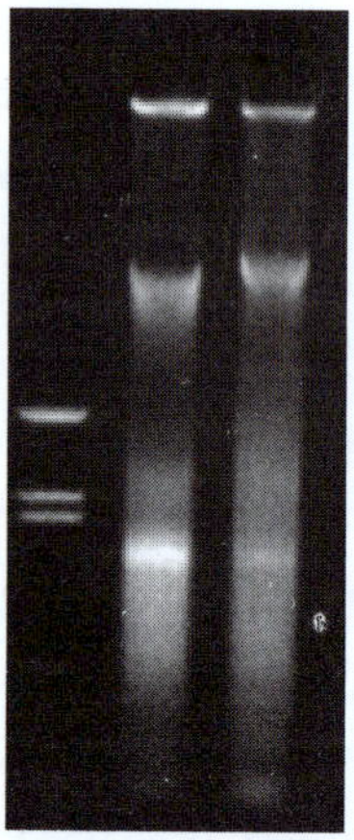

Fig. 3.1: Agarose gel stained with ethidium bromide and illuminated with UV light showing discrete DNA bands. Lane 1. Ladder, Lanes 2 and 3 isolated genomic DNA

When electropohresis is done preparatively, DNA is recovered from gel by cutting the DNA containing fragment of gel with scalpel followed by different treatments like crushing with glass rod in presence of agarase to digest agarose and setting DNA free or by process of electroelution in which the gel sealed in a dialysis tubing is placed between two electrodes in presence of buffer. Passage of an electrical current between electrodes causes DNA to migrate out of the gel piece but it remains trapped within the dialysis tubing and can, therefore, be recovered easily.

3.3 RESTRICTION MAPPING OF DNA FRAGMENTS

This involves the size analysis of restriction fragments produced by several restriction enzymes individually and in combination. For example, in Fig. 3.2 restriction sites of two enzymes A and B are being mapped. Cleavage with A gives fragments 2 and 7 kilo bases from a 9 kb molecule, hence we can have position of single A site from one end. Similarly, B gives fragments 3 kb and 6 kb away, so it has a single site 3 kb from one end; but it is still not clear whether this site is near A's site or is at opposite end of DNA. This can be resolved by double digestion. If the resultant fragments are 2 kb, 3 kb and 4 kb away, then A and B cut at opposite ends of the molecule; if they are 1 kb, 2 kb and 6 kb away, the sites are near to each other. It is worth stating here that the mapping of real molecules is rarely as simple as this.

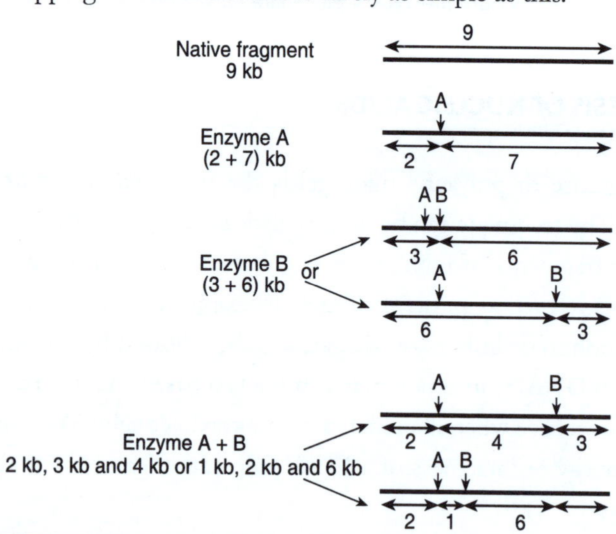

Fig. 3.2: Restriction mapping of DNA. It is worth noting that each experimental result and its interpretation should be considered in sequence.

3.4 NUCLEIC ACID BLOTTING METHOD

Electrophoresis allows separation on the basis of size but it does not allow the detection of specific fragment among the complex sample. This can be achieved by transferring the DNA from gel onto nylon or nitro cellulose membrane thus providing a more permanent record of the sample. The method is called southern blotting. In this, first the gel is soaked in alkali to render the DNA single stranded. It is then transferred to the membrane in the pattern exactly similar in the gel. This transfer can be brought electrophoretically or by drawing large volume of buffer through both gel and membrane, thus transferring DNA by capillary action. Now the transferred DNA molecule can be probed with labelled DNA molecule called gene probe. This single stranded gene probe will hybridise under right conditions to complimentary fragments immobilized onto the membrane. For hybridisation to take place effectively proper concentrations of salts and adequate temperature are very necessary. This is usually referred to as stringency of the hybridisation and is particular for each gene probe and sample of DNA. Then unbound probe is removed by washing with buffer and membrane is developed like photographic film after which the precise location of the probe and target can be visualized.

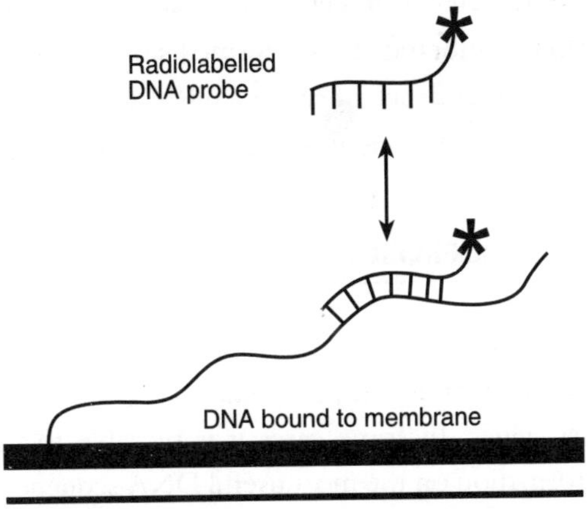

Fig. 3.3: Hybridisation during northern blotting

The same basic process of nucleic acid blotting can be used to transfer RNA from gels onto similar membrane. This allows identification of specific mRNA sequences of definite length by

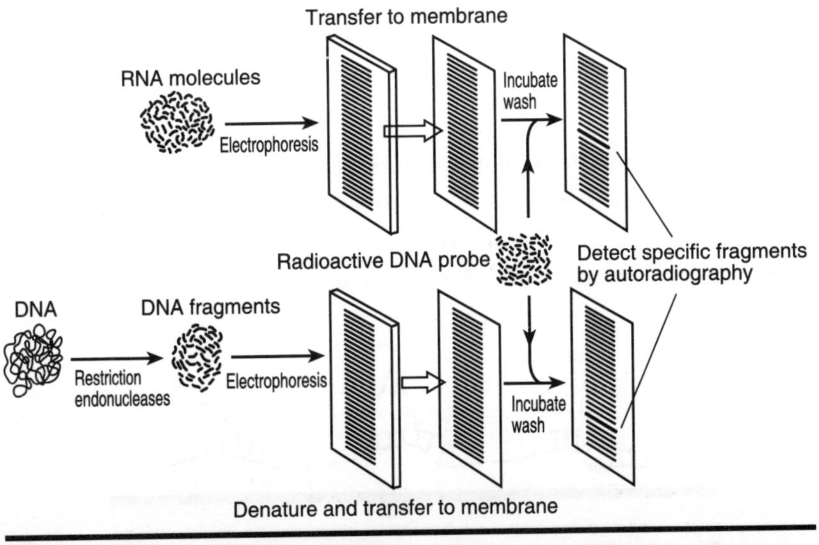

Fig. 3.4: Northern and Southern blotting procedure

hybridisation to a labelled gene probe and is known as "Northern blotting". By this one cannot only detect but also can quantify the relative amount of specific mRNA. For this mRNA transcripts are separated by electrophoresis under denaturing conditions which improve resolution and also aids in more accurate estimation of the sizes of transcripts. The format of the blotting may be altered from transfer from gel to direct application to slots on a specific blotting apparatus containing the nylon membrane. This is termed slot or dot blotting and provides a convenient means of measuring the abundance of specific mRNA transcripts without the need for gel electrophoresis; however, it does not provide information regarding the size of fragments.

3.5 GENE PROBES

The information needed to produce a gene probe may come from many sources but with the development and sophistication of genetic database, gaining this knowledge is usually one of the first stages. In some cases it is possible to use related proteins from same gene family to gain information on the most useful DNA sequences. Protein or DNA sequences that are similar but are from different species may also provide a starting point with which to produce a so-called heterologous gene probe. Also with the help of database, single stranded oligonucleotide probe can be synthesized chemically. This is done by computer controlled gene synthesizers that link dNTPs together on the basis of a desired sequence. Whatever way is used to obtain a probe but it should be assured that probe is unique, and is neither able to self-anneal nor is self-complimentary.

Where scant DNA information is available to prepare a gene probe it is possible in some cases to use the knowledge obtained form analysis of protein as from genetic code it is possible to predict the various DNA sequences which could code for the protein and thus then can synthesize oligonucleotide sequence chemically. However, due to degeneracy of genetic code there could be more than one oligonucleotide for a given polypeptide. Ideally, sequence that is specific for a given gene and is not longer than 20 bases that contains as many tryptophan and methionine as possible is sufficient as these have unique codons and therefore, fewer possible base sequence that can code for that part of protein.

Labelling of gene probes: An essential feature of gene probe is that it can be visualized by some means. Therefore, labeling of probe is a must and is done by two methods, traditionally by using radioactive labels and by non-radioactive labels. Mostly used radioactive labels are phosphorous 32 (^{32}P), sulphur 35 (^{35}S) and tritium (^{3}H) which are detected by process of autoradiography.

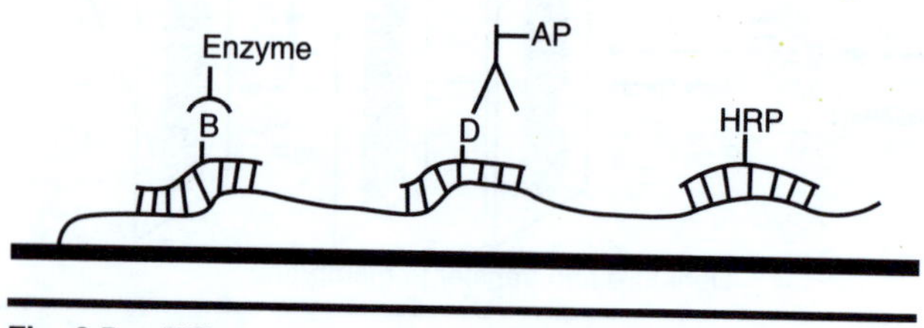

Fig. 3.5: Different types of non-radioactive probes

Non-radioactive probes, although less sensitive than radioactive probes but are safe to use. The labelling systems are termed either direct or indirect. Direct labelling allows an enzyme reporter such as alkaline phosphatase to couple directly to DNA. Indirect labelling method which is more popular involves binding of nucleotide that has label attached, e.g., Biotin, fluorescein and digoxigenin. These molecules are covalently linked to nucleotides. Specific binding proteins may then be used as a bridge between nucleotide and reporter protein such as enzyme. For example biotin incorporated into DNA molecule is recognized with a very high affinity by Streptavidin. This may be in turn is coupled with enzyme alkaline phosphatase which is able to convert colourless compound p-nitrophenol phosphate (PNPP) into yellow colour compound p-nitrophenol (PNP) and also offers a means of signal amplification. Thus, rather than the detection system relying on autoradiography, which is necessary for radiolabels, a series of reaction resulting in colour, a light or a chemiluminescence are much better, since autoradiography may take 1-3 days, where colour and chemiluminescence reaction takes a few minutes.

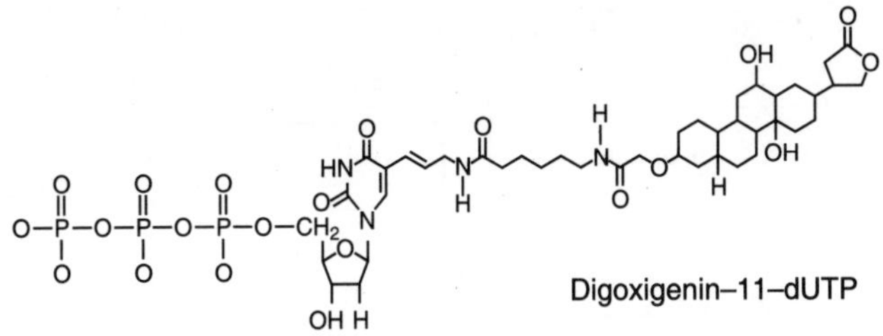

Fig. 3.6: Digoxigenin linked to dUTP (Enzyme linked non radioactive probe)

Table 3.1: Comparative chart of different gene probing methods

- random priming
  — used for cloned DNA fragments
- T4 nucleotide kinase
  — used primarily for synthetic oligonucleotides
- (nick translation)
  — earlier method replaced by random priming
  — quality control and reproducibility problems
- (terminal transferase)
  — adds nucleotides to 3′ end
  — used primarily for generation of homopolymer tails

3.5.1 End labelling of DNA molecules

The simplest form of labelling is by 5′- or 3′-end labelling. 5′-end labelling involves a phosphate transfer or exchange reaction where the 5′-phosphate of the DNA to be used is replaced by ^{32}P. This is carried by two enzymes: the first, alkaline phosphatase; removing phosphate group followed

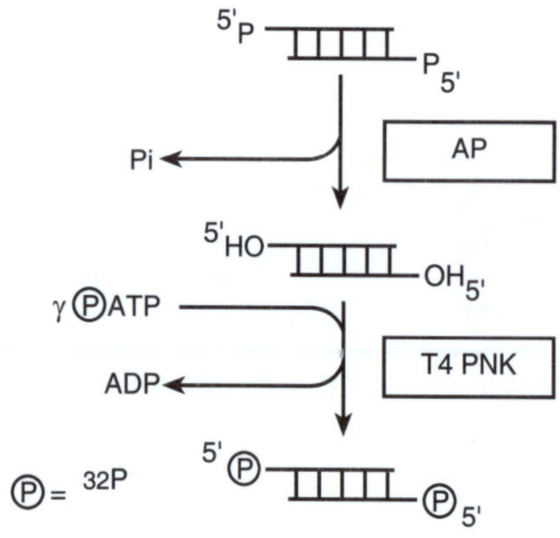

Fig. 3.7: End-labelling of a gene probe at 5′-end with alkaline phosphate and polynucleotide kinase

by poly nucleotide kinase catalyzing the transfer of phosphate group (^{32}P) to 5′ end of DNA. The newly labelled probe is then purified, usually by chromatography through a sephadex column to remove any unincorporated radiolabel.

Using other end of DNA molecule, the 3′-end is slightly less complex. Here a new, labelled dNTP ([α^{32}-P]ATP or biotin-labelled dNTP) is added to the 3′-end of DNA by enzyme terminal transferase. Although this is a simpler reaction, a potential problem exists because a new nucleotide is added to the existing sequence and so the complete sequence of the DNA is altered, which may affect its hybridization to its target sequence. End-labelling methods also suffer from the fact that only one label is added to DNA so such methods are of low activity in comparison to other who incorporate label through out the length of sequence of DNA.

3.5.2 Random primer labelling

The DNA to be labelled is first denatured and is then allowed to renatured in presence of random sequences of hexamers or hexanucleotide. These hexamers will by chance bind to DNA sample wherever they encounter a complementary sequence and so the DNA will acquire hexanucleotide annealing to it. Each of the hexamers can act as primer for synthesis of a fresh strand of DNA catalysed by DNA polymerase since it has free 3′- hydroxyl group. The Klenow fragment of DNA polymerase is used for random primer labelling since it lacks 5′ → 3′ exonuclease activity but still acts as 5′ → 3′ polymerase.

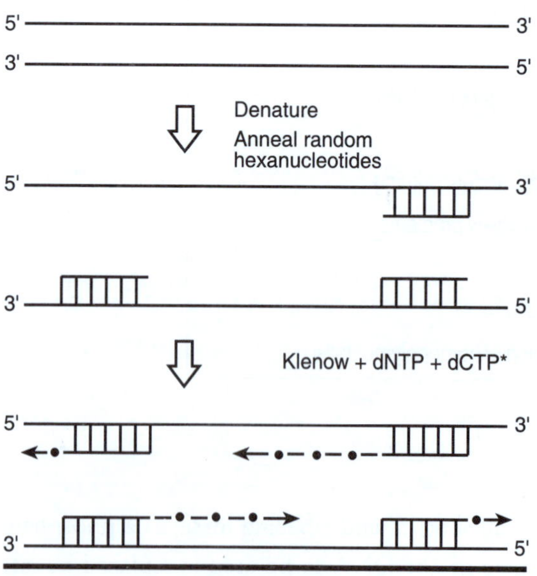

Fig. 3.8: Random primer gene probing

3.5.3 Nick translation

It is a traditional method of labelling DNA. Low concentrations of DNase I is used to create occasional single strand nick which is then filled in by DNA polymerase using an appropriate dNTP at the same time making a new nick at 3′-side of the previous one. If labelled dNTPs are used, the DNA can be labelled to a very high specific activity

3.6 THE POLYMERASE CHAIN REACTION

PCR has been one of the most important techniques developed in recent years. The reason behind is its simplicity of the reaction and relative case of the practical manipulation steps. The PCR is used to amplify a precise fragment of DNA from a complex mixture of starting material usually termed as template DNA. The reaction requires some information about sequence of flanking fragment of DNA to be amplified. From this information two oligonucleotide primers can be synthesized each complimentry to the stretch of DNA to the 3′-side of the target DNA, one for each of the two DNA strands. This technique in many cases has replaced traditional DNA cloning methods, since it fulfils the same function of producing large amount of DNA from small amount of starting materials.

Fig. 3.9: PCR machine (*Courtesy:* Dr D. Dash, Deptt. of Biochemistry, IMS, BHU)

3.6.1 Stages in PCR

It has three definite sets of times and temperature, termed as

Step I. Denaturation

Step II. Annealing and

Step III. Extension.

Each of the three steps are repeated 30-40 times or cycles. In first cycle the double stranded template DNA strand is first denatured by heating the reaction to above 90°C so that the region to be specifically amplified can be made accessible. The temperature is then cooled to between 40-60°C. The precise temperature is critical and each PCR system has to be defined and optimized. Reactions that are not optimized may give rise to other products along with amplified target sequence or even may not produce amplified product.

The second annealing step allows the hybridization of the two oligonucleotide primers, present in excess to bind to their complimentary sites which flank the DNA. The annealed oligonucleotide acts as a primer since they provide free 3′-end hydroxyl group to DNA polymerase. The third step, DNA synthesis or extension is carried out by a thermostable DNA polymerase, most commonly *Taq* DNA polymerase.

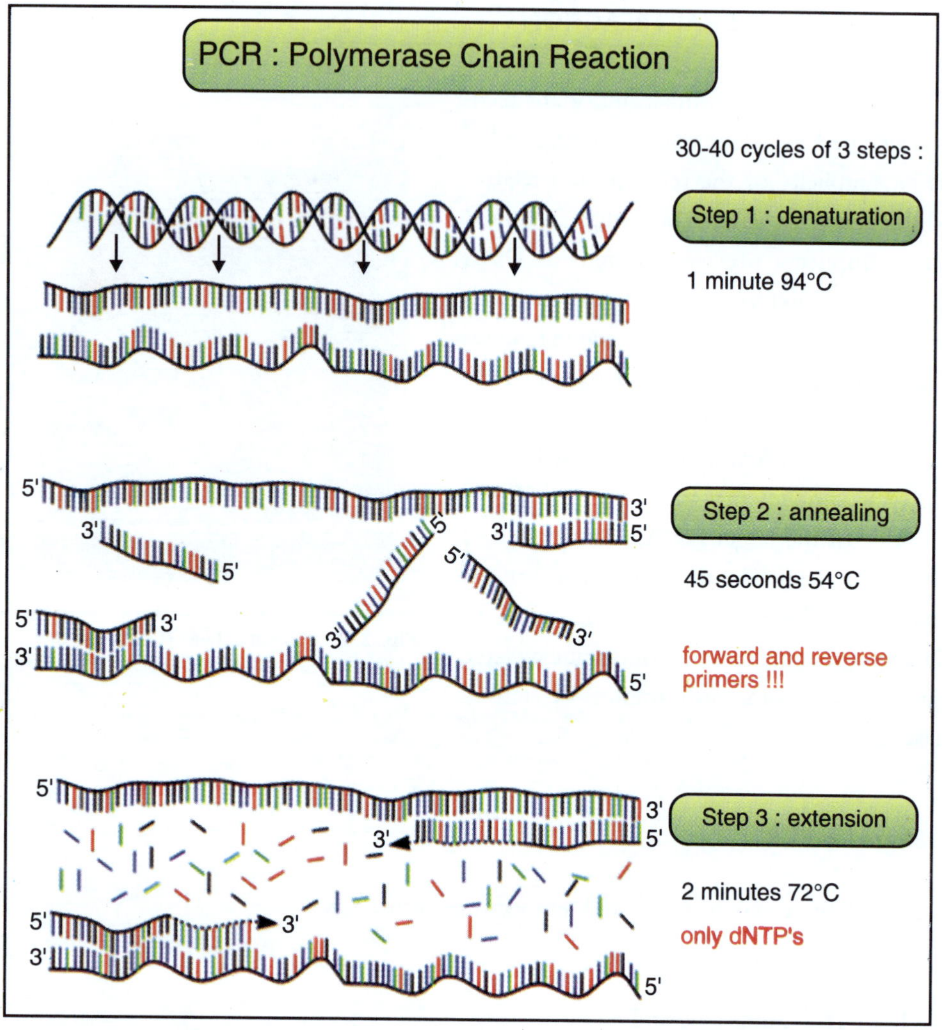

Fig. 3.10: Different Stages of PCR

DNA synthesis proceeds from both primers until new strands have been extended along and beyond the target DNA to be amplified. It is important to note that since the new strand extends beyond the target DNA, they will contain a region next to their 3′-end that is complimentary to the other primer. Thus in another round not only parental target strand but also newly synthesized strands also act as template DNA.

One problem with the early PCR reaction was the temperature needed to denature the DNA also denatures the DNA polymerase. Then thermostable DNA polymerase called Taq polymerase from thermophilic bacterium *Thermus aquaticus* with optimum temperature of 72°C and can also survive after prolonged exposure to 96°C provided the means to automate the reaction.

3.6.2 Primer design in PCR

The principle features o ideal primers are that these not only have to be complementary to the flanking sequence of the target DNA but must not be self-complementary otherwise they will form primer dimmers. Both primers have to be matched in their G+C content and should have similar annealing temperature. It is also possible to design primer with additional sequences at their 5′-ends, such as restriction endonuclease target site or promoter sequences. However, these require alteration in annealing temperature to compensate for areas of non-homology.

3.6.3 Types of PCR

The PCR may be used to amplify the templates from variety of sources. Unlike many manipulation methods used in current molecular biology the PCR technique is sensitive enough to require very little template preparation.

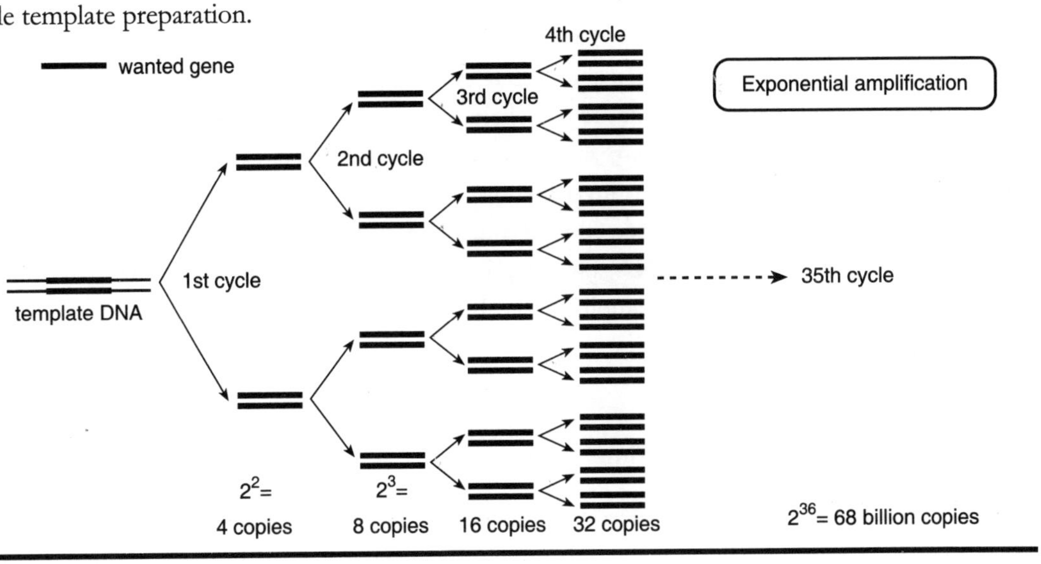

Fig. 3.11: Exponential amplification of genes in PCR

Extraction from many cells require just single boiling step; in fact, other ways involving SDS or proteinase K may affect adversely in PCR. The process may also be employed to amplify RNA, the process known as reverse transcriptase PCR (RT-PCR). This allows mRNA transcription products to be effectively analysed. It may also be used to differentiate latent viruses (detected by standard PCR) or active viruses, which replicate and thus produce transcription products and are, therefore, detectable by RT-PCR. In addition PCR may be extended to determine relative amount of transcription product.

3.6.4 Sensitivity of PCR

The enormous sensitivity of PCR is also its one of the main drawbacks, since the very large degree of amplification makes the system vulnerable to contamination. Even a trace of foreign DNA, such as that contained in dust particles, may be amplified to significant levels and may mislead the results. Hence, cleanliness is a paramount when carrying out PCR and dedicated equipment and in some cases laboratories are used. It is also possible that amplified products may also contaminate the PCR, although this may be overcome by ultraviolet irradiation to damage already amplified products so that they cannot be used as templates. Another way to overcome this kind of contaminatioin is the use of uracil in the PCR and then to treat the products with enzyme N-glycocylase (UNG) which degrades any PCR product with incorporated uracil, rendering them useless to be used as templates. The success of PCR process has given impetus to the development of other amplification based on either thermal or non-thermal (isothermal) cycling methods. The most popular alternative to the PCR is termed as "Ligase chain reaction or LCR". This operates in a fashion similar to PCR but a thermostable DNA ligase joins sets of primer together that are complimentary to target DNA. Following this, a similar exponential amplification reaction takes place, producing amounts of DNA that are similar to template DNA. Other alternative amplification methods are followed which all are based on one of the two strategies of either amplifying the target DNA/RNA molecule or by detecting the target and amplifying a signal molecule bound to it. Target amplification method includes Ligase chain reaction (LCR) employing thermostable DNA ligase used for mutation detection, etc. and other method under this category is nucleic acid sequence based amplification (NASBA) involving use of RNA, RNase H/reverse transcriptase, and T7 DNA polymerase used for viral detection, etc. (e.g., HIV). Signal amplification method includes branched DNA amplification (β-DNA) involving isothermal microwell format using hybridization or target /capture probe and signal amplification. It is used for mutation detection.

3.6.5 Factors Affecting the PCR

1. Denaturing Temperature and Time

The specific complementary association due to hydrogen bonding of single-stranded nucleic acids is referred to as "annealing": two complementary sequences will form hydrogen bonds between their complementary bases (G to C, and A to T or U) and form a stable double-stranded, anti-parallel "hybrid" molecule. One may make nucleic acid (NA) single-stranded for the purpose of annealing, if it is not single-stranded already, like most RNA viruses, by heating it to a point above the "melting temperature" of the double- or partially-double-stranded form, and then flash-cooling it; this ensures the "denatured" or separated strands do not re-anneal. Additionally, if the NA is heated in buffers of ionic strength lower than 150 mM NaCl, the melting temperature is generally less than 100°C — which is why PCR works with denaturing temperatures of 91-97°C.

Taq polymerase is given as having a half-life of 30 min at 95°C, which is partly why one should not do more than about 30 amplification cycles. However, it is possible to reduce the denaturation temperature after about 10 rounds of amplification, as the mean length of target DNA is decreased. For templates of 300 bp or less, denaturation temperature may be reduced to as low as 88°C for 50% (G+C) templates, which means one may do as many as 40 cycles without much decrease in enzyme efficiency.

"Time at temperature" is the main reason for denaturation / loss of activity of Taq. Thus, if one reduces this, one will increase the number of cycles that are possible, whether the temperature is reduced or not. Normally the denaturation time is 1 min at 94°C. It is possible, for short template sequences, to reduce this to 30 sec or less. Increase in denaturation temperature and decrease in time may also work: recommend 96°C for 15 sec.

2. Annealing Temperature and Primer Design

Primer length and sequence are of critical importance in designing the parameters of a successful amplification. The melting temperature of nucleic acid duplex increases both with its length, and with increasing (G+C) content. A simple formula for calculation of the T_m is

$$T_m = 4(G + C) + 2(A + T)°C.$$

Thus, the annealing temperature chosen for a PCR depends directly on length and composition of the primer(s). One should aim at using an annealing temperature (T_a) about 5°C below the lowest T_m of the pair of primers to be used. A more rigorous treatment of annealing temperature is given by Rychlik *et al.* (1990). They maintain that if the annealing temperature is increased by 1°C in every other cycle, specificity of amplification and yield of products <1kb in length are both increased. One consequence of having too low annealing temperature is that one or both primers will anneal to sequences other than the true target, as internal single-base mismatches or partial annealing may be tolerated. This is fine if one wishes to amplify similar or related targets. However, it can lead to "non-specific" amplification and consequent reduction in yield of the desired product, if the 3′-most base* is paired with a target.

A consequence of too high annealing temperature is that too little product will be made, as the likelihood of primer annealing is reduced; another and important consideration is that a pair of primers with very different annealing temperatures may never give appreciable yields of a unique product, and may also result in inadvertent "asymmetric" or single-strand amplification of the most efficiently primed product strand. Annealing does not take long, most primers will anneal efficiently in 30 sec or less, unless the annealing temperature (T_a) is too close to the T_m, or unless they are unusually long.

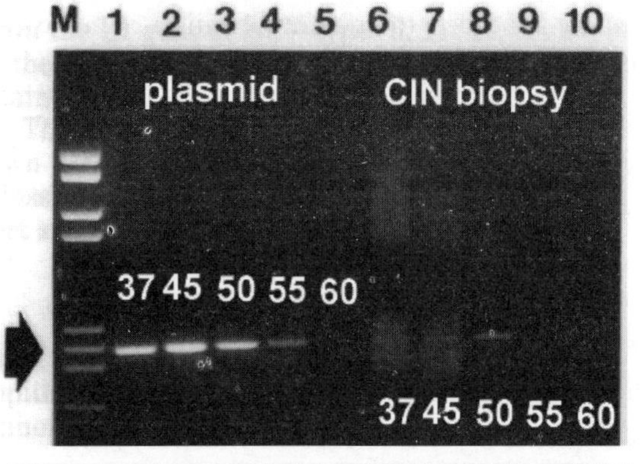

Fig. 3.12: Plasmid and biopsy sample DNA templates were amplified at different annealing temperatures as shown: Note that while plasmid is amplified from 37 to 55°C, HPV DNA is only specifically amplified at 50°C.

An illustration of the effect of annealing temperature on the specificity and on the yield of amplification of *Human papillomavirus type* 16 (HPV-16) is given below (Williamson and Rybicki, 1991: J Med Virol 33: 165-171).

* 3′-end base (towards 3′-end)

3. Primer Length

The optimum length of a primer depends upon its (A+T) content, and the T_m of its partner if one runs the risk of having problems as described above. Apart from the T_m, a prime consideration is that the primers should be complex enough so that the likelihood of annealing to sequences other than the chosen target is very low.

For example, there is a 1/4 chance (4^{-1}) of finding an A, G, C or T in any given DNA sequence; there is a 1/16 chance (4^{-2}) of finding any dinucleotide sequence (e.g., AG); a 1/256 chance of finding a given 4-base sequence. Thus, a sixteen base sequence will statistically be present only once in every 4^{16} bases (= 4 294 967 296 or 4 billion). This is about the size of the human or maize genome, and 1000 × greater than the genome size of *E. coli*. Thus, the association of a greater-than-17-base oligonucleotide with its target sequence is an extremely sequence-specific process, far more so than the specificity of monoclonal antibodies in binding to specific antigenic determinants. Consequently, 17-mer or longer primers are routinely used for amplification from genomic DNA of animals and plants. Too long a primer length may mean that even high annealing temperatures are not enough to prevent mismatch pairing and non-specific priming.

4. Degenerate Primers

For amplification of cognate sequences from different organisms, or for "evolutionary PCR", one may increase the chances of getting product by designing "degenerate" primers. These would in fact be a set of primers which have a number of options at several positions in the sequence so as to allow annealing to and amplification of a variety of related sequences. For example, Compton (1990) described the use of 14-mer primer sets with 4 and 5 degeneracies as forward and reverse primers, respectively, for the amplification of glycoprotein B (gB) from related herpesviruses. The reverse primer sequence was as follows:

TC<u>GAATTC</u>NCCYAAYTGNCCNT

where Y = T + C, and N = A + G + C + T, and the 8-base 5′-terminal extension comprises a *Eco*RI site (underlined) and flanking spacer to ensure that the restriction enzyme can cut the product (the New England Biolabs catalogue gives a good list of which enzymes require how long a flanking sequence in order to cut stub ends). Degeneracies obviously reduce the specificity of the primer(s), meaning mismatch opportunities are greater, and background noise increases; also, increased degeneracy means concentration of the individual primers decreases; thus, greater than 512-fold degeneracy should be avoided.

5. Elongation Temperature and Time

This is normally 70-72°C, for 0.5-3 min. Taq actually has a specific activity at 37°C which is very close to that of the Klenow fragment of *E. coli* DNA polymerase I, which accounts for the apparent paradox which results when one tries to understand how primers which anneal at an optimum temperature can then be elongated at a considerably higher temperature, the answer is that elongation occurs from the moment of annealing, even if this is transient, which results in considerably greater stability. At around 70°C the activity is optimal, and primer extension occurs at up to 100 bases/sec. About 1 min is sufficient for reliable amplification of 2 kb sequences. Longer products require longer times: 3 min is a good bet for 3 kb and longer products. Longer times may also be helpful in later cycles when product concentration exceeds enzyme concentration (>1 nM), and when dNTP and / or primer depletion may become limiting.

6. PCR Reaction Buffer

Recommended buffers generally contain:

- 10-50 mM Tris-HCl pH 8.3,
- up to 50 mM KCl, 1.5mM or higher $MgCl_2$,
- primers 0.2-1 μM each primer,
- 50-200 μM each dNTP,
- gelatin or BSA to 100 μg/ml,
- and/or non-ionic detergents such as Tween-20 or Nonidet P-40 or Triton X-100 (0.05-0.10% v/v)

Modern formulations may differ considerably; however, they are also generally proprietary. Following are some considerations which should be kept in mind during PCR.

(a) Higher than 50 mM KCl or NaCl inhibits Taq, but some is necessary to facilitate primer annealing.

(b) [Mg^{2+}] affects primer annealing; T_m of template, product and primer-template associations; product specificity; enzyme activity and fidelity. Taq requires *free* Mg^{2+}, so allowances should be made for dNTPs, primers and template, all of which chelate and sequester the cation; of these, dNTPs are the most concentrated, so [Mg^{2+}] should be 0.5-2.5 mM *greater* than [dNTP]. A titration should be performed with varying [Mg^{2+}] with all new template-primer combinations, as these can differ markedly in their requirements, even under the same conditions of concentrations and cycling times/temperatures.

(c) Some enzymes do not need added protein, others are dependent on it. Some enzymes work markedly better in the presence of detergent, probably because it prevents the natural tendency of the enzyme to aggregate.

(d) Primer concentrations should not go above 1 μM unless there is a high degree of degeneracy; 0.2 μM is sufficient for homologous primers.

(e) Nucleotide concentration need not be above 50 μM each; long products may require more, however.

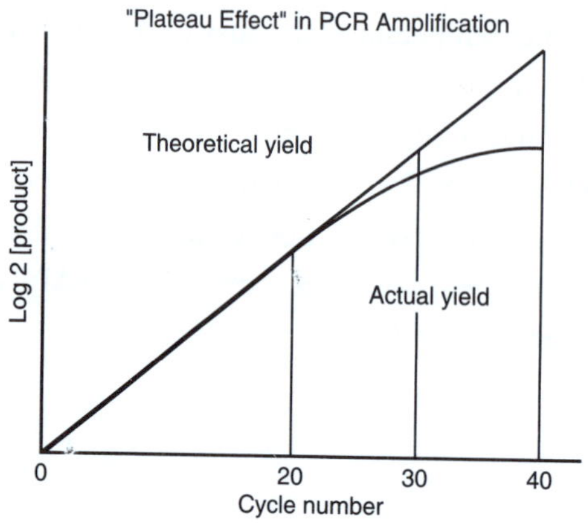

Fig. 3.13: Plateau effect in PCR

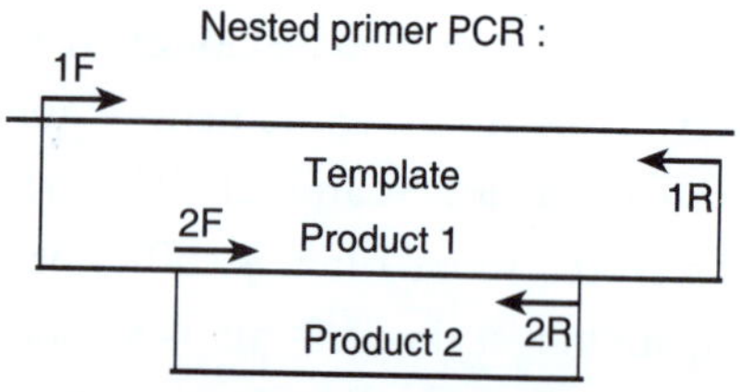

Fig. 3.14: Schematic representation of Nested primer PCR

7. Cycle Number

The number of amplification cycles necessary to produce a band visible on a gel depends largely on the starting concentration of the target DNA. Innis and Gelfand (1990) recommend from 40-45 cycles to amplify 50 target molecules, and 25-30 to amplify 3×105 molecules to the same concentration. This non-proportionality is due to a so-called *plateau effect*, which is the attenuation in the exponential rate of product accumulation in late stages of a PCR, when product reaches 0.3-1.0 nM. This may be caused by degradation of reactants (dNTPs, enzyme); reactant depletion (primers, dNTPs-former a problem with short products, latter for long products); end-product inhibition (pyrophosphate formation); competition for reactants by non-specific products; competition for primer binding by re-annealing of concentrated (10 nM) product.

If desired product is not made in 30 cycles, take a small sample (1 ul) of the amplified mix and re-amplify 20-30x in a new reaction mix rather than extending the run to more cycles. In some cases where template concentration is limiting, this can give good product where extension of cycling to 40x or more does not.

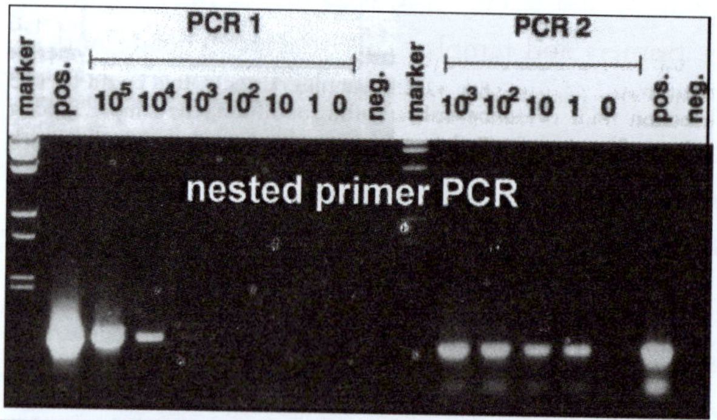

Fig. 3.15: The gel showing the effect of nested PCR amplification on the detectability of *Chicken anaemia virus* (CAV) DNA in a dilution series: the PCR1 just detects 1000 template molecules; PCR2 amplifies 1 template molecule (*Courtesy*: Soiné C *et al.* (1993) Avian Discovery)

A variant of this is nested primer PCR. PCR amplification is performed with one set of primers, and then some product is taken, with or without removal of reagents, for re-amplification with an internally-situated, "nested" set of primers. This process adds another level of specificity, meaning that all products non-specifically amplified in the first round will not be amplified in the second. This is illustrated in the left side:

8. Helix Destabilisers / Additives

With nucleic acids of high (G+C) content, it may be necessary to use harsher denaturation conditions. For example, one may incorporate up to 10% (w or v/v):

- dimethyl sulphoxide (DMSO),
- dimethyl formamide (DMF),
- urea or
- formamide

In the reaction mixture these additives are presumed to lower the T_m of the target nucleic acids, although DMSO at 10% and higher is known to decrease the activity of Taq by up to 50%. Additives may also be necessary in the amplification of long target sequences. DMSO often helps in amplifying products of >1kb. Formamide can apparently dramatically improve the specificity of PCR, while glycerol improves the amplification of high (G+C) templates. Polyethylene glycol (PEG) may be a useful additive when DNA template concentration is very low. It promotes macromolecular association by solvent exclusion, meaning the pol (DNA polymerase enzyme) can find the DNA.

3.6.6 Applications for PCR

1. The labelling of gene probes is one such area which has traditionally been undertaken by techniques such as nick translation. The nature of PCR makes it an ideal method for gene probe production and labelling.
2. Important modification of PCR known as quantitative PCR by which the initial concentration of template DNA can be estimated and is very useful for the measurement of, for example, a virus or an mRNA for protein expressed in abnormal amount in a disease process. Early quantitative PCR methods involved the comparision of a standard or control DNA template amplified with separate primers at the same time as the specific target DNA. These kind of quantification rely on the reaction being exponential and so any factor affecting this may also affect the result. Other method involves the incorporation of radiolabel through primer and their detection in subsequent purification of PCR products.
3. One of the most general application of development of PCR is direct PCR sequencing. This traditionally involves the cloning of sequences into vectors developed for chain termination sequencing. However, rapid accumulation of PCR products allows nucleotide sequence information to be obtained very quickly.

3.6.7 Variations on the basic PCR technique

1. **Allele-specific PCR:** AS-PCR is used to determine the genotype of single-nucleotide polymorphisms (SNPs) (single base differences in DNA) by using primers whose ends

overlap the SNP and differ by that single base. PCR amplification is less efficient in the presence of a mismatch, so the differences in amplification resulting from different primers can be used to quickly determine which primer matches the sample genotype.

2. **Assembly PCR:** Assembly PCR is the completely artificial synthesis of long gene products by performing PCR on a pool of long oligonucleotides with short overlapping segments. The oligonucleotides alternate between sense and antisense directions, and the overlapping segments serve to order the PCR fragments so that they selectively produce their final product.

3. **Asymmetric PCR:** Asymmetric PCR is used to preferentially amplify one strand of the original DNA more than the other. It finds use in some types of sequencing and hybridization probing where having only one of the two complementary stands is required. PCR is carried out as usual, but with a great excess of the primers for the chosen strand. Due to the slow (arithmetic) amplification later in the reaction after the limiting primer has been used up, extra cycles of PCR are required. A recent modification on this process, known as **L**inear-**A**fter-**T**he-**E**xponential-PCR (LATE-PCR), uses a limiting primer with a higher melting temperature (T_m) than the excess primer to maintain reaction efficiency as the limiting primer concentration decreases mid-reaction.

4. **Colony PCR:** Bacterial clones (E. coli) can be rapidly screened for correct DNA vector constructs. Selected bacterial colonies are picked with a sterile toothpick from an agarose plate and dabbed into the master mix or sterile water. Primers (and the master mix) are added, and the PCR is started with an extended time at 95°C when standard polymerase is used or with a shortened denaturation step at 100°C and special chimeric DNA polymerase.

5. **Helicase-dependent amplification:** Similar to traditional PCR, but maintains a constant temperature rather than cycling through denaturation and annealing/extension cycles. Helicase, an enzyme that unwinds DNA, is used in place of thermal denaturation.

6. **Hot-start PCR** is a technique that reduces non-specific amplification during the initial set up stages of the PCR. The technique may be performed manually by simply heating the reaction components briefly at the melting temperature (e.g., 95°C) before adding the polymerase. Specialized enzyme systems have been developed that inhibit the polymerase's activity at ambient temperature, either by the binding of an antibody or by the presence of covalently bound inhibitors that only dissociate after a high-temperature activation step. Hot-start/cold-finish PCR is achieved with new hybrid polymerases that are inactive at ambient temperature and are instantly activated at elongation temperature.

7. **Intersequence specific (ISSR) PCR:** A PCR method for DNA fingerprinting that amplifies regions between some simple sequence repeats to produce a unique fingerprint of amplified fragment lengths.

8. **Inverse PCR:** Inverse PCR is a method used to allow PCR when only one internal sequence is known. This is especially useful in identifying flanking sequences to various genomic inserts. This involves a series of DNA digestions and self-ligation, resulting in known sequences at either end of the unknown sequence.
9. **Ligation-mediated PCR**
10. **Methylation Specific PCR:** Methylation Specific PCR (MSP) is used to detect methylation of CpG islands in genomic DNA. DNA is first treated with sodium bisulphite, which converts unmethylated cytosine bases to uracil, which is recognized by PCR primers as thymine. Two PCR reactions are then carried out on the modified DNA, using primer sets identical except at any CpG islands within the primer sequences. At these points, one primer set recognizes DNA with cytosines to amplify methylated DNA, and one set recognizes DNA with uracil or thymine to amplify unmethylated DNA. MSP using qPCR can also be performed to obtain quantitative rather than qualitative information about methylation.
11. **Multiplex Ligation-dependent Probe Amplification (MLPA):** Permits multiple targets to be amplified with only a single primer pair, thus avoiding the resolution limitations of multiplex PCR (see below).
12. **Multiplex-PCR:** The use of multiple, unique primer sets within a single PCR reaction to produce amplicons of varying sizes specific to different DNA sequences. By targeting multiple genes at once, additional information may be gained from a single test run that otherwise would require several times the reagents and more time to perform. Annealing temperatures for each of the primer sets must be optimized to work correctly within a single reaction, and amplicon sizes, i.e., their base pair length, should be different enough to form distinct bands when visualized by gel electrophoresis.
13. **Nested PCR:** Nested PCR increases the specificity of DNA amplification, by reducing background due to non-specific amplification of DNA. Two sets of primers are being used in two successive PCR reactions. In the first reaction, one pair of primers is used to generate DNA products, which besides the intended target, may still consist of non-specifically amplified DNA fragments. The product(s) (sometimes after gel purification after electrophoresis of the PCR product) are then used in a second PCR reaction with a set of primers whose binding sites are completely or partially different from the primer pair used in the first reaction, but are completely within the DNA target fragment. Nested PCR is often more successful in specifically amplifying long DNA fragments than conventional PCR, but it requires more detailed knowledge of the target sequences.
14. **Quantitative PCR:** Q-PCR (**Q**uantitative PCR) is used to measure the quantity of a PCR product (preferably real-time). It is the method of choice to quantitatively measure starting amounts of DNA, cDNA or RNA. Q-PCR is commonly used to determine whether a DNA sequence is present in a sample and the number of its copies in the

sample. The method with currently the highest level of accuracy is **Quantitative real-time PCR**. It is often confusingly known as RT-PCR (**R**eal **T**ime PCR) or RQ-PCR. QRT-PCR or RTQ-PCR is more appropriate contractions. RT-PCR commonly refers to reverse transcription PCR (see below), which is often used in conjunction with Q-PCR. QRT-PCR methods use fluorescent dyes, such as Sybr Green, or fluorophore-containing DNA probes, such as TaqMan, to measure the amount of amplified product in real time.

15. **RT-PCR** (**R**everse **T**ranscription PCR) is a method used to amplify, isolate or identify a known sequence from a cellular or tissue RNA. The PCR reaction is preceded by a reaction using reverse transcriptase to convert RNA to cDNA. RT-PCR is widely used in expression profiling, to determine the expression of a gene or to identify the sequence of an RNA transcript, including transcription start and termination sites and, if the genomic DNA sequence of a gene is known, to map the location of exons and introns in the gene. The 5′ end of a gene (corresponding to the transcription start site) is typically identified by a RT-PCR method, named RACE-PCR, short for *Rapid Amplification of cDNA Ends*.
16. **TAIL-PCR: Thermal asymmetric interlaced** PCR is used to isolate unknown sequence flanking a known sequence. Within the known sequence TAIL-PCR uses a nested pair of primers with differing annealing temperatures; a degenerate primer is used to amplify in the other direction from the unknown sequence.
17. **Touchdown PCR:** Touchdown PCR is a variant of PCR that aims to reduce nonspecific background by gradually lowering the annealing temperature as PCR cycling progresses. The annealing temperature at the initial cycles is usually a few degrees above the T_m of the primers used, while at the later cycles, it is a few degrees below the primer T_m. The higher temperatures give greater specificity for primer binding, and the lower temperatures permit more efficient amplification from the specific products formed during the initial cycle

3.7 NUCLEOTIDE SEQUENCING OF DNA

The precise usage of codons, information regarding mutations and polymorphisms and identification of gene regulatory control sequences can only be elucidated by analyzing DNA sequences. Two techniques have been developed for this, one based on an enzymic method frequently termed Sanger sequencing and chemical method called Maxam and Gilbert sequencing.

3.7.1 Sanger's sequencing or dideoxynucelotide chain terminators

In this the reaction mixture is divided into four groups, representing the four dNTPs A, C, G and T. In addition to all the dNTPs being present in the A tube, an analogue of dATP is added (2′, 3′

ddATP) is added that is similar to A but has no 3′ hydroxyl group and so will terminate the growing chain. Situation for other tubes ddC,ddG and ddT are identical except they contain ddCTP, ddGTP and ddTTp respectively.

Since the incorporation of ddNTP rather than dNTP is a random event, the reaction will produce new molecule varying widely in length, but all terminating at the same type of base. Thus four sets of DNA sequences are generated, each terminating at a different type of base, but all have a common 5′- end. The four labelled and chain-terminated samples are then denatured by heating and loaded next to each other for electrophoresis. Electrophoresis is performed at 70°C in presence of urea, to prevent renaturation of DNA. Very thin and long gels are used for maximum resolution over a wide range of fragment lengths. After electrophoresis, the position of radioactive DNA bands on the gel is determined by autoradiography. Since every band in the track from dideoxyadenosine triphosphate must contain molecules that terminate at adenine, and those in ddCTP terminate at cytosine, etc, it is possible to read the sequence of the newly synthesized strand from autoradiograph, provided that the gel can resolve differences in length equal to single nucleotide. Under ideal conditions, sequences up to about 400 bases length can be read from one gel.

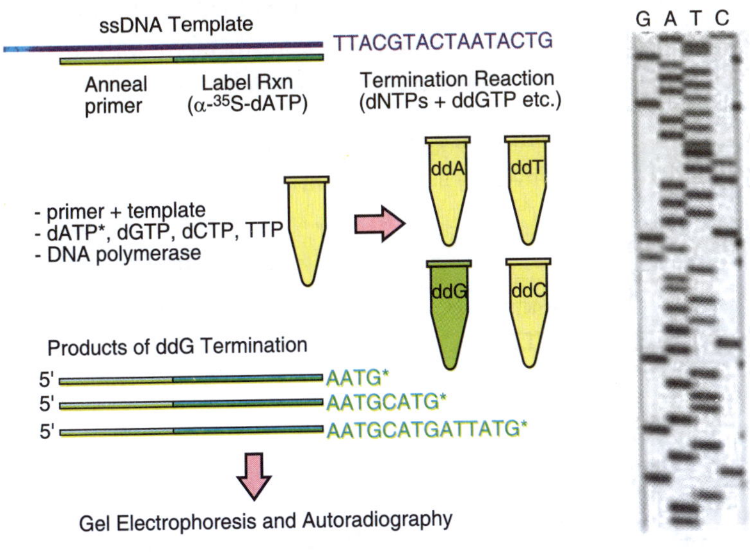

Fig. 3.16: Sanger sequencing

3.7.2 Direct PCR sequencing

It is possible to undertake nucleotide sequencing from double-staranded molecules such as plasmid cloning vectors and PCR products but the double-stranded DNA must be denatured prior to annealing with primer. In case of plasmids, an alkaline denaturation step is sufficient. However, for PCR products this is more problematic and a focus of much research. Unlike plasmids, PCR products are short and reannealed rapidly, so preventing the reannealing process or biasing the

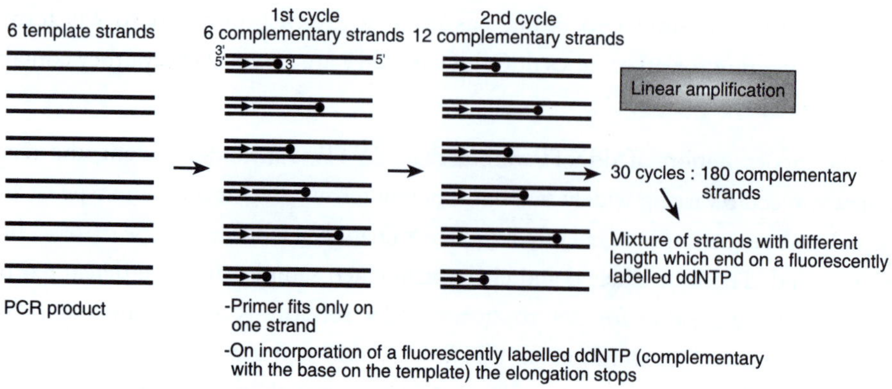

Fig. 3.17: Linear amplification of gene in sequencing

amplification towards one strand by using a primer ratio of 100:1 can overcome this problem to a certain extent. Denaturants such as formamide or dimethylsulphooxide (DMSO) are usually employed to prevent renaturation of PCR starnds after their separation.it is possible to physicaaly separate and retain one PCR starnd by incorporating a molecule such as biotin into one of the primer, which can be recover after PCR by affinity chromatography with strepavidin, leaving the complimentary PCR strand. Thus it provided high quality single stranded DNA for sequencing.

One of the most useful methods of sequencing PCR products is termed PCR cycle sequencing. This is not strictly a PCR, since it involves linear amplification with a single primer in about 20 PCR cycles. Radiolabelled or fluorescent–labelled dideoxynucleotides are then introduced into the final stages of reaction to generate chain termination extention products. Automated direct PCR sequencing is increasingly being refined, allowing greater lengths of DNA to be analysed in one sequencing run.

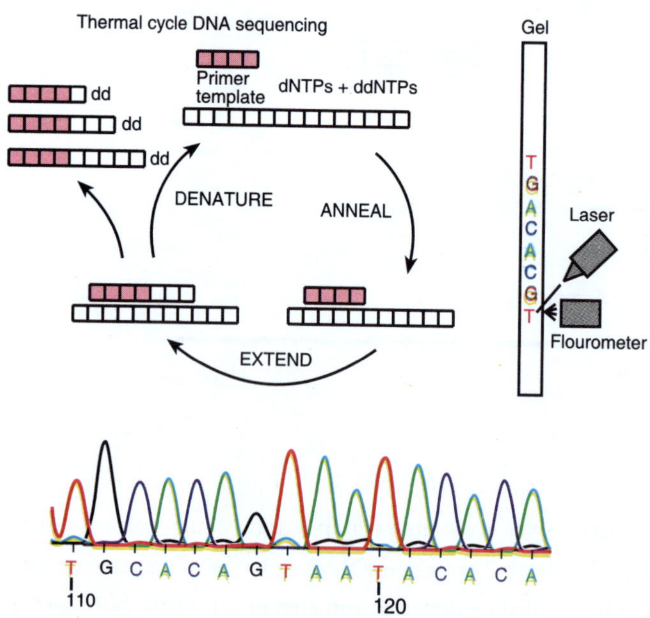

Fig. 3.18: Automated PCR sequencing and Fluorescent sequencing

3.7.3 Automated fluorescent DNA sequencing

This involves dideoxynucleotides labelled with different fluro-chromes and are used to carry out chain termination as in standard reactions. The advantage of this modification is that, sice different labellel is incorporated in each ddNTP all the products are run on same denaturing electrophoresis gel. Each product with their base specific dye is excited by a laser and dye then emits light at its characterstic wavelength. A diffraction grating separates the emissions, which are detected by charge couple device (CCD) and the sequence is interpreted by computer. In addition to real time sequencing, the length of the sequence that may be analysed is in excess of 500 bp.

3.7.4 Maxam and Gilbert sequencing

This is a chemical method of sequencing developed by Maxam and Gilbert and the method is often used for sequencing of small fragments of DNA such as oligonucleotides. A radioactive label is added to either the 3′ or the 5′ end of a double stranded DNA preparation. The strands are then separated by electrophoresis under denaturating conditions and analysed separately. DNA labelled at one end is divided into four aliquots; each is treated with chemicals that act on specific bases by methylation or removal of bases. Conditions are chosen such that each molecule is modified at only one position along its length and every base in the DNA strand has equal chances of being modified. After modification reactions separate samples are cleaved by piperidine, which breaks phosphodiester bonds exclusively at the 5′-side of nucleotides whose base has been modified. The result is similar to that produced by Sanger method, since each sample now contains radiolabelled molecules of various length, all with one end common (the labelled end), and with the other end cut at the same type of base. Analysis of the reaction products by electrophoresis is as already described for Sanger method.

3.8 TIPS TO REMEMBER

Isolation of nucleic acids

Types of Methods:

- differential solubility
- 'adsorption' methods
- density gradient centrifugation.

Isolation of RNA : Special Considerations

- RNAse inhibitors
- extraction in guanidine salts
- phenol extractions at pH 5-6 (pH 8 for DNA)
- treatment with RNase-free DNase
- selective precipitation of high MW forms (rRNA, mRNA) with LiCl oligo-dT column.

Isopycnic/CsCl (density)

- DNA ~1.7 g/cm3
- Protein ~1.3 g/cm3
- RNA > DNA
- ssDNA > dsDNA
- GC content.

Evaluation of nucleic acids

DNA	A_{260}	1.0 ≈ 50 μg/ml
	A_{260}/A_{280}	1.6 – 1.8
RNA	A_{260}	1.0 ≈ 40 μg/ml
	A_{260}/A_{280}	~ 2.0

Design of Primers

- analyze sequence with computer
- amplicon length (250-1000 bp)
- uniqueness (18-28 bases)
- T_m > 55°C
- 50% GC composition
- 3′-GC 'caps'
- no internal complementarity
- no 'primer dimers'
- HPLC purification (optional).

Factors Affecting Hybridization

- temperature
- ionic strength
- chaotropic agents
- probe length
- probe mismatch
- % GC.

Labelling DNA Probes

- Random priming: Used for cloned DNA fragments.
- T4 nucleotide kinase: Used primarily for synthetic oligonucleotides.
- Nick translation: Earlier method replaced by random priming. It has quality control and reproducibility problems.
- Terminal transferase: Adds nucleotides to 3′-end used primarily for generation of homopolymer tails.

3.9 VIVA VOCE

Q.1. What are the precautions that has to be followed while isolating RNA?

Q.2. How can you check the quality of the isolated nucleic acid?

Q.3. Among RNA and DNA which is having greater density and why?

Q.4. What are T_m and T_a?

Q.5. What is dot / slot blot?

Q.6. What are the applications of restriction mapping?

Q.7. What is quantitative PCR?

Q.8. What are gene probes and where are they used?

Q.9. What are the factors affecting PCR amplification?

Q.10. How primers are designed?

Q.11. What do you understand by direct PCR sequencing?

Q.12. What are taqman probes?

Q.13. What is SYBR green and where is it used?

Q.14. Differentiate between touch down and multiplex PCR.

Q.15. What is the advantage of nested PCR?

Q.16. Explain hot-start PCR.

Q.17. Differentiate between chemical and enzymatic method of DNA sequencing

Q.18. What is Plateau effect in PCR?

Q.19. Differentiate between polynucleotide kinase and terminal transferase enzymes.

Q.20. What are adaptors?

CHAPTER 4

Concepts of Cloning

In classical biology, a *clone* is a population of identical organisms derived from a single parental organism. For example, the members of a colony of bacterial cells that arise from a single cell on a petri plate are clones. Molecular biology has borrowed the term to mean a collection of molecules or cells all identical to an original molecule or cell. So, if the original cell on the petri plate harboured a recombinant DNA molecule in the form of a plasmid, the plasmids within the millions of cells in a bacterial colony represent a clone of the original DNA molecule, and these molecules can be isolated and studied. Furthermore, if the cloned DNA molecule is a gene (or part of a gene), that is, it encodes a functional product, a new avenue to isolating and studying this product has opened. Recombinant DNA methodology offers exciting new vistas in biochemistry.

4.1 CLONING VECTORS

For the cloning of any molecule of DNA it is necessary for the DNA to be incorporated into a cloning vector. These are DNA elements that may be stably maintained and propagated in a host organism for which the vector has replication functions. A typical host organism is a bacterium, such as *Escherichia coli*, that grows and divides rapidly. Any vector with a replication origin in *E. coli* will replicate efficiently (together with any incorporated DNA). Thus, any DNA cloned into a vector will enable the amplification of the inserted foreign DNA fragment and also allow any subsequent analysis to be undertaken. In this way the cloning process resembles the PCR, although there are some major differences between the two techniques. By cloning, it is possible not only to store a copy of any particular fragment of DNA but also to produce unlimited amounts of it.

The vectors used for cloning vary in their complexity, ease of manipulation, selection and the amount of DNA sequence they can accommodate (the insert capacity). Vectors have in general been developed from naturally occurring molecules such as bacterial plasmids, bacteriophages or combinations of the elements that make them up, such as cosmids. For gene library constructions there is a choice and trade-off between various vector types, usually related to the ease of the manipulations needed to construct the library and the maximum size of foreign DNA insert of the vector. Thus vectors with the advantage of large insert capacities are usually more difficult to manipulate, although there are many more factors to be considered, which are indicated in the following treatment of vector systems.

4.1.1 Plasmids

Plasmids are naturally occurring; circular, extrachromosomal DNA molecules. Natural strains of the common colon bacterium *Escherichia coli* isolated from various sources harbour diverse plasmids. Often these plasmids carry genes specifying novel metabolic activities that are advantageous to the host bacterium. These activities range from catabolism of unusual organic substances to metabolic functions that endow the host cells with resistance to antibiotics, heavy metals, or bacteriophages. Plasmids that are able to perpetuate themselves in *E. coli,* the bacterium favoured by bacterial geneticists and molecular biologists, have become the darlings of recombinant DNA technology.

Because restriction endonuclease digestion of plasmids can generate fragments with overlapping or "sticky" ends, artificial plasmids can be constructed by ligating different fragments together. Such artificial plasmids were among the earliest recombinant DNA molecules. These recombinant molecules can be autonomously replicated, and hence propagated, in suitable bacterial host cells, provided they still possess a site signalling where DNA replication can begin (also called origin of replication or *ori* sequence).

I. Plasmids as Cloning Vectors

The idea arose that "foreign" DNA sequences could be inserted into artificial plasmids and that these foreign sequences would be carried into *E. coli* and propagated as part of the plasmid. That is, these plasmids could serve as cloning vectors to carry genes. (The word *vector* is used here in the sense of "a vehicle or carrier.") Plasmids useful as cloning vectors possess three common features: a replicator, a selectable marker, and a cloning site (Fig. 4.1). A *replicator* is an origin of replication, or *ori*. The *selectable marker* is

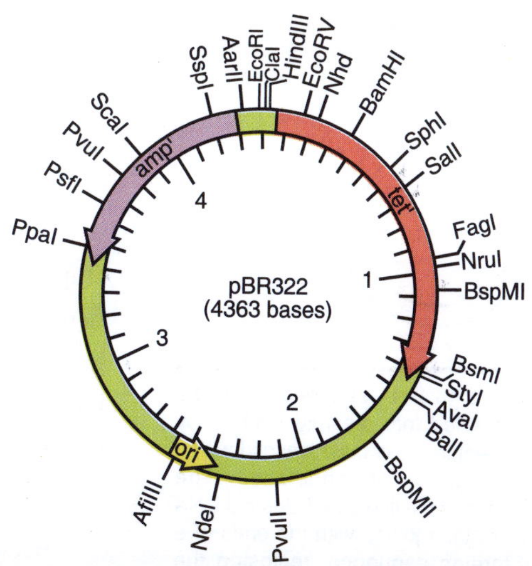

Fig. 4.1: One of the first widely used cloning vectors, the plasmid *pBR*322. This 4363-bp plasmid contains an origin of replication (*ori*) and genes encoding resistance to the drugs ampicillin (*amp*^{r}) and tetracycline (*tet*^{r}). The locations of restriction endonuclease cleavage sites are indicated.

typically a gene conferring resistance to an antibiotic. Only those cells containing the cloning vector will grow in the presence of the antibiotic. Therefore, growth on antibiotic containing media "selects for" plasmid-containing cells. Typically, the cloning *site* is a sequence of nucleotides representing one or more restrictions in endonuclease cleavage sites. Cloning sites are located where the insertion of foreign DNA neither disrupts the plasmid's ability to replicate nor inactivates essential markers.

II. Virtually any DNA Sequence can be Cloned

Nuclease cleavage at a restriction site opens, or linearizes, the circular plasmid so that a foreign DNA fragment can be inserted. The ends of this linearized plasmid are joined to the ends of the fragment so that the circle is closed again, creating a recombinant plasmid (Fig. 4.2). Recombinant plasmids are hybrid DNA molecules consisting of plasmid DNA sequences plus inserted DNA elements (called inserts). Such hybrid molecules are also called chimeric constructs or chimeric plasmids (The term chimera is borrowed from mythology and refers to a beast composed of the body and head of a lion, the heads of a goat and a snake, and the wings of a bat.). The presence of foreign DNA sequences does not adversely affect replication of the plasmid, so chimeric plasmids can be propagated in bacteria just like the original plasmid. Bacteria often harbour several hundred copies of common cloning vectors per cell. Hence, large amounts of a cloned DNA sequence can be recovered from bacterial cultures. The enormous power of recombinant DNA technology stems in part from the fact that virtually any DNA sequence can be selectively cloned and amplified in this manner. DNA sequences that are difficult to clone include inverted repeats, origins of replication, centromeres, and telomeres. The only practical limitation is the size of the foreign DNA segment: most plasmids with inserts larger than about 10 kbp are not replicated efficiently.

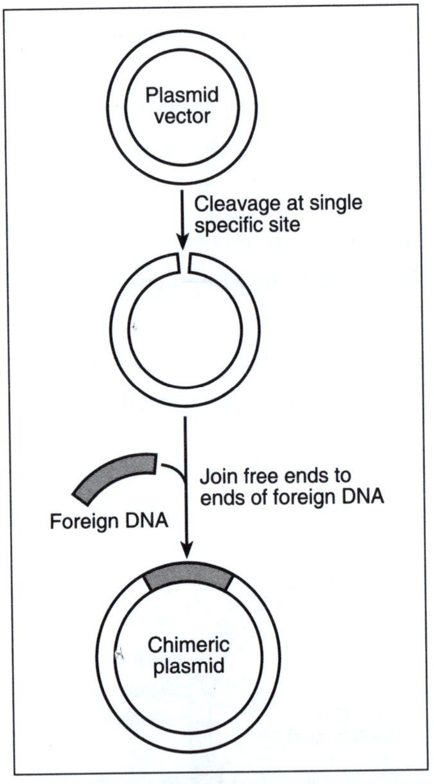

Fig. 4.2: Foreign DNA sequences can be inserted into plasmid vectors by opening the circular plasmid with a restriction endonuclease. The ends of the linearized plasmid DNA are then joined with the ends of a foreign sequence, reclosing the circle to create a chimeric plasmid.

Bacterial cells may harbour one or many copies of a particular plasmid, depending on the nature of the plasmid replicator. That is, plasmids are classified as high copy number or low copy number. The copy number of most genetically engineered plasmids is high (200 or so), but some are lower.

Table 4.1: List of restriction endonucleases with their recognition sites

Restriction Endonucleases

About 1000 restriction enzymes have been characterized. They are named by *italicized* three-letter codes, the first one is a capital letter denoting the genus of the organism of origin, while the next two letters are the abbreviation of the particular species. Because prokaryotes often contain more than one restriction enzymes, the various representatives are assigned letter and number codes as they are identified. Thus, *EroRI* is the initial restriction endonuclease isolated from *Escherichia coli,* strain R. With one exception (*Nci*I), all known type II restriction endonucleases generated fragments with 5′-PO_4 and 3′–OH ends.

Enzyme	Common Isoschizomers	Recognition Sequence	Compatible Cohesive Ends
*Alu*I		AG↓CT	Blunt
*Apy*I	*Atu*I, *Eco*RII	CC↓$\binom{A}{T}$GG	
*Asu*II		TT↓CGAA	*Cla*I, *Hpa*II, *Taq*I
*Ava*I		G↓PyCGPuG	*Sal*I, *Xho*I, *Xma*I
*Avr*II		C↓CTAGG	
*Bal*I		TGG↓CCA	Blunt
*Bam*HI		G↓GATCC	*Bcl*I, *Bgl*II, *Mbo*I, *Sau*3A, *Xho*II
*Bcl*I		T↓GATCA	*Bam*HI, *Bgl*II, *Mbo*I, *Sau*3A, *Xho*II
*Bgl*II		A↓GATCT	*Bam*HI, *Bcl*I, *Mbo*I, *Sau*3A, *Xho*II
*Bst*EII		G↓GTNACC	
*Bst*XI		CCANNNNN↓NTGG	
*Cla*I		AT↓CGAT	*Acc*I, *Acy*I, *Asy*II, *Hpa*II, *Taq*I
*Dde*I		C↓TNAG	
*Eco*RI		G↓AATTC	
*Eco*RII	*Atu*I, *Apy*I	↓CC$\binom{A}{T}$GG	
*Fnu*DII	*Tha*I	CG↓CG	Blunt
*Hac*I		$\binom{A}{T}$GG↓CC$\binom{T}{A}$	Blunt
*Hac*II		PuGCGC↓Py	
*Hac*III		GG↓CC	Blunt
*Hinc*II		GTPy↓PuAC	Blunt
*Hind*III		A↓AGCTT	
*Hpa*I		GTT↓AAC	Blunt
*Hpa*II		C↓CGG	*Acc*I, *Acy*I, *Asu*II, *Cla*I, *Taq*I
*Kpn*I		GGTAG↓G	*Bam*HI, *Bcl*I, *Bgl*II, *Xho*II
*Mbo*I	*Sau*3A	↓GATC	
*Msp*I		C↓CGG	
*Msl*I		TGC↓GCA	Blunt
*Nol*I		GC↓GGCCGC	

Contd.

Enzyme	Common Isoschizomers	Recognition Sequence	Compatible Cohesive Ends
PstI		CTGCA↓G	
SacI	*SstI*	GAGCT↓C	
SalI		G↓TCGAC	*AvaI*, *XhoI*
Sau3A		↓GATC	*Bam*HI, *BclI*, *BglII*, *MboI*, *XhoII*
SfiI		GGCCNNNN↓NGGCC	
SmaI	*XmaI*	CCC↓GGG	Blunt
SphI		GCATG↓G	
SstI	*SacI*	GAGCT↓G	
TaqI		T↓CGA	*AccI*, *AcyI*, *AsuII*, *ClaI*, *HpaII*
XbaI		T↓CTAGA	
XhoI		C↓TCGAG	*AvaI*, *SalI*
XhoII		$\binom{A}{G}$↓GATC$\binom{T}{G}$	*Bam*HI, *BclI*, *BglII*, *MboI*, *Sau3A*
XmaI	*SmaI*	C↓CCGGG	*AvaI*

III. Construction of Chimeric Plasmids

Creation of chimeric plasmids requires joining the ends of the foreign DNA insert to the ends of a linearized plasmid (Fig. 4.2). This ligation is facilitated if the ends of the plasmid and the insert have complementary, singlestranded overhangs. Then these ends can base-pair with one another, annealing the two molecules together. One way to generate such ends is to cleave the DNA with restriction enzymes that make staggered cuts; many such restriction endonucleases are available. For example, if the sequence to be inserted is an *Eco*RI fragment and the plasmid is cut with *Eco*RI, the single stranded sticky ends of the two DNAs can anneal (Fig. 4.3). The interruptions in the sugar–phosphate backbone of DNA can then be sealed with DNA ligase to yield a covalently closed, circular chimeric plasmid. DNA ligase is an enzyme that covalently links adjacent 3′-OH and 5′-PO_4 groups. An inconvenience of this strategy is that *any* pair of *Eco*RI sticky ends can anneal with each other. So, plasmid molecules can reanneal with themselves, as can the foreign DNA restriction fragments. These DNAs can be eliminated by selection schemes designed to identify only those bacteria containing chimeric plasmids. Blunt-end ligation is an alternative method for joining different DNAs. This method depends on the ability of phage T4 DNA ligase to covalently join the ends of any two DNA molecules (even those lacking 3′- or 5′-overhangs) (Fig. 4.4). Some restriction endonucleases cut DNA so that blunt ends are formed. Because there is no control over which pair of DNAs are blunt end and ligated by T4 DNA ligase, strategies to identify the desired products must be applied.

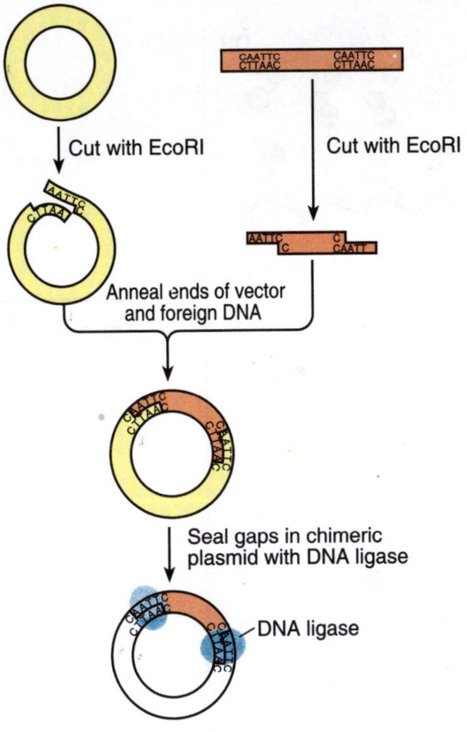

Fig. 4.3: Restriction endonuclease *Eco*RI cleaves double-stranded DNA. The recognition site for *Eco*RI is the hexameric sequence GAATTC:

5′ . . . NpNpNpNp**GpApApTpTpC**pNpNpNpNp . . . 3′

3′ . . . NpNpNpNp**CpTpTpApApG**pNpNpNpNp . . . 5′

Cleavage occurs at the G residue on each strand so that the DNA is cut in a staggered fashion, leaving 5'-overhanging single-stranded ends (sticky ends):

5′ . . . NpNpNpNp**G pApApTpTpC**pNpNpNpNp . . . 3′

3′ . . . NpNpNpNp**CpTpTpApAp G**pNpNpNpNp . . . 5′

An *Eco*RI restriction fragment of foreign DNA can be inserted into a plasmid having an *Eco*RI cloning site by (a) cutting the plasmid at this site with *Eco*RI, annealing the linearized plasmid with the *Eco*RI foreign DNA fragment, and (b) sealing the nicks with DNA ligase.

A great number of variations on these basic themes have emerged. For example, short synthetic DNA duplexes whose nucleotide sequence consists of little more than a restriction site can be blunt-end ligated onto any DNA. These short DNAs are known as linkers. Cleavage of the ligated DNA with the restriction enzyme then leaves tailor-made sticky ends useful in cloning reactions (Fig. 4.5). Similarly, many vectors contain a polylinker cloning site, a short region of DNA sequence bearing numerous restriction sites.

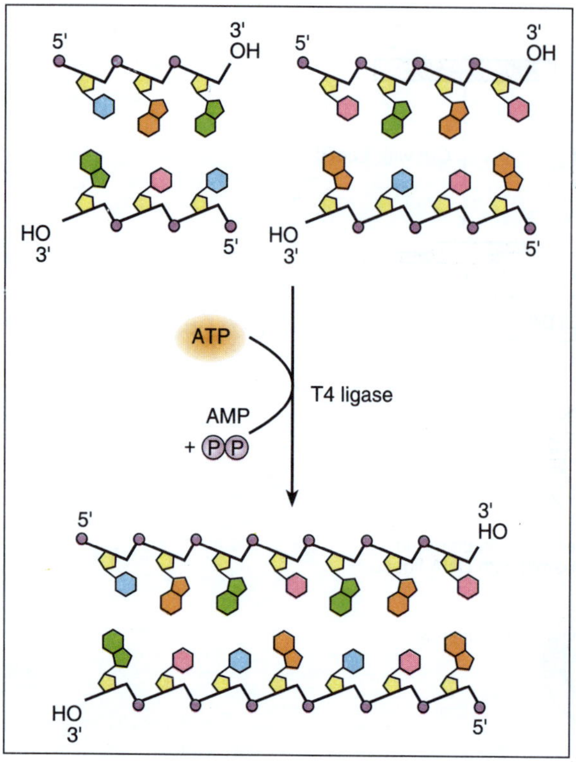

Fig. 4.4: Blunt-end ligation using phage T4 DNA ligase, which catalyzes the ATP dependent ligation of DNA molecules. AMP and PPi are by-products.

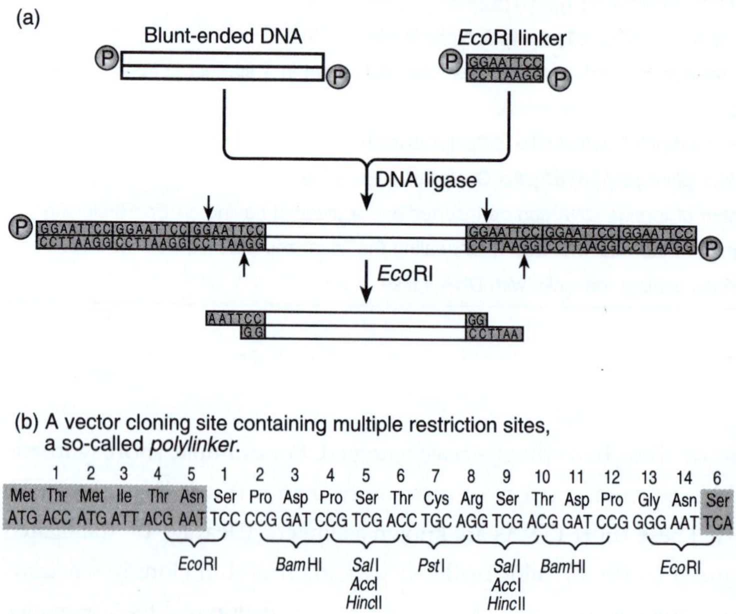

Fig. 4.5: (a) The use of linkers to create tailor-made ends on cloning fragments. Synthetic oligonucleotide duplexes whose sequences represent *Eco*RI restriction sites are blunt-end ligated to a DNA molecule using T4 DNA ligase. Note that the ligation reaction can add multiple linkers on each end of the blunt-ended DNA. *Eco*RI digestion removes all but the terminal one, leaving the desired 5'-overhangs. (b) Cloning vectors often have polylinkers consisting of a multiple array of restriction sites at their cloning sites, so restriction fragments generated by a variety of endonucleases can be incorporated into the vector. Note that the polylinker is engineered not only to have multiple restriction sites but also to have an uninterrupted sequence of codons, so this region of the vector has the potential for translation into protein. The sequence shown is the cloning site for the vectors M13mp7 and pUC7; the shaded amino acid residues are contiguous with the coding sequence of the *lacZ* gene carried by this vector.

IV. Promo2ters and Directional Cloning

Note that the strategies discussed thus far create hybrids in which the orientation of the DNA insert within the chimera is random. Sometimes it is desirable to insert the DNA in a particular orientation. For example, an experimenter might wish to insert a particular DNA (a gene) in a vector so that its gene product is synthesized. To do this, the DNA must be placed downstream from a promoter. A promoter is a nucleotide sequence lying upstream of a gene that controls expression of the gene. RNA polymerase molecules bind specifically at promoters and initiate transcription of adjacent genes, copying template DNA into RNA products. One way to insert DNA so that it will be properly oriented with respect to the promoter is to create DNA molecules whose ends have different overhangs. Ligation of such molecules into the plasmid vector can only take place in one orientation, to give directional cloning (Fig. 4.6).

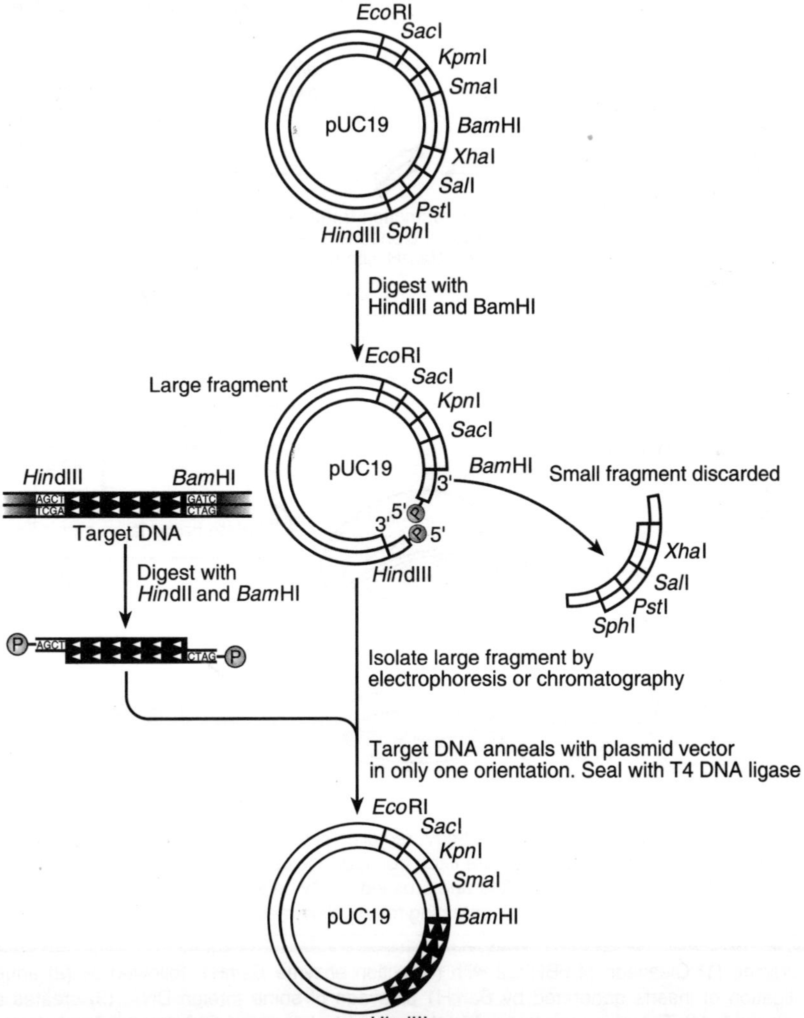

Fig. 4.6: Directional cloning. DNA molecules whose ends have different overhangs can be used to form chimeric constructs in which the foreign DNA can enter the plasmid in only one orientation. The foreign DNA is digested with two different restriction enzymes (*Hind*III and *Bam*HI), and the plasmid is digested with the same two enzymes. Note that pUC19 has a polylinker or universal cloning site.

V. Biologically Functional Chimeric Plasmids

The first biologically functional chimeric DNA molecules constructed *in vitro* were assembled from parts of different plasmids in 1973 by Stanley Cohen, Annie Chang, Herbert Boyer, and Robert Helling. These plasmids were used to transform recipient *E. coli* cells (transformation means the uptake and replication of exogenous DNA by a recipient cell). The bacterial cells were rendered somewhat permeable to DNA by Ca^{2+} treatment and a brief 42°C heat shock. Although less than 0.1% of the Ca^{2+} treated bacteria became competent for transformation following such treatment, transformed bacteria could be selected by their resistance to certain antibiotics (Fig. 4.7). Consequently, the chimeric plasmids must have been biologically functional in at least two aspects: they replicated stably within their hosts and they expressed the drug resistance markers they carried.

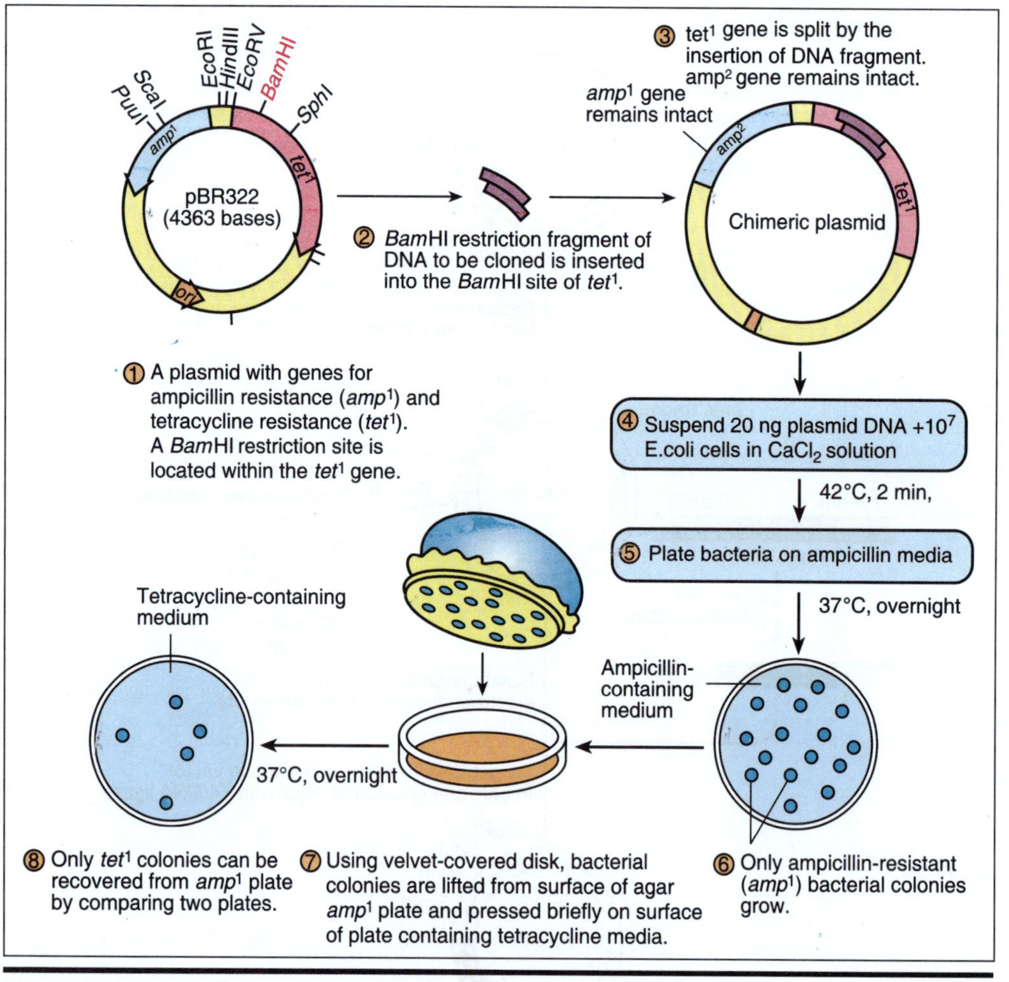

Fig. 4.7: Vector. (1) Cleavage of pBR322 with restriction enzyme *Bam*H1, followed by (2) annealing and ligation of inserts generated by *Bam*H1 cleavage of some foreign DNA, (3) creates a chimeric plasmid. (4) The chimeric plasmid is then used to transform Ca^{2+}-treated heat-shocked *E. coli* cells, and the bacterial sample is plated on a Petri plate. (5) Following incubation of the petri plate overnight at 37°C, (6) colonies of *ampr* bacteria are evident. (7) Replica plating of these bacteria on plates of tetracycline-containing media (8) reveals which colonies are *tetr* and which are tetracycline sensitive *(tets)*. Only the *tets* colonies possess plasmids with foreign DNA inserts.

In general, plasmids used as cloning vectors are engineered to be small, 2.5 kbp to about 10 kbp in size, so that the size of the insert DNA can be maximized. These plasmids have only a single origin of replication, so the time necessary for complete replication depends on the size of the plasmid. Under selective pressure in a growing culture of bacteria, overly large plasmids are prone to delete any nonessential "genes," such as any foreign inserts. Such deletion would thwart the purpose of most cloning experiments. The useful upper limit on cloned inserts in plasmids is about 10 kbp. Many eukaryotic genes exceed this size.

4.1.2 Bacteriophage l as a Cloning Vector

The genome of bacteriophage λ (lambda) is a 48.5 kbp linear DNA molecule that is packaged into the head of the bacteriophage. The middle one-third of this genome is not essential to phage infection, so λ phage DNA has been engineered so that foreign DNA molecules up to 16 kbp can be inserted into this region for cloning purposes. *In vitro* packaging systems are then used to package the chimeric DNA into phage heads which, when assembled with phage tails, form infective phage particles. Bacteria infected with these recombinant phage produce large numbers of phage progeny before they lyse, and large amounts of recombinant DNA can be easily purified from the lysate.

4.1.3 Cosmids

The DNA incorporated into phage heads by bacteriophage λ packaging systems must satisfy only a few criteria. It must possess a 14-bp sequence known as *cos* (which stands for cohesive end site) at each of its ends, and these *cos* sequences must be separated by no fewer than 36 kbp and no more than 51 kbp of DNA. Essentially any DNA satisfying these minimal requirements will be packaged and assembled into an infective phage particle. Other cloning features such as an *ori,* selectable markers, and a polylinker are joined to the *cos* sequence so that the cloned DNA can be propagated and selected in host cells. These features have been achieved by placing *cos* sequences on either side of cloning sites in plasmids to create cosmid vectors that are capable of carrying DNA inserts about 40 kbp in size (Fig. 4.8). Because cosmids lack essential phage genes, they reproduce in host bacteria as plasmids.

4.1.4 Fosmids

Vector containing the single copy E.coli F-factor replicon, developed as an improved method for constructing libraries of cosmid-sized (approximately 40 kb) clones. The stability of inserts cloned into fosmid vectors has been shown to be substantially greater than in high copy vectors. Copy control fosmids, e.g., pCC1fos, contain both the *E. coli* F-factor replicon and the oriV highcopy origin of replication, thus providing the user the clone stability afforded by single-copy fosmid cloning and the high yields of DNA that can be realized from cosmid clones.

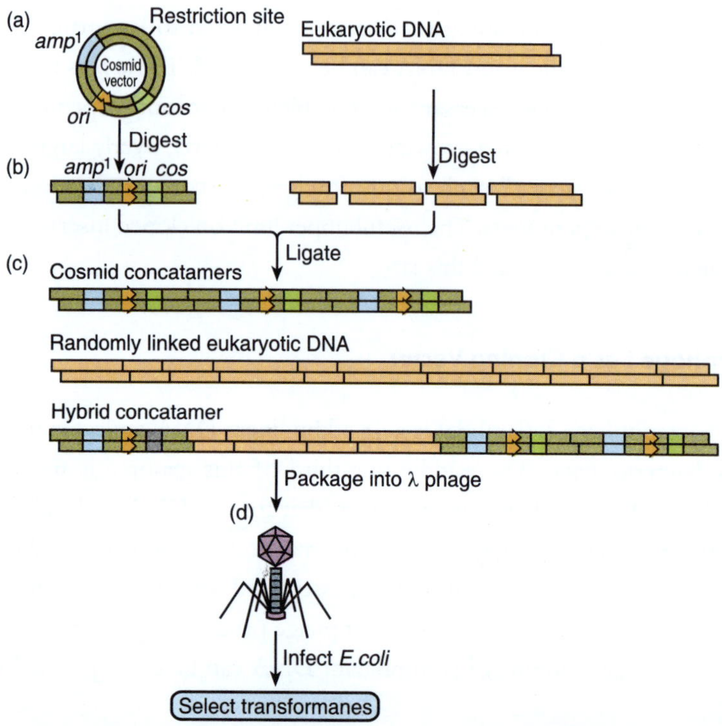

Fig. 4.8: Cosmid vectors for cloning large DNA fragments. (a) Cosmid vectors are plasmids that carry a selectable marker such as *ampr,* an origin of replication (*ori*), a polylinker suitable for insertion of foreign DNA, and (b) a *cos* sequence. Both the plasmid and the foreign DNA to be cloned are cut with a restriction enzyme, and the two DNAs are then ligated together. (c) The ligation reaction leads to the formation of hybrid concatamers, molecules in which plasmid sequences and foreign DNAs are linked in series in no particular order. The bacteriophage packaging extract contains the restriction enzyme that recognizes *cos* sequences and cleaves at these sites. (d) DNA molecules of the proper size (36 to 51 kbp) are packaged into phage heads, forming infective phage particles. (e) The *cos* sequence is

5′-TACG**GGGCGGCGACCT**CGCG-3′

3′-ATGC**CCCGCCGCTGGA**GCGC-5′

Endonuclease cleavage at the sites indicated by arrows leaves 12-bp cohesive ends.

4.1.5 Phagemid

A phage-plasmid vector is able to replicate as single- or double-stranded DNA. Phagemids can be induced to produce phage particles containing single-stranded DNA. Example: pBluescript series (Stratagene), which contains a filamentous **f1 phage** intergenic region including the origin of replication.

4.1.6 Shuttle Vectors

Plasmids used in eukaryotes require a eukaryotic origin of replication and marker genes that will be expressed by eukaryotic cells. At present plasmids are used for cloning in yeast and in plants. Although the yeast has a natural plasmid called the 2 μ circle, this is too large for use in cloning.

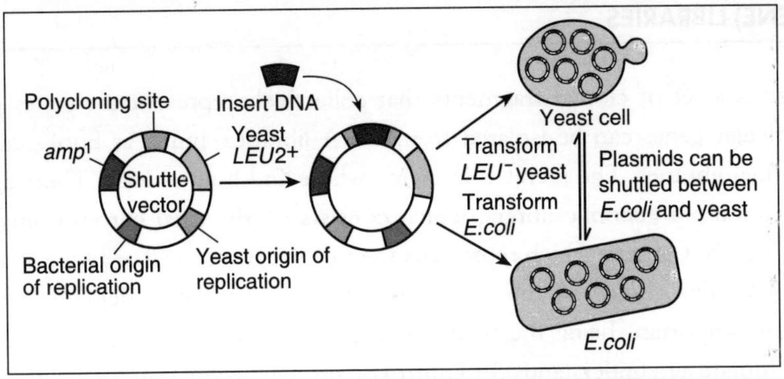

Fig. 4.9: A typical shuttle vector. This vector has yeast and bacterial origins of replication, *ampr* (ampicillin resistance gene for selection in *E. coli*) and *LEU2*$^+$, a gene in the yeast pathway for leucine biosynthesis. The recipient yeast cells are *LEU2*$^-$ (defective in this gene) and thus require leucine for growth. *LEU2*$^-$ yeast cells transformed with this shuttle vector can be selected on medium lacking any leucine supplement.

Plasmids such as yeast episomal plasmids (Yep) have been created by genetic manipulation using replication origins from the 2 μ circle, and by incorporating a gene that will complement a defective gene in the host yeast cell. Strain of such yeast with defective gene can be used as a selectable marker for the presence of that plasmid. Yeast, like bacteria, can be grown rapidly, hence well suited for use in cloning. Particular use has been the creation of shuttle vectors which has origin of replication for yeast and bacteria like *E.coli.*

Shuttle vectors are plasmids capable of propagating and transferring ("shuttling") genes between two different organisms, one of which is typically a prokaryote (*E. coli*) and the other a eukaryote (for example, yeast). Shuttle vectors must have unique origins of replication for each cell type as well as different markers for selection of transformed host cells harbouring the vector (Fig. 4.9). Shuttle vectors have the advantage that eukaryotic genes can be cloned in bacterial hosts, yet the expression of these genes can be analyzed in appropriate eukaryotic backgrounds.

4.1.7 Artificial Chromosomes

DNA molecules 2 megabase pairs in length have been successfully propagated in yeast by creating yeast artificial chromosomes or YACs. Further, such YACs have been transferred into transgenic mice for the analysis of large genes or multigenic DNA sequences *in vivo,* that is, within the living animal. For these large DNAs to be replicated in the yeast cell, YAC constructs must include not only an origin of replication (known in yeast terminology as an autonomously replicating sequence or ARS) but also a centromere and telomeres. Recall that centromeres provide the site for attachment of the chromosome to the spindle during mitosis and meiosis, and telomeres are nucleotide sequences defining the ends of chromosomes. Telomeres are essential for proper replication of the chromosome.

4.2 DNA (GENE) LIBRARIES

A DNA library is a set of cloned fragments that collectively represent the genes of a particular organism. Particular genes can be isolated from DNA libraries, much as books can be obtained from conventional libraries. The secret is knowing where and how to look. There are two general types of gene library: a genomic library, which consists of the total chromosomal DNA of an organism; and a cDNA library, which represents the mRNA from a cell or tissue at a specific point of time (Fig. 3.3). The choice of the particular type of gene library depends on a number of factors, the most important being the final application of any DNA fragment derived from the library. If the ultimate aim understands the control of protein production for a particular gene or its architecture, then genomic libraries must be used. However, if the goal is the production of new or modified proteins, or the determination of tissue-specific expression of timing patterns, cDNA libraries are more appropriate. The main consideration in the construction of genomic or cDNA libraries is, therefore, the nucleic acid starting material. Since the genome of an organism is fixed, chromosomal DNA may be isolated from almost any cell type in order to prepare genomic DNA. In contrast, however, cDNA libraries represent only mRNA being produced from a specific cell type at a particular time in the cell's development. Thus, it is important to consider carefully the cell or tissue type from which the mRNA is to be deriver in the construction of cDNA libraries.

There are a variety of cloning vectors available, many based on naturally occurring molecules such as bacterial plasmids or bacteria-infecting viruses. The choice of vector also depends on whether a genomic library or cDNA library is constructed

4.3 CONSTRUCTING GENE LIBRARIES

4.3.1 Digesting genomic DNA molecules

After genomic DNA has been isolated and purified, it is digested with restriction endonucleases. These enzymes are the key to molecular cloning because of the specificity they have for particular DNA sequences. It is important to note that every copy of a given DNA molecule from a specific organism will give the same set of fragments when digested with a particular enzyme. DNA from different organisms will, in general, give different sets of fragments when treated with the same enzyme. By digesting complex genomic DNA from an organism it is possible to reproducibly divide its genome into a large number of small fragments, each approximately the size of a single gene. Some enzymes cut straight across the DNA to give flush or blunt ends. Other restriction enzymes make staggered single-strand cuts, producing short single-stranded projections at each end of the digested DNA. These ends are not only identical but complementary and will base-pair with each other; they are, therefore, known as cohesive or sticky ends. In addition, the 5′-end projection of the DNA always retains the phosphate groups.

Over 500 restriction enzymes, recognizing more than 200 different sites, have been characterized. The choice of which enzyme to use depends on a number of factors. For example, the recognition sequence of 6 bp will occur, on average, every 4096 (4^6) bases, assuming a random sequence of each of the four bases. This means that digesting genomic DNA with *Eco*RI, which recognizes the sequence 5′-GAATTC-3′, will produce fragments each of which is, on average, just over 4 kb. Enzymes with 8 bp recognition sequences produce much longer fragments. Therefore, very large genomes, such as human DNA, are usually digested with enzymes that produce long DNA fragments. This makes subsequent steps more manageable, since a smaller number of those fragments need to be cloned and subsequently analyzed.

4.3.2 Ligating DNA molecules

The DNA products resulting from restriction digestion to form sticky ends may be joined to any other DNA fragments treated with the same restriction enzyme. Thus, when the two sets of fragments are mixed; base-pairing between sticky ends will result in the annealing of fragments that were derived from different starting DNA. There will, of course, also be pairing of fragments derived from the same starting DNA molecules, termed reannealing. All these pairing are transient, owing to the weakness of hydrogen bonding between the few bases in the sticky ends, but they can be stabilized by use of an enzyme, DNA ligase, in a process termed ligation. This enzyme, usually isolated from bacteriophage T4 and called T4 DNA ligase, forms a covalent bond between the 5′-phosphate at the end of one strand and the 3′-hydoxyl of the adjacent strand. The reaction which is ATP dependent is often carried out at 10°C to lower the kinetic energy of the molecules, and so reduce the chances of base-paired sticky ends parting before they have been stabilized by ligation. However, long reaction times are needed to compensate for the low activity of DNA ligase in the cold. It is also possible to join blunt ends of DNA molecules, although the efficiency of this reaction is much lower than in sticky-ended ligations.

Since ligation reconstructs the site of cleavage, recombinant molecules produced by ligation of sticky ends can be cleaved again at the 'joins', using the same restriction enzyme that was used to generate the fragments initially. In order to propagate digested DNA from an organism it is necessary to join or ligate that DNA with a specialized DNA carrier molecule termed a vector. Each DNA fragment is inserted by ligation into vector DNA molecule, which allows the whole recombinant DNA to then be replicated indefinetly within microbial cells. In this way a DNA fragment can be cloned to provide sufficient material for further detailed analysis or for further manipulations. Thus, all of the DNA extracted from an organism and digested with a restriction enzyme will result in a collection of clones. This collection of clones is known as a gene library.

4.4 GENOMIC LIBRARIES

Any particular gene constitutes only a small part of an organism's genome. For example, if the organism is a mammal whose entire genome encompasses some 106 kbp and the gene is 10 kbp, then the gene represents only 0.001% of the total nuclear DNA. It is impractical to attempt to recover such rare sequences directly from isolated nuclear DNA because of the overwhelming amount of extraneous DNA sequences. Instead, a genomic library is prepared by isolating total DNA from the organism, digesting it into fragments of suitable size, and cloning the fragments into an appropriate vector. This approach is called shotgun cloning because the strategy has no way of targeting a particular gene but instead seeks to clone all the genes of the organism at one time. The intent is that at least one recombinant clone will contain at least part of the gene of interest. This can be achieved by partial restriction digestion with an enzyme that recognizes tetranucleotide sequences. Complete digestion with such an enzyme would produce a large number of very short fragments, but, if the enzyme is allowed to cleave only a few of its potential restriction sites before the reaction is stopped, each DNA molecule will be cut into relatively large fragments. Average fragment size will depend on the relative concentrations of DNA and restriction enzyme and, in particular, on the conditions and durations of incubation. It is also possible to produce fragments of DNA by physical shearing, although the ends of the fragments may need to be repaired to make them flush ended. This is achieved by using a modified DNA polymerase termed *Klenow polymerase*. This is prepared by cleavage of DNA polymerase with subtilisin, giving a large enzyme fragment which has no 5′→3′ exonuclease activity, but which still acts as 5′→3′ polymerase. Using the appropriate dNTPs, this will fill in any recessed 3′ ends on the sheared DNA. The mixture of DNA fragments is then ligated with a vector, and subsequently cloned. If enough clones are produced there will be a very high chance that any particular DNA fragment, such as a gene, will be present in at least one of the clones. To keep the number of clones to a manageable size, fragments about 10 kb in length are needed for prokaryotic libraries, but the length must be increased to about 40 kb for mammalism libraries.

Genomic libraries have been prepared from hundreds of different species. Many clones must be created to be confident that the genomic library contains the gene of interest. The probability, P, that some number of clones, N, contains a particular fragment representing a fraction, f, of the genome is

$$P = 1 - (1 - f)^{N}.$$

Thus,
$$N = \text{In}\ (1 - P)/\ \text{In}\ (1 - f).$$

For example, if the library consists of 10 kbp fragments of the *E. coli* genome (4640 kbp total), over 2000 individual clones must be screened to have a 99% probability (P = 0.99) of finding a particular fragment. Since f =10/4640 = 0.0022 and P = 0.99, N = 2093. For a 99% probability of finding a particular sequence within the 3×10^6 kbp human genome, N would equal almost 1.4 million if the cloned fragments averaged 10 kbp in size. The need for cloning vectors capable of carrying very large DNA inserts becomes obvious from these numbers.

4.4.1 Combinatorial Libraries

Specific recognition and binding of other molecules is a defining characteristic of any protein or nucleic acid. Often, target ligands of a particular protein are unknown, or, in other instances, a unique ligand for a known protein may be sought in the hope of blocking the activity of the protein or otherwise perturbing its function. Combinatorial libraries are the products of emerging strategies to facilitate the identification and characterization of possible ligands for a protein. These strategies are also applicable to the study of nucleic acids. Unlike genomic libraries, combinatorial libraries consist of synthetic oligomers. Arrays of synthetic oligonucleotides printed as tiny dots on miniature solid supports are known as DNA chips. Specifically, combinatorial libraries contain very large numbers of chemically synthesized molecules (such as peptides or oligonucleotides) with randomized sequences or structures. Such libraries are designed and constructed with the hope that one molecule among a vast number will be recognized as a ligand by the protein (or nucleic acid) of interest. If so, perhaps that molecule will be useful in a pharmaceutical application, for instance as a drug to treat a disease involving the protein to which it binds. An example of this strategy is the preparation of a synthetic combinatorial library of hexapeptides. The maximum number of sequence combinations for hexapeptides is 20^6 or 64,000,000. One approach to simplify preparation and screening possibilities for such a library is to specify the first two amino acids in the hexapeptide while the next four are randomly chosen. In this approach, 400 libraries (20^2) are synthesized, each of which is unique in terms of the amino acids at positions 1 and 2 but random at the other four positions (as in AAXXXX, ACXXXX, ADXXXX, etc.) so that each of the 400 libraries contains 20^4 or 1,60,000 different sequence combinations. Screening these libraries with the protein of interest reveals which of the 400 libraries contains a ligand with high affinity. This library is then systematically expanded by specifying the first 3 amino acids (knowing from the chosen 1-of-400 libraries which amino acids are best as the first 2); only 20 synthetic libraries (each containing 20^3 or 8000 hexapeptides) are made here (one for each third-position possibility, the remaining three positions being randomized). Selection for ligand binding, again with the protein of interest, reveals the best of these 20, and this particular library is then varied systematically at the fourth position, creating 20 more libraries (each containing 20^2 or 400 hexapeptides). This cycle of synthesis, screening, and selection is repeated until all six positions in the hexapeptide are optimized to create the best ligand for the protein. A variation on this basic strategy using synthetic oligonucleotides rather than peptides identified a unique 15-mer (sequence GGTTGGTGTGGTTGG) with high affinity (K_D = 2.7 nM) toward thrombin, a serine protease in the blood coagulation pathway. Thrombin is a major target for the pharmacological prevention of clot formation in coronary thrombosis.

4.4.2 Screening Libraries

A common method of screening plasmid-based genomic libraries is to carry out a colony hybridization experiment. The protocol is similar for phage-based libraries except that bacteriophage plaques, not bacterial colonies, are screened. In a typical experiment, host bacteria containing either a plasmid based or bacteriophage-based library are plated out on a petri dish and allowed to grow overnight to form colonies (or in the case of phage libraries, plaques) (Fig 4.10). A replica of the bacterial colonies (or plaques) is then obtained by overlaying the plate with a nitrocellulose disc. The disc is removed, treated with alkali to dissociate bound DNA duplexes into single-stranded DNA, dried,

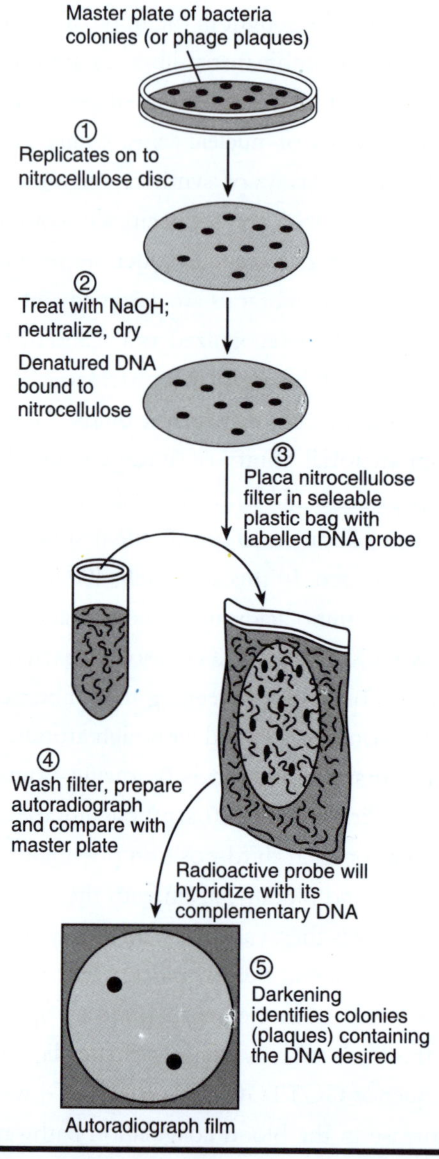

Fig. 4.10: Screening a genomic library by colony hybridization (or plaque hybridization). Host bacteria transformed with a plasmid-based genomic library or infected with a bacteriophage-based genomic library are plated on a petri plate and incubated overnight to allow bacterial colonies (or phage plaques) to form. A replica of the bacterial colonies (or plaques) is then obtained by overlaying the plate with a nitrocellulose disc. (1) Nitrocellulose strongly binds nucleic acids; single-stranded nucleic acids are bound more tightly than double-stranded nucleic acids. (Nylon membranes with similar nucleic acid – and protein-binding properties are also used.) Once the nitrocellulose disc has taken up an impression of the bacterial colonies (or plaques), it is removed and the petri plate is set aside and saved. The disc is treated with 2 M NaOH, neutralized, and dried. (2) NaOH both lyse any bacteria (or phage particles) and dissociates the DNA strands. When the disc is dried, the DNA strands become immobilized on the filter. The dried disc is placed in a sealable plastic bag, and a solution containing heat-denatured (single-stranded), labelled probe is added. (3) The bag is incubated to allow annealing of the probe DNA to any target DNA sequences that might be present on the nitrocellulose. The filter is then washed, dried, and placed on a piece of X-ray film to obtain an autoradiogram. (4) The position of any spots on the X-ray film reveals where the labelled probe has hybridized with target DNA. (5) The location of these spots can be used to recover the genomic clone from the bacteria (or plaques) on the original petri plate.

and placed in a sealed bag with labelled probe. If the probe DNA is duplex DNA, it must be denatured by heating at 70°C. The probe and target DNA complementary sequences must be in a single stranded form if they are to hybridize with one another. Any DNA sequences complementary to probe DNA will be revealed by autoradiography of the nitrocellulose disc. Bacterial colonies (phage plaques) containing clones bearing target DNA are identified on the film and can be recovered from the master plate.

4.4.2.1 Probes for Southern Hybridization

Clearly, specific probes are essential reagents if the goal is to identify a particular gene against a background of innumerable DNA sequences. Usually, the probes that are used to screen libraries are nucleotide sequences that are complementary to some part of the target gene. To make useful probes requires some information about the gene's nucleotide sequence. Sometimes such information is available. Alternatively, if the amino acid sequence of the protein encoded by the gene is known, it is possible to work backward through the genetic code to the DNA sequence (Fig. 4.11). Because the genetic code is degenerate (that is, several codons may specify the same amino acid), probes designed by this approach are usually degenerate oligonucleotides about 17 to 50 residues long (such oligonucleotides are so-called 17- to 50-mers). The oligonucleotides are synthesized so that different bases are incorporated at sites where degeneracies occur in the codons. The final preparation thus consists of a mixture of equal-length oligonucleotides whose sequences vary to accommodate the degeneracies. Presumably, one oligonucleotide sequence in the mixture will hybridize with the target gene. These oligonucleotide probes are at least 17-mers because shorter degenerate oligonucleotides might hybridize with sequences unrelated to the target sequence.

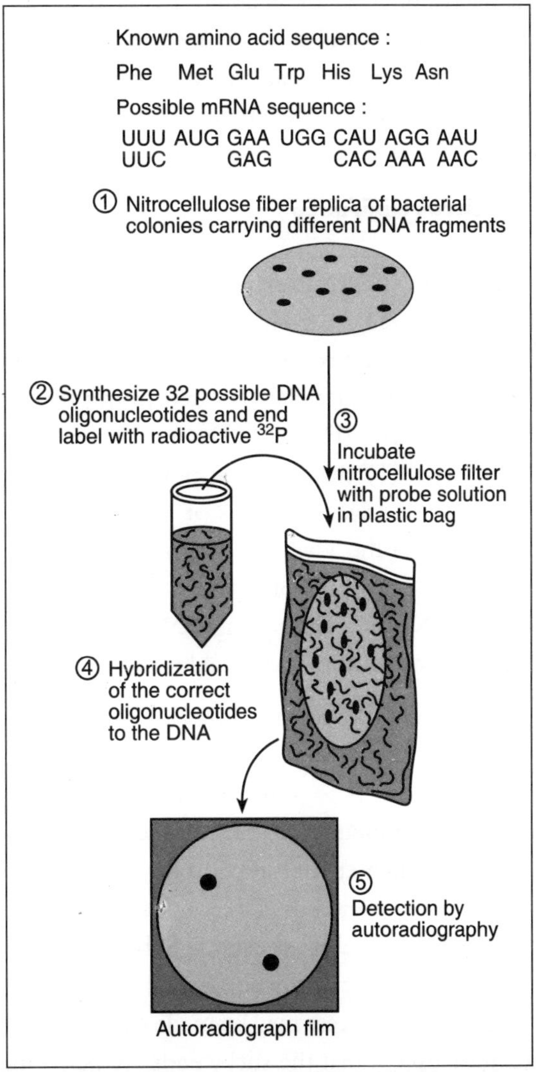

Fig. 4.11: sequence. A radioactively labelled set of DNA (degenerate) oligonucleotides representing all possible mRNA coding sequences is synthesized. (In this case, there are 25 or 32.) The complete mixture is used to probe the genomic library by colony hybridization.

A piece of DNA from the corresponding gene in a related organism can also be used as a probe in screening a library for a particular gene. Such probes are termed heterologous probes because they are not derived from the homologous (same) organism. Problems arise if a complete eukaryotic gene is the cloning target; eukaryotic genes can be tens or even hundreds of kilobase pairs in size. Genes of this size are fragmented in most cloning procedures. Thus, the DNA identified by the probe may represent a clone that carries only part of the desired gene. However, most cloning strategies are based on a partial digestion of the genomic DNA, a technique that generates an overlapping set of genomic fragments. This being so, DNA segments from the ends of the identified clone can now be used to probe the library for clones carrying DNA sequences that flanked the original isolate in the genome. Repeating this process ultimately yields the complete gene among a subset of overlapping clones.

4.5 cDNA LIBRARIES

cDNAs are DNA molecules copied from mRNA templates. cDNA libraries are constructed by synthesizing cDNA from purified cellular mRNA. These libraries present an alternative strategy for gene isolation, especially eukaryotic genes. Because most eukaryotic mRNAs carry 3′-poly(A) tails, mRNA can be selectively isolated from preparations of total cellular RNA by oligo(dT)-cellulose chromatography (Fig. 4.12). DNA copies of the purified mRNAs are synthesized by first annealing short oligo(dT) chains to the poly(A) tails. These oligo(dT) chains serve as primers for reverse transcriptase-driven synthesis of DNA (Fig. 4.13). (Random oligonucleotides can also be used as primers, with the advantages being less dependency on poly(A) tracts and increased likelihood of creating clones representing the 5′-ends of mRNAs.) Reverse transcriptase is an enzyme that synthesizes a DNA strand, copying RNA as the template. DNA polymerase is then used to copy the DNA strand and form a double-stranded (duplex DNA) molecule.

Ligation of blunt-ended DNA fragments is not as efficient as ligation of sticky ends; therefore, with cDNA molecules additional procedures are undertaken before ligation with cloning vectors. One approach is to add cDNA small, double stranded molecules with one internal site for a restriction endonuclease; these are termed nucleic acid linkers. Numerous linkers are commercially available with internal restriction for many of the most commonly used restriction enzymes. Linkers are blunt end ligated to cDNA but since they are added much in excess of the cDNA, the ligation process is reasonably successful. Subsequently the linkers are digested with the appropriate restriction enzyme, which provides the sticky ends for efficient ligation to a vector digested with the same enzyme. This process may be made easier by the addition of adaptors rather than linkers, which are identical except that the sticky ends are performed and so there is no need of restriction digestion following ligation.

Therefore, lastly Linkers are added to the DNA duplexes rendered from the mRNA templates, and the cDNA is cloned into a suitable vector. Once a cDNA derived from a particular gene has been identified, the cDNA becomes an effective probe for screening genomic libraries for isolation of the gene itself.

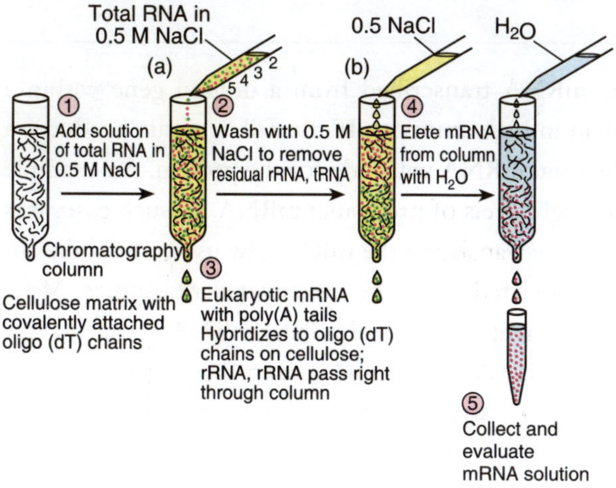

Fig. 4.12: Isolation of eukaryotic mRNA via oligo(dT)-cellulose chromatography. (a) In the presence of 0.5 M NaCl, the poly(A) tails of eukaryotic mRNA anneal with short oligo(dT) chains covalently attached to an insoluble chromatographic matrix such as cellulose. Other RNAs, such as rRNA (green), pass right through the chromatography column. (b) The column is washed with more 0.5 M NaCl to remove residual contaminants. (c) Then the poly(A) mRNA is recovered by washing the column with water because the base pairs formed between the poly(A) tails of the mRNA and the oligo(dT) chains are unstable in solutions of low ionic strength.

Because different cell types in eukaryotic organisms express selected subsets of genes, RNA preparations from cells or tissues in which genes of interest are selectively transcribed are enriched for the desired mRNAs. cDNA libraries prepared from such mRNA are representative of the pattern and extent of gene expression that uniquely define particular kinds of differentiated cells. cDNA libraries of many normal and diseased human cell types are commercially available, including cDNA libraries of many tumour cells. Comparison of normal and abnormal cDNA libraries, in conjunction with two dimensional gel electrophoretic analysis of the proteins produced in normal and abnormal cells is a promising new strategy in clinical medicine to understand disease mechanisms.

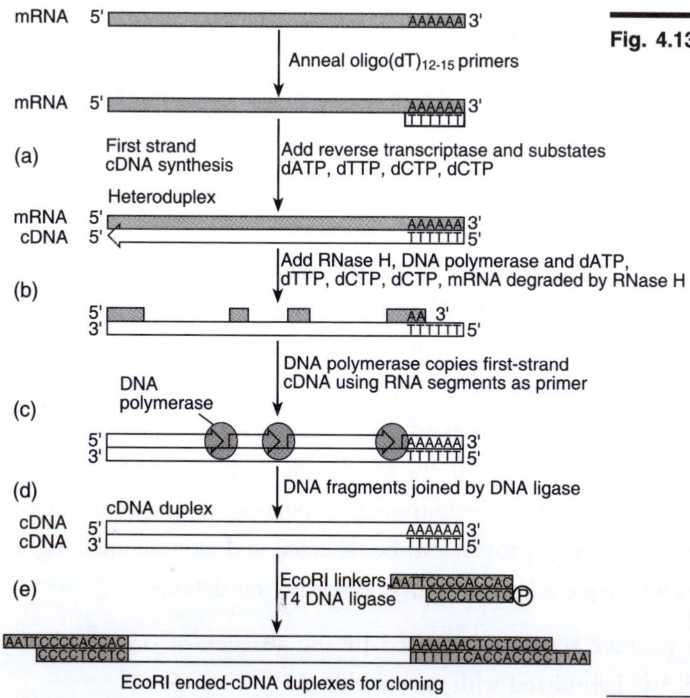

Fig. 4.13: Reverse transcriptase–driven synthesis of cDNA from oligo(dT) primers annealed to the poly(A) tails of purified eukaryotic mRNA. (a) Oligo(dT) chains serve as primers for synthesis of a DNA copy of the mRNA by reverse transcriptase. Following completion of first-strand cDNA synthesis by reverse transcriptase, RNase H and DNA polymerase are added (b). RNase H specifically digests RNA strands in DNA:RNA hybrid duplexes. DNA polymerase copies the first-strand cDNA, using as primers the residual RNA segments after RNase H has created nicks and gaps (c). DNA polymerase has a 5′→3′ exonuclease activity that removes the residual RNA as it fills in with DNA. The nicks remaining in the second-strand DNA are sealed by DNA ligase (d), yielding duplex cDNA. *Eco*RI adapters with 5′-overhangs are then ligated onto the cDNA duplexes (e) using phage T4 DNA ligase to create *Eco*RI ended cDNA for insertion into a cloning vector.

4.6 ENRICHMENT METHODS FOR RNA

Frequently an attempt is made to isolate the mRNA transcribed from a desired gene within a particular cell or tissue that produces the protein in high amounts. If the cell or tissue produces a major protein of the cell, a large fraction of the total mRNA will code for the protein. An example of this is B cells of the pancreas, which contain high levels of proinsulin mRNA. In such cases it is possible to precipitate polysomes, which are actively translating the mRNA, by using antibodies to the ribosomal proteins; mRNA can then be dissociated from the precipitated ribosomes. More usually, the mRNA required is only a minor component of the total cellular mRNA. In such cases total mRNA may be fractioned by size using sucrose density gradient centrifugation. Then each fraction is used to direct the synthesis of proteins using an *in vitro* translation system.

4.6.1 Subtractive hybridization

It is often the case that genes are transcribed in a specific cell type or differentially activated during a particular stage of cellular growth, often at very low levels. It is possible to isolate those mRNA transcripts by subtractive hybridization. Usually the mRNA species common to the different cell types are removed, leaving the cell type or tissue-specific mRNA for analysis. This may be undertaken by isolating the mRNA from the so called subtractor cells and producing a first strand cDNA. The original mRNA from the subtractor cells is then degraded and the mRNA from the target cells isolated and mixed with the cDNA. All the complementary mRNA-cDNA molecules common to both cell types will hybridize, leaving the unbound mRNA, which may be isolated and further analyzed. A more rapid approach for the analyzing the differential expression of genes has been developed using the polymerase chain reaction (PCR).

4.7 HYBRID SELECT/ARREST TRANSLATION

The difficulty of characterizing clone and detecting desired DNA from a mixed cDNA library have been made simpler by two useful techniques namely hybrid select (release) translation or hybrid arrest transltion. After preparation of cDNA library in a plasmid vector, the plasmid is extracted from the part of each colony, and each preparation is then denatured and immobilized on nylon membrane. The membrane is soaked in total cellular mRNA under stringent conditions. The temperature is usually a few degrees below T_m, in which hybridization will occur only between complimentary strands of nucleic acid. Hence membrane will bind to just one species of mRNA, since it has only one type of cDNA immobilized on it. Unbound mRNA is washed off the membrane by Guffer. And then the bound mRNA is used to direct *in vitro* translation. Then by immunoprecipitation or electrophoresis of the protein, the mRNA coding for one particular protein can be detected and clone containing its corresponding cDNA is isolated. This technique is known as hybrid release translation.

In hybrid arrest translation the positive result is indicated by the absence of one particular translation product when total mRNA is hybridized with excess cDNA.

4.8 CLONING PCR PRODUCTS

While PCR has to some extent replaced cloning as a method for the generation of large quantities of a desired DNA fragment, there is, in certain circumstances, still a requirement for the cloning of PCR amplified DNA. For example, certain techniques such as *in vitro* protein synthesis are best achieved with a DNA fragment inserted into an appropriate plasmid or phage cloning vector. Cloning methods for PCR follow closely the cloning of DNA fragments derived from the conventional manipulation of DNA. This may be achieved through one of the two ways: blunt-ended or cohesive-ended cloning. Certain thermostable DNA polymerases such as *Taq* DNA polymerase and *Tth* DNA polymerase give rise to PCR products having a 3′ overhanging A residue. It is possible to clone the PCR products into dT vectors, termed dA:dT cloning. This makes the use of the fact that the terminal additions of A residues may be successfully ligated to vectors prepared with T residue overhangs to follow efficient ligation of PCR product. This reaction is catalyzed by DNA ligase, as in conventional ligation reaction.

It is possible to carry out cohesive-ended cloning with PCR products. In this case, oligonucleotide primers are designed with a restriction endonuclease site incorporated into them. Since the complementarity of the primers needs to be absolute at the 3′ end, the 5′ end of the primer is usually the region for the location of the restriction site. This needs to be designed with care, since the efficiency of digestion of certain restriction endonucleases decreases if extra nucleotides not involved in recognition are absent at the 5′-end. In this case the digestion and ligation reactions are the same as those undertaken for conventional reactions.

4.9 EXPRESSION VECTORS

Frequently it is the product of the cloned gene, rather than the gene itself, that is of primary interest—particularly when the protein has commercial, therapeutic, or research value. With an increased understanding of the fundamentals of DNA, RNA, and protein metabolism and their regulation, investigators can now manipulate cells to express cloned genes in order to study their protein products.

Expression vectors are engineered so that any cloned insert can be transcribed into RNA, and, in many instances, even translated into protein. cDNA expression libraries can be constructed in specially designed vectors derived from either plasmids or bacteriophage l. Proteins encoded by the various cDNA clones within such expression libraries can be synthesized in the host cells, and if suitable assays are available to identify a particular protein, its corresponding cDNA clone can be identified and isolated. Expression vectors designed for RNA expression or protein expression, or both, are available.

I. RNA Expression

A vector for *in vitro* expression of DNA inserts as RNA transcripts can be constructed by putting a highly efficient promoter adjacent to a versatile cloning site. Figure 4.14 depicts such an expression

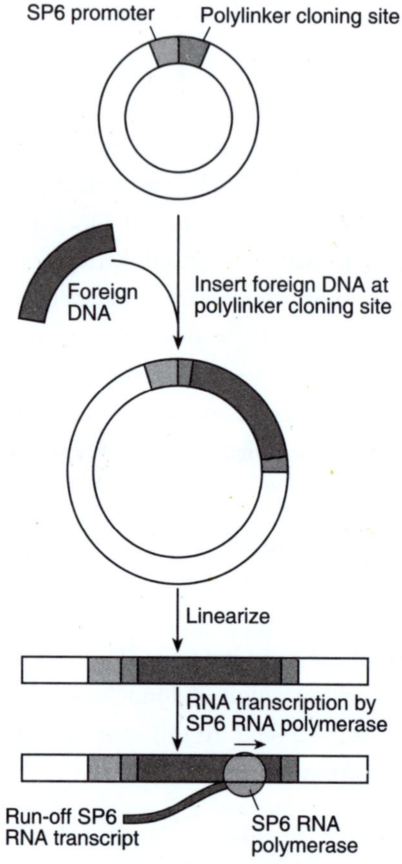

Fig. 4.14: Expression vectors carrying the promoter recognized by the RNA polymerase of bacteriophage SP6 are useful for making RNA transcripts *in vitro*. SP6 RNA polymerase works efficiently *in vitro* and recognizes its specific promoter with high specificity. These vectors typically have a polylinker adjacent to the SP6 promoter. Successive rounds of transcription initiated by SP6 RNA polymerase at its promoter lead to the production of multiple RNA copies of any DNA inserted at the polylinker. Before transcription is initiated, the circular expression vector is linearized by a single cleavage at or near the end of the insert so that transcription terminates at a fixed point.

vector. Linearized recombinant vector DNA is transcribed *in vitro* using SP6 RNA polymerase. Large amounts of RNA product can be obtained in this manner; if radioactive ribonucleotides are used as substrates, labelled RNA molecules useful as probes are made

II. Protein Expression

Because cDNAs are DNA copies of mRNAs, cDNAs are uninterrupted copies of the exons of expressed genes. Because cDNAs lack introns, it is feasible to express these cDNA versions of eukaryotic genes in prokaryotic hosts that cannot process the complex primary transcripts of

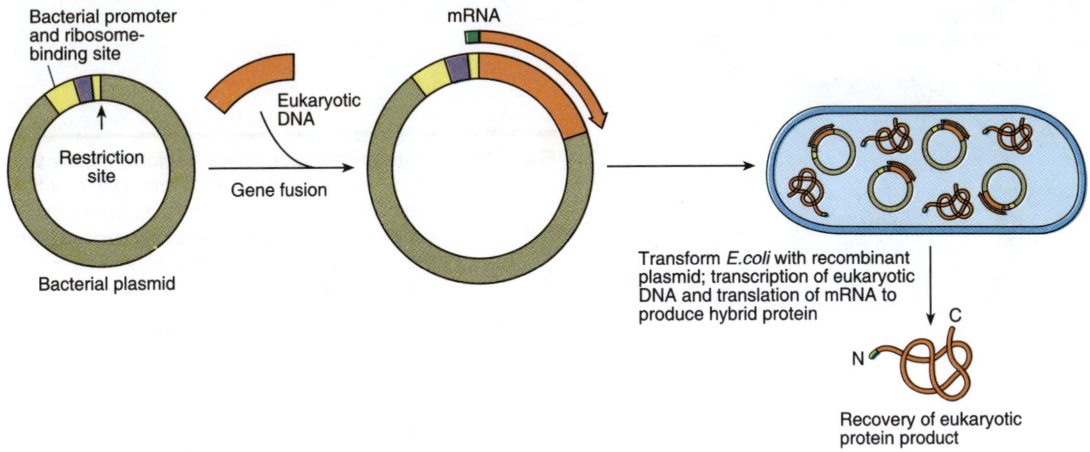

Fig. 4.15: A typical expression-cloning vector. Eukaryotic coding sequences are inserted at the restriction site just downstream from a promoter region where RNA polymerase binds and initiates transcription. Transcription proceeds through a region encoding a bacterial ribosome-binding site and into the cloned insert. The presence of the bacterial ribosome-binding site in the RNA transcript ensures that the RNA can be translated into protein by the ribosomes of the host bacteria.

eukaryotic genes. To express a eukaryotic protein in *E. coli,* the eukaryotic cDNA must be cloned in an expression vector that contains regulatory signals for both transcription and translation.

Accordingly, a promoter where RNA polymerase initiates transcription as well as a ribosome binding site to facilitate translation are engineered into the vector just upstream from the restriction site for inserting foreign DNA. The AUG initiation codon that specifies the first amino acid in the protein (the translation start site) is contributed by the insert (Fig. 4.15). Strong promoters have been constructed that drive the synthesis of foreign proteins to levels equal to 30% or more of total *E. coli* cellular protein. An example is the hybrid promoter, *ptac,* which was created by fusing part of the promoter for the *E. coli* genes encoding the enzymes of lactose metabolism (the *lac* promoter) with part of the promoter for the genes encoding the enzymes of tryptophan biosynthesis (the *trp* promoter) (Fig. 4.16). In cells carrying *ptac* expression vectors, the *ptac* promoter is not induced to drive transcription of the foreign insert until the cells are exposed to *inducers* that lead to its activation. Analogs of lactose (a β-galactoside) such as isopropyl- β-thiogalactoside, or **IPTG,** are excellent inducers of *ptac.* Thus, expression of the foreign protein is easily controlled. The bacterial production of valuable eukaryotic proteins represents one of the most important uses of recombinant DNA technology.

For example, human insulin for the clinical treatment of diabetes is now produced in bacteria. Analogous systems for expression of foreign genes in eukaryotic cells include vectors carrying promoter elements derived from mammalian viruses, such as *simian virus 40 (SV40),* the *Epstein–Barr virus,* and the human *cytomegalovirus (CMV).* A system for high-level expression of foreign genes uses insect cells infected with the baculovirus expression vector. Baculoviruses infect *lepidopteran* insects (butterflies and moths). In engineered baculovirus vectors, the foreign gene is cloned downstream of the promoter for polyhedrin, a major viral-encoded structural protein, and the recombinant vector is incorporated can lead to accumulation of the foreign gene product to levels as high as 500 mg/l.

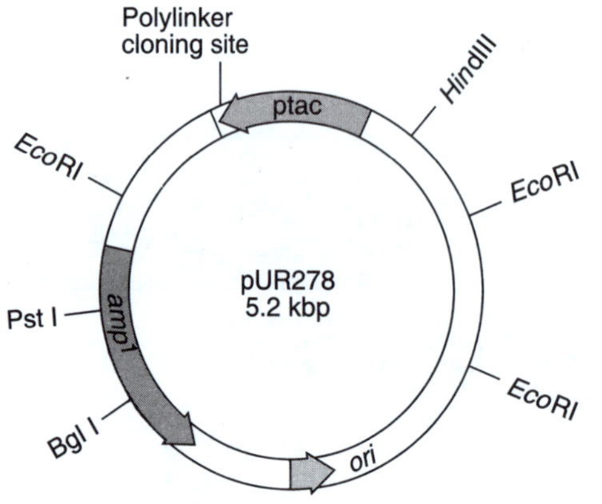

Fig. 4.16: A *ptac* protein expression vector contains the hybrid promoter *ptac* derived from fusion of the *lac* and *trp* promoters. Expression from *ptac* is more than 10 times greater than expression from either the *lac* or *trp* promoter alone. Isopropyl-β-D-thiogalactoside, or IPTG, induces expression from *ptac* as well as *lac.*

III. Screening cDNA Expression Libraries with Antibodies

Antibodies that specifically cross-react with a particular protein of interest are often available. If so, these antibodies can be used to screen a cDNA expression library to identify and isolate cDNA clones encoding the protein. The cDNA library is introduced into host bacteria, which are plated out and grown overnight, as in the colony hybridization scheme previously described. DNA binding nylon membranes are placed on the plates to obtain a replica of the bacterial colonies. The nylon membrane is then incubated under conditions that induce protein synthesis from the cloned cDNA inserts, and the cells are treated to release the synthesized protein. The synthesized protein binds tightly to the nylon membrane, which can then be incubated with the specific antibody. Binding of the antibody to its target protein product reveals the position of any cDNA clones expressing the protein, and these clones can be recovered from the original plate. Like other libraries, expression libraries can be screened with oligonucleotide probes, too.

IV. Fusion Protein Expression

Some expression vectors carry cDNA inserts cloned directly into the coding sequence of a vector-borne protein-coding gene (Fig. 4.17). Translation of the recombinant sequence leads to synthesis of a hybrid protein or fusion protein. The N-terminal region of the fused protein represents amino acid sequences encoded in the vector, whereas the remainder of the protein is encoded by the foreign insert. Keep in mind that the triplet codon sequence within the cloned insert must be in phase with codons contributed by the vector sequences to make the right protein. The N-terminal protein sequence contributed by the vector can be chosen to suit purposes. Furthermore, adding an N-terminal signal sequence that targets the hybrid protein for secretion from the cell simplifies recovery of the fusion protein. A variety of gene fusion systems have been developed to facilitate isolation of a specific protein encoded by a cloned insert. The isolation procedures are based on affinity chromatography purification of the fusion protein through exploitation of the unique ligand-binding properties of the vector-encoded protein (Table 4.2).

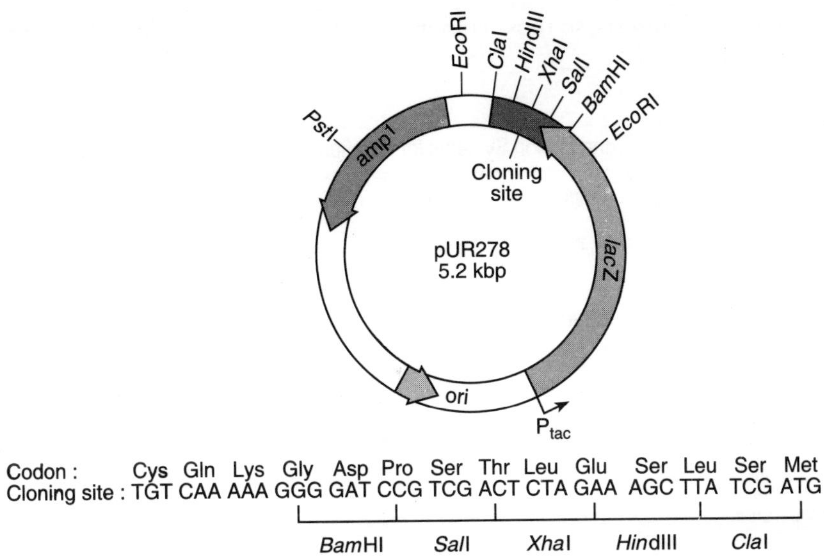

Fig. 4.17: A typical expression vector for the synthesis of a hybrid protein. The cloning site is located at the end of the coding region for the protein β-galactosidase. Insertion of foreign DNAs at this site fuses the foreign sequence to the β-galactosidase coding region (the *lacZ* gene). IPTG induces the transcription of the *lacZ* gene from its promoter *plac*, causing expression of the fusion protein.

V. 1-β-Galactosidase and Blue or White Selection

One version of these fusion protein expression vectors places the cloning site at the end of the coding region of the protein β-galactosidase, so that among other things the fusion protein is attached to β-galactosidase and can be recovered by purifying the β-galactosidase activity. Alternatively, placing the cloning site within the β-galactosidase coding region means that cloned inserts disrupt the β-galactosidase amino acid sequence, inactivating its enzymatic activity. This property has been exploited in developing a visual screening protocol that distinguishes those clones in the library that bear inserts from those that lack them.

Cells that have been transformed with a plasmid-based β-galactosidase expression cDNA library (or infected with a similar library constructed in a bacteriophage λ-based β-galactosidase fusion vector) are plated on media containing 5-bromo-4-chloro-3-indolyl-β-D-galactopyranoside, or **X-gal** (Fig. 4.18). X-gal is a chromogenic substrate, a colourless substance that upon enzymatic reaction yields a coloured product. Following induction with IPTG, bacterial colonies (or plaques) harbouring vectors in which the β-galactosidase gene is intact (those vectors lacking inserts) express an active β-galactosidase that cleaves X-gal, liberating 5-bromo-4-chloro-indoxyl, which dimerizes to form an indigo blue product. These blue colonies (or plaques) represent clones that lack inserts. The β-galactosidase gene is

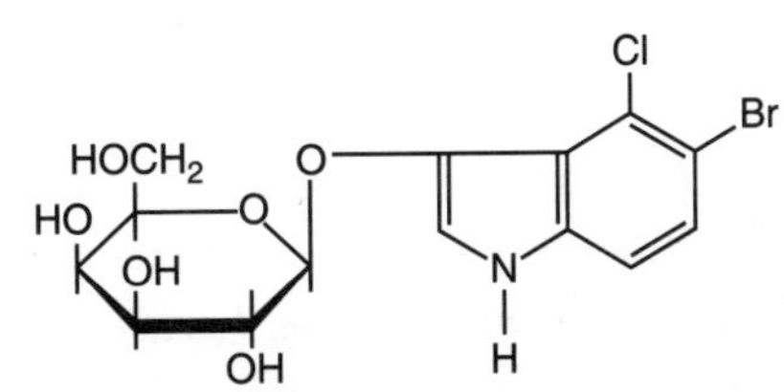

X-gal

Fig. 4.18: The structure of 5-bromo-4-chloro-3-indolyl-β-D-galactopyranoside, or X-gal.

inactivated in clones with inserts, so those colonies (or plaques) that remain "white" (actually, colourless) are recombinant clones.

Table 4.2: Gene Fusion Systems for Isolation of Cloued Fusion Proteins

Gene Product	Origin	Molecular Mass	Secreted	Affinity Ligand
β-Galactosidase	*E. coli*	116	No	β-Aminophenyl-β-o-thiogalactoside (APTG)
Protein A	*S. aunus*	31	Yes	Immunoglobulin G (IgG)
Chloramphenicol acctyltransferase (CAT)	*E. coli*	24	Yes	Chloramphenicol
Streptavidin	*Streptomyces*	13	Yes	Biotin
Glutathione-S-transferase (GST)	*E. coli*	26	No	Glutathione
Maltose-binding protein (MBP)	*E. coli*	40	Yes	Starch

4.10 ALTERATIONS IN CLONED GENES PRODUCE MODIFIED PROTEINS

Cloning techniques can be used not only to overproduce proteins but to produce protein products subtly altered from their native forms. Specific amino acids may be replaced individually by site-directed mutagenesis. This powerful approach to studying protein structure and function changes the amino acid sequence of a protein by altering the DNA sequence of the cloned gene. If appropriate restriction sites flank the sequence to be altered, researchers can simply remove a DNA segment and replace it with a synthetic one that is identical to the original except for the desired change (Fig. 4.19a). When suitably located restriction sites are not present, an approach called oligonucleotide-directed mutagenesis (Fig. 4.19b) can create a specific DNA sequence change. A short synthetic DNA strand with a specific base change is annealed to a single-stranded copy of the cloned gene within a suitable vector. The mismatch of a single base pair in 15 to 20 bp does not prevent annealing if it is done at an appropriate temperature. The annealed strand serves as a primer for the synthesis of a strand complementary to the plasmid vector. This slightly mismatched duplex recombinant plasmid is then used to transform bacteria, where the mismatch is repaired by cellular DNA repair enzymes. About half of the repair events will remove and replace the altered base and restore the gene to its original sequence; the other half will remove and replace the normal base, retaining the desired mutation. Transformants are screened (often by sequencing their plasmid DNA) until a bacterial colony containing a plasmid with the altered sequence is found.

Changes can also be introduced that involve more than one base pair. Large parts of a gene can be deleted by cutting out a segment with restriction endonucleases and ligating the remaining portions to form a smaller gene. Parts of two different genes can be ligated to create new combinations. The product of such a fused gene is called a fusion protein.

Researchers now have ingenious methods to bring about virtually any genetic alteration in vitro. Reintroduction of the altered DNA into the cell permits investigation of the consequences of the alteration. Site-directed mutagenesis has greatly facilitated research on proteins by allowing investi-

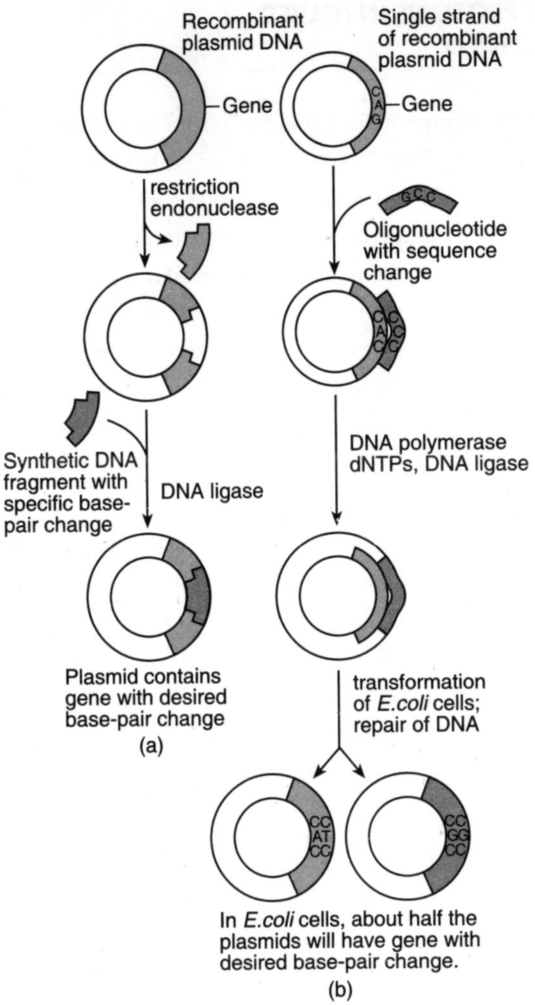

Fig. 4.19: Two approaches to site-directed mutagenesis. (a) A synthetic DNA segment replaces a DNA fragment that has been removed by cleavage with a restriction endonuclease. (b) A synthetic oligonucleotide with a desired sequence change at one position is hybridized to a single-stranded copy of the gene to be altered. This acts as primer for synthesis of a duplex DNA (with one mismatch), which is then used to transform cells. Cellular DNA repair systems will convert about 50% of the mismatches to reflect the desired sequence change.

gators to make specific changes in the primary structure of a protein and to examine the effects of these changes on the folding, three dimensional structure, and activity of the protein.

4.11 REPORTER GENE CONSTRUCTS

Potential regulatory regions of genes (such as promoters) can be investigated by placing these regulatory sequences into plasmids upstream of a gene, called a reporter gene, whose expression is easy to measure. Such chimeric plasmids are then introduced into cells of choice (including eukaryotic cells) to assess the potential function of the nucleotide sequence in regulation because expression of the reporter gene serves as a report on the effectiveness of the regulatory element. A number of different genes have been used as reporter genes, such as the *lacZ* gene. A reporter gene with many inherent advantages is encoding of the green fluorescent protein (or GFP). Unlike the protein expressed by other reporter gene systems, GFP does not require any substrate to measure its activity, nor is it dependent on any cofactor or prosthetic group. Detection of GFP requires only irradiation with near UV or blue light (400-nm light is optimal), and the green fluorescence (light of 500 nm) that results are easily observed with the naked eye, although it can also be measured precisely with a fluorometer.

4.12 THE TWO-HYBRID SYSTEM TO IDENTIFY PROTEINS INVOLVED IN SPECIFIC PROTEIN-PROTEIN INTERACTIONS

Specific interactions between proteins (so-called protein-protein interactions) lie at the heart of many essential biological processes. Stanley Fields, Cheng-Ting Chien, and their collaborators have invented a method to identify specific protein-protein interactions *in vivo* through expression of a reporter gene whose transcription is dependent on a functional transcriptional activator, the *GAL4* protein. The *GAL4* protein consists of two domains: a DNA-binding (or DB) domain and a transcriptional activation (or TA) domain. Even if expressed as separate proteins, these two domains will still work, provided they can be brought together. The method depends on two separate plasmids encoding two hybrid proteins, one consisting of the *GAL4* DB domain fused to protein X, and the other consisting of the *GAL4* TA domain fused to protein Y (Fig. 4.19, part A). If proteins X and Y interact in a specific protein-protein interaction, the *GAL4* DB and TA domains are brought together so that transcription of a reporter gene driven by the *GAL4* promoter can take place (Fig. 4.19, part B). Protein X, fused to the *GAL4*-DNA binding domain (DB), serves as the "bait" to fish for the protein Y "target" and its fused *GAL4* TA domain. This method can be used to screen cells for protein "targets" that interact specifically with a particular "bait" protein. To do so, cDNAs encoding proteins from the cells of interest are inserted into the TA-containing plasmid to create fusions of the cDNA coding sequences with the *GAL4* TA domain coding sequences, so a fusion protein library is expressed.

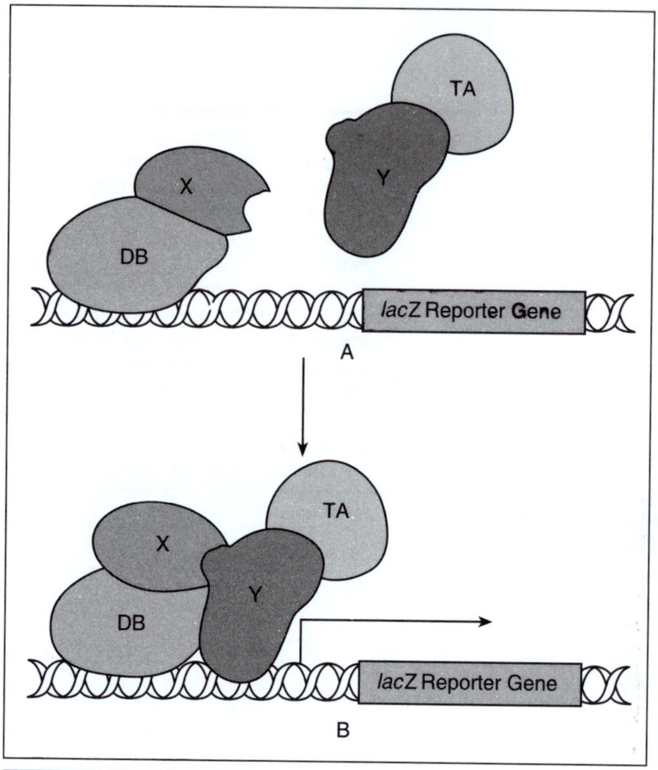

Fig. 4.20: Example of two hybrid system

Identification of a target of the "bait" protein by this method also yields directly a cDNA version of the gene encoding the "target" protein.

4.13 DNA FINGERPRINTING

DNA fingerprinting is based on sequence polymorphisms, slight sequence differences (usually single base-pair changes) between individuals, 1 bp in every 1,000 bp, on average. Each difference from the prototype human genome, sequence (the first one obtained) occurs in some fraction of the

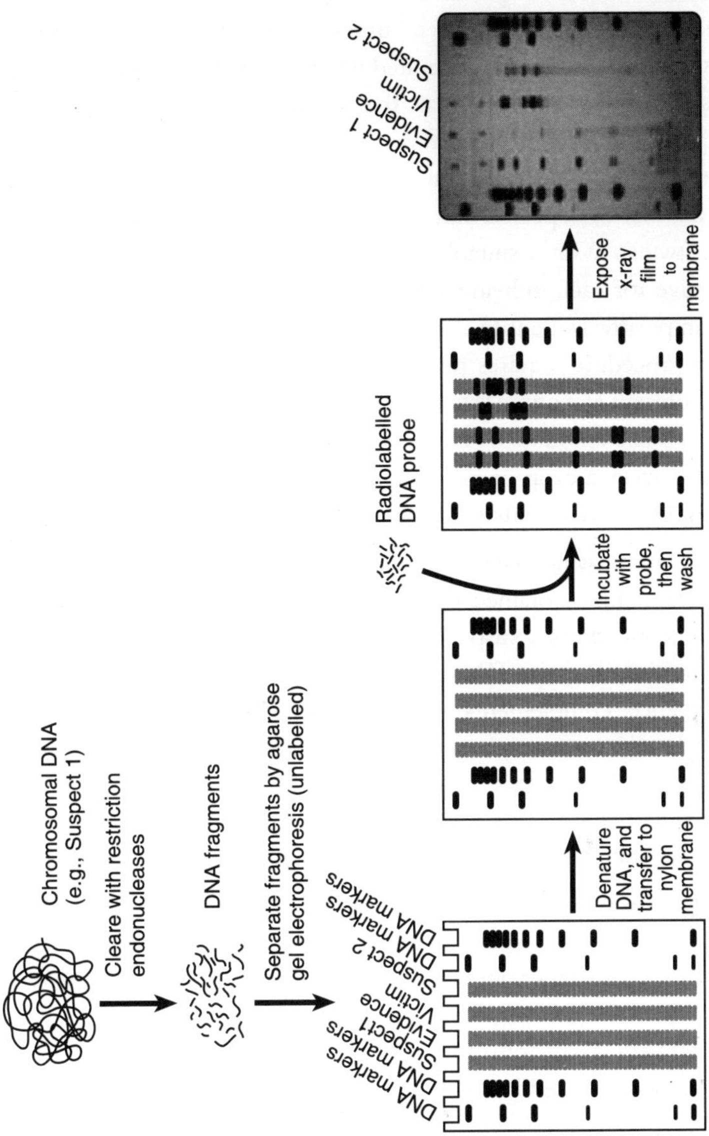

Fig. 4.21: DNA fingerprinting procedure

human population; every individual has some differences. Some of the sequence changes affect recognition sites for restriction enzymes, resulting in variation in the size of DNA fragments produced by digestion with a particular restriction enzyme. These variations are restriction fragment length polymorphisms (RFLPs). The detection of RFLPs relies on a specialized hybridization procedure called Southern blotting. DNA fragments from digestion of genomic DNA by restriction endonucleases are separated by size electrophoretically, denatured by soaking the agarose gel in alkali, and then blotted onto a nylon membrane to reproduce the distribution of fragments in the gel. The membrane is immersed in a solution containing a radioactively labelled DNA probe. A probe for a sequence that is repeated several times in the human genome generally identifies a few of the thousands of DNA fragments generated when the human genome is digested with a restriction endonuclease. Autoradiography reveals the fragments to which the probe hybridizes. The genomic DNA sequences used in these tests are generally regions containing repetitive DNA (short sequences repeated thousands of times in tandem), which are common in the genomes of higher eukaryotes. The number of repeated units in these DNA regions varies among individuals (except between identical twins). With a suitable probe, the pattern of bands produced by DNA fingerprinting is distinctive for each individual. Combining the use of several probes makes the test so selective that it can positively identify a single individual in the entire human population. However, the Southern blot procedure requires relatively fresh DNA samples and larger amounts of DNA than are generally present at a crime scene. RFLP analysis sensitivity is augmented by using PCR to amplify vanishingly small amounts of DNA. This allows investigators to obtain DNA fingerprints from a single hair follicle, a drop of blood, a small semen sample from a rape victim, or samples that might be months or even many years old.

These methods are proving decisive in court cases worldwide. In the example in Fig. 4.21, the DNA from a semen sample obtained from a rape and murder victim was compared with DNA samples from the victim and two suspects. Each sample was cleaved into fragments and separated by gel electrophoresis. Radioactive DNA probes were used to identify a small subset of fragments that contained sequences complementary to the probe. The sizes of the identified fragments varied from one individual to the next, as seen here in the different patterns for the three individuals (victim and two suspects) tested. One suspect's DNA exhibits a banding pattern identical to that of a semen sample taken from the victim. This test used a single probe, but three or four different probes would be used (in separate experiments) to achieve an unambiguous positive identification.

Such results have been used to both convict and acquit suspects and, in other cases, to establish paternity with an extraordinary degree of certainty. The impact of these procedures on court cases will continue to grow as societies agree on the standards and as formal methods become widely established in forensic laboratories. Even decades-old murder mysteries can be solved: in 1996, DNA fingerprinting helped to confirm the identification of the bones of the last Russian czar and his family, who were assassinated in 1918.

4.14 TIPS TO REMEMBER

Recombinant DNA

- prepare foreign (target) DNA
  - (a) gDNA (fragments)
  - (b) cDNA (copy of RNA)
- prepare vector (host)
- recombine target and vector DNA
- introduce rDNA to host
- screen for DNA of interest

Preparation of cDNA

1. Isolate mRNA
2. Synthesize DNA-RNA hybrid
   - reverse transcriptase
     - (a) RNA dependent
     - (b) DNA polymerase
   - oligo-dT primer
   - random priming
3. Synthesize 2nd DNA strand
   - self-priming
   - replacement synthesis
   - primed synthesis
4. Add termini (linkers or adaptors)

Plasmids

- extra-chromosomal elements
- 1-200 kb size range
- transmitted during conjugation
- antibiotic resistance
- low copy number vs. high copy number

Termini	Cloning Requirements	Comments
Identical Overhangs	Phosphatase treatment of linear plasmid improves efficiency.	Restriction sites at junctions preserved. Both orientations of insert DNA possible. Tandem copies of insert possible.
Blunt-end	High concentrations of DNA and ligase needed. Phosphatase treatment.	Restriction sites at junctions often eliminated. Tandem copies of insert DNA possible. Both orientations possible.
Different Overhangs	Purification of double-cut plasmid increases efficiency.	Restriction sites at junctions preserved. Background of non-recombinants is low. One possible orientation of insert. Tandem copies unlikely.

Classes of Restriction Enzymes

Type I cleavage occurs 400-7000 bp from recognition site.

Type II cleavage occurs adjacent or within recognition site.

Type III cleavage occurs 25-27 bp from recognition site.

- Eukaryotic DNA differs from bacterial (prokaryotic) DNA; in that it has introns (intervening sequences) and exons (expressed or translated sequences).

- In order for a eukaryotic gene to be expressed, the introns are 'edited' out of mRNA after transcription.

Screening by Hybridization

- Plaque/colony lifting
- Making labelled probes
- Hybridization
  - Pre-hybridization
  - Hybridization
  - Washing
- Detection
  - Radioactivity
  - Luminescence
  - Fluorescence
  - Chemiluminescence
- Rescuing clones and further analysis

4.15 VIVA VOCE

Q.1. What is cloning?

Q.2. How can you clone a man?

Q.3. Differentiate between cloning vector and expression vector?

Q.4. Name one eukaryotic cloning vector.

Q.5. Explain shuttle vectors.

Q.6. What are the qualities that plasmid possesses to be used as a cloning vector?

Q.7. What do you understand by high copy number and low copy number plasmids?

Q.8. Explain the principle of DNA fingerprinting.

Q.9. What are DNA/gene libraries?

Q.10. Differentiate between genomic DNA library and cDNA library.

Q.11. How can you prepare cDNA library?

Q.12. Colony assay is used for............................

Q.13. What are adaptors and where are they used?

Q.14. Differentiate between adaptors and linkers.

Q.15. What are combinational libraries?

Q.16. What is hybrid arrest translation and how does it differ from hybrid release translation?

Q.17. What is transformation and how is it achieved?

Q.18. What are the qualities of good cloning vector?

Q.19. What do you understand by fusion proteins?

Q.20. What is side directed mutagenesis?

Q.21. Differentiate between side directed mutagenesis and oligonucleotide directed mutagenesis.

CHAPTER 5

Centrifugation

5.1 INTRODUCTION

A centrifuge is a device for separating particles from a solution according to their size, shape, density, viscosity of the medium and rotor speed. In a solution, particles whose density is higher than that of the solvent sink (sediment), and particles that are lighter than it float to the top. The greater the difference in density, the faster they move. If there is no difference in density (isopycnic conditions), the particles stay steady. To take advantage of even tiny differences in density to separate various particles in a solution, gravity can be replaced with the much more powerful "centrifugal force" provided by a centrifuge. This technique plays crucial role in biochemistry or biotechnology as it is non-dispensable part of one or the other step in every method involved in biological study right from the separation of cell organelles to complex experiments involving separation of sub-cellular fractions.

5.2 BASIC PRINCIPLES OF SEDIMENTATION

The rate of sedimentation is dependent upon the applied centrifugal field (G) being directed readily outwards; this is determined by the square of the angular velocity of the rotor (ω in radians s^{-1}) and the radians (r, in centimetres) of the particle from the axis of the rotation, according to the equation

$$G = \omega^2 r.$$

Since one revolution of the rotor is equal to 2 radians, its angular velocity, in radians s^{-1}, can be readily expressed in terms of revolutions per minute (rev min^{-1}), the common way of expressing rotor speed being

$$\omega = \frac{2\pi \text{ rev min}^{-1}}{60}.$$

The centrifugal field (G) in terms of rev min^{-1} is then

$$G = \frac{4\pi^2 (\text{rev min}^{-1})^2 r}{3600}$$

and is generally expressed as a multiple of the earth's gravitational field (g = 981 cm s^{-1}), i.e., the ratio of the weight of the particle in the centrifugal field to the weight of the same particle when acted on by gravity alone, and is then referred to as the relative centrifugal field (RCF) or more commonly as the 'number times g'.

Hence

$$RCF = \frac{4\pi^2 (\text{rev min}^{-1})^2 r}{3600 \times 981},$$

which may be shortened to give

$$RCF = (1.118 \times 10^{-5})\ (\text{rev min}^{-1})^2\ r.$$

Because rotors are different from various manufactures, we use RCF to represent the centrifugation force.

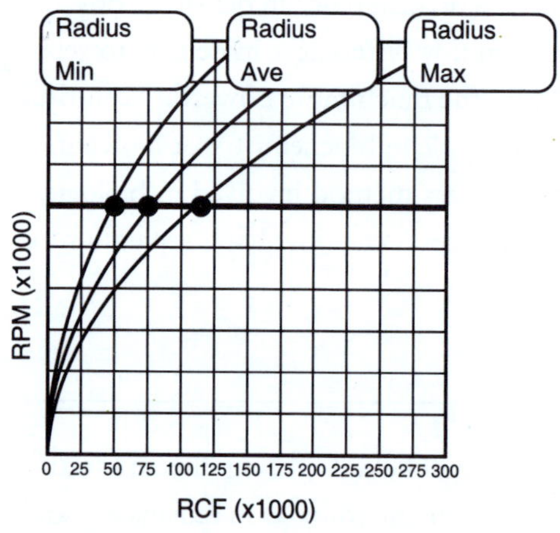

Fig. 5.1: Relation between RCF and RPM

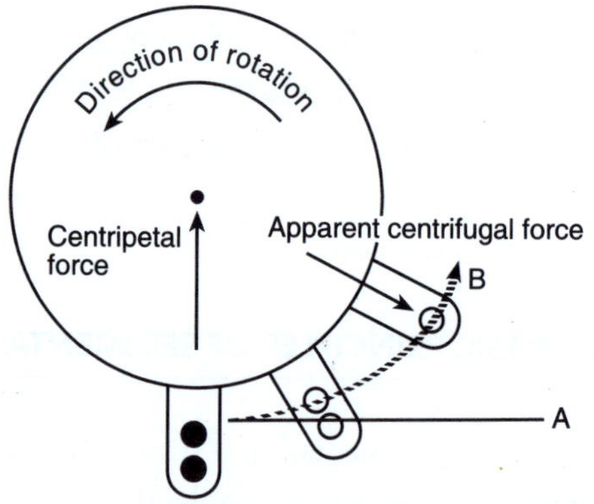

Fig. 5.2: Different forces during centrifugation

When conditions for the centrifugal separation of particles are reported, therefore, rotor speed, radial dimensions and time of operation of the rotor must all be quoted. Since biochemical experiments are usually conducted with particles dissolved or suspended in solution, the rate of sedimentation of a particle is dependent not only upon the applied centrifugal field but also upon the mass of the particle, which may be expressed as the product of its volume and density, the density and viscosity of the medium in which it is sedimenting and the extent to which its shape deviates from spherical. When particle sediments it must displace some of the solution in which it is suspended, resulting in an apparent upthrust on the particle equal to the weight of the liquid displaced. If a particle is assumed to be spherical and of known volume and density, the latter being corrected for the buoyancy due to the density of the medium, then the net outward force (F) it experiences when centrifuged at an angular velocity of ω radians s^{-1} is given by

$$F = \frac{4}{3}\pi r_p^3 (\rho_p - \rho_m)\omega^2 r,$$

where $\frac{4}{3}\pi r_p^3$ is the volume of a sphere of radius r_p, ρ_p is the density of the particle, ρ_m is the density of the suspending medium, and r is the distance of the particle from the centre of rotation. Particles, however, generate friction as they migrate through the solution. If a particle is rigid and spherical and moving at a known velocity, then the frictional force (F_0) opposing motion is given by

$$F_0 = vf,$$

where v is the velocity or sedimentation rate of the particle, and f is the frictional coefficient of the particle in the solvent.

The frictional coefficient of a particle is the function of its size, shape and hydration, and of the viscosity of the medium, and according to the Stokes equation, for an unhydrated spherical particle, is given by

$$F = 6\pi\eta r_p,$$

where η is the viscosity coefficient of the medium.

For asymmetric and/or hydrated particles, the actual radius of the particle in equation 5.8 is replaced by the effective of Stokes radius, r_{eff}. An unhydrated, spherical particle of known volume and density, and present in a medium of constant density, therefore accelerates in a centrifugal field, its velocity increasing until the net force of sedimentation equals the frictional force resisting its motion through the medium, i.e.,

$$F = F_0 \text{ or } \frac{4}{3}\pi r_p^3 (\rho_p - \rho_m)\omega^2 r = 6\pi\eta r_p v.$$

In practice, the balancing of these forces occurs quickly and the particle reaches a constant velocity because the frictional resistance increases with the velocity of the particle. Under these conditions, the net force acting on the particle is zero. Hence, the particle no longer accelerates but achieves a maximum velocity, with the result that it now sediments at a constant rate. Its rate of sedimentation (v) is then given by

$$v = dr/dt = 2r_p^2 (\rho_p - \rho_m)\omega^2 r / 9\eta.$$

It is evident from this equation that the sedimentation rate of a given particle is proportional to its size, to the difference in density between the particle and the medium and to the applied centrifugal field. It is zero when the densities of the particle and medium are equal; it decreases when the viscosity of the medium increases, and increases as the force field increases. However, since the equation involves the square of the particle radius, it is apparent that the size of the particle has the greatest influence upon its sedimentation rate.

5.2.1 Sedimentation Coefficient (S)

When the frictional force balances the driving force, $\frac{dv}{dt} = 0$

$$\omega^2 r(m_p - m_s) - f \cdot v = 0$$

$$S = \frac{v_t}{\omega^2 x} = \frac{m(1 - \bar{v}_2 \rho)}{f},$$

where S ≡ terminal velocity / unit acceleration

Sedimentation coefficients have units of sec. 10^{-13} sec is called 1 svedberg (or 1 S).

m = particle mass
f = frictional coefficient of the particle in the solvent
ρ = density of solution
v = particle velocity.

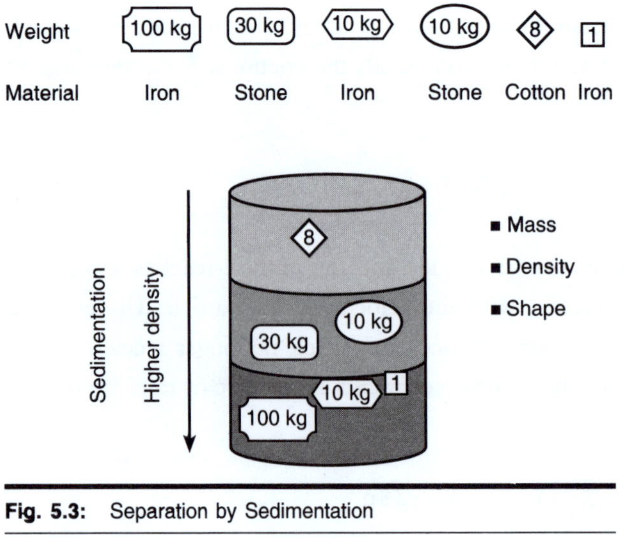

Fig. 5.3: Separation by Sedimentation

Characters of Sediment Coefficient (S)

1. S is increased for particle of larger mass (because sedimenting force in α man (1 – vr)).
2. S is increased for particle of larger density (equal volume).
3. S is increased for more compact structures (shape) of equal particle mass (frictional coefficient is less).
4. S is increased with rotational speed.

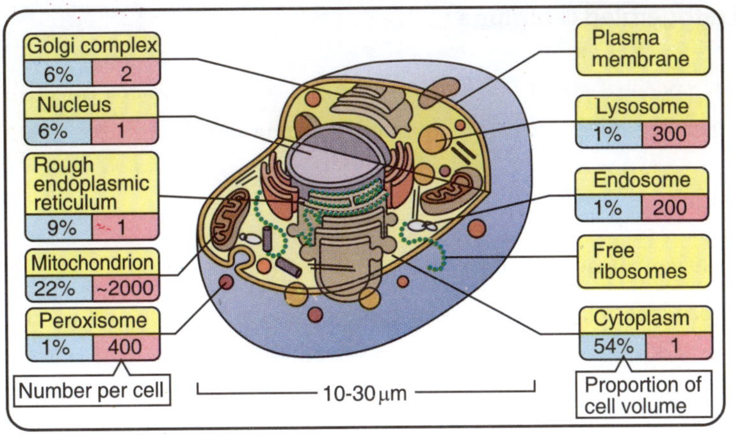

Fig. 5.4: Distribution of various organelles in cells

5.3 TYPES OF CENTRIFUGES AND THEIR USES

There are four major types of centrifuges. They are—

1. Small bench centrifuges

They are used to collect small amount of material that rapidly sediment like yeast cells, erythrocytes etc. They have maximum relative centrifugal field of 3000-7000 g.

Fig. 5.5: Small bench centrifuge (*Courtesy:* Dr D Dash, Deptt. of Biochemistry, IMS, BHU

2. Large capacity refrigerated centrifuges

They have refrigerated rotor chamber and have capacity to change rotor chambers for varying size. They can go up to maximum of 6500 g and use to sediment or collect the substances that sediment rapidly like erythrocytes, yeast cell, nuclei and chloroplast.

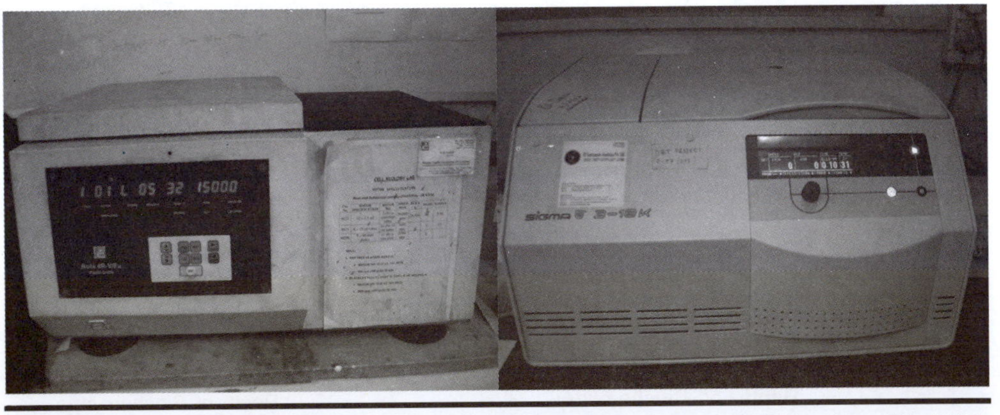

Fig. 5.6: Table top large capacity centrifuges (*Courtesy:* Dr D Dash, Deptt. of Biochemistry, IMS, BHU)

3. High speed refrigerated centrifuge

They can generate speed of about 60000g and are used to collect micro-organism, cellular debris, larger cellular organelles and proteins precipitated by ammonium sulphate.

4. Ultra centrifuges

(a) **Preparative ultracentrifuge:**

It can produce relative centrifugal force of about 600000g and its chamber is refrigerated, sealed and evacuated. It is employed for separation of macromolecues/ligand binding kinetic studies, separation of various lipoprotein fraction from plasma and deprotonisation of physiological fluids for amino acid ananlysis.

(b) **Analytical ultracentrifuge:**

It is capable of operating at 500000 g. Three kinds of optical systems are available in analytical ultracentrifuges: a light absorption system, and the alternative Schlieren system and Rayleigh interferometric system, both of which detect changes in the refractive index of the solution.

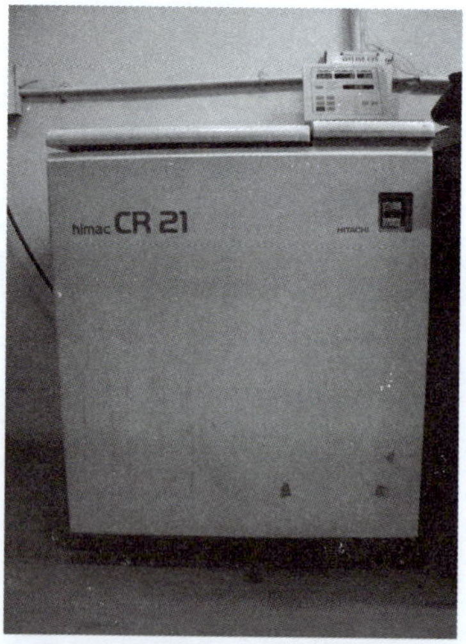

Fig. 5.7: High speed centrifuge (*Courtesy:* Dr D Dash, Deptt. of Biochemistry, IMS, BHU)

Fig. 5.8: Ultracentrifuge (*Courtesy:* Dr D Dash, Deptt. of Biochemistry, IMS, BHU)

Table 5.1: Classes of centrifuges and their applications

	Centrifuge Classes		
	Low speed	High speed	Ultra/micro-ultra
Maximum Speed (rpm × 10^3)	10	20	100/150
Maximum RCF (× 10^3)	7	100	800/900
Pelleting applications			
Bacteria	Yes	Yes	Yes
Animal and plant cells	Yes	Yes	Yes
Nuclei	Yes	Yes	Yes
Precipitates	Some	Most	Yes
Membrane fractions	Some	Some	Yes
Ribosomes/Polysomes	—	—	Yes
Macromolecules	—	—	Yes
Viruses	—	Most	Yes

5.4 DESIGN AND TYPES OF ROTORS

1. Swinging bucket rotors

The swinging bucket rotor has buckets that start off in a vertical position but during acceleration of the rotor swing out to a horizontal position so that during centrifugation the tube and hence the solution in the tube, is aligned perpendicular to the axis of rotation and parallel to the applied centrifugal field, the tube returning to its original position during deceleration of the rotor.

Fig. 5.9: Swinging bucket rotors (*Courtesy:* Dr D Dash, Deptt. of Biochemistry, IMS, BHU)

Fig. 5.10: Fixed angle rotors (*Courtesy:* Dr D Dash, Deptt. of Biochemistry, IMS, BHU)

2. Fixed angle rotors

In fixed angles the tubes are located in holes in the rotor body set at a fixed angle between 14° and 40° to the vertical. Under the influence of centrifugal field, particles move radially outward and have only a short distance to travel before colliding with, and precipitating on, the outer wall of the centrifuge tube. A region of high concentrion is formed that has a density greater than surrounding medium, with the result that the precipitate sinks and collects as a small compact pellet at the outermost point of the tube.

3. Vertical tube rotors

They are considered as zero angle fixed angle rotors in which the tubes are alinged vertically in the body of the rotors at all times.

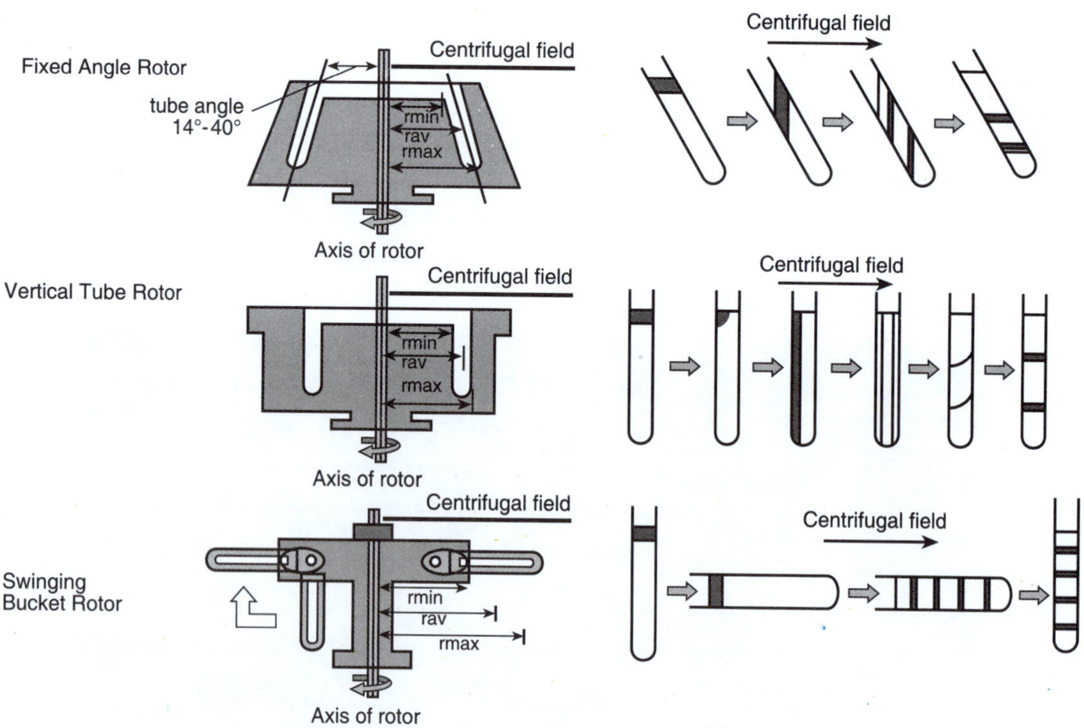

Fig. 5.11: Types of centrifugation rotors

4. Zonal rotors

The zonal rotors may be of the batch or continuous flow type. The former being more extensively used than the latter, and are designed to minimize the wall effect that is encountered in swinging-bucket and fixed angle rotors, and to increase sample size.

5. Elutriator rotors

The elutriator is a kind of continuous flow rotor that contains recesses to hold a single conical shaped separation chamber, the apex of which points away from the axis of rotation, and a bypass chamber on the opposite side of the rotor that serves as a counter balance and to provide the fluid outlet.

5.5 CARE OF ROTORS

The protective anodized coating on aluminium rotor is very thin and does not provide a high degree of protection against corrosion; thus rotors should always be handled with care to prevent scratching. Rotors should always be thoroughly washed preferably with deionised water and since moisture is a potential source of corrosion, they should be allowed to dry upside down in a warm atmosphere; they should then be stored in a clean, dry environment. Rotors' outer surface only can be given a protective coat of lanolin or silicone polish. Swinging-bucket rotors, however, should never be completely immersed in water as the bucket hanging system is difficult to dry. Titanium rotors are essentially resistant to corrosion. To prevent possible damage to the drive shaft of the centrifuge due to vibration caused by rotor imbalance, sample loads should be balanced within the limits specified by the manufacturer. Swinging-bucket rotors should not be run with any bucket or caps removed or individual rotor buckets interchanged as they form integral part of the balance of the rotor.

During acceleration and deceleration of the rotor, cyclic stretching and relaxing of metal can cause metal fatigue, leading to eventual failure of rotor. To avoid overstressing the rotor and to ensure its continued safe operation, an accurate record should be kept of its total usage, i.e, number of runs (at any speed) and time of each run so that rotor can either be derated after a certain number of runs or replaced after a set period of time as specified by manufacturer.

5.6 SAMPLE CONTAINERS

Centrifuge tubes and bottles are available in different range of sizes, thickness and rigidity from different variety of materials including glass, cellulose, esters, polyallomer, polycarbonate, polyethylene, polypropylene, kynar, nylon and stainless steel.

The type of container used will depend upon nature and volume of sample to be centrifuged along with centrifugal forces to be withstood.

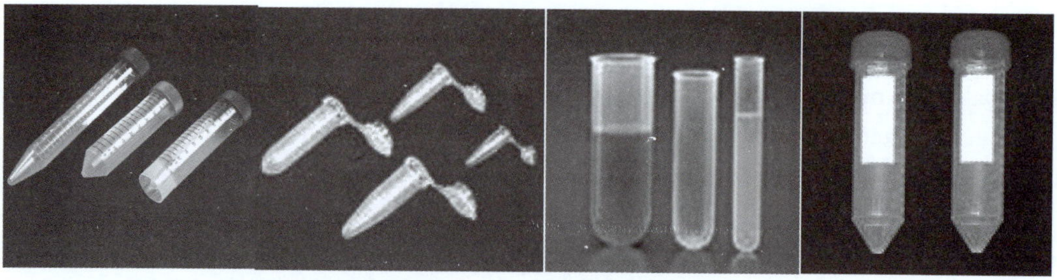

Fig. 5.12: Different types of centrifugation tubes

Glass centrifuge tubes are suitable only for centrifugation at low speeds as they disintegrate at higher centrifugal fields. Thin walled tubes may be used in swinging bucket rotors because the tube is protected by the surrounding bucket; however, thick walled tubes are required with fixed angle and vertical tube rotors. The centrifuge tubes should be filled to accurate level and need to cap the tube or bottle depends upon the speed and type of the container used.

Major precautionary measures in selection of a tube (plastic) material:

1. Clarity
2. Chemical resistance
3. Sealing mechanism (if needed)
4. Check product guide pages or tube packaging for notes on recommended sample volume and maximum speed.
5. Always run thin-walled, sealed tubes full in a fixed angle or vertical rotor. Autoclave tubes only if absolutely necessary and only at 121°C for 15 min.
6. Avoid cleaning plastic tubes in automated dishwashers or glassware washers, which may produce excessively hot temperatures.
7. Tube must be carefully matched with rotor type to prevent sample loss and/or failure as illustrated in Table below.
8. To prolong tube life and avoid breakage or collapse:

Table 5.2: Tube type and rotor compatibility

	Rotor Type		
Tube type	Fixed angle	Swinging-bucket	Vertical
Thin-walled open top	No	Yes	
Thick-walled open top	Yes	Yes	No
Thin-walled sealed	Yes	Some tube types	Yes
Oak ridge	Yes	No	No

5.7 SEPARATION METHODS

1. Differential centrifugation

This depends upon the sedimentation rate of particles of different size and density. Centrifugations will initially sediment the largest particles. For particles with same mass but with different densities, the one with highest density will sediment first. Particles having similar banding densities can usually be efficiently separated one from another by differential centrifugation or rate zonal method, provided that there are at least 10-fold differences in their masses.

In differential centrifugation the material to be separated is divided centrifugally into number of fractions by increasing the applied centrifugal field. The centrifugal field at each step is chosen so that particular type of material sediments. Any type of particle originally present in homogenate may be found in pellet or the supernatant or both fractions, depending upon the time and speed of centrifugation and size and density of particles. At the end of each stage the pellet and supernatant are separated and pellet washed several times by resuspension and recentrifugation in homogenation medium.

Initially all particles of homogenate are homogenously distributed throughout the centrifuge tube. During centrifugation particles move down the centrifuge tubes at their respective sedimen-

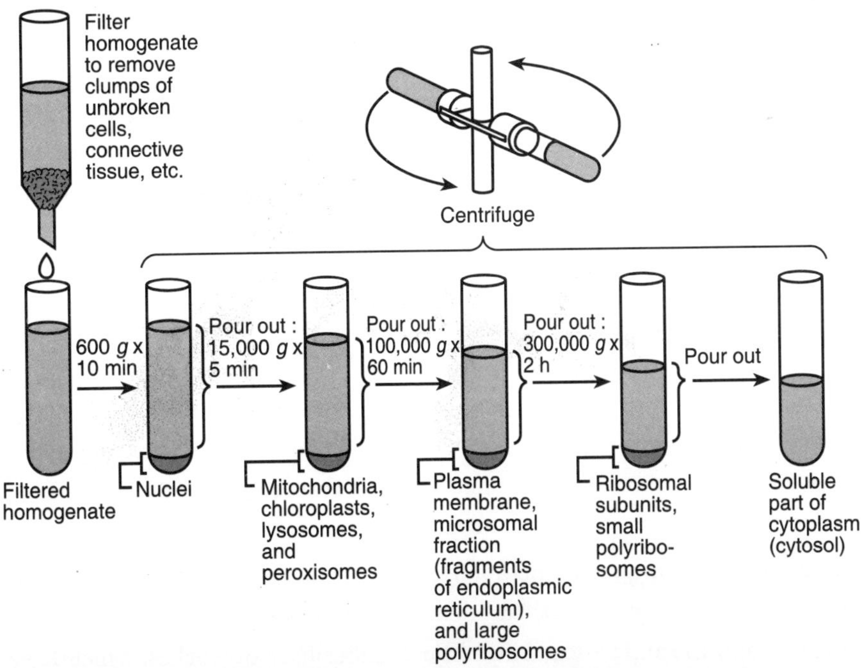

Fig. 5.13: Isolation of different cell organelles by differential centrifugation

tation rates and start to form a pellet on the bottom of centrifuge tube. Ideally centrifugation is continued enough to pellet all the largest class of particles, the resulting supernatant then being centrifuged at a higher speed to separate medium-sized particles and so on. However, since particles of varying sizes and densities were distributed homogenously at the commencement of centrifugation, it is evident that the pellet will not be homogenous but will contain a mixture of all the sedimented components, being enriched with fastest sedimenting particles. In the time required for complete sedimentation of heavier particles, some of the lighter and medium sized particles, originally suspended near the bottom of the tube, will also sediment and thus contaminate the fraction. Pure preparation of the pellet of the heaviest particle cannot be, therefore, obtained in single centrifugation step. It is only the most slowly sedimenting component of mixture remaining in the supernatant after all the larger particles have been sedimented that can be purified by single centrifugation step.

The separation achieved by differential centrifugation can be improved by repeated resuspension and recentrifugation under similar condition. Further centrifugation of the supernatant with gradually increasing centrifugal fields results in sedimentation of intermediate and finally the smallest and least dense particles. In spite of its inherent limitations, differential centrifugation is probably the most commonly employed method for isolation of cell organelles from homogenized tissue.

2. Centrifugal elutriation

In this technique the separation and purification of a large variety of cells from different tissues and species can be achieved by a gentle "washing action" using an elutriator rotor. The technique is based upon the differences in the set-up in separation chamber of rotor, between the opposing centripetal liquid flow and applied centrifugal field being used to separate particles mainly on the basis of differences in their size. The technique does not employ the density gradient and have advantage that any medium totally compatible with the particles can be used. By this separation can be achieved very quickly, giving high cell concentrations and a very good recovery yield.

3. Density gradient centrifugation

There are two methods of density gradient centrifugation, the rate zonal technique and the isopycnic (isodensity or equal density) technique, and both can be used when quantitative separation of all the components of mixture of particles is required. They are also used for the determination of buoyant densities and for the estimation of sedimentation coefficient.

(a) **Rate zonal technique:** Particle separation by the rate zonal technique is based upon differences in the size, shape and density of particles, the density and viscosity of the medium and the applied centrifugal field. Subcellular organelles, which have different densities but are similar in size, do not separate efficiently using this method, but separation of proteins of similar densities and differing only 3 folds in relative molecular

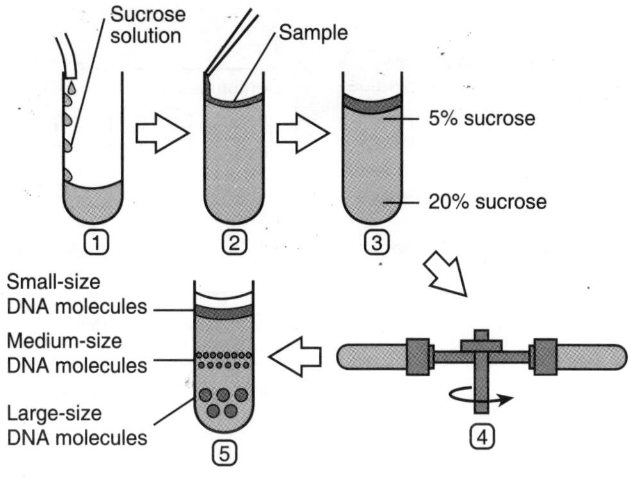

Fig. 5.14: Principle of rate zonal centrifugation technique

mass can be achieved easily. The technique involves carefully layering a sample solution on top of preformed liquid density gradient, the highest density of which does not exceed that of densest particle to be separated. The function of gradient is primarily to stabilize the liquid column in the tube against the movements resulting from conventional currents and secondarily to produce a gradient that helps to improve the resolution of gradient. The sample is then centrifuged until the desired degree of separation is achieved. Since the technique is time dependent, centrifugation must be terminated before any of the separated zone pellets at he bottom of tube. The technique is employed for the separation of enzymes, RNA-DNA hybrids, ribosomal subunit, subcellular organelle, etc.

(b) **Isopycnic centrifugation technique:** Isopycnic centrifuigation depends solely upon the buyont density and not on its shape, size and time, the size of the particle affecting only the rate at which it reaches its isopycnic position in the gradient. The technique is used to separate particles of similar size but of different density. Hence soluble proteins which have very similar densities cannot be usually separated by this method, where as sub cellular organelles can be effectively separated.

The methods are a combination of sedimentation and flotation and involve layering the sample on top of a density gradient that spans the whole range of the particle densities that are to be separated. The maximum density of the gradient, therefore, must always exceed the density of the densest particle. During centrifugation, sedimentation of the particle occurs until the buoyant density of the particle and density of the gradient are equal. At this point of isodensity no further sedimentation occurs, irrespective of how long centrifugation continues, because the particles are

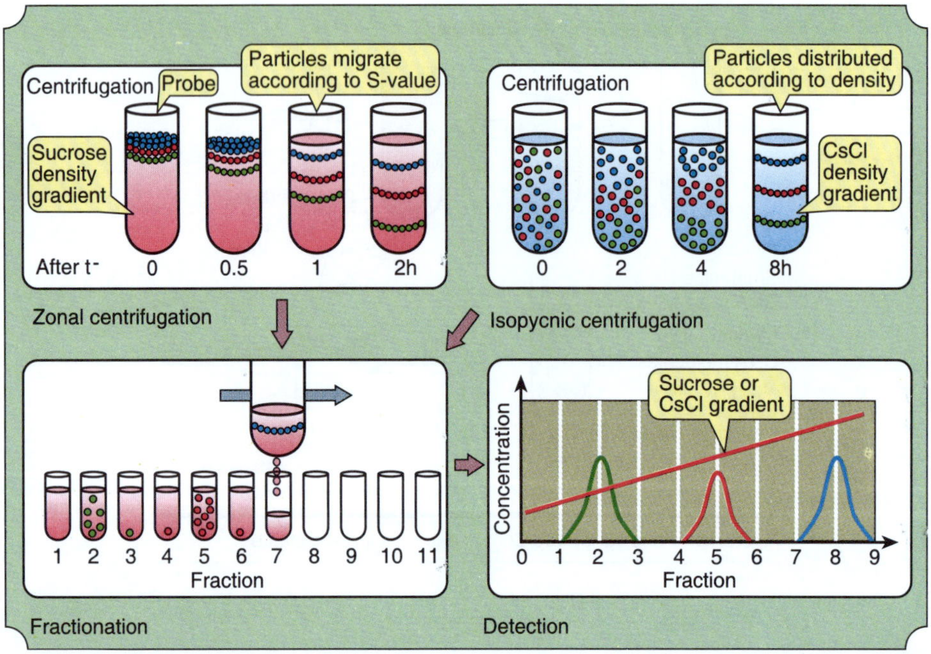

Fig. 5.15: Density gradient centrifugation

floating on the cushion of material that has density greater than their own. Isopycnic centrifugation, in contrast to the rate zonal technique, is an equilibrium method, the particle banding to form zones each at their own characteristic buoyant density. In case when not all components in a mixture of particle are required, a gradient range can be selected in which unwanted materials will be sediment at the bottom of the tube and whole of the particles of interest will float at their respective isopynic positions. Such a technique involves a combination of both the rate zonal and isopynic approaches.

Table 5.3: Densities of various important biological molecules

Material	Density (g/cm^3)
Microbial cells	1.05–1.15
Mammalian cells	1.04–1.10
Organelles	1.10–1.60
Proteins	1.30
DNA	1.70
RNA	2.00

5.7.1 Features of gradient material

There is no ideal all purpose gradient material; the choice of solute depends upon the nature of the particles to be fractionated. The gradient material should:

- permit the desire type of separation
- be stable in solution
- be inert towards biological material and not react with the centrifuge , rotor, tubes or caps:
- not absorb light at wavelengths appropriate for spectrophotometric monitoring (visible or ultraviolet range), or otherwise interfere with assaying procedures;
- be sterilisable, non-toxic or flammable;
- have negligible osmotic pressure and cause minimum changes in ionic strength, pH and viscosity;
- be inexpensive and readily available in pure form and capable of forming a solution covering the density range needed for a particular application without overstressing the rotor;
- allow easy separation of the sample material from the gradient medium with loss of the sample or its activity.

Generally used gradient materials are salts of alkali metals (e.g., caesium and rubidium chloride), small neutral hydrophilic organic molecules (e.g., sucrose), hydrophilic macromolecules (e.g., proteins and polysaccharides), and a number of miscellaneous compounds such as colloidal silica (e.g., percoll and ludox) and non-ionic iodinated aromatic compounds (e.g., metrizamide, nycodenz and renograffin).

Sucrose solution while suffering from disadvantages of being very viscous at densities greater than 1.1 to 1.2 g cm^{-3} and exerting very high osmotic effects even at very low concentrations have been found to be most convenient gradient material for rate zonal separation. Glycerol is also used as gradient material especially for the separation of enzymes, or alternative media such as ficoll, metrizamide or percoll gradients may be utilized. Non-ionic media, such as sucrose, glycerol, metrizamide, ficoll and percoll are generally considered to be gentler than ionic salts like caesium chloride and potassium bromide and require a lower centrifugal fields to achieve adequate separation of particles. In case of isopycnic separation, no one medium has proved satisfactory for the isolation of all type of biological particles. Ficoll has successfully used for the separation of whole cell and subcellular organelles by rate zonal and isopycnic centrifugations. Caesium and rubidium salts are used exclusively for isopycnic separations and have been used most frequently for separation of high density solutes like nucleic acids. However, at high concentrations their high ionic strength and osmolarity tends to disrupt intra- and inter-molecular bonds. Generally, ionic media have been used for the separation of nucleic acids, proteins and viruses and non-ionic iodinated aromatic compounds, because of their increased versatility, have been used for a much wider variety of applications.

Table 5.4: Application of density gradient media for isopycnic separations

Gradient media	Cells	Viruses	Organelles	Nucleoproteins	Macromolecules
Sugars (e.g., sucrose)	+	+++	+++	+	–
Polysaccharides (e.g., ficoll)	++	++	++	–	–
Colloidal silica (e.g., percoll)	+++	+	+++	–	–
Iodinated media (e.g., nycodenz)	++++	++	++++	+++	+
Alkali metal salts (e.g., CsCl)	–	++	–	++	++++

++++ excellent, +++ good, ++ good for some applications, + limited use, – unsatisfactory

5.8 PRECAUTIONARY MEASURES WITH CENTRIFUGES

Centrifuges are extremely dangerous instruments if not properly maintained and correctly used. It is, therefore, always advisable that one must read and understand the operating manual for particular centrifuge.

Manufacturers of centrifuges should ensure effective lid locks. Access to the rotor chamber of centrifuges should always be avoided when spin is in progress. Centrifuges should have imbalance detectors, over speed detectors and devices and ability to contain any failure of rotor. To prevent possible physical injury when rotors are filled and emptied, care must be taken to ensure that the moving rotor is not touched and that long hair and loose clothing (e.g., Ties) do not get caught in any rotating part. This is especially important with older centrifuges where lid can be opened before the rotor has stopped rotating. It is important when is centrifuging hazardous materials like pathogenic microorganisms, infectious viruses, carcinogenic, corrosive or toxic chemicals, radioactive materials), especially in low speed non-refrigerated centrifuges in which rotor temperature is controlled by air-flow through the rotor bowl , samples should be kept in air-tight, leak-proof containers. This is to prevent aerosol formations arising from accidental spillage of sample which would contaminate the rotor, centrifuge and possibly the whole laboratory.

5.9 TIPS TO REMEMBER

Sedimentation coefficient (S)

- **S** can be considered "Sedimentation Rate" of a particle under centrifugation force.
- S is increased for particle of larger mass (because sedimenting force a m (1 - vr).
- S is increased for particle of larger density (equal volume).

- S is increased for more compact structures (shape) of equal particle mass (frictional coefficient is less).
- S is increased with rotational speed.

Fixed Angle Rotor

Sedimenting particles have only short distance to travel before pelleting. **Shorter run time is sufficient for this kind of rotor.** The most widely used rotor type.

Swinging Bucket Rotor

Longer distance of travel may allow better separation, such as in density gradient centrifugation. Easier to withdraw supernatant without disturbing pellet.

Differential Centrifugation

Based on the differences in the **sedimentation rate** of the biological particles of different size, shape and density.

Density gradient centrifugation

Molecules separated on equilibrium position, NOT by rates of sedimentation. After centrifugation, each molecule floats or sinks (=redistribution) to position where density equals density of CsCl or sucrose solution. Then no net sedimenting force on molecules and separation is on basis of different densities of the particles.

Relative Centrifugal Force

- expressed as 'x gravity'
- $RCF = Fc/Fg = \omega 2r/980$ $[\omega = \pi(rpm)/30]$
- $RCF = 1.119 \times 105\ (rpm)2r$ [radius in cm]

5.10 VIVA VOCE

Q.1. What are the forces affecting the separation of particles in centrifugation?

Q.2. What is sedimentation coefficient?

Q.3. Which rotors will be suitable for isolation of nucleic acids?

Q.4. Differentiate between isopycnic and zonal centrifugation techniques.

Q.5. What is the principle of differential centrifugation?

Q.6. Which type of gradient materials are best suited for zonal centrifugation and why?

Q.7. What is "1s"?

Q.8. Which type of sample containers are recommended for ultracentrifugation?

Q.9. How many types of centrifuges are there?

Q.10. What are the precautions one should take for centrifugation techniques?

CHAPTER 6

Immunochemical Techniques

In this chapter we will be dealing with immunochemical techniques which are employed in molecular biology. These techniques are not only employed for research but also as diagnostic tools in many of the pathological conditions. First we will start with antibody production.

6.1 ANTIBODY PRODUCTION

(a) Polyclonal antibodies

It is possible to produce antibodies which bind to proteins, peptides, carbohydrate and nucleic acids. In general, most immunochemical methods are devised for use with antibodies that recognize proteins and peptides.

Choice of species for antibody production depends upon the amount of antigen available, amount of antiserum required and quality of antiserum desired. In some case use of closely related species or different strains within single species for derivation of antigen and production of antibody can provide antibodies with specific properties. The most important consideration for immunogenicity is differences in amino acid sequences. Immunogenicity tends to increase with size. Peptides with molecular weight less than 2000 are poor antigens while with more than 10000 are good antigens as long as they are recognized as foreign in responding animal.

Antibodies that bind to smaller peptides can be produced by linking (conjugating) these substances to larger proteins called as carrier proteins. Thus the Antisera produce will recognize carrier protein too along with peptide. Substances that are not immunogenic alone, but are when conjugated are known as heptanes. Such substances potentiate the immune response by forming a slow –release depot of antigen, by stimulating T cell help or by aiding antigen presentation. Single immunization results in production of antisera but it is usually suboptimal, containing antibodies of low avidity and a high proportion of IgM, therefore several "booster" immunizations with the antigen is done so as to boost the antibody production by making the animal hyperimmune. Such "hyperimmune sera" are usually the polyclonal reagent of choice for immunochemical techniques.

(b) Monoclonal antibodies

Monoclonal antibodies are secreted by cloned, i.e., monoclonal cells. Mature, antibody secreted lymphocytes from immunized animals can be cloned, but these survive only for very few period in culture, and therefore, do not provide useful amount of antibody. However, procedures have been developed to separate antibodies from continuously growing (immortal) cells. These involve generation of hybrid cells, transformation of lymphocytes with virus or recombinant DNA.

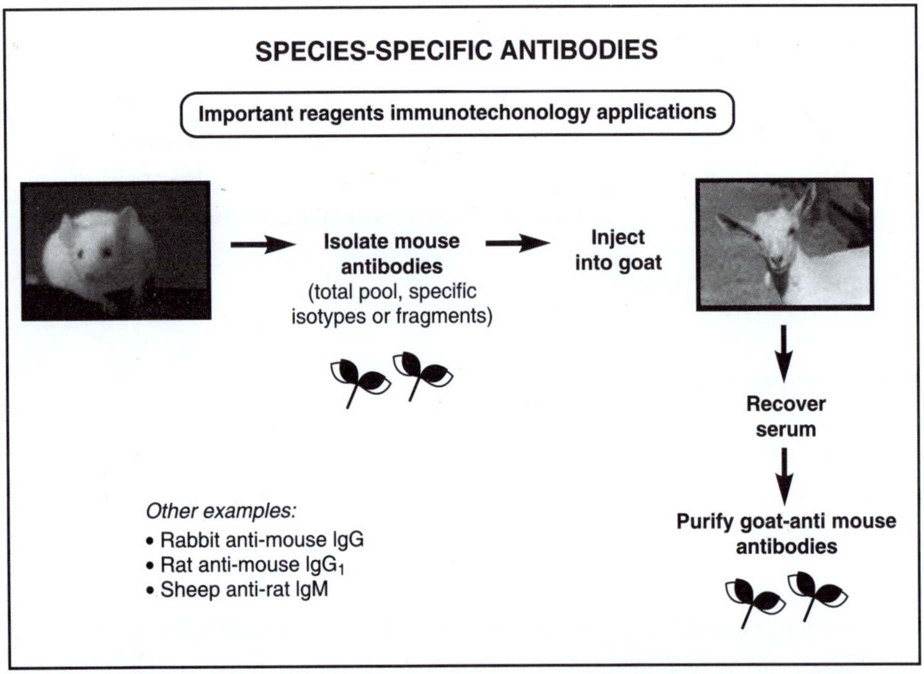

Fig. 6.1: Figure showing the development of species specific antibodies

6.2 IMMUNOASSAY

An immunoassay is a biochemical test that measures the concentration of a substance in a biological liquid, typically serum or urine, using the reaction of an antibody or antibodies to its antigen. The assay takes advantage of the specific binding of an antibody to its antigen. Monoclonal antibodies are often used as they only usually bind to one site of a particular molecule, and therefore, provide a more specific and accurate test, which is less easily confused by the presence of other molecules. The antibodies selected must have a high affinity for the antigen (if there is antigen available, a very high proportion of it must bind to the antibody).

The presence of both antigen and antibody can be measured. For instance, when detecting infection the presence of antibody against the pathogen is measured. For measuring hormones such as insulin, the insulin acts as the antigen.

For numerical results, the response of the fluid being measured must be compared to standards of a known concentration. This is usually done through the plotting of a standard curve on a graph, the position of the curve at response of the unknown is then examined, and so the quantity of the unknown is found.

Detecting the quantity of antibody or antigen can be achieved by a variety of methods. One of the most common is to label either the antigen or antibody. The label may consist of an enzyme (see enzyme-linked immunosorbent assay (ELISA)), radioisotopes such as I-125 Radioimmunoassay (RIA) or fluorescence. Other techniques include agglutination, nephelometry, turbidimetry and Western Blot. Immunoassays have a particularly important role in the diagnosis of HIV through the HIV test

Types

Immunoassays can be divided into those that involve labelled reagents and those which involve non-labelled reagents. Those which involve labelled reagents are divided into homogenous and heterogenous (which involved a separation step) immunoassays. Heterogenous immunoassays can be competitive or non-competitive.

- In a competitive immunoassay, the antigen in the unknown sample competes with labelled antigen to bind with antibodies. The amount of labelled antigen bound to the antibody site is then measured. In this method, the response will be inversely proportional to the concentration of antigen in the unknown. This is because the greater the response, the less antigen in the unknown was available to compete with the labelled antigen.
- In noncompetitive immunoassays, also referred to as the "sandwich assay," antigen in the unknown is bound to the antibody site, and then labelled antibody is bound to the antigen. The amount of labelled antibody on the site is then measured. Unlike the competitive method, the results of the noncompetitive method will be directly proportional to the concentration of the antigen. This is because labelled antibody will not bind if the antigen is not present in the unknown sample.

Immunoassays can be homogeneous or heterogeneous:

- A heterogeneous immunoassay will require an extra step to remove unbound antibody or antigen from the site, usually using a solid phase reagent.
- Because homogeneous assays do not require this step, they are typically faster and easier to perform.

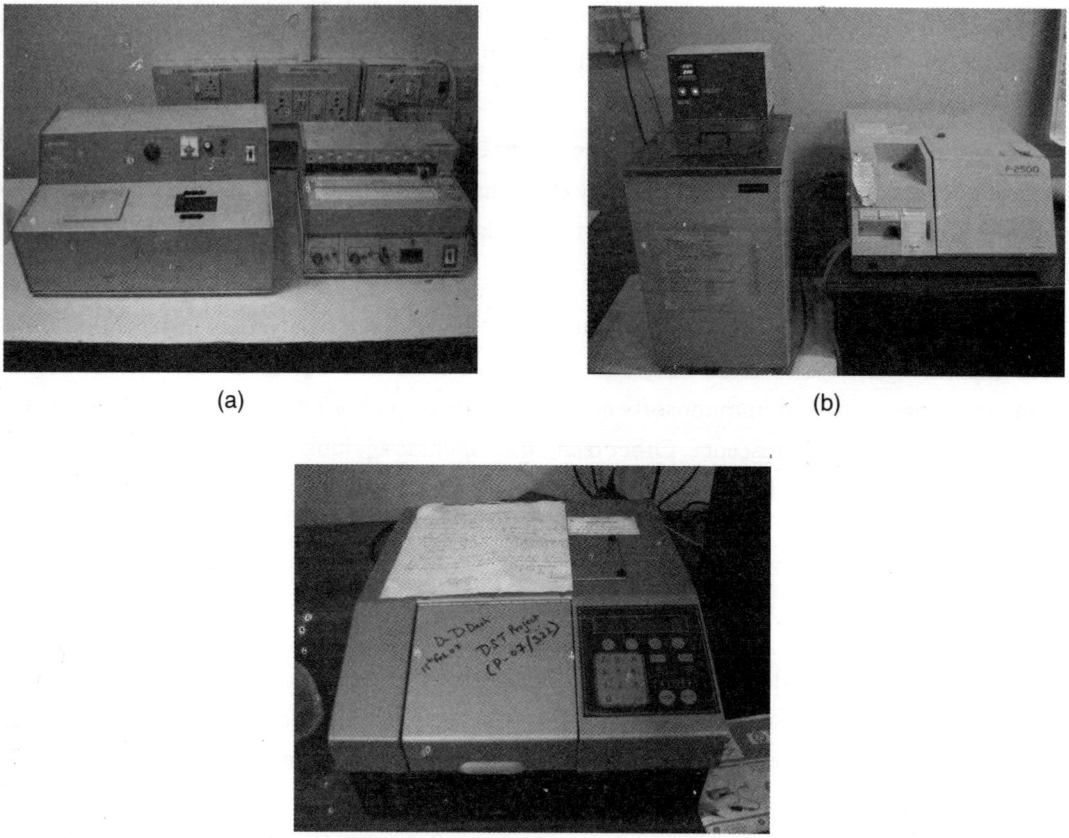

(a) (b) (c)

Fig. 6.2: Different instruments employed for various immunoassays; (a) Aggregometer, (b) Fluorescence spectrophotometer and (c) Fluorescence plate reader (*Courtsey:* Dr D Dash, Deptt. of Biochemistry, IMS, BHU)

6.3 RADIOIMMUNOASSAY (RIA)

One of the most sensitive techniques for detecting antigen or antibody is radioimmunoassay (RIA). The technique was first developed in 1960 by two endocrinologists, S. A. Berson and Rosalyn Yalow, to determine levels of insulin-anti-insulin complexes in diabetics. Although their technique encountered some skepticism, it soon proved its value for measuring hormones, serum proteins, drugs, and vitamins at concentrations of 0.001 micrograms per millilitre or less. In 1977, some years after Berson's death, the significance of the technique was acknowledged by the award of a Nobel Prize to Yalow. The principle of RIA involves competitive binding of radiolabelled antigen and unlabelled antigen to a high-affinity antibody. The labelled antigen is mixed with antibody at a concentration that saturates the antigen-binding sites of the antibody. Then test samples of unlabelled antigen of unknown concentration are added in progressively larger amounts. The antibody does not distinguish labelled antigen from unlabelled antigen, so the two kinds of antigen compete for available binding sites on the antibody. As the concentration of unlabelled antigen increases, more labelled antigen will be displaced from the binding sites. The decrease in the amount of radiolabelled antigen bound to specific antibody in the presence of the test sample is measured in order to determine the amount of antigen present in the test sample. The antigen is generally labeled with a gamma-emitting isotope such as 125I, but beta-emitting isotopes such as tritium (3H) are also routinely used as labels. The radiolabelled antigen is part of the assay mixture; the test sample may be a complex mixture, such as serum or other body fluids, that contains the unlabelled antigen. The first step in setting up an RIA is to determine the amount of antibody needed to bind 50%–70% of a fixed quantity of radioactive antigen (Ag*) in the assay mixture. This ratio of antibody to Ag* is chosen to ensure that the number of epitopes presented by the labelled antigen always exceeds the total number of antibody binding sites. Consequently, unlabelled antigen added to the sample mixture will compete with radiolabelled antigen for the limited supply of antibody. Even a small amount of unlabelled antigen added to the assay mixture of labelled antigen and antibody will cause a decrease in the amount of radioactive antigen bound, and this decrease will be proportional to the amount of unlabelled antigen added. To determine the amount of labelled antigen bound, the Ag-Ab complex is precipitated to separate it from free antigen (antigen not bound to Ab), and the radioactivity in the precipitate is measured. A standard curve can be generated using unlabelled antigen samples of known concentration (in place of the test sample), and from this plot the amount of antigen in the test mixture may be precisely determined. Several methods have been developed for separating the bound antigen from the free antigen in RIA. One method involves precipitating the Ag-Ab complex with a secondary anti-isotype antiserum. For example, if the Ag-Ab complex contains rabbit IgG antibody, then goat anti-rabbit IgG will bind to the rabbit IgG and precipitate the complex. Another method makes use of the fact that protein A of Staphylococcus aureus has high affinity for IgG. If the Ag-Ab complex contains an IgG antibody, the complex can be precipitated by mixing with formalin-killed S. aureus. After removal of the complex by either of these methods, the amount of free labelled antigen remaining in the supernatant can be measured in a radiation counter; subtracting this value from the total amount of labelled antigen added yields the amount of labelled antigen bound. Various solid-phase RIAs have been developed that make it easier to separate the Ag-Ab complex from the unbound antigen. In some cases, the antibody is covalently crosslinked to Sepharose beads. The amount of radiolabelled antigen bound to the beads can be measured after the beads have been centrifuged and washed. Alternatively, the antibody can be immobilized on polystyrene or polyvinylchloride wells and the amount of free labelled antigen in the supernatant can be determined in a radiation counter. In another approach, the antibody is immobilized on the walls of microtiter wells and the amount of bound antigen is determined. Because the procedure requires only small amount of sample and can be conducted in small 96-well microtiter plates, this procedure is well suited for determining the concentration of a particular antigen in large numbers of samples. For example, a microtiter RIA has been widely used to screen for the presence of the hepatitis B virus. RIA screening of donor blood has sharply reduced the incidence of hepatitis B infections in recipients of blood transfusions.

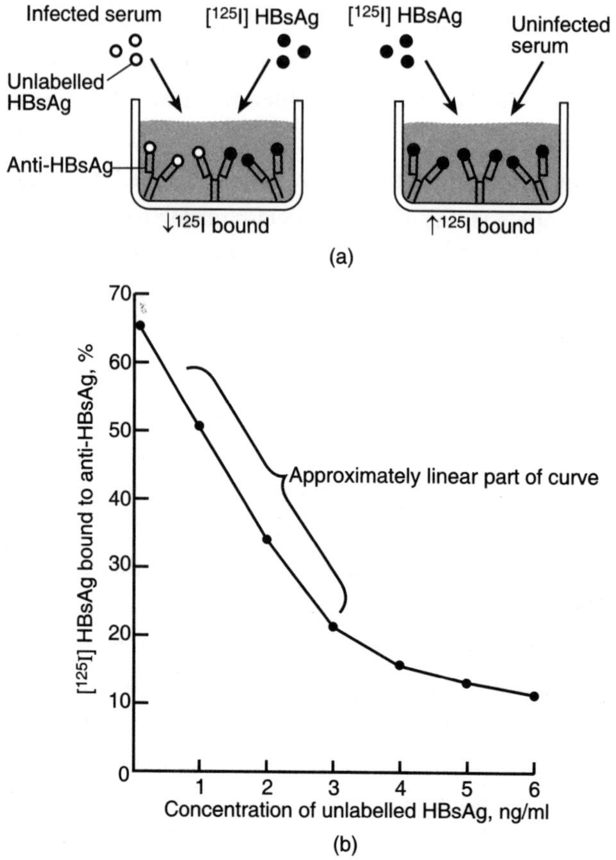

Fig. 6.3: A solid-phase radioimmunoassay (RIA) to detect hepatitis B virus in blood samples. (a) Microtiter wells are coated with a constant amount of antibody specific for HBsAg, the surface antigen on hepatitis B virus. A serum sample and [^{125}I]HBsAg are then added. After incubation, the supernatant is removed and the radioactivity of the antigen-antibody complexes is measured. If the sample is infected, the amount of label bound will be less than in controls with uninfected serum. (b) A standard curve is obtained by adding increasing concentrations of unlabelled HBsAg to a fixed quantity of [^{125}I]HBsAg and specific antibody. From the plot of the percentage of labelled antigen bound versus the concentration of unlabelled antigen, the concentration of HBsAg in unknown serum samples can be determined by using the linear part of the curve.

6.4 ENZYME-LINKED IMMUNOSORBENT ASSAY (ELISA)

Enzyme-linked immunosorbent assay, commonly known as ELISA (or EIA), is similar in principle to RIA but depends on an enzyme rather than a radioactive label. An enzyme conjugated with an antibody reacts with a colourless substrate to generate a coloured reaction product. Such a substrate is called a chromogenic substrate. A number of enzymes have been employed for ELISA, including alkaline phosphatase, horseradish peroxidase, and β-galactosidase. These assays approach the sensitivity of RIAs and have the advantage of being safer and less costly.

Because the ELISA can be performed to evaluate either the presence of antigen or the presence of antibody in a sample, it is a useful tool both for determining serum antibody concentrations (such as with the HIV test or West Nile Virus) and also for detecting the presence of antigen. It has also found applications in the food industry in detecting potential food allergens such as milk, peanuts, walnuts, almonds, and eggs.

6.4.1 General method and instrumentation of ELISA

The steps of the general, "indirect," ELISA for determining serum antibody includes:

1. A sample of known antigen is applied to a surface, often the well of a microtiter plate. The antigen is fixed to the surface to render it immobile. Simple adsorption of the protein to the plastic surface is usually sufficient. These samples of known antigen concentrations will constitute a standard curve used to calculate antigen concentrations of unknown samples. Note that the antigen itself may be an antibody.
2. The plate wells or other surfaces are then coated with serum samples of unknown antigen concentration, diluted into the same buffer used for the antigen standards. Since antigen immobilization in this step is due to non-specific adsorption, it is important for the total protein concentration to be similar to that of the antigen standards.
3. A concentrated solution of non-interacting protein, such as Bovine Serum Albumin (BSA) or casein, is added to all plate wells. This step is known as blocking, because the serum proteins block non-specific adsorption of other proteins to the plate.
4. The plate is washed, and a detection antibody specific to the antigen of interest is applied to all plate wells. This antibody will only bind to immobilized antigen on the well surface, not to other serum proteins or the blocking proteins.
5. The plate is washed to remove any unbound detection antibody. After this wash, only the antibody-antigen complexes remain attached to the well.
6. Secondary antibodies, which will bind to any remaining detection antibodies, are added to the wells. These secondary antibodies are conjugated to the substrate-specific enzyme. This step may be skipped if the detection antibody is conjugated to an enzyme.
7. Wash the plate, so that excess unbound enzyme-antibody conjugates are removed.
8. Apply a substrate which is converted by the enzyme to elicit a chromogenic or fluorogenic or electrochemical signal.
9. View/quantify the result using a spectrophotometer, spectrofluorometer, or other optical/electrochemical device.

The enzyme acts as an amplifier; even if only a few enzyme-linked antibodies remain bound, the enzyme molecules will produce many signal molecules. A major disadvantage of the indirect ELISA is that the method of antigen immobilization is non-specific; any proteins in the sample will stick to the microtiter plate well, so small concentrations of analyte in serum must compete with other serum proteins when binding to the well surface. The sandwich ELISA provides a solution to this problem.

ELISA may be run in a qualitative or quantitative format. Qualitative results provide a simple positive or negative result for a sample. The cut-off between positive and negative is determined by the analyst and may be statistical. Two or three times the standard deviation is often used to distinguish positive and negative samples. In quantitative ELISA, the optical density or fluorescent units of the sample is interpolated into a standard curve, which is typically a serial dilution of the target.

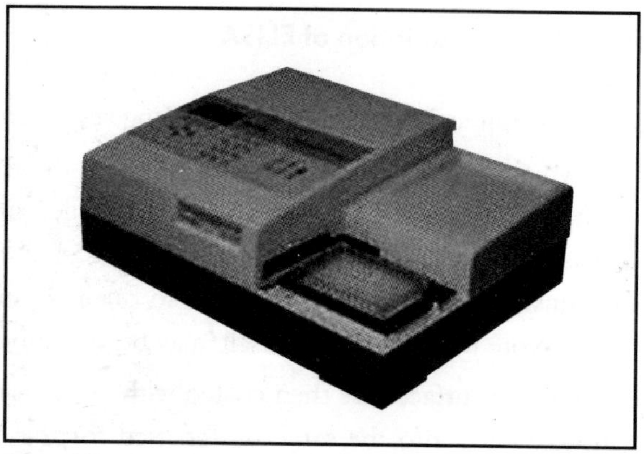

Fig. 6.4: ELISA plate reader (*Courtesy:* Dr D Dash, Deptt. of Biochemistry, IMS, BHU)

6.4.2 Numerous Variants of ELISA

A number of variations of ELISA have been developed, allowing qualitative detection or quantitative measurement of either antigen or antibody. Each type of ELISA can be used qualitatively to detect the presence of antibody or antigen. Alternatively, a standard curve based on known concentrations of antibody or antigen is prepared, from which the unknown concentration of a sample can be determined.

6.4.2.1 Indirect ELISA

Antibody can be detected or quantitatively determined with an indirect ELISA. Serum or some other sample containing primary antibody (Ab_1) is added to an antigen-coated microtiter well and allowed to react with the antigen attached to the well. After any free Ab_1 is washed away, the presence of antibody bound to the antigen is detected by adding an enzyme-conjugated secondary anti-isotype antibody (Ab_2), which binds to the primary antibody. Any free Ab_2 then is washed away, and a substrate for the enzyme is added. The amount of coloured reaction product that forms is measured by specialized spectrophotometric plate readers, which can measure the absorbance of all of the wells of a 96-well plate in seconds. Indirect ELISA is the method of choice to detect the presence of serum antibodies against human immunodeficiency virus (HIV), the causative agent of AIDS. In this assay, recombinant envelope and core proteins of HIV are adsorbed as solid-phase antigens to microtiter wells. Individuals infected with HIV will produce serum antibodies to epitopes on these viral proteins. Generally, serum antibodies to HIV can be detected by indirect ELISA within 6 weeks of infection.

6.4.2.2 Sandwich ELISA

Antigen can be detected or measured by a sandwich ELISA. In this technique, the antibody (rather than the antigen) is immobilized on a microtiter well. A sample containing antigen is added and allowed to react with the immobilized antibody. After the well is washed, a second enzyme-linked antibody specific for a different epitope on the substrate is added, and the coloured reaction product is measured.

6.4.2.3 Competitive ELISA

Another variation for measuring amounts of antigen is competitive ELISA. In this technique, antibody is first incubated in solution with a sample containing antigen. The antigen-antibody mixture is then added to an antigen-coated microtiter well. The more antigen present in the sample, the less free antibody will be available to bind to the antigen-coated well. Addition of an enzyme-conjugated secondary antibody (Ab_2) specific for the isotype of the primary antibody can be used to determine the amount of primary antibody bound to the well as in an indirect ELISA. In the competitive assay, however, the higher the concentration of antigen in the original sample, the lower the absorbance.

6.4.2.4 Chemiluminescence in ELISA

Measurement of light produced by chemiluminescence during certain chemical reactions provides a convenient and highly sensitive alternative to absorbance measurements in ELISA assays. In versions of the ELISA using chemiluminescence, a luxogenic (light-generating) substrate takes the place of the chromogenic substrate in conventional ELISA reactions. For example, oxidation of the compound luminol by H_2O_2 and the enzyme horseradish peroxidase (HRP) produces light:

$$\text{Ab-HRP} + \text{Ag} \longrightarrow \text{Ab-HRP-Ag} \xrightarrow{\text{luminol} + H_2O_2} \text{light}$$

The advantage of chemiluminescence assays over chromogenic ones is enhanced sensitivity. In general, the detection limit can be increased at least ten-fold by switching from a chromogenic to a luxogenic substrate, and with the addition of enhancing agents, more than 200-fold. In fact, under ideal conditions, as little as 5×10^{-18} moles (5 attomoles) of target antigen have been detected.

6.4.2.5 ELISPOT Assay

A modification of the ELISA assay called the ELISPOT assay allows the quantitative determination of the number of cells in a population that are producing antibodies specific for a given antigen or an antigen for which one has a specific antibody. In this approach, the plates are coated with the antigen (capture antigen) recognized by the antibody of interest or with the antibody (capture antibody) specific for the antigen whose production is being assayed. A suspension of the cell population under investigation is then added to the coated plates and incubated. The cells settle onto the surface of the plate, and secreted molecules reactive with the capture molecules are bound by the capture molecules in the vicinity of the secreting cells, producing a ring of antigen-antibody complexes around each cell that is producing the molecule of interest. The plate is then washed and an enzyme-linked antibody specific for the secreted antigen or specific for the species (e.g., goat anti-rabbit) of the secreted antibody is added and allowed to bind. Subsequent development of the assay by addition of a suitable chromogenic or chemiluminescence-producing substrate reveals the position of each antibody- or antigen-producing cell as a point of colour or light

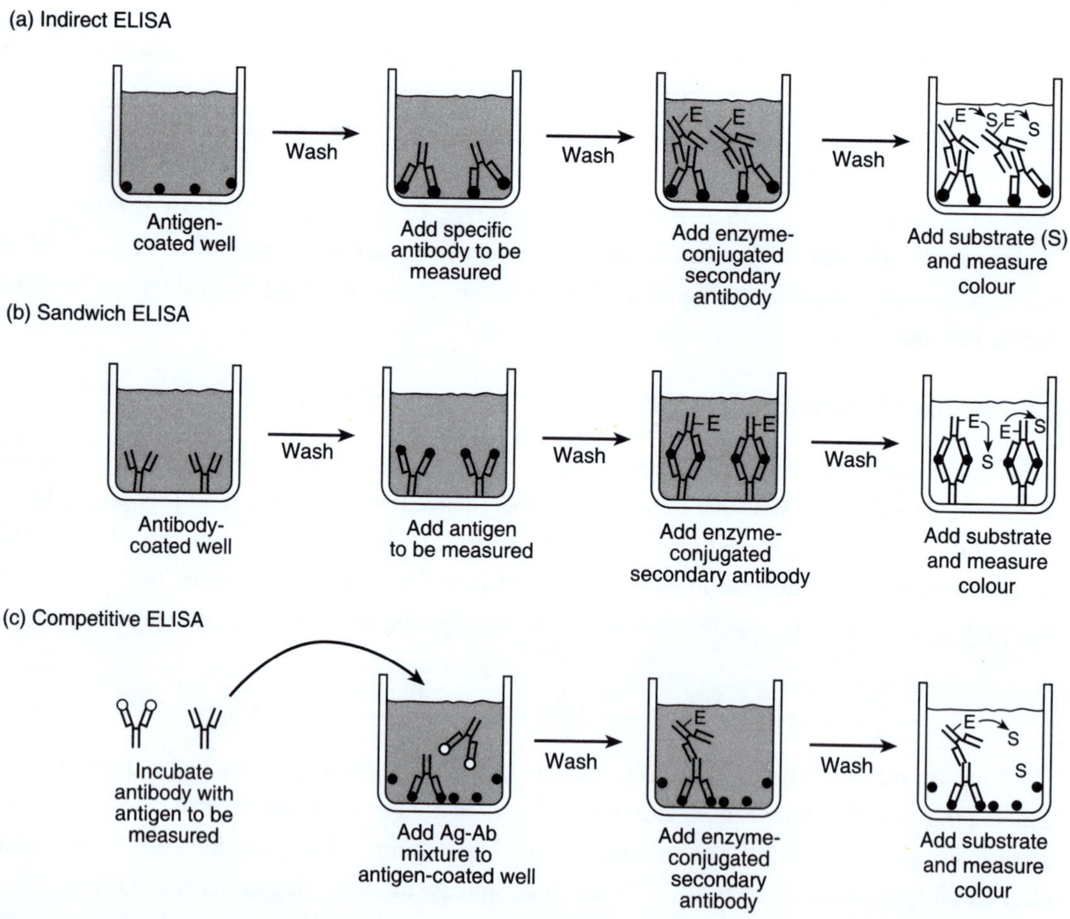

Fig. 6.5: Variations in the enzyme-linked immunosorbent assay (ELISA) technique allow determination of antibody or antigen. Each assay can be used qualitatively, or quantitatively by comparison with standard curves prepared with known concentrations of antibody or antigen. Antibody can be determined with an indirect ELISA (a), whereas antigen can be determined with a sandwich ELISA (b) or competitive ELISA (c). In the competitive ELISA, which is an inhibition-type assay, the concentration of antigen is inversely proportional to the colour produced.

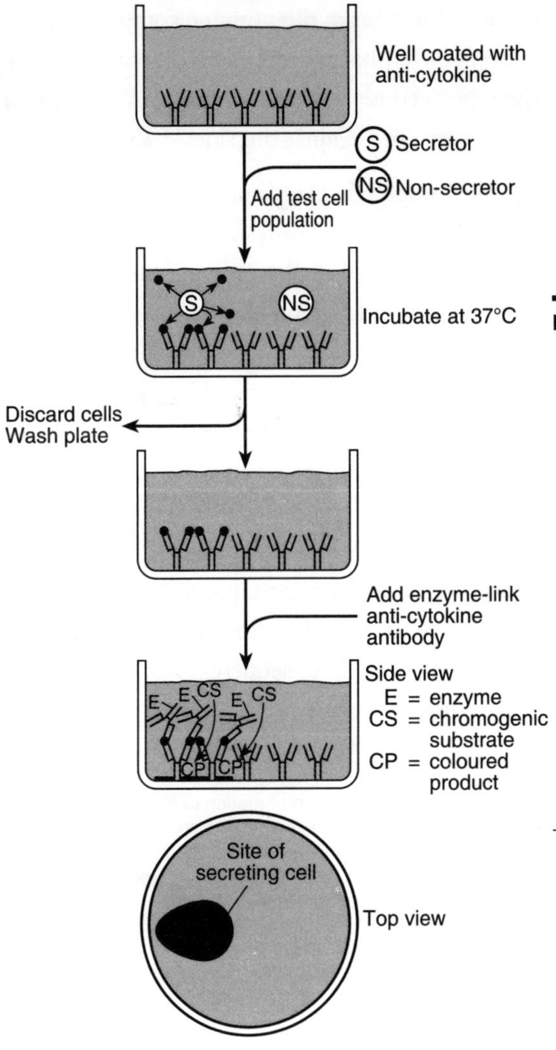

Fig. 6.6: In the ELISPOT assay, a well is coated with antibody against the antigen of interest, a cytokine in this example, and then a suspension of a cell population thought to contain some members synthesizing and secreting the cytokine are layered onto the bottom of the well and incubated. Most of the cytokine molecules secreted by a particular cell react with nearby well-bound antibodies. After the incubation period, the well is washed and an enzyme-labelled anti-cytokine antibody is added. After washing away unbound antibody, a chromogenic substrate that forms an insoluble coloured product is added. The coloured product precipitates and forms a spot only on the areas of the well where cytokine-secreting cells had been deposited. By counting the number of coloured spots, it is possible to determine how many cytokine-secreting cells were present in the added cell suspension.

6.5 IMMUNOPRECIPITATION

The immunoprecipitation technique has the advantage of allowing the isolation of the antigen of interest for further analysis. It also provides a sensitive assay for the presence of a particular antigen in a given cell or tissue type. An extract produced by disruption of cells or tissues is mixed with an antibody against the antigen of interest in order to form an antigen-antibody complex that will precipitate. However, if the antigen concentration is low (often the case in cell and tissue extracts), the assembly of antigen-antibody complexes into precipitates can take hours, even days, and it is difficult to isolate the small amount of immunoprecipitate that forms. Fortunately, there are a number of ways to avoid these limitations. One is to attach the antibody to a solid support, such as a synthetic bead, which allows the antigen-antibody complex to be collected by centrifugation. Another is to add a secondary antibody specific for the primary antibody to bind the antigen-antibody complexes. If the secondary antibody is attached to a bead, the immune complexes can be collected by centrifugation. A particularly ingenious version of this procedure involves the coupling of the secondary antibody to magnetic beads. After the secondary antibody binds to the primary antibody, immunoprecipitates are collected by placing a magnet against the side of the tube.

When used in conjunction with biosynthetic radioisotope labelling, immunoprecipitation can also be used to determine whether a particular antigen is actually synthesized by a cell or tissue. Radiolabelling of proteins synthesized by cells of interest can be done by growing the cells in cell-culture medium containing one or more radiolabelled amino acids. Generally, the amino acids used for this application are those most resistant to metabolic modification, such as leucine, cysteine, or methionine. After growth in the radioactive medium, the cells are lysed and subjected to a primary antibody specific for the antigen of interest. The Ag Ab complex is collected by immunoprecipitation, washed free of unincorporated radiolabelled amino acid and other impurities, and then analyzed. The complex can be counted in a scintillation counter to obtain a quantitative determination of the amount of the protein synthesized. Further analysis often involves disruption of the complex, usually by use of SDS and heat, so that the identity of the immunoprecipitated antigen can be confirmed by checking that its molecular weight is that expected for the antigen of interest. This is done by separation of the disrupted complex by SDS-PAGE and subsequent autoradiography to determine the position of the radiolabelled antigen on the gel.

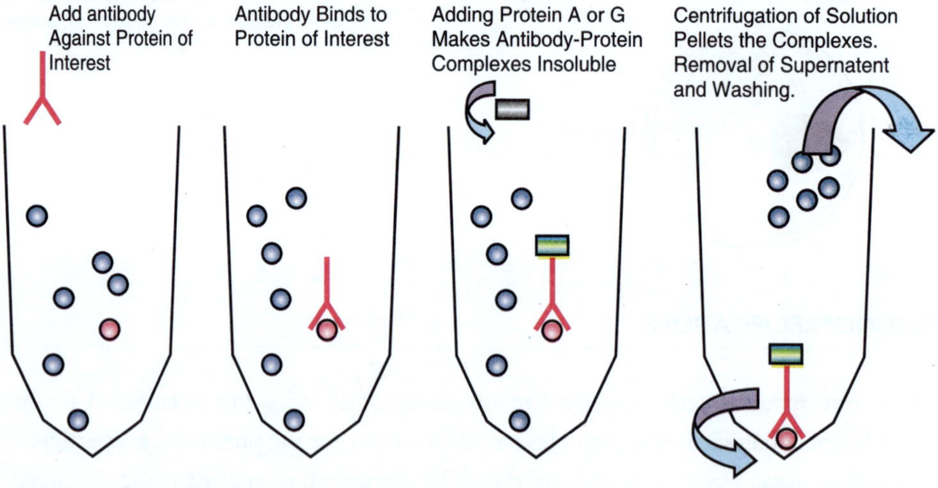

Fig. 6.7: Forms of immunoprecipitation reaction

Antibody and soluble antigen interacting in aqueous solution form a lattice that eventually develops into a visible precipitate. Antibodies that aggregate soluble antigens are called precipitins. Although formation of the soluble Ag-Ab complex occurs within minutes, formation of the visible precipitate occurs more slowly and often takes a day or two to reach completion. Formation of an Ag-Ab lattice depends on the valency of both the antibody and antigen. The antibody must be bivalent; a precipitate will not form with monovalent Fab fragments.The antigen must be either bivalent or polyvalent; that is, it must have at least two copies of the same epitope, or have different epitopes that react with different antibodies present in polyclonal antisera. Experiments with myoglobin illustrate the requirement that protein antigens be bivalent or polyvalent for a precipitin reaction to occur. Myoglobin precipitates well with specific polyclonal antisera but fails to precipitate with a specific monoclonal antibody because it contains multiple, distinct epitopes but only a single copy of each epitope. Myoglobin thus can form a crosslinked lattice structure with polyclonal antisera but not with monoclonal antisera. The principles that underlie precipitation reactions are presented because they are essential for an understanding of commonly used immunological assays. Although various modifications of the precipitation reaction were at one time the major types of assay used in immunology, they have been largely replaced by methods that are faster and, because they are far more sensitive, require only very small quantities of antigen or antibody.

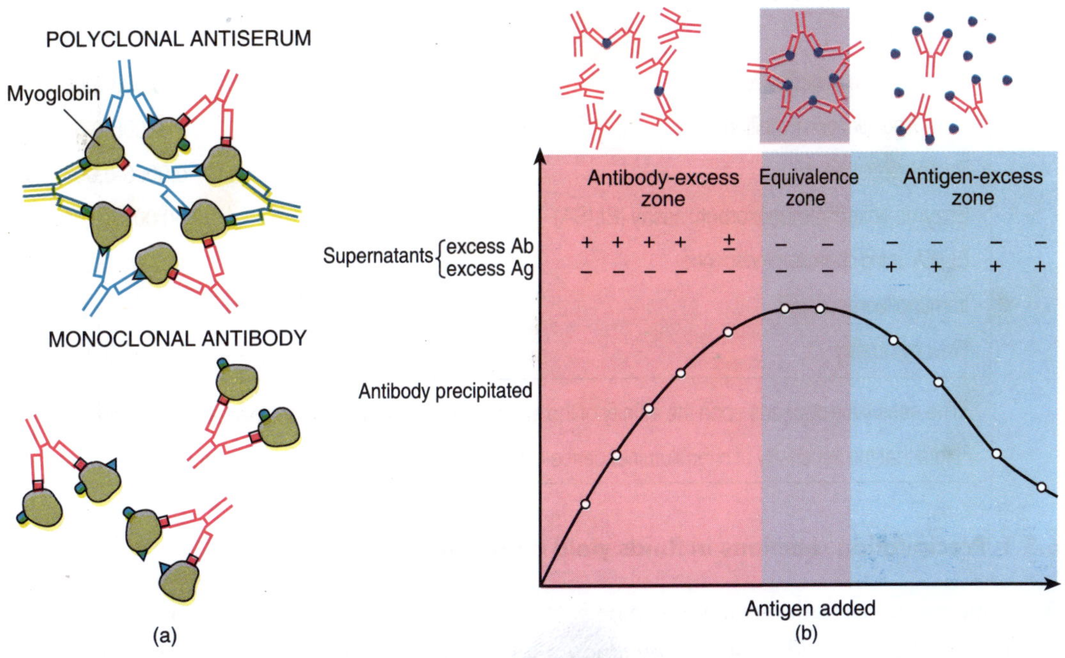

Fig. 6.8: Precipitation reactions. (a) Polyclonal antibodies can form lattices, or large aggregates, that precipitate out of solution. However, if each antigen molecule contains only a single epitope recognized by a given monoclonal antibody, the antibody can link only two molecules of antigen and no precipitate is formed. (b) A precipitation curve for a system of one antigen and its antibodies. This plot of the amount of antibody precipitated versus increasing antigen concentrations (at constant total antibody) reveals three zones: a zone of antibody excess, in which precipitation is inhibited and antibody not bound to antigen can be detected in the supernatant; an equivalence zone of maximal precipitation in which antibody and antigen form large insoluble complexes and neither antibody nor antigen can be detected in the supernatant; and a zone of antigen excess in which precipitation is inhibited and antigen not bound to antibody can be detected in the supernatant.

Also, these modern assay methods are not limited to antigen-antibody reactions that produce a precipitate. Table 6.1 presents a comparison of the sensitivity, or minimum amount of antibody detectable, by a number of immunoassays.

Table 6.1: Sensitivity of various immunoassays

Assay	Sensitivity (mg antibody/ml)
Precipitation reaction in fluids	20-200
Precipitation reactions in gels	
Mancini radial immunodiffusion	10-50
Ouchterlony double immunodiffusion	20-200
Immunoelectrophoresis	20-200
Rocket electrophoresis	2
Agglutination reactions	
Direct	0.3
Passive agglutination	0.006-0.06
Agglutination inhibition	0.006-0.06
Radioimmunoassay	0.0006-0.006
Enzyme-linked immunosorbent assay (ELISA)	< 0.0001-0.01
ELISA using chemiluminescence	< 0.0001-0.01[1]
Immunofluorescence	1.0
Flow cytometry	0.06-0.006

[1]The sensitivity depends upon the affinity of the antibody as well as the epitope density and distribution.

[1]Note that the sensitivity of chemiluminescence-based ELISA assays can be made to match that of RIA.

6.5.1 Precipitation reactions in fluids yield a precipitin curve

A quantitative precipitation reaction can be performed by placing a constant amount of antibody in a series of tubes and adding increasing amounts of antigen to the tubes. At one time this method was used to measure the amount of antigen or antibody present in a sample of interest. After the precipitate forms, each tube is centrifuged to pellet the precipitate, the supernatant is poured off, and the amount of precipitate is measured. Plotting the amount of precipitate against increasing antigen concentrations yields a precipitin curve. As Figure 6.8b shows, excess of either antibody or antigen interferes with maximal precipitation, which occurs in the so-called equivalence zone, within which the ratio of antibody to antigen is optimal. As a large multimolecular lattice is formed at equivalence, the complex increases in size and precipitates out of solution. As shown in Fig. 6.8, under conditions of antibody excess or antigen excess, extensive lattices do not form and precipitation is inhibited. Although the quantitative precipitation reaction is seldom used experimentally today, the principles of antigen excess, antibody excess, and equivalence apply to many Ag-Ab reactions.

6.5.2 Precipitation reactions in gels yield visible precipitin lines

Immune precipitates can form not only in solution but also in an agar matrix. When antigen and antibody diffuse toward one another in agar, or when antibody is incorporated into the agar and antigen diffuses into the antibody-containing matrix, a visible line of precipitation will form. As in a precipitation reaction in fluid, visible precipitation occurs in the region of equivalence, whereas no visible precipitate forms in regions of antibody or antigen excess. Two types of immunodiffusion reactions can be used to determine relative concentrations of antibodies or antigens, to compare

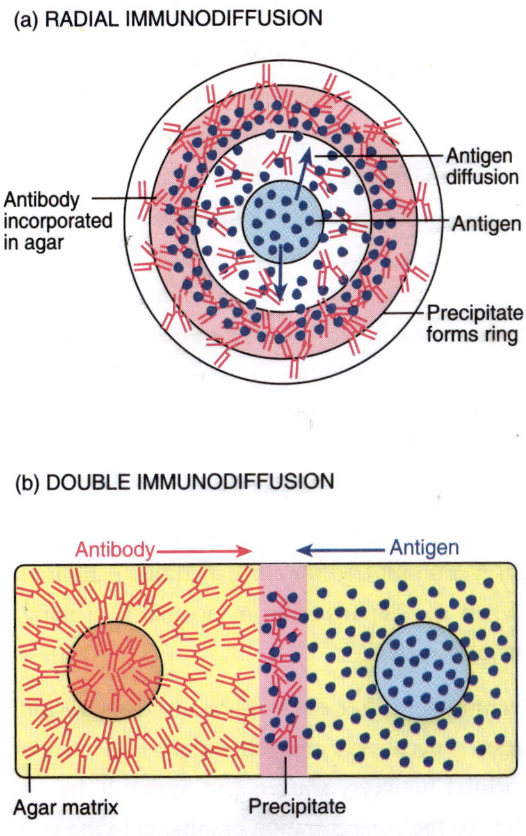

Fig. 6.9: Diagrammatic representation of radial immunodiffusion (Mancini method) and double immunodiffusion (Ouchterlony method) in a gel. In both cases, large insoluble complexes form in the agar in the zone of equivalence, visible as lines of precipitation. Only the antigen diffuses in radial immunodiffusion, whereas both the antibody and antigen diffuse in double immunodiffusion.

antigens, or to determine the relative purity of an antigen preparation. They are radial immunodiffusion (the Mancini method) and double immunodiffusion (the Ouchterlony method); both are carried out in a semisolid medium such as agar. In radial immunodiffusion, an antigen sample is placed in a well and allowed to diffuse into agar containing a suitable dilution of an antiserum. As the antigen diffuses into the agar, the region of equivalence is established and a ring of precipitation, a precipitin ring, forms around the well (Fig. 6.9, upper panel). The area of the precipitin ring is proportional to the concentration of antigen. By comparing the area of the precipitin ring with a standard curve (obtained by measuring the precipitin areas of known concentrations of the antigen), the concentration of the antigen sample can be determined. In the Ouchterlony method, both antigen and antibody diffuse radially from wells toward each other, thereby establishing a concentration gradient. As equivalence is reached, a visible line of precipitation, a precipitin line, forms (Fig. 6.9, lower panel).

6.6 IMMUNOELECTROPHORESIS

In immunoelectrophoresis, the antigen mixture is first electrophoresed to separate its components by charge. Troughs are then cut into the agar gel parallel to the direction of the electric field, and antiserum is added to the troughs. Antibody and antigen then diffuse toward each other and produce lines of precipitation where they meet in appropriate proportions. Immunoelectrophoresis is used in clinical laboratories to detect the presence or absence of proteins in the serum. A sample of serum is electrophoresed, and the individual serum components are identified with antisera specific for a given protein or immunoglobulin class. This technique is useful in determining whether a patient produces abnormally low amounts of one or more isotypes, characteristic of certain immunodeficiency diseases. It can also show whether a patient overproduces some serum protein, such as albumin, immunoglobulin, or transferrin. The immunoelectrophoretic pattern of serum from patients with multiple myeloma, for example, shows a heavy distorted arc caused by the large amount of myeloma protein, which is monoclonal Ig and therefore uniformly charged. Because immunoelectrophoresis is a strictly qualitative technique that only detects relatively high antibody concentrations (greater than several hundred mg/ml), its utility is limited to the detection of quantitative abnormalities only when the departure from normal is striking, as in immunodeficiency states and immunoproliferative disorders.

A related quantitative technique, rocket electrophoresis, does permit measurement of antigen levels. In rocket electrophoresis, a negatively charged antigen is electrophoresed in a gel containing antibody. The precipitate formed between antigen and antibody has the shape of a rocket, the height of which is proportional to the concentration of antigen in the well. One limitation of rocket electrophoresis is the need for the antigen to be negatively charged for electrophoretic movement within the agar matrix. Some proteins, immunoglobulins for example, are not sufficiently charged to be quantitatively analyzed by rocket electrophoresis; nor is it possible to measure the amounts of several antigens in a mixture at the same time.

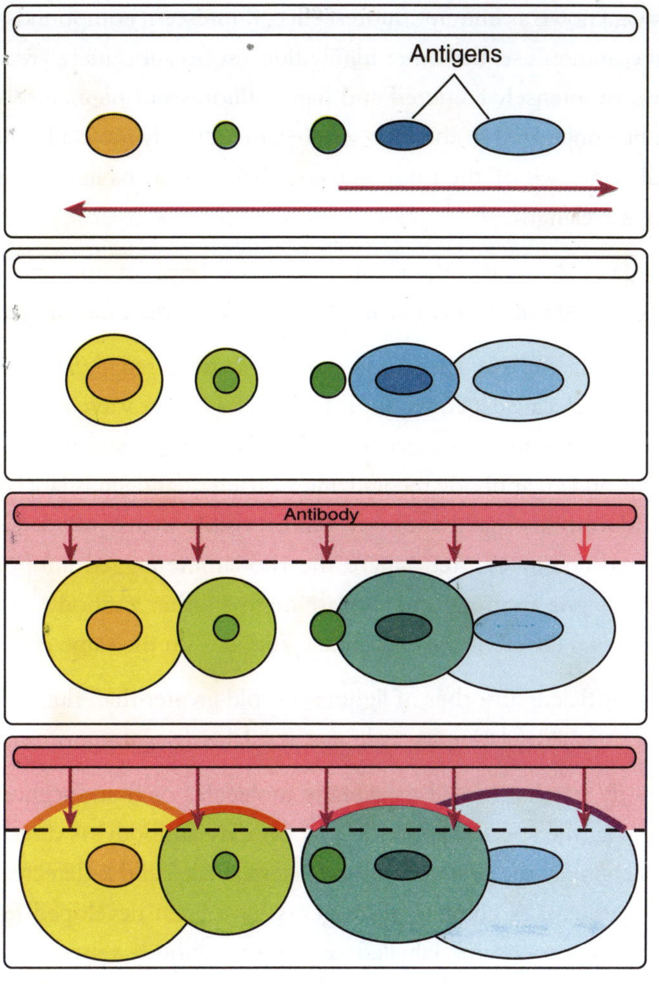

Fig. 6.10: Immunoelectrophoresis of an antigen mixture; an antigen preparation (orange) is first electrophoresed, which separates the component antigens on the basis of charge. Antiserum (blue) is then added to troughs on one or both sides of the separated antigens and allowed to diffuse; in time, lines of precipitation (coloured arcs) form where specific antibody and antigen interact.

6.7 IMMUNOFLUORESCENCE

In 1944, Albert Coons showed that antibodies could be labelled with molecules that have the property of fluorescence. Fluorescent molecules absorb light of one wavelength (excitation) and emit light of another wavelength (emission). If antibody molecules are tagged with a fluorescent dye, or fluorochrome, immune complexes containing these fluorescently labelled antibodies (FA) can be detected by coloured light emission when excited by light of the appropriate wavelength. Antibody molecules bound to antigens in cells or tissue sections can similarly be visualized. The emitted light can be viewed with a fluorescence microscope, which is equipped with a UV light source. In this technique, known as immunofluorescence, fluorescent compounds such as fluorescein and rhodamine are in common use, but other highly fluorescent substances are also routinely used, such as phycoerythrin, an intensely coloured and highly fluorescent pigment obtained from algae. These molecules can be conjugated to the Fc region of an antibody molecule without affecting the specificity of the antibody. Each of the fluorochromes below absorbs light at one wavelength and emits light at a longer wavelength.

Fluorescein, an organic dye that is the most widely used label for immunofluorescence procedures, absorbs blue light (490 nm) and emits an intense yellow-green fluorescence (517 nm).

Rhodamine, another organic dye, absorbs in the yellow-green range (515 nm) and emits a deep red fluorescence (546 nm). Because it emits fluorescence at a longer wavelength than fluorescein, it can be used in two-colour immunofluorescence assays. An antibody specific to one determinant is labeled with fluorescein, and an antibody recognizing a different antigen is labelled with rhodamine. The location of the fluorescein-tagged antibody will be visible by its yellow-green colour, easy to distinguish from the red colour emitted where the rhodamine-tagged antibody has bound. By conjugating fluorescein to one antibody and rhodamine to another antibody, one can, for example, visualize simultaneously two different cell-membrane antigens on the same cell.

Phycoerythrin is an efficient absorber of light (~30-fold greater than fluorescein) and a brilliant emitter of red fluorescence, stimulating its wide use as a label for immunofluorescence.

Fluorescent-antibody staining of cell membrane molecules or tissue sections can be direct or indirect. In direct staining, the specific antibody (the primary antibody) is directly conjugated with fluorescein; in indirect staining, the primary antibody is unlabelled and is detected with an additional fluorochrome-labelled reagent. A number of reagents have been developed for indirect staining. The most common is a fluorochrome-labelled secondary antibody raised in one species against antibodies of another species, such as fluorescein-labelled goat anti-mouse immunoglobulin. Indirect immunofluorescence staining has two advantages over direct staining. First, the primary antibody does not need to be conjugated with a fluorochrome. Because the supply of primary antibody is often a limiting factor, indirect methods avoid the loss of antibody that usually occurs during the conjugation reaction. Second, indirect methods increase the sensitivity of staining because multiple molecules of the fluorochrome reagent bind to each primary antibody molecule, increasing the amount of light emitted at the location of each primary antibody molecule. Immunofluorescence has been applied to identify a number of subpopulations of lymphocytes, notably the $CD4^+$ and $CD8^+$ T-cell subpopulations. The technique is also suitable for identifying bacterial species, detecting Ag-Ab complexes in autoimmune disease, detecting complement components in tissues, and localizing hormones and other cellular products stained in situ. Indeed, a major application of the fluorescent-antibody technique is the localization of antigens in tissue sections or in subcellular compartments. Because it can be used to map the actual location of target antigens, fluorescence microscopy is a powerful tool for relating the molecular architecture of tissues and organs to their overall gross anatomy.

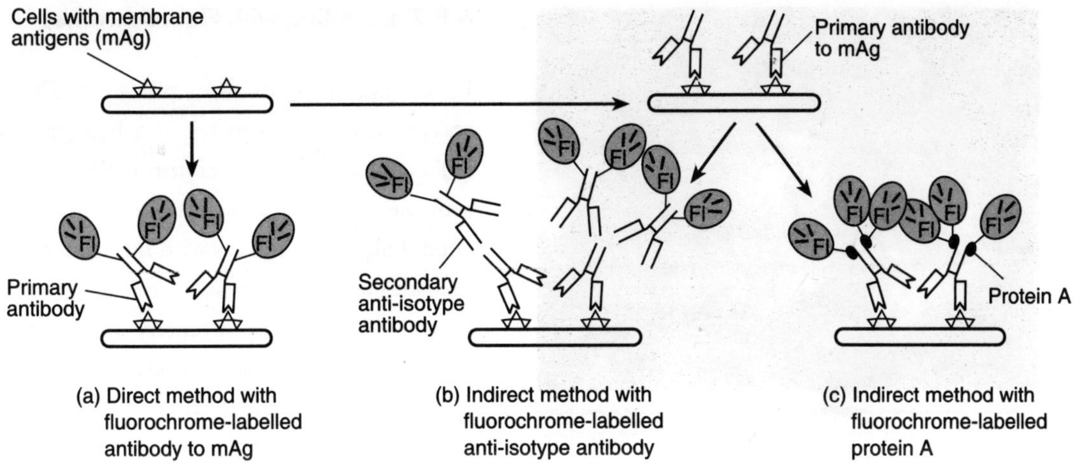

Fig. 6.11: Direct and indirect immunofluorescence staining of membrane antigen (mAg). Cells are affixed to a microscope slide. In the direct method (a), cells are stained with anti-mAg antibody that is labelled with a fluorochrome (Fl). In the indirect methods (b and c), cells are first incubated with unlabelled anti-mAg antibody and then stained with a fluorochrome-labelled secondary reagent that binds to the primary antibody. Cells are viewed under a fluorescence microscope to see if they have been stained.

6.8 ANTIBODY LABELLING

Like the fluorochrome as described in above section, antibodies can also be tagged with enzymes, or carbohydrate or biotin for different purposes. Labelling antibodies with enzymes, fluorochromes, or biotin provides a signal for visualization or quantitation of the target molecule. Antibody bound to agarose is useful for separating a target antigen from a complex mixture. To avoid excessive background staining and to improve sensitivity, only purified antibodies should be used for staining. At the minimum, the IgG fraction, which contains naturally occurring immunoglobulin as well as specific antibody, should be isolated from the antiserum. This can be done by various methods, usually involving a combination of fractionation and chromatography. Preferably, the antibody should be affinity isolated on a column containing the antigen bound to a solid support. This will eliminate all serum proteins, including immunoglobulins that do not specifically bind to the antigen.

6.8.1 Labelling with Enzymes

Antibodies may be labelled with various enzymes to provide highly specific probes that both visualize the target and amplify the signal by acting on a substrate to produce a coloured or chemiluminescent product. Horseradish peroxidase and alkaline phosphatase are the most commonly used enzymes for this purpose. Antibody-enzyme conjugates may be used for detecting proteins in immunohistology and immunocytology, immunoblotting, and ELISA.

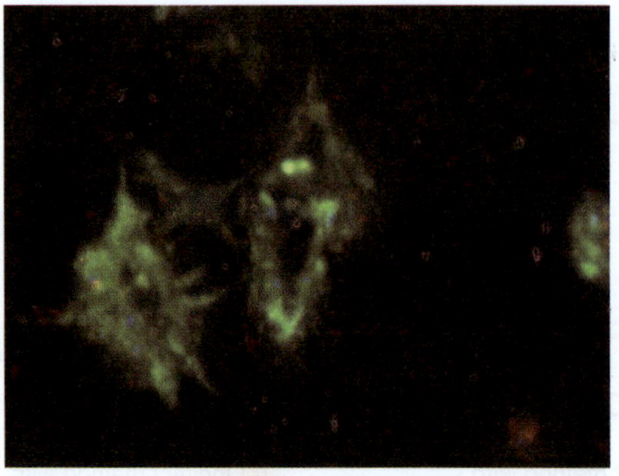

Fig. 6.12: Direct immunofluorescence of cytoskeletal protein actin in human platelets by phalloidin labelled with FITC (Courtesy: Dr D Dash, Deptt. of Biochemistry, IMS, BHU)

6.8.2 Labelling with Fluorochromes

Fluorescein isothiocyanate (FITC), tetramethylrhodamine isothiocyanate (TRITC), and R-phycoerythrin (PE), represent the fluorescent probes most commonly cited. Labelling proteins with fluorochromes provides a coloured reagent that can be observed directly. The intensity and narrow wavelength of the fluorochromes make them useful in immunohistochemistry and immunocytochemisty, using both visible and fluorescence microscopy, and in flow cytometry.

6.8.3 Labelling with Biotin

Biotinylated antibodies are widely used in systems where signal amplification is desired. Biotin binds avidin with a high degree of affinity and specificity. Avidin, ExtrAvidin®, or streptavidin labelled with enzymes or fluorochromes can bind biotinylated antibodies, amplifying the signal and allowing detection of antigens present in small amounts. This system may be used in immunohistochemistry and immunocytochemistry, immunoblotting, ELISA, and flow cytometry.

6.8.4 Labelling with Gold

Antibodies conjugated to colloidal gold are used primarily in electron microscopy (EM) because of the electron-dense nature of the gold particles. Gold-antibody conjugates may also be used for immunohistochemistry and immunocytochemistry, and immunoblotting. In some cases silver enhancement may be used to amplify the signal. In general, 5 nm particles are recommended for intracellular staining because they are able to penetrate the cell membranes more easily, and for high resolution EM because the small size allows more exact localization of antigen. 10 nm particles are recommended for cell surface staining and for light microscopy because the larger size makes the stain more visible. 20 nm particles are recommended for blotting and for some histochemical applications. These are suggestions only — the end user will have to determine the best particle size for each application.

6.8.5 Attaching Antibodies to Agarose

Antibodies attached to agarose have a variety of applications. They can be used to isolate proteins and other compounds from sera or from cell and tissue homogenates by affinity chromatography for quantitation or further analysis. They have also been used for immunoprecipitation of proteins from cell lysates, and for reduction of serum immunoglobulin in autoimmune diseases and transplantation experiments.

Table 6.2 : The following table shows different modes of antibody labelling and their applications

ANTIBODY LABELLING

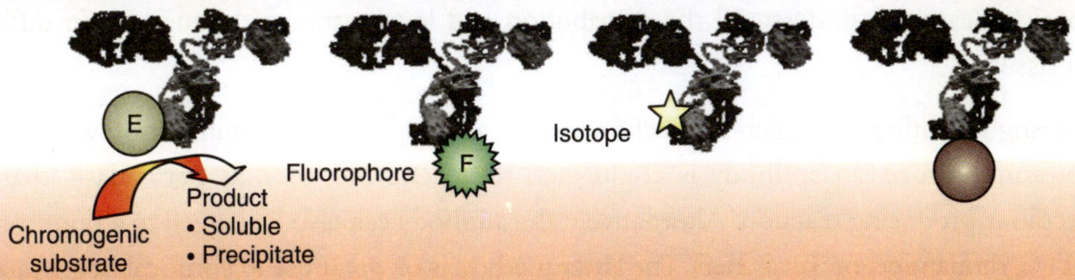

Horseradish Peroxidase (HRP) Alkaline Phosphatase (AP) β-Galactosidase (β-gal)	Fluorescein (FITC) Rhodamine Phycoerythrin Cy-3 Cy-5 Alexa Fluor	^{125}Iodine	Gold Ferritin Wolfram
ELISA Immunohistochemistry Immunoblotting	Immunohistochemistry Immunocytochemistry Flow cytometry Immunoblotting Protein arrays	Radioimmunoassays (RIA)	Immunoelectron microscopy

6.8.6 Antibody Fragmentation

In some assays it is preferable to use only the antigen binding (Fab) portion of the antibody. For these applications, antibodies may be enzymatically digested to produce either an Fab or an F $(ab')_2$ fragment of the antibody. To produce an F $(ab')_2$ fragment, IgG is digested with pepsin, which cleaves the heavy chains near the hinge region. One or more of the disulphide bonds that join the heavy chains in the hinge region are preserved, so the two Fab regions of the antibody remain joined together, yielding a divalent molecule (containing two antibody binding sites), hence the designation F $(ab')_2$. The light chains remain intact and attached to the heavy chain. The Fc fragment is digested into small peptides. Fab fragments are generated by cleavage of IgG with papain instead of pepsin. Papain cleaves IgG above the hinge region containing the disulphide bonds that join the heavy chains, but below the site of the disulphide bond between the light chain and heavy chain. This generates two separate monovalent (containing a single antibody binding site) Fab fragments and an intact Fc fragment. The fragments can be purified by gel filtration, ion exchange, or affinity chromatography.

6.9 IMMUNOHISTOCHEMISTRY

Immunohistochemistry or IHC refers to the process of localizing proteins in cells of a tissue section exploiting the principle of antibodies binding specifically to antigens in biological tissues. It takes its name from the roots "immuno," in reference to antibodies used in the procedure, and "histo," meaning tissue. Immunohistochemical staining is widely used in the diagnosis and treatment of cancer. Specific molecular markers are characteristic of particular cancer types. IHC is also widely used in basic research to understand the distribution and localization of biomarkers in different parts of a tissue.

Visualising an antibody-antigen interaction can be accomplished in a number of ways. In the most common instance, an antibody is conjugated to an enzyme, such as peroxidase, that can catalyse a colour-producing reaction. Alternatively, the antibody can also be tagged to a fluorophore, such as FITC, rhodamine, or Texas Red. The latter method is of great use in confocal laser scanning microscopy, which is highly sensitive and can also be used to visualise interactions between multiple proteins.

6.9.1 Antibody types

The antibodies used for specific detection can be polyclonal or monoclonal. Monoclonal antibodies are generally considered to exhibit greater specificity. Polyclonal antibodies are made by injecting animals with peptide antigens, and then after a secondary immune response is stimulated, isolating antibodies from whole serum. Thus, polyclonal antibodies are a heterogeneous mix of antibodies that recognize several epitopes.

Antibodies can also be classified as primary or secondary reagents. Primary antibodies are raised against an antigen of interest and are typically unconjugated (unlabelled), while secondary antibodies are raised against primary antibodies. Hence, secondary antibodies recognize immunoglobulins of a particular species and are conjugated to either biotin or a reporter enzyme such as alkaline phosphatase or horseradish peroxidase. Some secondary antibodies are conjugated to fluorescent agents, such as the Alexa-Fluor family, are also frequently used for detection of proteins in IHC procedures. Protein concentration is generally measured by densitometry analysis, where the intensity of staining correlates with the amount of the protein of interest.

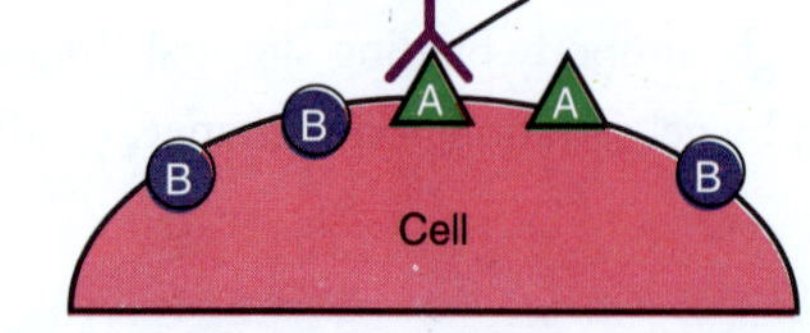

Fig. 6.13: The direct method of immunohistochemical staining uses one labelled antibody, which binds directly to the antigen being stained for.

6.9.2 Sample preparation

In the procedure, depending on the purpose and the thickness of the experimental sample, either thin (about 4-40 μm) slices are taken of the tissue of interest, or if the tissue is not very thick and is penetrable it is used whole. The slicing is usually accomplished through the use of a microtome, and slices are mounted on slides. "Free-floating IHC" uses slices that are not mounted; these slices are normally produced using a vibrating microtome.

Direct and indirect IHC: There are two strategies used for the immmunohistochemical detection of antigens in tissue, the direct method and the indirect method.In both cases, the tissue is treated to rupture the membranes, usually by using a kind of detergent, called Triton X-100.

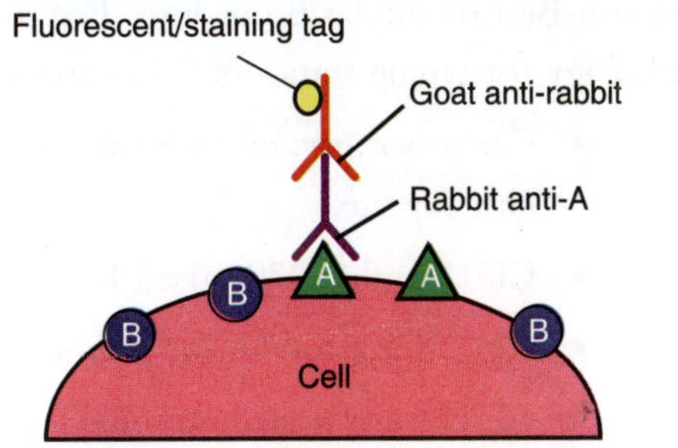

Fig. 6.14: The indirect method of immunohistochemical staining uses one antibody against the antigen being probed for, and a second, labelled, antibody against the first.

The direct method is a one-step staining method, and involves a labelled antibody (e.g., FITC conjugated antiserum) reacting directly with the antigen in tissue sections. This technique utilizes only one antibody and the procedure is therefore simple and rapid. However, it can suffer problems with sensitivity due to little signal amplification and is in less common use than indirect methods.

The indirect method involves an unlabelled primary antibody (first layer) which reacts with tissue antigen, and a labelled secondary antibody (second layer) which reacts with the primary antibody. (The secondary antibody must be against the IgG of the animal species in which the primary antibody has been raised.) This method is more sensitive due to signal amplification through several secondary antibody reactions with different antigenic sites on the primary antibody. The second layer antibody can be labelled with a fluorescent dye or an enzyme.

In a common procedure, a biotinylated secondary antibody is coupled with streptavidin-horseradish peroxidase. This is reacted with 3, 3'-Diaminobenzidine (DAB) to produce a brown staining wherever primary and secondary antibodies are attached in a process known as DAB staining. The reaction can be enhanced using nickel, producing a deep purple/gray staining.

The indirect method, apart from its greater sensitivity, also has the advantage that only a relatively small number of standard conjugated (labelled) secondary antibodies need to be generated. For example, a labelled secondary antibody raised against rabbit IgG, which can be purchased "off the shelf," is useful with any primary antibody raised in rabbit. With the direct method, it would be necessary to make custom labelled antibodies against every antigen of interest.

6.9.3 Diagnostic IHC markers

IHC is an excellent detection technique and has the tremendous advantage of being able to show exactly where a given protein is located within the tissue examined. This has made it a widely-used technique in the neurosciences, enabling researchers to examine protein expression within specific brain structures. Its major disadvantage is that, unlike immunoblotting techniques where staining is checked against a molecular weight ladder, it is impossible to show in IHC that the staining corresponds with the protein of interest. For this reason, primary antibodies must be well-validated in a

Western Blot or similar procedure. The technique is even more widely used in diagnostic surgical pathology for typing tumours (e.g., carcinoma vs. melanoma).

- Carcinoembryonic antigen (CEA): used for identification of adenocarcinomas. Not specific for site.
- CD15 and CD30 : used for Hodgkin's disease.
- Alpha fetoprotein: for yolk sac tumours and hepatocellular carcinoma.
- CD117: for gastrointestinal stromal tumours (GIST).
- Prostate specific antigen (PSA): for prostate cancer.
- Estrogens and progesterone staining for tumour identification.
- Identification of B-cell lymphomas using CD20.

6.9.4 Immunotherapy

Many proteins shown to be highly upregulated in pathological states by immunohistochemistry are potential targets for therapies utilising monoclonal antibodies. For example, Her-2/neu (also known as Erb-B2) is highly expressed in a variety of cancer cell types. As such, antibodies against Her-2/neu have been FDA approved for clinical treatment of cancer under the drug named Herceptin.

6.10 IMMUNOBLOTTING (WESTERN BLOTTING)

Identification of a specific protein in a complex mixture of proteins can be accomplished by a technique known as Western blotting, named for its similarity to Southern blotting, which detects DNA fragments, and Northern blotting, which detects mRNAs. In Western blotting, a protein is electrophoretically separated on gel (denaturing conditions or native/ non-denaturing conditions). The proteins are then transferred to a membrane (typically nitrocellulose or PVDF), where they are probed (detected) using antibodies specific to the target protein. The Ag-Ab complexes that form on the band containing the protein recognized by the antibody can be visualized in a variety of ways. If the protein of interest was bound by a radioactive antibody, its position on the blot can be determined by exposing the membrane to a sheet of X-ray film, a procedure called autoradiography. However, the most generally used detection procedures employ enzyme-linked antibodies against the protein. After binding of the enzyme-antibody conjugate, addition of a chromogenic substrate that produces a highly coloured and insoluble product causes the appearance of a coloured band at the site of the target antigen. The site of the protein of interest can be determined with much higher sensitivity if a chemiluminescent compound along with suitable enhancing agents is used to produce light at the antigen site.

Western blotting can also identify a specific antibody in a mixture. In this case, known antigens of well-defined molecular weight are separated by SDS-PAGE and blotted onto nitrocellulose. The separated bands of known antigens are then probed with the sample suspected of containing antibody specific for one or more of these antigens. Reaction of an antibody with a band is de-

Fig. 6.15: Electroblotter (*Courtesy:* Dr D Dash, Deptt. of Biochemistry, IMS, BHU)

tected by using either radiolabelled or enzyme-linked secondary antibody that is specific for the species of the antibodies in the test sample. The most widely used application of this procedure is in confirmatory testing for HIV, where Western blotting is used to determine whether the patient has antibodies that react with one or more viral proteins.

The method originated from the laboratory of George Stark at Stanford. The name Western blot was given to the technique by W. Neal Burnette.

6.10.1 Steps in a Western Blot

6.10.1.1 Tissue preparation

Samples may be taken from whole tissue or from cell culture or from cell lysate. In most cases, solid tissues are first broken down mechanically using a blender (for larger sample volumes), using a homogenizer (smaller volumes), or by sonication. Cells may also be broken open by one of the above mechanical methods.

Assorted detergents, salts, and buffers may be employed to encourage lysis of cells and to solubilize proteins. Protease and phosphatase inhibitors are often added to prevent the digestion of the sample by its own enzymes.

A combination of biochemical and mechanical techniques – including various types of filtration and centrifugation – can be used to separate different cell compartments and organelles.

6.10.1.2 Gel electrophoresis

The proteins of the sample are separated using gel electrophoresis. Separation of proteins may be by isoelectric point (pI), molecular weight, electric charge, or a combination of these factors. The nature of the separation depends on the treatment of the sample and the nature of the gel (for details please refer to "electrophoretic technique").

By far the most common type of gel electrophoresis employs polyacrylamide gels and buffers loaded with sodium dodecyl sulphate (SDS). SDS-PAGE (SDS polyacrylamide gel electrophoresis) maintains polypeptides in a denatured state once they have been treated with strong reducing agents to remove secondary and tertiary structure (e.g., S-S disulphide bonds to SH and SH) and thus allows separation of proteins by their molecular weight. Sampled proteins become covered in the negatively charged SDS and move to the positively charged electrode through the acrylamide mesh of the gel. Smaller proteins migrate faster through this mesh and the proteins are thus separated according to size (usually measured in kilo daltons, kDa). The concentration of acrylamide determines the resolution of the gel – the greater the acrylamide concentration the better the resolution of higher molecular weight proteins. Proteins travel only in one dimension along the gel for most blots.

Samples are loaded into wells in the gel. One lane is usually reserved for a marker or ladder, a commercially available mixture of proteins having defined molecular weights, typically stained so as to form visible, coloured bands. When voltage is applied along the gel, proteins migrate into it at different speeds. These different rates of advancement (different electrophoretic mobilities) separate into bands within each lane.

It is also possible to use a two-dimensional (2-D) gel which spreads the proteins from a single sample out in two dimensions. Proteins are separated according to isoelectric point (pH at which they have neutral net charge) in the first dimension, and according to their molecular weight in the second dimension.

6.10.1.3 Transfer

In order to make the proteins accessible to antibody detection, they are moved from within the gel onto a membrane made of nitrocellulose or PVDF. The membrane is placed on top of the gel resting on stack of paper towels, and a stack of tissue papers placed on top of that. The entire set-up is placed in a buffer solution which moves up the paper by capillary action, bringing the proteins with it. Another method for transferring the proteins is called electroblotting and uses an electric current to pull proteins from the gel into the PVDF or nitrocellulose membrane. The proteins have now moved from within the gel onto the membrane while maintaining the organization they had within the gel. As a result of this "blotting" process, the proteins are exposed on a thin surface layer for detection. Both varieties of membrane are chosen for their non-specific protein binding properties (i.e., binds all proteins equally well). Protein binding is based upon hydrophobic interactions as well as charged interactions between the membrane and protein. Nitrocellulose membranes are cheaper than PVDF, but are far more fragile and do not stand up well to repeated probings.

The uniformity and overall effectiveness of transfer of protein from the gel to the membrane can be checked by staining the membrane with Coomassie or Ponceau S dyes. Coomassie is the more sensitive of the two, although Ponceau S's water solubility makes it easier to subsequently destain and probe the membrane.

6.10.1.4 Blocking

Since the membrane has been chosen for its ability to bind protein, and both antibodies and the target are proteins, steps must be taken to prevent interactions between the membrane and the antibody used for detection of the target protein. Blocking of non-specific binding is achieved by placing the membrane in a dilute solution of protein - typically Bovine serum albumin (BSA) or non-fat dry milk (both are inexpensive), with a minute percentage of detergent such as Tween 20. The protein in the dilute solution attaches to the membrane in all places where the target proteins have not attached. Thus, when the antibody is added, there is no room on the membrane for it to attach other than on the binding sites of the specific target protein. This reduces "noise" in the final product of the Western blot, leading to clearer results, and eliminates false positives.

6.10.1.5 Detection

During the detection process the membrane is "probed" for the protein of interest with a modified antibody which is linked to a reported enzyme, which when exposed to an appropriate substrate drives a colorimetric reaction and produces a colour. For a variety of reasons, this traditionally takes place in a two-step process, although there are now one-step detection methods available for certain applications.

(A) Two-step processing

1. Primary antibody probing

Antibodies are generated when a host species or immune cell culture is exposed to the protein of interest (or a part thereof). Normally, this is part of the immune response, whereas here they are harvested and used as sensitive and specific detection tools that bind the protein directly.

After blocking, a dilute solution of primary antibody (generally between 0.5 and 5 micrograms/ml) is incubated with the membrane under gentle agitation. Typically, the solution is comprised of buffered saline solution with a small percentage of detergent, and sometimes with powdered milk or BSA. The antibody solution and the membrane can be sealed and incubated together for anywhere from 30 minutes to overnight. It can also be incubated at different temperatures, with warmer temperatures being associated with more binding, both specific (to the target protein, the "signal") and non-specific ("noise").

2. Secondary antibody probing

After rinsing the membrane to remove unbound primary antibody, the membrane is exposed to another antibody, directed at a species-specific portion of the primary antibody. This is known as a secondary antibody, and due to its targeting properties, tends to be referred to as "anti-mouse," "anti-goat," etc. Antibodies come from animal sources (or animal sourced hybridoma cultures); an anti-mouse secondary will bind to just about any mouse-sourced primary antibody. This allows some cost savings by allowing an entire lab to share a single source of mass-produced antibody, and provides far more consistent results. The secondary antibody is usually linked to biotin or to a reported enzyme such as alkaline phosphatase or horseradish peroxidase. This means that several secondary antibodies will bind to one primary antibody and enhances the signal.

Most commonly, a horseradish peroxidase-linked secondary is used in conjunction with a chemiluminescent agent, and the reaction product produces luminescence in proportion to the amount of protein. A sensitive sheet of photographic film is placed against the membrane, and exposure to the light from the reaction creates an image of the antibodies bound to the blot.

As with the ELISPOT and ELISA procedures, the enzyme can be provided with a substrate molecule that will be converted by the enzyme to a coloured reaction product that will be visible on the membrane.

A third alternative is to use a radioactive label rather than an enzyme coupled to the secondary antibody. Since other methods are safer, quicker and cheaper this method is now rarely used.

(B) One step processing

Historically, the probing process was performed in two steps because of the relative ease of producing primary and secondary antibodies in separate processes. This gives researchers and corporations huge advantages in terms of flexibility, and adds an amplification step to the detection process. Given the advent of high-throughput protein analysis and lower limits of detection, however, there has been interest in developing one-step probing systems that would allow the process to occur faster and with less consumables. This requires a probe antibody which both recognizes the protein of interest and contains a detectable label probes which are often available for known protein tags. The primary probe is incubated with the membrane in a manner similar to that for the primary antibody in a two-step process, and then is ready for direct detection after a series of wash steps.

6.10.1.6 Detection

1. Colorimetric detection

The colorimetric detection method depends on incubation of the Western blot with a substrate that reacts with the reporter enzyme (such as peroxidase) that is bound to the secondary antibody. This converts the soluble dye into an insoluble form of a different colour that precipitates next to the enzyme and thereby stains the nitrocellulose membrane. Development of the blot is then stopped by washing away the soluble dye. Protein levels are evaluated through densitometry (how intense the stain is) or spectrophotometry.

2. Chemiluminescence

Chemiluminescent detection methods depend on incubation of the Western blot with a substrate that will luminesce when exposed to the reporter on the secondary antibody. The light is then detected by photographic film, and more recently by CCD cameras which captures a digital image of the Western blot. The image is analysed by densitometry, which evaluates the relative amount of protein staining and quantifies the results in terms of optical density. Newer software allows further data analysis such as molecular weight analysis if appropriate standards are used. So-called "enhanced chemiluminescent" (ECL) detection is considered to be among the most sensitive detection methods for blotting analysis.

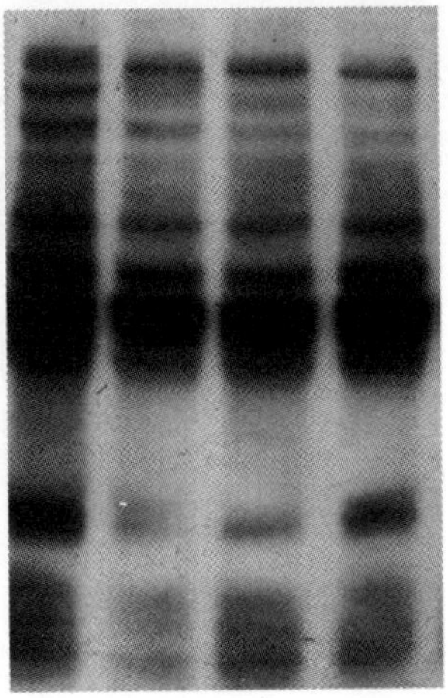

Fig. 6.16: Western blotting of platelets protein for tyrosine phosphorylation profile developed by ECL detection method (*Courtesy:* Dr D Dash, Deptt. of Biochemistry, IMS, BHU)

3. Radioactive detection

Radioactive labels do not require enzyme substrates, but rather allow the placement of medical X-ray film directly against the Western blot which develops as it is exposed to the label and creates dark regions which correspond to the protein bands of interest (see image to the right). The importance of radioactive detection methods is declining, because it is very expensive, health and safety risks are high and ECL provides a useful alternative.

4. Fluorescent detection

The fluorescently labelled probe is excited by light and the emission of the excitation is then detected by a photosensor such as CCD camera equipped with appropriate emission filters which captures a digital image of the Western blot and allows further data analysis such as molecular weight analysis and a quantitative Western blot analysis. Fluorescence is considered to be among the most sensitive detection methods for blotting analysis.

6.10.1.7 Analysis

After the unbound probes are washed away, the western blot is ready for detection of the probes that are labelled and bound to the protein of interest. In practical terms, not all westerns reveal protein only at one band in a membrane. Size approximations are taken by comparing the stained bands to that of the marker or ladder loaded during electrophoresis. The process is repeated for a structural protein, such as actin or tubulin, that should not change between samples. The amount of target protein is indexed to the structural protein to control between groups. This practice ensures correction for the amount of total protein on the membrane in case of errors or incomplete transfers.

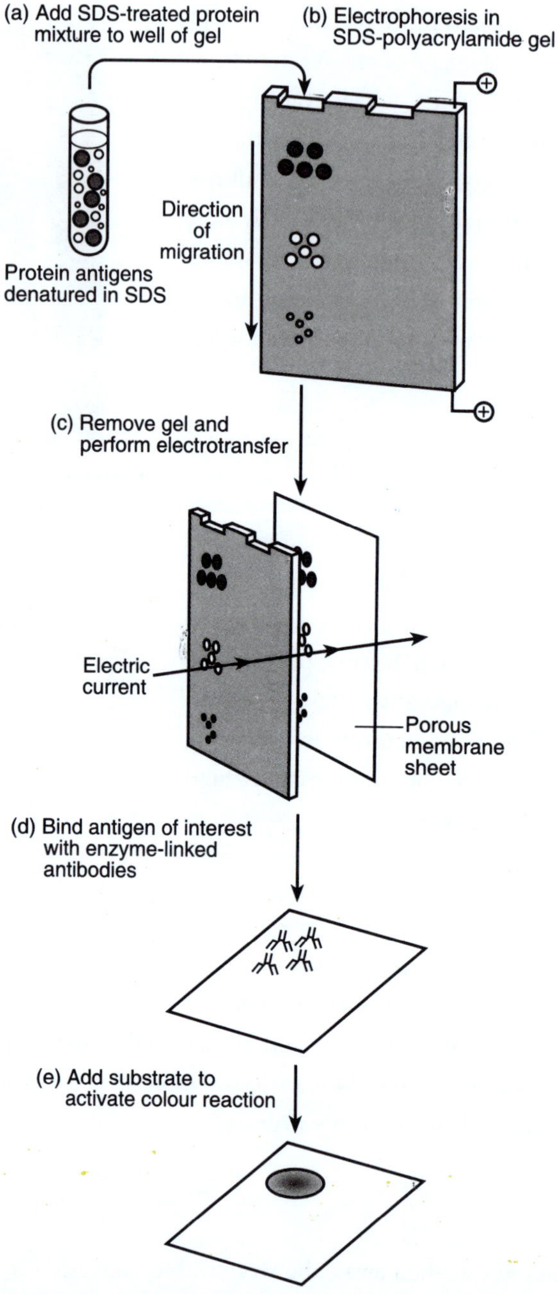

Fig. 6.17: Stern blotting, a protein mixture is (a) treated with SDS, a strong denaturing detergent, (b) then separated by electrophoresis in an SDS polyacrylamide gel (SDS-PAGE) which separates the components according to their molecular weight; lower molecular weight components migrate farther than higher molecular weight ones. (c) The gel is removed from the apparatus and applied to a protein-binding sheet of nitrocellulose or nylon, and the proteins in the gel are transferred to the sheet by the passage of an electric current. (d) Addition of enzyme-linked antibodies detects the antigen of interest, and (e) the position of the antibodies is visualized by means of an ELISA reaction that generates a highly coloured insoluble product that is deposited at the site of the reaction. Alternatively, a chemiluminescent ELISA can be used to generate light that is readily detected by exposure of the blot to a piece of photographic film.

6.10.2 Secondary probing (Stripping)

One major difference between nitrocellulose and PVDF membranes relates to the ability of each to support "stripping" antibodies off and reusing the membrane for subsequent antibody probes. While there are well-established protocols available for stripping nitrocellulose membranes, the sturdier PVDF allows for easier stripping, and for more reuse before background noise limits experiments. Avoid overstripping the membrane, as target proteins may be lost from the blot during extended incubations. Membrane scanning 9 is done regularly to determine when stripping is complete. If overstripping is a problem, try reducing the amount of SDS in the stripping buffer. Depending on the strength of the antibody-antigen interactions, one needs to optimize the stringency of stripping protocol. Another difference is that, unlike nitrocellulose, PVDF must be soaked in 95% ethanol, isopropanol or methanol before use. PVDF membranes also tend to be thicker and more resistant to damage during use.

6.10.3 Medical diagnostic applications

1. The confirmatory HIV test employs a Western blot to detect anti-HIV antibody in a human serum sample. Proteins from known HIV-infected cells are separated and blotted on a membrane as above. Then, the serum to be tested is applied in the primary antibody incubation step; free antibody is washed away, and a secondary anti-human antibody linked to an enzyme signal is added. The stained bands then indicate the proteins to which the patient's serum contains antibody.
2. A Western blot is also used as the definitive test for Bovine spongiform encephalopathy (BSE, commonly referred to as 'mad cow disease').
3. Some forms of Lyme disease testing employ Western blotting

6.11 AFFINITY AND AVIDITY

6.11.1 Antibody Affinity is a Quantitative Measure of Binding Strength

The combined strength of the noncovalent interactions between a single antigen-binding site on an antibody and a single epitope is the affinity of the antibody for that epitope.

Low-affinity antibodies bind antigen weakly and tend to dissociate readily, whereas high-affinity antibodies bind antigen more tightly and remain bound longer. The association between a binding site on an antibody (Ab) with a monovalent antigen (Ag) can be described by the equation:

$$\text{Ag} + \text{Ab} \underset{k_{-1}}{\overset{k_1}{\rightleftharpoons}} \text{Ag-Ab}$$

where k_1 is the forward (association) rate constant and k_{-1} is the reverse (dissociation) rate constant. The ratio k_1/k_{-1} is the association constant K_a (i.e., $k_1/k_{-1} = K_a$), a measure of affinity. Because K_a

is the equilibrium constant for the above reaction, it can be calculated from the ratio of the molar concentration of bound Ag-Ab complex to the molar concentration of unbound antigen and antibody at equilibrium as follows:

$$K_a = \frac{[Ag - Ab]}{[Ab][Ag]}.$$

The value of K_a varies for different Ag-Ab complexes and depends upon both k_1, which is expressed in units for litres/mole/second (l/mol/s), and k_{-1}, which is expressd in units of 1/second (inverse second). For small haptens, the forward rate constant can be extremely high; in some cases, k_1 can be as high as 4×10^8 l/mol/s, approaching the theoretical upper limit of diffusion-limited reactions (10^9 l/mol/s). For larger portein antigens, however, k_1 is smaller, with values in the range of 10^5 l/mol/s.

The rate at which bound antigen leaves as antibody's binding site (i.e., the dissociation rate constant, k_{-1}) plays a major role in determining the antibody's affinity for an antigen.

For some purposes, the dissociation of the antigen-antibody complex is of interest:

$$\text{Ag-Ab} \rightleftharpoons \text{Ab} + \text{Ag}$$

The equilibrium constant for that reactions is K_d, the reciprocal of K_a

$$K_d = [Ab]\ [Ag]/[Ab\text{-}Ag] = 1/K_a$$

and is a quantitative indicator of the stability of an Ag-Ab complex; very stable complexes have very low values of K_d, and less stable ones have higher values.

The affinity constant, K_a, can be determined by equilibrium dialysis or by various newer methods. Because equilibrium dialysis remains for many the standard against which other methods are evaluated, it is described here. This procedure uses a dialysis chamber containing two equal compartments separated by a semipermeable membrane. Antibody is placed in one compartment, and a radioactively labelled ligand that is small enough to pass through the semipermeable membrane is placed in the other compartment (Fig. 6.17). Suitable ligands include haptens, oligosaccharides, and oligopeptides.

In the absence of antibody, ligand added to compartment B will equilibrate on both sides of the membrane (Fig. 6.17a). In the presence of antibody, however, part of the labelled ligand will be bound to the antibody at equilibrium, trapping the ligand on the antibody side of the vessel, whereas unbound ligand will be equally distributed in both compartments. Thus the total concentration of ligand will be greater in the compartment containing antibody (Fig. 6.17b). The difference in the ligand concentration in the two compartments represents the concentration of ligand bound to the antibody (i.e., the concentration of Ag-Ab complex). The higher the affinity of the antibody, the more ligand is bound.

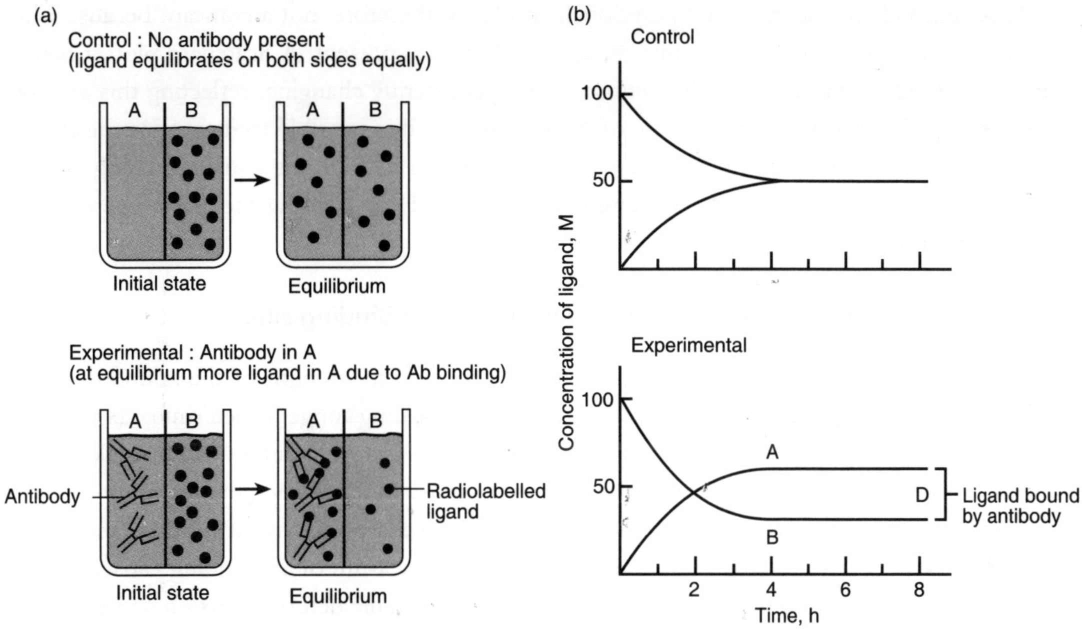

Fig. 6.18: Determination of antibody affinity by equilibrium dialysis. (a) The dialysis chamber contains two compartments (A and B) separated by a semipermeable membrane. Antibody is added to one compartment and a radiolabelled ligand to another. At equilibrium, the concentration of radioactivity in both compartments is measured. (b) Plot of concentration of ligand in each compartment with time. At equilibrium, the difference in the concentration of radioactive ligand in the two compartments represents the amount of ligand bound to antibody.

Since the total concentration of antibody in the equilibrium dialysis chamber is known, the equilibrium equation can be rewritten as:

$$K_a = [\text{Ab-Ag}]/[\text{Ab}][\text{Ag}] = \frac{r}{c(n-r)},$$

where r equals the ratio of the concentration of bound ligand to total antibody concentration, c is the concentration of free ligand, and n is the number of binding sites per antibody molecule. This expression can be rearranged to give the Scatchard equation:

$$\frac{r}{c} = K_a n - K_a r.$$

Values for r and c can be obtained by repeating the equilibrium dialysis with the same concentration of antibody but with different concentrations of ligand. If K_a is a constant, that is, if all the antibodies within the dialysis chamber have the same affinity for the ligand, then a Scatchard plot of r/c versus r will yield a straight line with a slope of $-K_a$ (Fig. 6.18a). As the concentration of unbound ligand c increases, r/c approaches 0, and r approaches n, the valency, equal to the number of binding sites per antibody molecule.

Most antibody preparations are polyclonal, and K_a is, therefore, not a constant because a heterogeneous mixture of antibodies with a range of affinities is present. A Scatchard plot of heterogeneous antibody yields a curved line whose slope is constantly changing, reflecting this antibody heterogeneity (Fig. 6.19b).With this type of Scatchard plot, it is possible to determine the average affinity constant, K_0, by determining the value of K_a, when half of the antigen-binding sites are filled. This is conveniently done by determining the slope of the curve at the point where half of the antigen binding sites are filled.

6.11.2 Antibody avidity incorporates affinity of multiple binding sites

The affinity at one binding site does not always reflect the true strength of the antibody-antigen interaction. When complex antigens containing multiple, repeating antigenic determinants are mixed with antibodies containing multiple binding sites, the interaction of an antibody molecule with an antigen molecule at one site will increase the probability of reaction between those two molecules at a second site. The strength of such multiple interactions between a multivalent antibody and antigen is called the avidity. The avidity of an antibody is a better measure of its binding capacity within biological systems (e.g., the reaction of an antibody with antigenic determinants on a virus or bacterial cell) than the affinity of its individual binding sites. High avidity can compensate for low affinity. For example, secreted pentameric IgM often has a lower affinity than IgG, but the high avidity of IgM, resulting from its higher valence, enables it to bind antigen effectively.

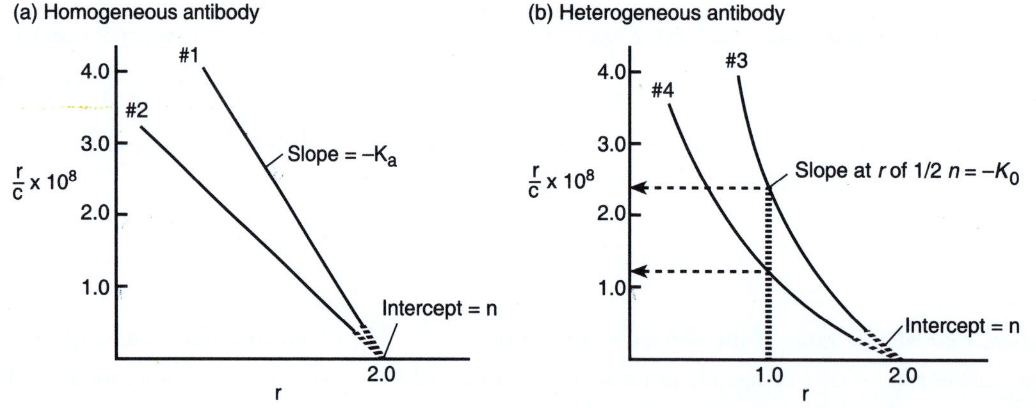

Fig. 6.19: Scatchard plots are based on repeated equilibrium dialyses with a constant concentration of antibody and varying concentration of ligand. In these plots, r equals moles of bound ligand/ mole antibody and c is the concentration of free ligand. From a Scatchard plot, both the equilibrium constant (K_a) and the number of binding sites per antibody molecule (n), or its valency, can be obtained. (a) If all antibodies have the same affinity, then a Scatchard plot yields a straight line with a slope of $-K_a$. The x intercept is n, the valency of the antibody, which is 2 for IgG and other divalent Igs. For IgM, which is pentameric, n =10, and for dimeric IgA, n = 4. In this graph, antibody #1 has a higher affinity than antibody #2. (b) If the antibody preparation is polyclonal and has a range of affinities, a Scatchard plot yields a curved line whose slope is constantly changing.The average affinity constant K_0 can be calculated by determining the value of K_a when half of the binding sites are occupied (i.e., when r =1 in this example). In this graph, antiserum #3 has a higher affinity ($K_0 = 2.4 \times 10^8$) than antiserum #4 ($K_0 = 1.25 \times 10^8$). Note that the curves shown in (a) and (b) are for divalent antibodies such as IgG.

6.12 TIPS TO REMEMBER

Monoclonal Antibodies Preparation

1. Immunization
2. Fusion
3. Selection
4. Screening
   Cloning
5. Production

Technique	General Procedure	Typical Application
ELISA	• adsorb Ag to solid support • incubate with Ab • detect bound Ab	• quantify Ab • quantify Ag • process large # of samples
BLOTTING	• SDS-PAGE and transfer • incubate with Ab • detect bound Ab	• identify subunit MW • quantify Ag • qunatify Ab?
IFA	• fix cells on slide • incubate with Ab • detect bound Ab	• subcellular localization • quantify Ab • quantify Ag?

Protein size (kDa)	Gel percentage (%)
4 - 40	20
12 - 45	15
10 - 70	12.5
15 - 100	10
25 - 200	8

Fluorochromes

- fluorescein
- rhodamine

Enzyme Crosslinking

- AP
- HRP

Radiolabelling

- iodination
- metabolically (mAbs)

Biotinylation

Immunogen vs. Antigen

- immunogen induces immune response

- antigen reacts with products of the immune response
- immunogenicity
  - nature of the immunogen
  - ability of immune system to react
  - possible to manipulate

Immunoprecipitation

- affinity purification based on isolation of Ag-Ab complexes
- analyze by gel electrophoresis
- initially based on centrifugation of large supramolecular complexes
- [high] and equal amounts
- isolation of Ag-Ab complexes
- fixed *S. aureus*
- protein A-agarose
- protein G-agarose

6.13 VIVA VOCE

Q.1. What are monoclonal antibodies and how are they produced?
Q.2. What is species specific reactivity?
Q.3. Explain direct immunostaining.
Q.4. Explain various fluorescence based immunoassay.
Q.5. How antibodies can be labelled?
Q.6. What are secondary antibodies?
Q.7. What is the role of secondary antibodies in immunoassays?
Q.8. Differentiate between RIA and ELISA.
Q.9. What is sandwich ELISA?
Q.10. What are different variants of ELISA?
Q.11. What are flurochromes and how are they employed in immunological techniques?
Q.12. How can you determine the co-localization of two proteins within the cells?
Q.13. What is the purpose of blocking in Western blotting?
Q.14. What is ELISA spot assay?
Q.15. What is affinity?
Q.16. Differentiate between affinity and avidity.
Q.17. What is the force responsible for transfer of proteins from gel to the membrane in immunoblotting?
Q.18. What are the qualities of nitrocellulose and PVDF membranes?
Q.19. What is immunoprecipitation?
Q.20. How can you detect or measure the expression of one particular protein of a cell?
Q.21. What are the precautions one should take during immunoprecipitation assay?

CHAPTER 7

Enzymes: Experimental Measures of Activity

7.1 INTRODUCTION

Enzyme kinetics offer a wealth of information on the mechanisms of enzyme catalysis and on the interactions of enzymes with ligands, such as substrates and inhibitors. We know the basis for determining the kinetic constants k_{cat} and K_m from initial velocity measurements taken at varying substrate concentrations during steady state catalysis. The determination of these kinetic constants rests on the ability to measure accurately the initial velocity of an enzymatic reaction under well-controlled conditions.

In this chapter we describe some of the experimental methods used to determine reaction velocities. We shall see that numerous strategies have been developed for following over time the loss of substrate or the appearance of products that result from enzyme turnover. The velocity of an enzymatic reaction is sensitive to many solution conditions, such as pH, temperature, and solvent isotopic composition; these conditions must be well controlled if meaningful data are to be obtained. Controlled changes in these solution conditions and measurement of their effects on the reaction velocity can provide useful information about the mechanism of catalysis as well. Like all proteins, enzymes are sensitive to storage conditions and can be denatured easily by mishandling. Therefore, we also discuss methods for the proper handling of enzymes to ensure their maximum catalytic activity and stability.

7.2 INITIAL VELOCITY MEASUREMENTS: DIRECT, INDIRECT, AND COUPLED ASSAYS

To measure the velocity of a reaction, it is necessary to follow a signal that reports product formation or substrate depletion over time. The type of signal that is followed varies from assay to assay but usually relies on some unique physicochemical property of the substrate or product, and/or the analyst's ability to separate the substrate from the product. Generally, most enzyme assays rely on one or more of the following broad classes of detection and separation methods to follow the course of the reaction:

Spectroscopy
Polarography
Radioactive decay
Electrophoretic separation
Chromatographic separation
Immunological reactivity

These methods can be used in *direct assay*: the direct measurement of the substrate or product concentration as a function of time. For example, the enzyme cytochrome c oxidase catalyzes the oxidation of the heme-containing protein cytochrome c. In its reduced (ferrous iron) form, cytochrome c displays a strong absorption band at 550 nm, which is significantly diminished in intensity when the heme iron is oxidized (ferric form) by the oxidase. One can thus measure the change in light absorption at 550 nm for a solution of ferrous cytochrome c as a function of time after addition of cytochrome c oxidase; the diminution of absorption at 550 nm that is observed is a direct measure of the loss of substrate (ferrous cytochrome c) concentration (Fig. 7.1).

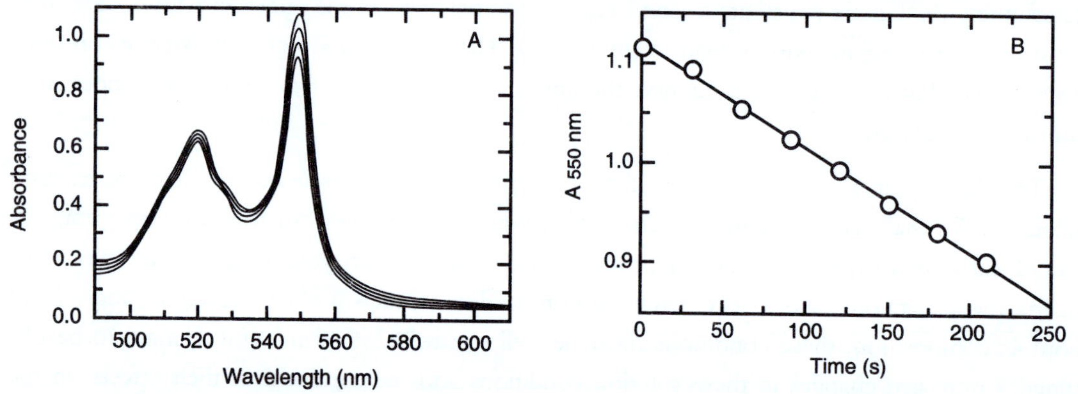

Fig. 7.1: Absorption of ferrocytochrome c as a function of time after addition of the enzyme cytochrome c oxidase. As the cytochrome c iron is oxidized by the enzyme, the absorption feature at 550 nm decreases. (B) Plot of the absorption at 550 nm for the spectra in (A), as a function of time. Note that in this early stage of the reaction (10% of the substrate has been converted), the plot yields a linear relationship between absorption and time. The reaction velocity can thus be determined from the slope of this linear function

In some cases the substrate and product of an enzymatic reaction do not provide a distinct signal for convenient measurement of their concentrations. Often, however, product generation can be coupled to another, nonenzymatic, reaction that does produce a convenient signal; such a strategy is referred to as an *indirect* assay. Dihydroorotate dehydrogenase (DHODase) provides an example of the use of indirect assays. This enzyme catalyzes the conversion of dihydroorotate to orotic acid in the presence of the exogenous cofactor ubiquinone. During enzyme turnover, electrons generated by the conversion of dihydroorotate to orotic acid are transferred by the enzyme to a ubiquinone cofactor to form ubiquinol. It is difficult to measure this reaction directly, but the reduction of ubiquinone can be coupled to other nonenzymatic redox reactions. Several redoxactive dyes are known to change colour upon oxidation or reduction. Among these, 2,6-dichlorophenolindophenol (DCIP) is a convenient dye with which to follow the DHODase reaction. In its oxidized form DCIPs bright blue, absorbing light strongly at 610 nm. Upon reduction, however, this absorption band is completely lost. DCIP is reduced stoichiometrically by ubiquinol, which is formed during DHODase turnover. Hence, it is possible to measure enzymatic turnover by having an excess of DCIP present in a solution of substrate (dihydroorotate) and cofactor (ubiquinone), then following the loss of 610 nm absorption with time after addition of enzyme to initiate the reaction.

A third way of following the course of an enzyme-catalyzed reaction is referred to as the *coupled* assays method. Here the enzymatic reaction of interest is paired with a second enzymatic reaction, which can be conveniently measured. In a typical coupled assay, the product of the enzyme reaction of interest is the substrate for the enzyme reaction to which it is coupled for convenient measurement. An example of this strategy is the measurement of activity for hexokinase, the enzyme that catalyzes the formation of glucose 6-phosphate and ADP from glucose and ATP. None of these products or substrates provide a particularly convenient means of measuring enzymatic activity. However, the product glucose 6-phosphate is the substrate for the enzyme glucose 6-phosphate dehydrogenase, which, in the presence of NADP, converts this molecule to 6-phosphogluconolactone. In the course of the second enzymatic reaction, NADP is reduced to NADPH, and this cofactor reduction can be monitored easily by light absorption at 340 nm.

This example can be generalized to the following scheme:

$$A \xrightarrow{v_1} B \xrightarrow{v_2} C$$

where A is the substrate for the reaction of interest, v_1 is the velocity for this reaction, B is the product of the reaction of interest and also the substrate for the coupling reaction, v_2 is the velocity for the coupling reaction, and C is the product of the coupling reaction being measured. Although we are measuring C in this scheme, it is the steady state velocity v_1 that we wish to study. To accomplish this we must achieve a situation in which v_1 is rate limiting (i.e., $v_1 >> v_2$) and B has reached a steady state concentration. Under these conditions B is converted to C almost instantaneously, and the rate of C production is a reflection of v_1. The measured rate will be less than the steady state rate v_1, however, until [B] builds up to its steady state level. Hence, in any coupled assay there will be a lag phase prior to steady state production of C (Fig. 7.2), which can interfere with the measurement of the initial velocity. Thus to measure the true initial velocity of the reaction of interest, conditions must be sought to minimize the lag phase that precedes steady state product formation, and care must be taken to ensure that the velocity is measured during the steady state phase.

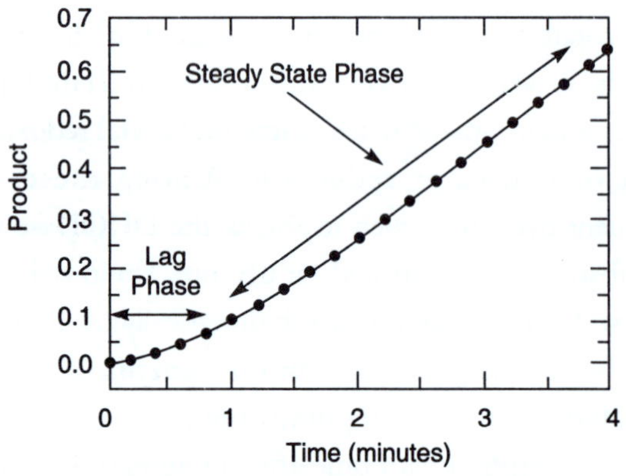

Fig. 7.2: Typical data for a coupled enzyme reaction illustrating the lag phase that precedes the steady state phase of the time course

The velocity of the coupled reaction, v_2, follows simple Michaelis-Menten kinetics as follows:

$$v_2 = \frac{V_2[B]}{K_m^2[B]},$$

where K_m^2 refers to the Michaelis constant for enzyme E_2, not the square of the K_m. Early in the reaction, v_1 is constant for a fixed concentration of E_1. Hence, the rate of B formation is given by:

$$\frac{d[B]}{dt} = v_1 - v_2 = v_1 - \frac{V_2[B]}{K_m^2 + [B]},$$

This equation was evaluated by integration by Storer and Cornish-Bowden (1974), who showed that the time required for [B] to reach some percentage of its steady state level $[B]_{ss}$ can be defined by the following equation:

$$t_{99\%} = \frac{\phi K_m^2}{v_1},$$

where $t_{99\%}$ is the time required for [B] to reach 99% $[B]_{ss}$ and Φ is a dimensionless value that depends on the ratio v_1/V_2 and v_2/v_1. It is known that maximal velocity V_2 is the product of k_{cat} for the coupling enzyme and the concentration of coupling enzyme $[E_2]$. The values of k_{cat} and K_m for the coupling enzyme are constants that cannot be experimentally adjusted without changes in reaction conditions. The maximal velocity V_2, however, can be controlled by the researcher by adjusting the concentration $[E_2]$. Thus by varying $[E_2]$ one can adjust V_2, hence the ratio v_1/V_2, and also the lag time for the coupled reaction.

Table 7.1: Values of Φ, for [B]= 99% $[B]_{ss}$, useful in designing coupled assays

v_1/V_2	ϕ
0.1	0.54
0.2	1.31
0.3	2.42
0.4	4.12
0.5	6.86
0.6	11.70
0.7	21.40
0.8	45.50
0.9	141.00

Source: Adapted from Storer and Cornish-Bowden (1974)

Let us say that we can measure the true steady state velocity v_1 after [B] has reached 99% of $[B]_{ss}$. How much time is required to achieve this level of $[B]_{ss}$? We can calculate this if we know the values of v_1 and Φ. Storer and Cornish-Bowden tabulated the ratios v_1/V_2 that yield different values of Φ for reaching different percentages of $[B]_{ss}$. Table 7.1 lists the values for $[B]=99\%\ [B]_{ss}$. This percentage is usually considered to be optimal for measuring v_1 in a coupled assay. In certain cases this requirement can be relaxed. For example, $[B]=90\%\ [B]_{ss}$ would be adequate for use of a coupled assay to screen column fractions for the presence of the enzyme of interest. In this situation we are not attempting to define kinetic parameters, but merely wish a relative measure of primary enzyme concentration among different samples. The reader should consult the original paper by Storer and Cornish-Bowden (1974) for additional tables of Φ for different percentages of $[B]_{ss}$.

Easterby (1973) and Cleland (1979) have presented a slightly different method for determining the duration of the lag phase for a coupled reaction. From their treatments we find that as long as the coupling enzyme(s) operate under first-order conditions (i.e., $[B]_{ss} << K_m^2$), we can write:

$$[B]_{ss} = v_1\left(\frac{K_m^2}{V_2}\right).$$

And

$$\tau = \frac{K_m^2}{V_2},$$

where τ is the lag time. The time required for [B] to approach $[B]_{ss}$ is exponentially related to τ so that [B] is 92% $[B]_{ss}$ at 2.5τ, 95% $[B]_{ss}$ at 3τ, and 99% $[B]_{ss}$ at 4.6τ (Easterby, 1973). Product (C) formation as a function of time (t) is dependent on the initial velocity v_1 and the lag time (τ) as follows:

$$[C] = v_1\left(t + \tau e^{-t/\tau} - \tau\right).$$

If more than one enzyme is used in the coupling steps, the overall lag time can be calculated as $\Sigma\left(K_m^i / V_i\right)$. For example, if one uses two consecutive coupling enzymes, A and B to follow the reaction of the primary enzyme of interest, the overall lag time would be given by:

$$\tau = \frac{K_m^A}{V_A} + \frac{K_m^n}{V_B}.$$

Because coupled reactions entail multiple enzymes, these assays present a number of potential problems that are not encountered with direct or indirect assays. For example, to obtain meaningful data on the enzyme of interest in coupled assays, it is imperative that the reaction of interest remains rate limiting under all reaction conditions. Otherwise, any velocity changes that accompany changes in reaction conditions may not accurately reflect effects on the target enzyme. For example, to use a coupled reaction scheme to determine k_{cat} and K_m for the primary enzyme of interest, it is neces-

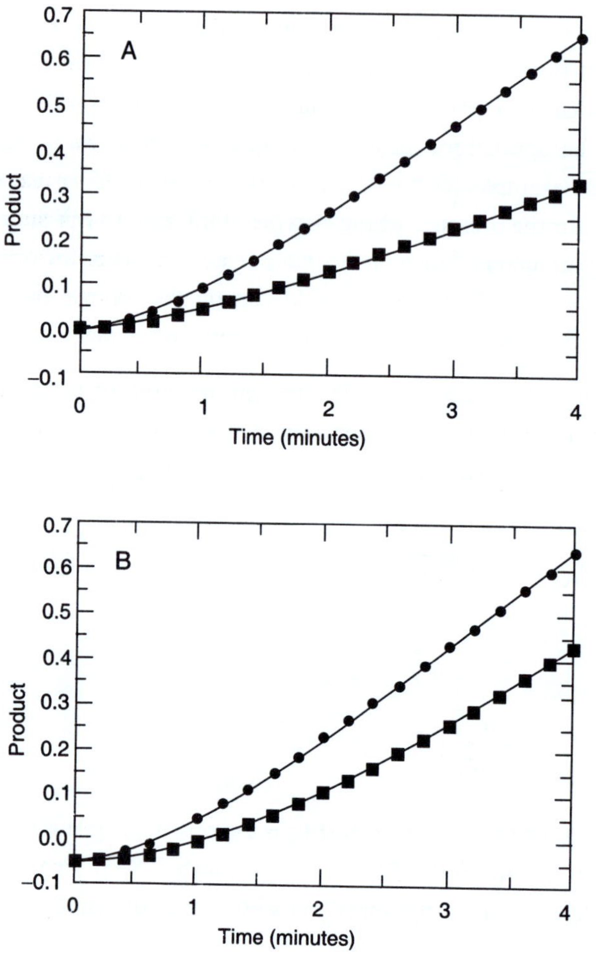

Fig. 7.3: Effects of inhibitors on a coupled enzyme reaction: circles, the data points for the enzyme in the absence of inhibitor; squares, the data points when inhibitor is present as some fixed concentration. (A) Effect of an inhibitor of the primary enzyme of interest: the lag phase is unaffected, but the steady state rate (slope) is diminished. (B) Effect of an inhibitor of the coupling enzyme: the lag phase is extended, but the steady state rate is unaffected.

sary to ensure that v_1 is still rate limiting at the highest values of [A] (i.e., substrate for the primary enzyme of interest).

Use of a coupled assay to study inhibition of the primary enzyme might also seem problematic. The presence of multiple enzymes could introduce ambiguities in interpreting the results of such experiments: for example, which enzyme(s) are really being inhibited? Easterby (1973) points out, however, that using coupled assays to screen for inhibitors makes it relatively easy to distinguish between inhibitors of the primary enzyme and the coupling enzyme(s). Inhibitors of the primary enzyme would have the effect of diminishing the steady state velocity v_1 (Fig. 7.3A), while inhibitors of the coupling enzyme(s) would extend the lag phase without affecting v_1 (Fig. 7.3B). In practice these distinctions are clear-cut only when one measures product formation over a range of time points covering significant portions of both the lag phase and the steady state phase of the progress curves (Fig. 7.3). Rudolph et al. (1979), Cleland (1979), and Tipton (1992) provide more detailed discussions of coupled enzyme assays.

7.3 ANALYSIS OF PROGRESS CURVES: MEASURING TRUE STEADY STATE VELOCITY

The full progress curve of an enzymatic reaction is rich in kinetic information. Throughout the progress curve, the velocity is changing as the substrate concentration available to the enzyme continues to diminish. Hence, throughout the progress curve the instantaneous velocity approximates the initial velocity for the instantaneous substrate concentration at a particular time point. Modern computer graphing programs often provide a means of computing the derivative of the data points from a plot of *y* versus *x*, hence one could determine the instantaneous velocity ($v = d[P]/dt = -d[S]/dt$) from derivitization of the enzyme progress curve. At each time point for which the instantaneous velocity is determined, the instantaneous value of [S] can be determined, and so the instantaneous velocity can be replotted as a function of [S]. This replot is hyperbolic and can be fit to the Michaelis-Menten equation to extract estimates of k_{cat} and K_m from a single progress curve. This approach and its limitations have been described by several authors (Cornish-Bowden, 1972; Duggleby and Morrison, 1977; Waley, 1982; Kellershohn and Laurent, 1985; Duggleby, 1985, 1994).

7.4 CONTINUOUS VERSUS END POINT ASSAYS

Once an appropriate linear time period has been established, the researcher has two options for obtaining a velocity measurement. First, the signal might be monitored at discrete intervals over the entire linear time period, or some convenient portion thereof. This strategy, referred to as a *continuous assay*, provides the safest means of accurately determining reaction velocity from the slope of a plot of signal versus time.

It is not always convenient to assay samples continuously, however, especially when one is using separation techniques, such as high performance liquid chromatography (HPLC) or electrophoresis. In such cases a second strategy, called *end point* or *discontinuous assay*, is often employed. Having established a linear time period for an assay, one measures the signal at a single specific time point within the linear time period (most preferably, a time point near the middle of the linear phase). The reaction velocity is then determined from the difference in signal at that time point and at the initiation of the reaction, divided by the time

$$v = \frac{\Delta I}{\Delta t} = \frac{I_t - I_0}{t_{reading}},$$

where the intensity of the signal being measured at time t and time zero is given by I_t and I_0, respectively, and $t_{reading}$ is the time interval between initiation of the reaction and measurement of the signal.

In many instances it is much easier to take a single reading than to make multiple measurements during a reaction. Inherent in the use of end point readings, however, is the danger of assuming that the signal will track linearly with time over the period chosen, under the conditions of the measurement.

Changes in temperature, pH, substrate, and enzyme concentrations, as well as the presence of certain types of inhibitor can dramatically change the linearity of the signal over a fixed time window. Hence, end point readings can be misleading. Whenever feasible, then, one should use continuous assays to monitor substrate loss or product formation. When this is impractical, end point readings can be used, but cautiously, with careful controls.

7.5 INITIATING, MIXING, AND STOPPING REACTIONS

In a typical enzyme assay, all but one of the components of the reaction mixture are added to the reaction vessel, and the reaction is started at time zero by adding the missing component, which can either be the enzyme or the substrate. The choice of the initiating component will depend on the details of the assay format and the stability of the enzyme sample to the conditions of the assay. In either case, the other components should be mixed well and equilibrated in terms of pH, temperature, and ionic strength. The reaction should then be initiated by addition of a small volume of a concentrated stock solution of the missing component. A small volume of the initiating component is used to ensure that its addition does not significantly perturb the conditions (temperature, pH, etc.) of the overall reaction volume. Unless the reaction mixture and initiating solutions are well matched in terms of buffer content, pH, temperature, and other factors, the initiating solution should not be more than about 5-10% of the total volume of the reaction mixture.

Samples should be mixed rapidly after addition of the initiating solution, but vigorous shaking or vortex mixing is denaturing to enzymes and should be avoided. Mixing must, however, be complete; otherwise there will be artifactual deviations from linear initial velocities as mixing continues during the measurements.

One way to rapidly achieve gentle but complete mixing is to add the initiating solution to the side of the reaction vessel as a "hanging drop" above the remainder of the reaction mixture, as illustrated in Fig. 7.4. With small volumes (say, < 50 µL), the surface tension will hold the drop in place above the reaction mixture. At time zero the reaction is initiated by gently inverting the closed vessel two or three times to mix the solutions. Fig. 7.4 illustrates this technique for a reaction taking place in a microcentrifuge tube. It is also convenient to place the initiating solution in the tube cap, which then can be closed, permitting the solutions to be mixed by inversion as illustrated in Fig. 7.4. For optical spectroscopic assays, the reaction can be initiated directly in the spectroscopic cuvette by the same technique, using a piece of Parafilm and one's thumb to seal the top of the cuvette during the inversions.

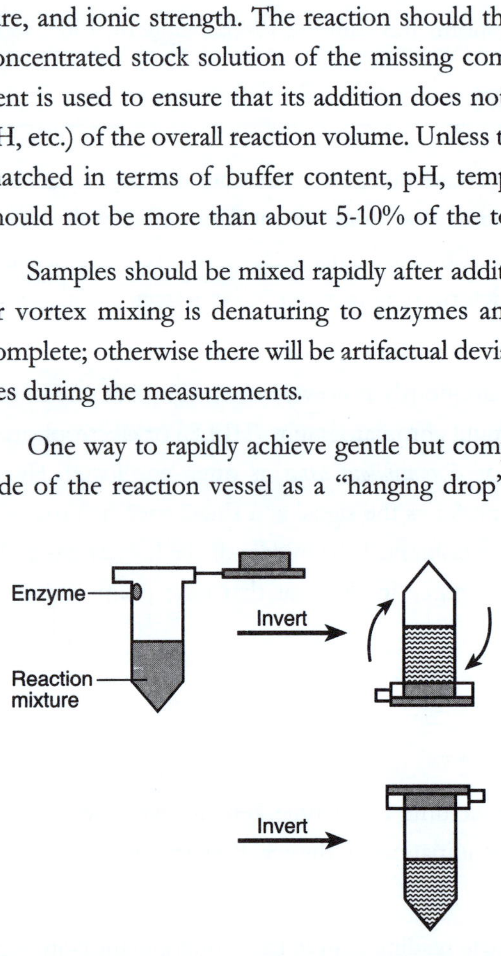

Fig. 7.4: A common strategy for initiating an enzymatic reaction in a microcentrifuge tube

Regardless of how the reacting and initiating solutions are mixed, the mixing must be achieved in a short period of time relative to the time interval between measurements of the reaction's progress. With a little practice one can use the inversion method just described to achieve this mixing in 10 seconds or less.

This is usually fast enough for assays in which measurements are to be made in intervals of 1 minute or longer time. A number of parameters, such as temperature and enzyme concentration, can be adjusted to ensure that the reaction velocity is slow enough to allow mixing of the solutions and making of measurements on a convenient time scale. In some rare cases, the enzymatic velocity is so rapid that it cannot be measured conveniently in this way. Then one must resort to specialized rapid mixing and detection methods, such as stopped-flow techniques (Roughton and Chance, 1963; Kyte, 1995); these methods are also used to measure pre-steady state enzyme kinetics.

For assays in which samples are removed from the reaction vessel at specific times for measurement, one can start the timer at the point of mixing and make measurements at known time intervals after the initiation point. In many spectroscopic assays, however, one measures changes in absorption or fluorescence with time. For most modern spectrometers, the detection is initiated by pressing a button on an instrument panel or depressing a key on a computer keyboard. Thus to start an assay one must mix the solutions, place the cuvette, or optical cell, in the holder of the spectrometer, and start the detection by pressing the appropriate button. The delay between mixing and actually starting a measurement can be as much as 20 seconds. Thus the time point recorded by the spectrometer as zero will not be the true zero point (i.e., mixing point) of the reaction. Again, with practice one can minimize this delay time, and in most cases the assay can be set up to render this error insignificant.

As we shall see, there should always be two control measurements: one in which all the reaction components except the enzyme are present, and a separate one in which everything but the substrate is present. (In these controls, buffer is added to make up for the volumes that would have been contributed by the enzyme or substrate solutions.) With these two control measurements one can calculate what the absorption or fluorscence should be for the reaction mixture at the true time zero. If the first spectrometer reading (i.e., the time point recorded as time zero by the spectrometer) is significantly different from this calculated value, it is necessary to correct the time points recorded by the spectrometer for the time delay between the start of mixing and the initiation of the detection device. A laboratory timer or stopwatch can be used to determine the time gap.

Many nonspectroscopic assays require measurement times that are long in comparison to the rate of the enzymatic reaction being monitored. Suppose, for example, that we wish to measure the amount of product formed every 5 minutes over the course of a 30-minute reaction and assay for product by an HPLC method. The HPLC measurement itself might take 20-30 minutes to complete. If the enzymatic reaction is continuing during the measurement time, the amount of product produced during specific time intervals cannot be determined accurately. In such cases it is necessary to quench or stop the reaction at a specific time, to prevent further enzymatic production of product or utilization of substrate.

Methods for stopping enzymatic reactions usually involve denaturation of the enzyme by some means, or rapid freezing of the reaction solution. Examples of quenching methods include immersion in a dry ice—ethanol slurry to rapidly freeze the solution, and denaturation of the enzyme by addition of strong acid or base, addition of electrophoretic sample buffer, or immersion in a boiling water bath. In addition to these methods, reagents can be added that interfere in a specific way with a particular enzyme. For example, the activity of many metalloenzymes can be quenched by adding an excess of a metal chelating agent, such as ethylenediaminetetraacetic acid (EDTA). Three points must be considered in choosing a quenching method for an enzymatic reaction. First, the technique used to quench the reaction must not interfere with the subsequent detection of product or substrate. Second, it must be established experimentally that the quenching technique chosen does indeed completely stop the reaction. Finally, the volume change that occurs upon addition of the quenching reagent to the reaction mixture must be accounted for. Similarly, measurement of product or substrate concentration must be corrected to compensate for the dilution effects of quencher addition.

7.6 THE IMPORTANCE OF RUNNING CONTROLS

Regardless of the detection method used to follow an enzymatic reaction, it is always critical to perform control measurements in which enzyme and substrate are separately left out of the reaction mixture. These control experiments permit the analyst to correct the experimental data for any time-dependent changes in signal that might occur independent of the action of the enzyme under study, and to correct for any static signal due to components in the reaction mixture. To illustrate these points, let us follow a hypothetical enzymatic reaction by tracking light absorption decrease at some wavelength, as substrate is converted to product. Let us say that there is some low rate of spontaneous product formation in the absence of the enzyme, and that the enzyme itself imparts a small, but measurable absorption at the analytical wavelength. We might set up an experiment in which all the reaction components are placed in a cuvette, and the reaction is initiated by the addition of a small volume of enzyme stock solution. For this illustration, let us say that the reaction mixture is prepared by addition of the volumes of stock solutions listed in Table 7.2. The strategy for preparing the reaction mixtures in Table 7.2 is typical of what one might use in a real experimental situation.

Table 7.2: Volumes of stock solutions to prepare experimental and control samples for a hypothetical enzyme-catalyzed reaction

	Volumes Added (μl)		
Stock Solution	Experimental	No Substrate Control	No Enzyme Control
10 × Substrate	100 μl	0 μl	100 μl
10 × Buffer	100 μl	100 μl	100 μl
Distilled water	790 μl	890 μl	800 μl
Enzyme	10 μl	10 μl	0 μl
Total volume	1.0 l	1.0 l	1.0 l

Fig. 7.5 A illustrates the time courses we might see for the hypothetical solutions from Table 7.2. For our experimental run, the true absorption readings are displaced by about 0.1 unit, as a result of the absorption of the enzyme itself ("No substrate" trace in Fig. 7.5A). To correct for this, we subtract this constant value from all our experimental data points. If now we were to determine the slope of our corrected experimental trace, however, we would be overestimating the velocity of our reaction because such a slope would reflect both the catalytic conversion of substrate to product and the spontaneous absorption change seen in our "No enzyme" control trace. To correct for this, we subtract these control data points from the experimental points at each measurement time to yield the difference plot in Fig. 7.5B.

Measuring the slope of this difference plot yields the true reaction velocity.

As illustrated in Fig. 7.5A, the correction for the spontaneous absorption change may appear at first glance to be trivial. However, the velocity measured for the uncorrected data differs from the corrected velocity by more than 10% in this example. In some cases the background signal change is even more substantial. Hence, the types of control measurement discussed here are essential for obtaining meaningful velocity measurements for the catalyzed reaction under study.

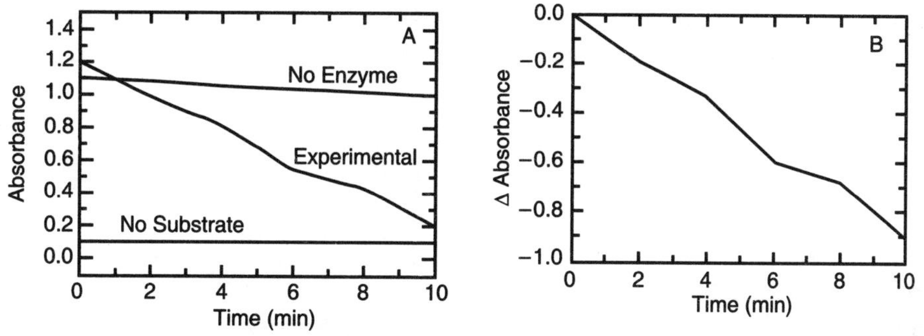

Fig. 7.5: The importance of running blank controls. (A) Time courses of absorption for a hypothetical enzymatic reaction (experimental trace), along with two control samples: one with all the reaction mixture components except enzyme; and the other with all of the reaction mixture components except substrate. In this example, the enzyme absorbs minimally at the analytical wavelength, but enough to displace the time zero measurement by about 0.1 absorption unit. (B) The two control readings at each time point have been subtracted from the experimental measurement to yield the true reaction time course of the reaction.

7.7 DETECTION METHODS

A wide variety of physicochemical methods have been used to follow the course of enzymatic reactions. Some of the more common techniques are described here, but our discussion is far from comprehension. Any signal that differentiates the substrates or products of the reaction from the other components of the reaction mixture can, in principle, form the basis of an enzyme assay; the only limit is the creativity of the investigator

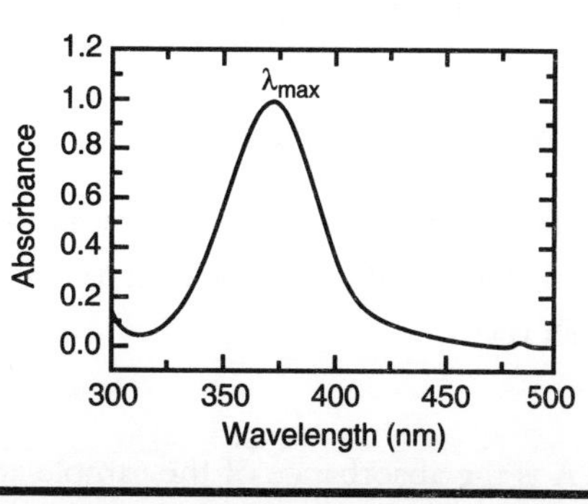

Fig. 7.6: A typical absorption spectrum of a molecule with an absorption maximum at 375 nm.

7.8 ASSAYS BASED ON OPTICAL SPECTROSCOPY

Two very common means of following the course of an enzymatic reaction are absorption spectroscopy and fluorescence spectroscopy. Both methods are based on the changes in electronic configuration of molecules that result from their absorption of light energy of specfic wavelengths. Molecules can absorb electromagnetic radiation, such as light, causing transitions between various energy levels. Energy in the infrared region, for example, can cause transitions between vibrational levels of a molecule. Microwave energy can induce transitions among rotational energy levels, while radiofrequency energy, which forms the basis of NMR spectroscopy, can cause transitions among nuclear spin levels. The energy differences between electronic levels of a molecule are so large that light energy in the ultraviolet (UV) and visible regions of the electromagnetic spectrum is required to induce transitions among these states.

If a molecule is irradiated with varying wavelengths of light (of similar intensity), only light of specific wavelengths will be strongly absorbed by the sample. At these wavelengths, the energy of the light matches the energy gap between two electronic states of the molecule, and the light is absorbed to induce such a transition. Fig. 7.6 is a hypothetical absorption spectrum of a molecule in which light of wavelength l_{max} induces an electron redistribution to bring the molecule from its ground π state to an excited π^* state. This process is illustrated as a potential energy diagram in Fig. 7.7A.

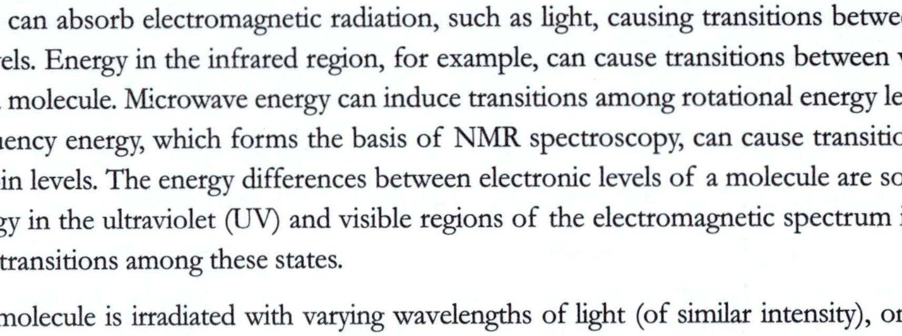

Fig 7.7: Energy level diagrams for (A) a light-induced transition from a π electronic ground state to an excited π^* state of a molecule, and (B) the relaxation of the excited state molecule back to the ground state by photon emission (fluorescence).

7.8.1 Absorption measurements

The value of absorption spectroscopy as an analytical tool is that the absorption of a molecule at a particular wavelength can be related to the concentration of that molecule in solution, as described by Beer's law:

$$A = \varepsilon cl,$$

where A is the absorbance of the sample at some wavelength, c is the concentration of sample in molarity units, l is the path length of sample the light beam traverses (in centimetres), and ε is an intrinsic constant of the molecule, known as the extinction coefficient, or molar absorptivity.

Since absorbance is a unitless quantity, ε must have units of reciprocal molarity times reciprocal path length (typically expressed as $M^{-1}\cdot cm^{-1}$ or $mM^{-1}\cdot cm^{-1}$). Thus, if we know the value of ε for a particular molecule, and the path length of the cuvette, we can calculate the concentration of that molecule in a solution by measuring the absorption of that solution. As we have seen in the examples of Figs. 7.1 and 7.5, we can use the unique absorption features of a substrate or product of an enzymatic reaction to follow the course of such a reaction.

Using Beer's law, we can convert our measured ΔA values at different time points to the concentration of the molecule being followed, and thus report the reaction velocity in terms of change in molecular concentration as a function of time. Note that the units here, molarity per unit time (i.e., moles per litre per unit time) are commonly used in reporting enzyme velocities. Some workers instead report velocity in units of moles per unit time. The two units are easily interconverted by making note of the total volume of the reaction mixture for which the velocity is being measured.

7.8.2 Choosing an Analytical Wavelength

The wavelength used for following enzyme kinetics should be one that gives the greatest difference in absorption between the substrate and product molecules of the reaction. In many cases, this will correspond to the wavelength maximum of the substrate or product molecule. However, when there is significant spectral overlap between the absorption bands of the substrate and product, the most sensitive analytical wavelength may *not* be the same as the wavelength maximum. This concept is illustrated in Fig. 7.8. In inspecting the spectra for the substrate and product of the hypothetical enzymatic reaction of Fig. 7.8A, note that at the wavelength maximum for each molecule, the other molecule displays significant absorption. The wavelength at which the largest difference in signal is observed can be determined by calculating the difference spectrum between these two spectra, as illustrated in Fig. 7.8B. Thus, in our hypothetical example, the wavelength maxima for the substrate and product are 374 nm and 385 nm, respectively, but the most sensitive wavelengths for following loss of substrate or formation of product would be 362 nm and 402 nm, respectively.

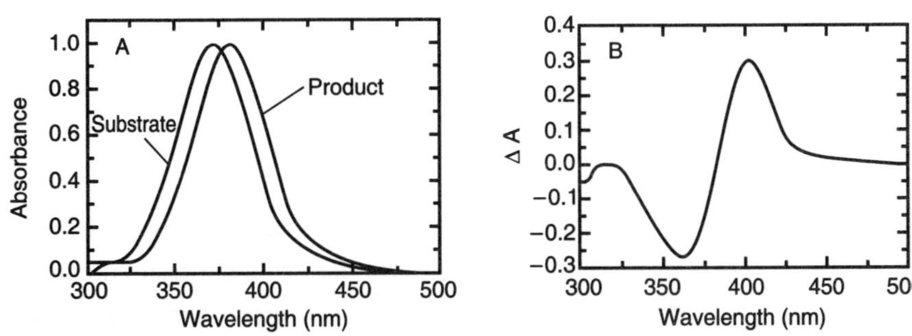

Fig. 7.8: Example of the use of difference spectroscopy. (A) Absorption spectra of the substrate and product of some enzymatic reaction. Because of the high degree of spectral overlap between these two molecules, it would be difficult to quantify changes in one component of a mixture of the two species. (B) The mathematical difference spectrum (product minus substrate) for the two spectra in (A). The difference spectrum highlights the differences between the two spectra, making quantitation of changes more straightforward.

7.8.3 Optical Cells

Absorption measurements are most commonly performed with a standard spectrophotometer, and the samples are contained in specialized cells known as cuvettes. These specialized cells come in a range of path lengths and are constructed of various optical materials. Disposable plastic cuvettes that hold 1 ml or 3 ml samples are commercially available. Although these disposable units are very convenient and reduce the chances of sample-to-sample cross-contamination, their use should be restricted to the visible wavelength range (350-800 nm). For measurements at wavelengths less than 350 nm, high quality quartz cuvettes must be used, since both plastic and glass absorb too much light in the ultraviolet. Quartz cuvettes are available from numerous manufacturers in a variety of sizes and configurations. Regardless of the type of cuvette selected, its path length must be known to ensure correct application of Beer's law to the measurements. Usually, the manufacturer provides this information at the time of purchase. If, for any reason, the path length of a particular cuvette is not known with certainty, it can be determined experimentally by measuring the absorption of any stable chromophoric solution in a standard 1 cm path length cuvette and then measuring the absorption of the same sample in the cuvette of unknown path length. The ratio of the two absorption readings ($A_{unknown}/A_{1\ cm}$) yields the path length of the unknown cuvette in centimetres. A good standard solution for this purpose can be prepared as follows (Haupt, 1952). Weigh out 0.0400 g of potassium chromate and 3.2000 g of potassium hydroxide and dissolve both in 700 ml of distilled water. Transfer to a one-litre volumetric flask and bring the total volume to 1000 ml with additional distilled water. Mix the solution well. The resulting solution will have absorption of 0.991 at 375 nm in a 1 cm cuvette.

Many workers now perform absorption measurements with 96-well microtiter plate readers. These devices have remarkable sensitivity (many can measure changes in absorption of as little as 0.001) and can greatly increase productivity by allowing up to 96 samples to be assayed simultaneously. Most plate readers are based on optical filters rather than monochromators, hence providing only a finite number of analytical wavelengths for measurements. It is usually not difficult, however, to purchase an optical filter for a particular wavelength and to install it in the plate reader. Because most commercially available microtiter plates are constructed of plastic, plate reader measurements are normally restricted to the visible region of the spectrum.

Recently, however, monochromator-based plate readers have come on the market that allow one to make measurements at any wavelength between 250-750 nm. The manufacturers of these instruments also provide special quartz-bottomed plates for performing measurements in the ultraviolet. As with nonstandard cuvettes, it is necessary to determine the path length through which one is making measurements in a microtiter plate well. The path length in this situation will depend on the total volume of reaction mixture in the well.

For example, a 200 μl solution in a well of a 96-well plate has an approximate path length of 0.7 cm. The exact path length under given experimental conditions should be determined empirically, as described above.

Filter-based plate readers may show an apparent extinction coefficient of a molecule that differs from corresponding measurements made with a conventional spectrometer. Such discrepancies are due in part to the broader bandwidths of the optical filters used in plate readers. The absorption readings thus survey a wider range of wavelengths than the corresponding spectrometer measurements. Therefore, it is important to use the plate reader under the same conditions as the experimental measurements to determine a standard curve of absorption as a function of chromophore concentration.

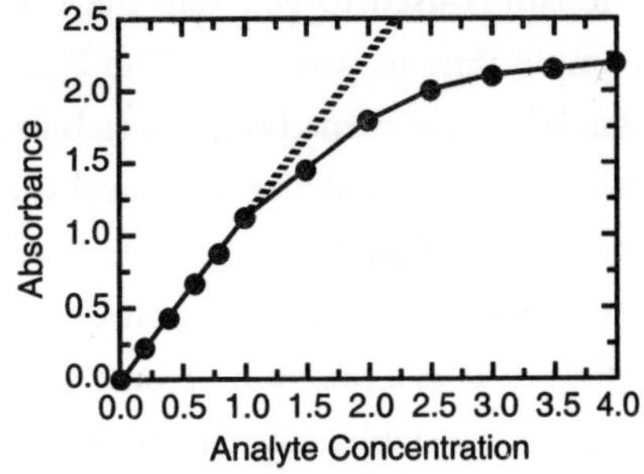

Fig. 7.9: Deviation from Beer's law. Over a small concentration range, the absorption at some analytical wavelength tracks linearly with analyte concentration, as expected from Beer's law. When the analyte concentration increases to the point at which A=1.0, however, significant deviations from this straight-line behaviour begin to appear.

Mixing of samples can be problematic in a 96-well plate, since the inversion method cannot conveniently be applied. Instead, many commercial plate readers have automatic built-in shaking devices that allow for sample mixing. Another way to ensure good mixing is to alter the general method just described as follows. The components of the reaction mixture, except for the initiating reagent, are mixed together in sufficient volume for all the wells of the plate to be used. This solution is placed in a multichannel pipette reservoir.

The initiating solution is added to the individual wells of the plate, and the remainder of the reaction solution is added to the wells by means of a multichannel pipet. The solutions are mixed by repeatedly pulling up and dispensing the reaction mixture with the pipeette. An entire row of 12 wells can be mixed in this way in less than 10 seconds.

7.8.4 Errors in absorption spectroscopy

A common error associated with absorption measurements is deviation from Beer's law. The form of Beer's law suggests that the absorption of a sample will increase linearly with the concentration of the molecule being analyzed, and indeed, this is the basis for the use of absorption spectroscopy as an analytical tool. Experimentally, however, one finds that this linear relationship holds only over a finite range of absorption values. As illustrated in Fig. 7.9, absorption readings greater than 1.0 in general should not be trusted to accurately reflect the concentration of analyte in solution. Thus, experiments should be designed so that the maximum absorption to be measured is less than 1.0. With a few preliminary trials, it usually is possible to adjust conditions so that the measurements fall safely below this limit. Additionally, the amount of instrumental noise in a measurement is affected by the overall absorption of the sample. For this reason it is more difficult to measure a small absorption change for a sample of high absorption. Empirically it turns out that the best compromise between minimizing this noise and having a reasonable signal follow occurs when the sample absorption is in the vicinity of 0.5. This is usually a good target absorption for following small absorption changes.

The lamps used to generate the UV and visible light for absorption spectrometers must be given ample time to warm up. The light intensity from these sources varies considerably shortly after the lamps are turned on, but stabilizes after about 30-90 minutes. Since the amount of warm-up time needed to stabilize the lamp output will vary from instrument to instrument, and from lamp to lamp within the same instrument, it is best to determine the required warm-up time for one's own instrument. This is easily done by measuring the signal from a sample of low absorption (say, ~0.05–0.1) as a function of time after turning the lamp on, and noting how long it takes for the signal to reach a stable, constant reading.

Another source of error in absorption measurements is sample turbidity. Particulate matter in a solution will scatter light that is detected as increased absorption by the sample. If settling of such particles occurs during kinetic measurements, significant noise in the data may result, and in severe cases there will appear to be an additional kinetic component of the data. The best way to avoid these complications is to ensure that the sample is free of particles by filtering all the solutions through 0.2 μm filters or by centrifugation (see Copeland, 1994, for further details).

7.9 FLUORESCENCE MEASUREMENTS

Light of an appropriate wavelength can be absorbed by a molecule to cause an electronic transition from the ground state to some higher lying excited state, as we have discussed. Because of its highly energetic nature, the excited state is short-lived (excited state lifetimes are typically less than 50 ns), and the molecule must find a means of releasing this excess energy to return to the ground state electronic configuration. Most of the time this excess energy is released through the dissipation of heat to the surrounding medium. Some molecules, however, can return to the ground state by emitting the excess energy in the form of light. Fluorescence, the most common and easily detected of these emissive processes, involves singlet excited and ground electronic states. The energetic processes depicted in Figure 7.7 are characteristic of molecular fluorescence. First, light of an appropriate wavelength is absorbed by the molecule, exciting it to a higher lying electronic state (Fig. 7.7A). The molecule then decays through the various high energy vibrational substates of the excited electronic state by heat dissipation, finally, relaxing from its lowest vibrational level to the ground electronic state with release of a photon (Fig. 7.7B).

Because of the differences in equilibrium interatomic distances between the ground and excited states, and because of the loss of energy during the decay through the higher energy vibrational substates, the emitted photon is far less energetic than the corresponding light energy required exciting the molecule in the first place. For these reasons, the fluorescence maximum of a molecule is always at a longer wavelength (less energy) than the absorption maximum; this difference in wavelength between the absorption and fluorescence maxima of a molecule is referred to as the Stokes shift. For example, the amino acid tryptophan absorbs light maximally at about 280 nm and fluoresces strongly between 325-350 nm (Copeland, 1994). To take advantage of this behaviour, fluorescence instruments are designed to excite a sample in a cuvette with light at the wavelength of maximal absorption and detect the emitted light at a different (longer) wavelength. To best detect the emitted light with minimal interference from the excitation light beam, most commercial fluorometers are designed to collect the emitted light at an angle of 90° from the excitation beam path. Thus, unlike cells for absorption spectroscopy, fluorescence cuvettes must have at least two optical quality widows at right angles to one another; all four sides of most fluorescence cuvettes have polished optical surfaces.

The strategies for following enzyme kinetics by fluorescence are similar to those just described for absorption spectroscopy. Many enzyme substrates-product pairs are naturally fluorescent and provide convenient signals with which to follow their loss or production in solution. If these molecules are not naturally fluorescent, it is often possible to covalently attach a fluorescent group without significantly perturbing the interactions with the enzyme under study. Fluorescence measurements offer two key advantages over absorption measurements for following enzyme kinetics. First, fluorescence instruments are very sensitive, permitting the detection of much lower concentration changes in substrate or product. Second, since many fluorophores have large Stokes shifts, the fluorescence signal is typically in an isolated region of the spectrum, where interferences from signals due to other reaction mixture components are minimal.

Fluorescence signals track linearly with the concentration of fluorophore in solution over a finite concentration range. In principle, fluorescence signals should vary with fluorophore concentration by a relationship similar to Beer's law, where the extinction coefficient is replaced by the molar quantum yield (Φ). In practice, however, it is difficult to calculate sample concentrations by means of applying tabulated values of Φ to experimental fluorescence measurement.

This limitation is in part due to the nature of the instrumentation and the measurements. Thus, to convert fluorescence intensity measurements into concentration units, it is necessary to prepare a standard curve of fluorescence signal as a function of fluorophore concentration, using a set of standard solutions for which the fluorophore concentration has been determined independently. The standard curve data points must be collected at the same time as the experimental measurements, however, since day-to-day variations in lamp intensity and other instrumental factors can greatly affect fluorescence measurements.

Sometimes the fluorophore is generated only as a result of the enzymatic reaction, and it is difficult to obtain a standard sample of this molecule for construction of a standard curve. In such cases it may not be possible to report velocity in true concentration units, and units of relative fluorescence per unit time must be used instead. It is still important to quantify this fluorescence relative to some standard fluorescent molecule, to permit comparisons of relative fluorescence measurements from one day to the next and from one laboratory to another. A good standard for this purpose is quinine sulphate. A dilute solution of quinine sulphate in an aqueous sulphuric acid solution can be excited at any wavelength between 240-400 nm to yield a strong fluorescence signal that maximizes at 453 nm (Fletcher, 1969). Russo (1969) suggests the following protocol for preparing a quinine sulphate solution as a standard for fluorescence spectroscopy:

- Weigh out 5 mg of quinine sulphate dihydrate and dissolve in 100 ml of 0.1 (N) H_2SO_4
- Measure the absorption of the sample at 366 nm, and adjust the concentration with 0.1 (N) H_2SO_4 so that the solution has absorption of 0.40 at this wavelength in a 1 cm cuvette.
- Dilute a sample of this solution 1/10 with 0.1 (N) H_2SO_4 and use the solution to record the fluorescence spectrum.

The relative fluorescence of a sample can then be reported as the fluorescence intensity of the sample at some wavelength, divided by the fluorescence intensity of the quinine sulphate standard at 453 nm, when the same fluorometer is used to excite both sample and standard, at the same wavelength. Of course, both sample and standard measurements must be made under the same set of experimental conditions (monochrometer slit width, lamp voltage, dwell time, etc.), and the second set should be made soon after the first.

7.9.1 Internal fluorescence quenching and energy transfer

If a molecule absorbs light at the same wavelength at which another molecule fluoresces, the fluorescence from the second molecule can be absorbed by the first molecule, leading to a diminution or *quenching* of the observed fluorescence intensity from the second molecule (Note that this is only one of numerous means of quenching fluorescence). The first molecule may then decay back to its ground state by radiationless decay (e.g., heat dissipation), or it may itself fluoresce at some characteristic wavelength. We refer to the first process as quenching because the net effect is a loss of fluorescence intensity. The second situation is described as "resonance energy transfer", because here excitation at the absorption maximum of one molecule leads to fluorescence by the other molecule (Fig. 7.10).

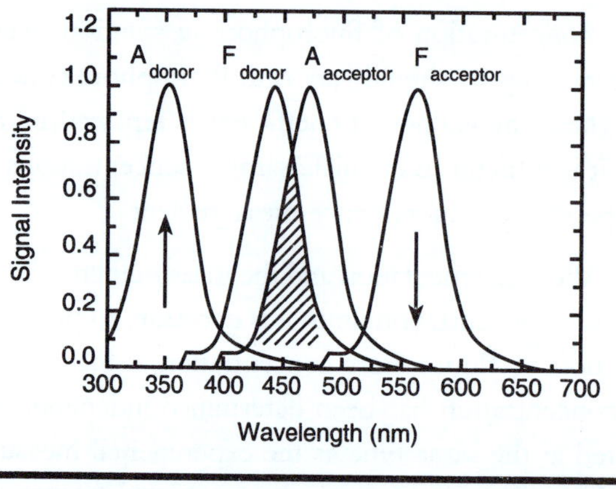

Fig. 7.10: Resonance energy transfer. In an energy transfer experiment, the sample is excited with light at the wavelength of the donor absorption band (A donor) to induce fluorescence of the donor molecule at the wavelength indicated by F donor. The absorption band of the acceptor molecule (A acceptor) occurs in a wavelength range where it overlaps with the fluorescence band of the donor; the area of overlap between these two features is shown by the hatched region.

Both these processes depend on several factors, including the spatial proximity of the two molecules. This property has been exploited to develop fluorescence assays for proteolytic enzymes based on synthetic peptide substrates. The basic strategy is to incorporate a fluorescent group (the donor into a synthetic peptide on either the N- or C-terminal side of the scissile peptide bond that is recognized by the target enzyme. A fluorescence quencher or energy acceptor molecule (both referred to hereafter as the acceptor molecule) is also incorporated into the peptide on the other side of the scissile bond. When the peptide is intact, the donor and acceptor molecules are covalently associated and remain apart at a relatively fixed distance, able to energetically interact. Once hydrolyzed by the enzyme, however, the two halves of the peptide will diffuse away from each other, thus eliminating the possibility of any interaction between the donor and acceptor. The observed effect of this hydrolysis will be an increase in the fluorescence from the donor molecule, and, in the case of energy transfer, a concomitant decrease in the fluorescence of the acceptor molecule with exication under the absorption maximum of the donor.

These approaches have been used to follow hydrolysis of peptide substrates for a large variety of proteases (e.g., see Matayoshi et al., 1990; Knight et al., 1992; Knight, 1995; and Packard et al., 1997). Table 7.3 summarizes some donor—acceptor pairs that are commonly used in synthetic peptide substrates for proteases. Another good source for information on donor—acceptor pairs is the Internet site of the Molecular Probes Company, a company specializing in fluorescence tools for biochemical and biological research.

Table 7.3: Donor-acceptor pairs[a] for quenching by resonance energy transfer in peptide substrates of proteolytic enzymes

		Wavelengths (nm)	
Quencher	Fluorophore	Excitation	Emission
Dabcyl	Edans	336	490
Dansyl	Trp	336	350
DNP	Trp	328	350
DNP	MCA	328	393
DNP	Abz	328	420
Tyr (NO_2)	Abz	320	420

[a]Dabcyl, 4-(4-dimethylaminophenylazo) benzoic acid; Edans, 5-[(2-aminoethyl) amino] naphthalene-1-sulphonic acid; Dansyl, (5-dimethylaminonaphthalene-1-sulphonyl); DNP, 2,4-dinitrophenyl; MCA, 7-methoxycoumarin-4 acetic acid; Abz, o-aminobenzyl; Tyr(NO_2), 3-nitrotyrosine.

Recently, fluorescence resonance energy transfer (FRET) has been applied to the study of enzymatic group transfer reactions, and to the study of protein-protein interactions in solution. Space does not permit a review of these applications. The interested reader is referred to the online handbook from the Molecular Probes Company for more information and literature examples of biochemical applications of FRET Technology.

7.9.2 Errors in Fluorescence Measurements

Most of the caveats described for absorption spectroscopy hold for fluorescence measurements as well. Samples must be free of particulate matter, since light scattering is a severe problem in fluorescence. Many of the commonly used fluorophores emit light in the visible region but must be excited at wavelengths in the near ultraviolet, necessitating the use of quartz cuvettes for these measurements. Also, any fluorescence due to buffer components and soon must be measured and corrected to ensure that meaningful data are obtained.

In addition to these, more common considerations are several sources of error unique to fluorescence measurements. First, many fluorescent molecules are prone to photodecomposition after long exposure to light. Hence, fluorescent substrates and reagents should be stored in amber glass or plastic, and the containers should be wrapped in aluminum foil to minimize exposure to environmental light. Second, the quantum yield of fluorescence for any molecule is highly dependent on sample temperature. We shall see shortly that temperature affects enzyme kinetics directly, but this is distinct from the general influence of temperature on fluorescence intensity. In general, the fluorescence signal increases with decreasing temperature, as competing nonradiative decay mechanisms for return to the ground state become less efficient. Hence, good temperature control of the sample must be maintained. Most commercial fluorometers provide temperature control by means of jacketed sample holders that attach to circulating water baths.

Finally, a major source of error in fluorescence measurements is light absorption by the sample at high concentrations. Individual molecules in a sample may be excited by the excitation light beam and caused to fluoresce.

To be detected, these emitted photons must traverse the rest of the sample and escape the cuvette to impinge on the surface of the detection device (typically a photomultiplier tube or diode array). Any such photon will be lost from detection, however, if before escaping the sample it encounters another molecule that is capable of absorbing light at that wavelength. As the sample concentration increases, the likelihood of such encounters and instances of light reabsorption increases exponentially. This phenomenon, referred to as the *inner filter effect* (Fig. 7.11), can dramatically reduce the fluorescence signal observed from a sample. The inner filter effect can be corrected if the absorption of the sample is known at the excitation and emission wavelengths used in the fluorescence measurement. The true, or corrected fluorescence F_{corr} can be calculated from the observed fluorescence F_{obs} as follows (Lackowicz, 1983):

$$F_{corr} = F_{obs} \times 10^{(A_{ex} \times A_{em})/2},$$

where A_{ex} and A_{em} are the sample absorptions at the excitation and emission wavelengths, respectively. This correction works only over a limited sample absorption range. If the sample absorption is greater than about 0.1, the correction will not be adequate. Hence, a good rule of thumb is to begin with samples that have absorption values of about 0.05 at the excitation wavelength. The sample concentration can be adjusted from this starting point to optimize the signal-to-noise ratio, with care taken not to introduce a significant inner filter effect.

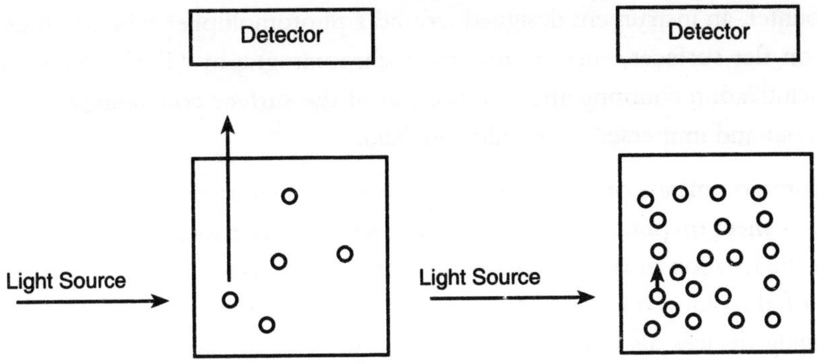

Fig. 7.11: Schematic diagram illustrating the inner filter effect. When a dilute sample (left) of a fluorescent molecule is excited at an appropriate wavelength, a detector stationed at 90° relative to the excitation source will detect the emitted light that emerges from the sample container. If, however, the sample is very concentrated, emitted light from one molecule in a sample may encounter and be reabsorbed by another molecule before emerging from the sample compartment (right). These reabsorbed photons, of course, will not be detected. The likelihood of this self-absorption, or inner filter effect, increases with increasing sample concentration.

7.10 RADIOISOTOPIC MEASUREMENTS

The basic strategy for the use of radioisotopes in enzyme kinetic measurements is to incorporate into the structure of the substrate a radioactive species that will be retained in the product molecule after catalysis. Using an appropriate technique for separating the substrate from the product, one can then measure the amount of radioactivity in the substrate and product fractions, and thus quantify substrate loss and product production. Most of the isotopes that are used commonly in enzyme kinetic measurements decay through emission of β-particles (Table 7.4).

Table 7.4: Properties of radioisotopes that are commonly used in enzyme kinetic assays

Isotope	Decay Process	Half-life
Carbon-14	${}^{14}_{6}C \rightarrow {}_{-1}\beta^0 + {}^{14}_{7}N$	5700 years
Phosphorus-32	${}^{32}_{15}P \rightarrow {}_{-1}\beta^0 + {}^{32}_{16}S$	14.3 days
Sulphur-35	${}^{35}_{16}S \rightarrow {}_{-1}\beta^0 + {}^{35}_{17}Cl$	87.1 days
Tritium-3	${}^{3}_{1}H \rightarrow {}_{-1}\beta^0 + {}^{3}_{2}He$	12.3 years

The decay process follows first-order kinetics, and the loss (or disintegration) of the starting material is thus associated with a characteristic half-life for the parent isotope. The standard unit of radioactivity is the curie (Ci), which originally defined the rate at which 1 gram of radium-226 decays completely. Relating this to other isotopes, a more useful working definition of the curie is that quantity of any substance that decays at a rate of 2.22×10^{12} disintegrations per minute (dpm).

Solutions of *p*-terphenyl or stilbene, in xylene or toluene, will emit light when in contact with a radioactive solute. This light emission, known as *scintillation*, is most commonly measured with a scintillation counter, an instrument designed around a photomultiplier tube or other light detector. Radioactivity on flat surfaces, such as thin-layer chromatography (TLC) plates and gels can be measured by scintillation counting after the portion of the surface containing the sample has been scraped or cut out and immersed in scintillation fluid.

Another common means of detecting radioactivity on such surfaces entails placing the surface in contact with a sheet of photographic film. The radioactivity darkens the film, making a permanent record of the location of the radioactive species on the surface. This technique, called *autoradiography*, is one of the oldest methods known for detecting radioactivity. Today computerinterfaced phosphor imaging devices also are commonly used for locating and quantifying radioactivity on two-dimensional surfaces (dried gels, TLC plates, etc.).

Radioactivity in a sample is quantified by measuring the dpm's of a sample using one of the methods just described. Since, however, no detector is 100% efficient, any instrumental reading obtained experimentally will differ from the true dpm of the sample. The experimental units of radioactivity are referred to as counts per minute (cmp's: events detected or counted by the instrument per minute). For example, a 1 μCi sample would display 2.22×10^6 dpm. If the detector used to measure this sample had an efficiency of 50%, the experimental value obtained would be 1.11×10^6 cpm. To convert this experimental reading into true dpm's, it would be necessary to measure a standard sample of the isotope of interest, of known dpm's. This information would permit the calibration of the efficiency of the instrument and the ready conversion of the cpm values of samples into dpm units.

When radiolabelled substrates are used in enzyme kinetic studies, the labelled substrate is usually mixed with "cold" (i.e., unlabelled) substrate to achieve a particular total substrate concentration without having to use high quantities of radioactivity. It is important, however, to quantify the proportion of radiolabelled molecules in the substrate sample. Quantification is commonly expressed in terms of the specific radioactivity of the sample. (*Note:* "Specific activity" in this case refers to the radioactivity of the sample and should not be confused with the specifc activity of an enzyme sample, which is defined later.) Specific radioactivity is given in units of radioactivity per mass or molarity of the sample. Common units of specific radioactivity include dpm/ mmol and mCi/mg. With the specific radioactivity of a substrate sample defined, one can easily convert into velocity units radioactivity measurements taken during an enzymatic reaction.

Good bookkeeping is essential in these assays. The amount of total substrate used will be dictated by the purpose of the experiment and its K_m for the enzyme. The specific radioactivity, on the other hand, should be adjusted to ensure that the amount of radioactivity used is the minimum that will provide good signal-over-background readings. Guidelines for sample preparation using different radioisotopes can be found in the reviews by Oldham (1968, 1992). The other point that should be kept in mind is that good post reaction separation of the labelled substrate and product molecules is critical to the use of radiolabels for following enzyme kinetics. Radiolabelled substrates are commonly used in conjunction with chromatographic and electrophoretic separation methods. When the substrate and/or the product is a protein, as in some assays for kinases and proteases, bulk precipitation or capture on nitrocellulose membranes can be used to separate the macromolecule from the other solution components.

7.10.1 Errors in Radioactivity Measurements

Aside from errors associated with bookkeeping, the most commonly encountered cause of inaccurate radioactivity measurements is self-absorption. When the separation method used in conjunction with the assay involves a solid separation medium, such as paper or thin-layer plate chromatography, gel electrophoresis, or capture on activated charcoal, the solid material in the sample may absorb some of the emitted radiation, preventing the signals from reaching the detection device. This self-absorption is best corrected for by measuring all samples and standards at a constant density in terms of milligrams of material per milliliter. Segel (1976) suggests using an inert material such a gelatin to adjust the density of all samples for this purpose. Because scintillation counting measures light emission, the same interferences discussed for fluorescence measurements can occur. In particular, if the sample is highly colored, quenching of the signal due to the equivalent of an inner filter effect may be observed When possible, this should be correct for by adjusting the optical density of the samples and standards with a similarly colored inert material (Segel, 1976).

7.11 OTHER DETECTION METHODS

Absorption, fluorescence, and radioactivity are by far the most common means of following enzyme kinetics, but a wide variety of other techniques have been utilized as well. Immunologic detection, for example, has been applied to follow proteolytic cleavage of a protein substrate by Western blotting, using antibodies raised against that protein substrate. Recently, antibodies have been developed that react exclusively with the phosphorylated forms of peptides and proteins; these reagents have been widely used to follow the enzymatic activity of the kinases and phosphatases using Western and dot blotting as well as ELISA-type assays. Reviews of immunologic detection methods can be found in Copeland (1994) and in Harlow and Lane (1988).

Polarographic methods have also been used extensively to follow enzyme reactions. Many oxidases utilize molecular oxygen during their turnover, and the accompanying depletion of dissolved O_2 from the solutions in which catalysis occurs can be monitored with an O_2-specific electrode. Very sensitive pH electrodes can be used to follow proton abstraction or release into solution during enzyme turnover. Enzymes that perform redoxchemistry as part of their catalytic cycle can also be monitored by electrochemical means. Reviews of these methods can be found in the text by Eisenthal and Danson (1992).

The variety of detection methods that have been applied to enzyme activity measurements is too broad to be covered comprehensively in any one volume. Our discussion should provide the reader with a good overview of the more common techniques employed in this field. The references given can provide more in-depth accounts. Another very good source for new and interesting enzyme assay methods, the journal *Analytical Biochemistry* (Academic Press), has historically been a repository for papers dealing with the development and improvement of enzyme assays. Finally, the series *Methods in Enzymology* (Academic Press) comprises volumes dedicated to in-depth reviews of varying topics in enzymology. This series details assay methods for many of the enzymes one is likely to work with and very frequently will indicate at least an assay for a related enzyme that can serve as a starting point for development of an individual assay method.

7.12 TIPS TO REMEMBER

Generally, most enzyme assays rely on one or more of the following broad classes of detection and separation methods to follow the course of the reaction:

Spectroscopy

Polarography

Radioactive decay

Electrophoretic separation

Chromatographic separation

Immunological reactivity.

Direct assay: The direct measurement of the substrate or product concentration as a function of time.

Indirect assay: In some cases the substrate and product of an enzymatic reaction do not provide a distinct signal for convenient measurement of their concentrations.

Often, however, product generation can be coupled to another, nonenzymatic, reaction that does produce a convenient signal; such a strategy is referred to as an ***indirect*** **assay**.

Coupled assays: Here the enzymatic reaction of interest is paired with a second enzymatic reaction, which can be conveniently measured.

Continuous assay: Here the signal might be monitored at discrete intervals over the entire linear time period, or some convenient portion thereof.

Discontinuous assay: Here after having established a linear time period for an assay, one measures the signal at a single specific time point within the linear time period

7.13 VIVA VOCE

Q.1. Explain Michaelis-Menten equation and kinetics.

Q.2. Explain with the help of example what non-competitive inhibition is and how it is different from uncompetitive inhibition.

Q.3. What are the different types of assay systems which are employed for analyzing enzyme kinetics?

Q.4. What are coupled assays?

Q.5. Explain how radioactive measurements of enzyme kinetics are done.

Q.6. What are the errors associated with fluorescent measurement of enzyme kinetics?

Q.7. What is the importance of running controls in enzyme kinetics study?

Q.8. What is inner filter effect?

Q.9. Explain assays based on optical spectroscopy.

Q.10. Differentiate between direct and indirect assays with examples.

Q.11. What is end point discontinuous assay and where is it used?

CHAPTER 8

Electrophoretic Techniques

8.1 GENERAL PROPERTIES

The term electrophoresis describes the migration of a charged particle under the influence of an electric field. Many important biological molecules, such as amino acids, peptides, proteins, nucleotides and nucleic acids, posses ionisable groups and, therefore, at any pH, exist in solution as electrically charged species either as cations (+) or anions (–). Under the influence of an electric field charged particles will migrate either to the cathode or to the anode, depending on the nature of their net charge.

In order to understand fully how charged species separate, it is necessary to look at some simple equations related to electrophoresis. When a potential difference (voltage) is applied across the electrodes, it generates a potential gradient, E, which is applied voltage, V, divided by the distance, d, between the electrodes. When this potential gradient E is applied, the force on a molecule bearing a charge of q coulombs is Eq newtons. It is this force that derives a charged molecule towards an electrode. However, there is also a frictional resistance that retards the movement of this charged molecule. This frictional force is a measure of the hydrodynamic size of the molecule, the shape of the molecule, the pore size of the medium in which electrophoresis is taking place and the viscosity of the buffer. The velocity, v, of a charged molecule in electrical field is, therefore, given by the equation:

$$v = Eq/f,$$

where f is the frictional coefficient.

More commonly the term electrophoretic mobility (μ) of an ion is used, which is the ratio of the velocity of the ion to field strength (v/E). When a potential difference is applied, therefore, molecules with different overall charges will begin to separate owing to their different electrophoretic mobilities. Even molecules of similar charges will begin to separate if they have different molecular sizes, since they will experience different frictional forces. As will be seen below, some forms of electrophoresis rely almost totally on the different charges on the molecules to effect separation, whilst other methods exploit differences in molecular size and, therefore, encourage frictional effects to bring about separation. In short, electrophoretic mobility can be given as:

$$\text{mobility} = \frac{\text{(voltage) (charge)}}{\text{(frictional coefficient)}}$$

Provided the electric field is removed before the molecules in the sample reach the electrodes, the components will have been separated according to their electrophoretic mobility. Electrophoresis is thus an incomplete form of electrolysis. The separated samples are then located by staining with an appropriate dye or by autoradiography if the sample is radiolabelled.

The current in the solution between the electrodes is conducted mainly by the buffer ions, a small proportion being conducted by the sample ions. Ohm's law expresses the relationship between current (I), voltage (V) and resistance (R):

$$V/I = R.$$

It, therefore, appears that it is possible to accelerate an electrophoretic separation by voltage, which would result in a corresponding increase in the current flowing. The distance migrated by the ions will be proportional to both current and time. However, this would ignore one of the major problems for most common forms of electrophoresis, namely the generation of the heat.

During electrophoresis, the power (W, watts) generated in the supporting medium is given by

$$W = I^2R.$$

Most of this power generated is dissipated as heat. Heating of the electrophoretic medium has the following effects:

- An increased rate of diffusion of sample and buffer ions leading to broadening of the separated samples.
- The formation of convection currents, which leads to mixing of the separated samples.
- Thermal instability of the samples that are rather sensitive to heat. This may include denaturation of the proteins (e.g., the loss of enzyme activity).
- A decrease of buffer viscosity, and hence a reduction in the resistance of the medium.

If a constant voltage is applied, the current increases during electrophoresis owing to the decrease in the resistance and the rise in current increases the heat output still further. For this reason, workers often use a stabilized power supply, which provides constant power and thus eliminates fluctuations in heating.

Constant heat generation is, however, a problem. One may think that electrophoresis may be run at a very low power (low current) to overcome any heating problem, but this can lead to poor separation as a result of the increased amount of the diffusion resulting from a long separation times. Compromise conditions, therefore, have to be found with reasonable power settings, to give acceptable separation times, and an appropriate cooling system, to remove liberated heat. While such systems work fairly well, the effects of heating are not always totally eliminated. For example, for electrophoresis carried out in cylindrical tubes or in slab gels, although heat is generated uniformly through the medium, heat is removed only from the edges, resulting in a temperature gradient within the gel, the temperature at the centre of the gel being higher than that at the edges. Since the warmer fluid at the centre is less viscous, electrophoretic mobility are, therefore, greater in the central region (electrophoretic mobilities increase by about 2% for each 1°C rise in the temperature), and electrophoretic zones develop a bowled shape, with the zone centre migrating faster than the edges.

A final factor that can effect electrophoresis separation is the phenomenon of electroendoosmosis (also known as electroosmotic flow), which is due to the presence of charged groups on the surface of the support medium. For example, paper has some carboxyl groups present, agarose (depending on the purity grade) contains sulphate group and the surface of the glass walls used in capillary electrophoresis contains silanol (Si–OH) groups. Mechanism of electroendoosmosis has been explained in details in the section covering capillary electrophoresis, although the principle is the same for any support medium that has charged groups on it. In short in a fused silica capillary tube, above a pH value of about 3, silanol groups on the silica capillary walls will ionise, generating negatively charged sites. It is these charges that generate electroendoosmosis. The ionized silanol groups create an electrical double layer, or region of charge separation, at the capillary wall/electrolyte interface. When a voltage is applied, cations in the electrolyte near the capillary wall migrate towards the cathode, pulling electrolyte solution with them. This creates a net electrosmotic flow towards the cathode.

However, the introduction of the use of gels as a support medium led to a rapid improvement in methods for analyzing macromolecules. The earliest gel system to be used was the starch gel and, although this has some uses, the vast majority of electrophoretic techniques used nowadays involve either agarose gel or polyacrylamide gel. Now we will discuss different types of electrophoretic methods in details

8.2. HORIZONTAL AND VERTICAL GEL ELECTROPHORESIS SYSTEMS

The equipment required for electrophoresis consists basically of two items, a power pack and an electrophoresis unit. Electrophoresis units are available for running either vertical or horizontal gel systems. Vertical slab gel units are commercially available and routinely used to separate proteins in acrylamide gel. The gel is formed between two glass plates that are clamped together but held apart by electrical spacers.

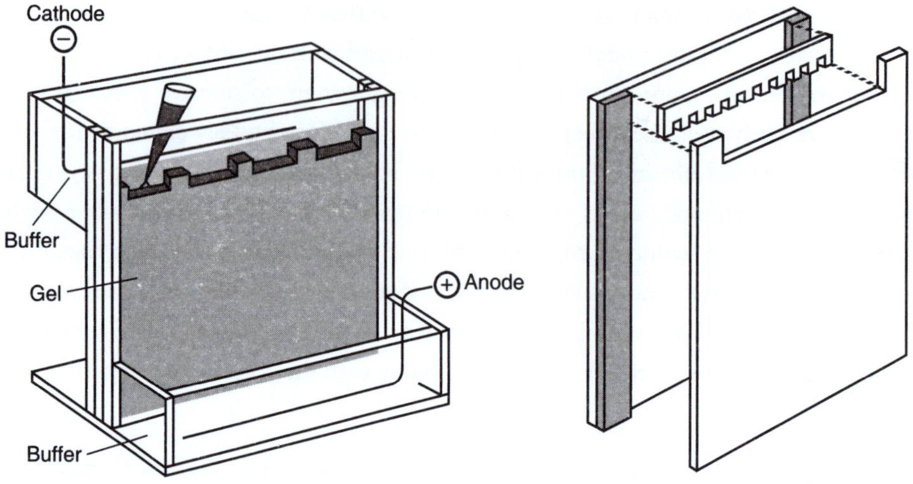

Fig. 8.1: A typical vertical gel apparatus

The most commonly used units are the so-called mini gel apparatus. Gel dimensions are typically 8.5 cm wide × 5 cm high, with a thickness of 0.5-1 mm. A plastic comb is placed in the gel solution and is removed after polymerization to provide loading wells up to 10 samples. When the apparatus is assembled, the lower electrophoresis tank buffer surrounds the gel plates and affords some cooling of the gel plates. A typical horizontal gel system is shown in Fig. 8.1.

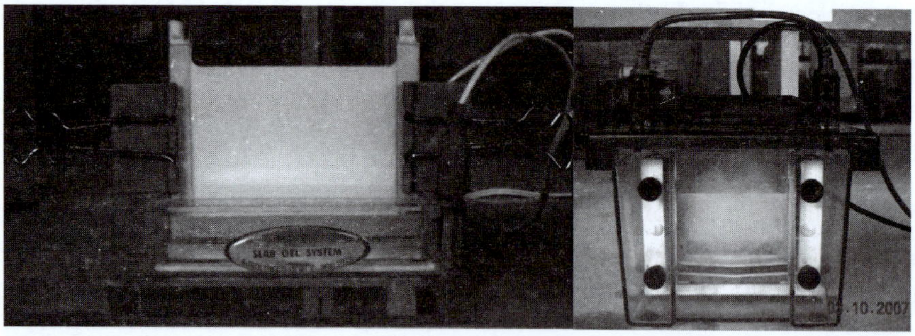

Fig. 8.2: Vertical gel electrophoresis system (*Courtesy:* Dr D Dash, Deptt. of Biochemistry, IMS, BHU)

The gel is cast on a glass or plastic sheet and placed on a cooling plate (an insulated surface through which cooling water is passed to conduct away generated heat). Connection between the gel and electrode buffer is made using a thick wad of wetted filter paper. Note, however, that agarose gel for DNA electrophoresis are run submerged in the buffer. The power pack supplies a direct current between the electrodes in the electrophoresis unit. All electrophoresis are carried out in an appropriate buffer, which is essential to maintain a constant state of ionization of the molecules being separated. Any variation in pH will alter the overall charge and hence the mobilities (rate of migration in the applied field) of the molecules are separated.

Fig. 8.3: Dual gel electrophoresis system attached to power pack (*Courtesy*: Dr D Dash, Deptt. of Biochemistry, IMS, BHU)

8.3 AGAROSE GEL ELECTROPHORESIS

Agarose is a linear polysaccharide (average relative molecular mass about 12000) made up of the basic repeat unit agarobiose, which comprises alternating units of the galactose and 3,6-anhydrogalactose (Fig. 8.4).

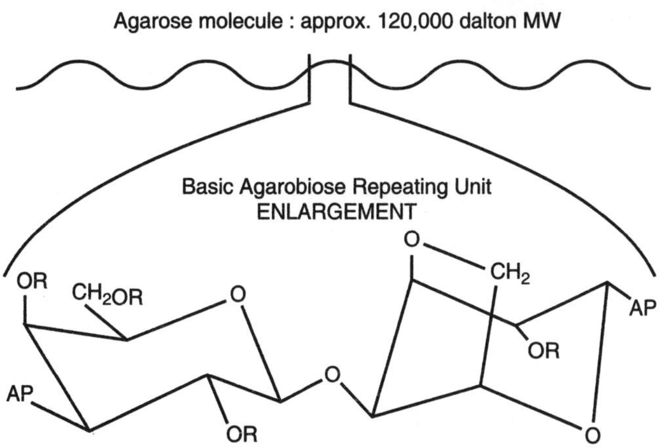

Fig. 8.4: Structure of agarose

Agarose is one of the components of agar that is mixture of polysaccharides isolated from certain seaweeds. Agarose is usually used at a concentration between 1% and 3%. Agarose gels are formed by suspending dry agarose in aqueous buffer, then boiling the mixture until a clear solution is formed. This is poured and allowed to cool to room temperature to form a rigid gel. The gelling properties are attributed to both inter- and intramolecular hydrogen bonding within and between the long agarose chains. This cross-linked structure gives the gel good anticonventional properties. The pore size in the gel is controlled by the initial concentration of agarose; large pore sizes are formed from low concentration and smaller pore sizes are formed from higher concentrations.

Although essentially free from charge, substitution of the alternating sugar residues with carboxyl, methoxyl, pyruvate and specially sulphate group occur to varying degrees. This substitution can result in electroendoosmosis during electrophoresis and ionic interactions between the gel and sample in all uses, both unwanted effects. Agarose is, therefore, sold in very different purity grades, based on sulphate concentration – the lower the sulphate content the higher the purity.

Agarose gels are used for the electrophoresis of both proteins and nucleic acids. For proteins, the pore sizes of a 1% agarose gel are large relative to the sizes of proteins. Agarose gels are therefore used in techniques such as immunoelectrophoresis or flat bed isoelectric focusing, where the proteins are required to move unhindered in the gel matrix according to their native charge. Such large pore gels are also used to separate much larger molecules such as DNA or RNA, because the pore sizes in the gel are still large enough for DNA or RNA molecule to pass through the gel. Now, however, the pore size and molecule size are more comparable and frictional effects begin to play a role in the separation of these molecules. A further advantage of using agarose is the availability of low melting temperature agarose (62-65°C). As the name suggests, these gels can be reliquified by heating to 65°C and thus, for example, DNA samples separated in a gel can be cut out of the gel, returned to solution and recovered.

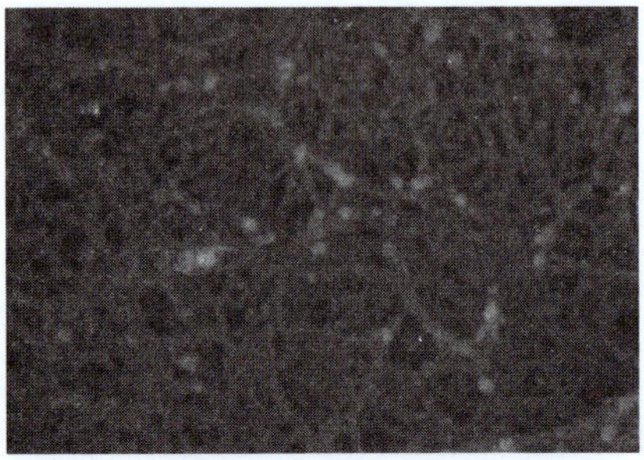

Fig. 8.5: SEM photograph of 1% agarose gel at 22000x magnification

Table 8.1: Range for fractionation of different compounds in different agarose percentages

Agarose Concentration	Protein Fractionation Range* (kDa)	Polysaccharide Fractionation* (kDa)	Nucleic Acid Exclusion Limit (bp)
1.0%	1,000 to 150,000	1,000 to 150,000	(> 3,000)
2.0%	80 to 40,000	90 to 20,000	1,340
4.0%	50 to 15,000	40 to 5,000	860
6.0%	10 to 5,000	10 to 1,000	180

Owing to the poor elasticity of agarose gels and the consequent problems of removing them from small tubes, the gel rod system is sometimes used, since acrylamide gel is not used. Horizontal slab gels are invariably used for isoelectric focusing or immunoelectrphoresis in agarose. Horizontal gels are also used routinely for DNA and RNA gels, although vertical systems have been used by some workers.

* The sentence means that since agarose gels have poor elasticity and thus they can not be properly removed from small tubes; therefore, small rod gel systems employed for acrylamide gels sometimes, can not be used for running agarose gels; i.e., they are not suitable for agarose gels

Table 8.2: Approximate gel concentrations for separation of DNA linear fragments of various sizes

Gel Concentration	Separation Range (bp)
agarose - 0.3%	60,000 - 5,000
agarose - 0.7%	20,000 - 800
agarose - 0.9%	7,000 - 500
agarose - 1.2%	6,000 - 400
agarose - 1.5%	4,000 - 200
agarose - 2.0%	3,000 - 100
agarose - 4.0%	500 - 10
acrylamide - 4%	1,000 - 800
acrylamide - 10%	500 - 25
acrylamide - 20%	50 - 1

8.3.1 Agarose Gel Electrophoresis of Nucleic Acids

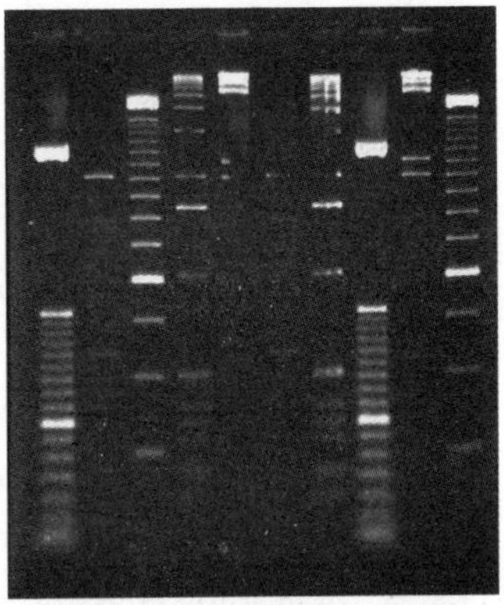

Fig. 8.6: DNA separated on 0.8% agarose gel (*Courtesy*: Dr D Dash, Deptt. of Biochemistry, IMS, BHU)

Nucleic acids are polymers composed of individual nucleotide units. The units are connected via phosphate diester linkages of the backbone sugars. The net effect of these linkages is to give the polymers a net negative charge. From the earliest days of electrophoresis it has been axiomatic that molecules carrying an electrical charge will migrate in an electrical field in a predictable manner. When subjected to an electrical field, a molecule carrying a net negative charge will migrate toward the positive pole and a molecule with a net positive charge will migrate toward the negative pole. In a semi-solid matrix like agarose, the equation describing mobility can be re-interpreted, at least heuristically, by defining ç as gel density or concentration and r as the length of the molecule. Thus, when they are placed in the semi-solid matrix of a gel, nucleic acids will migrate toward the positive pole in a predictable and reproducible manner that can be described as a negative exponential function of length. That is to say, shorter molecules will migrate faster and longer molecules will migrate slower. Indeed, in the case of a nucleic acid in a gel in an electrical field, every other element of the migration expression is a constant and mobility is completely determined by molecular length.

As a matter of practice, it is difficult to accurately resolve double-stranded nucleic acids smaller than about 100 bases in an agarose gel because the sieving properties of agarose are not fine enough. On the other end of the scale, molecules longer than about 25,000 bp but shorter than around 2,000,000 bp will all run at the same rate. This is called limiting mobility. Nucleic acid molecules longer than 2,000,000 bp will not even enter an agarose gel. Thus, the effective size range

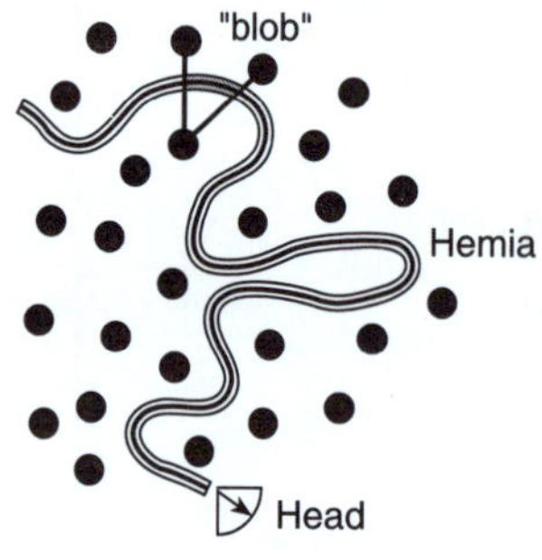

Fig. 8.7: Biased reptation

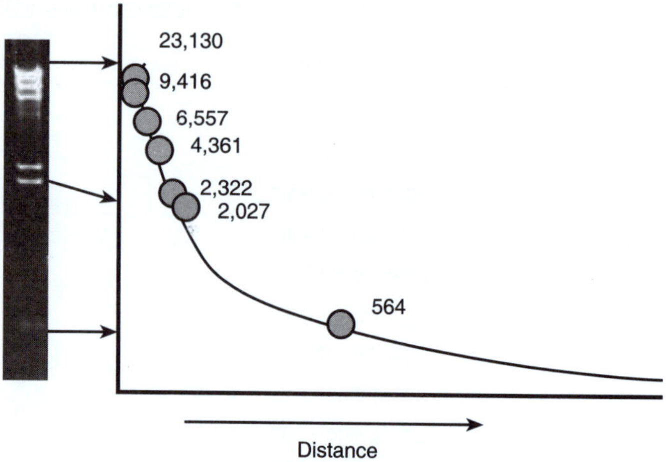

Fig. 8.8: Electrophoretic mobility of DNA restriction fragments in an agarose gel. The DNA is from bacteriophage lambda digested with the restriction enzyme *Hind III*. The graph shows migration distance in centimetres by size of the restriction fragment in base pairs.

for agarose gel electrophoresis of double stranded nucleic acids is between 100 bp and 25,000 bp. In this range the behaviour of the molecule is precise and predictable. This behaviour is shown in Fig. 8.8. As can be seen there is minimal separation of the larger fragments but resolution improves as the fragments get smaller. While this phenomenon has been known for many years, what was not known was how the nucleic acid molecules actually moved in the gel matrix. In the 1980s a theory was put forward that nucleic acids migrated through the gel much the same way that a snake moves. That is, the leading edge moves forward and pulls the rest of the molecule with it. In this model, as the molecule gets longer resistance being pulled along increases. This resistance is further increased by the interaction of the molecule with the gel matrix. As has been discussed above, the increase in resistance is non-linear. This model, called "biased reptation", is sufficient to explain all of the behaviour of a nucleic acid in a semi-solid matrix (Lerman et al., 1982; Lumpkin et al., 1985).

In 1989 a group at the university of Washington put this theory to the test. They filmed DNA molecules moving through an agarose gel. Their films showed both reptation and the nucleic acid/ gel matrix interactions (Smith et al., 1989).

In the mid-1980s a number of methods were developed to electrophoretically analyze nucleic acid molecules in the limiting mobility size range. The solution involved artificially introducing a size dependent mobility on nucleic acid molecules by altering the electrophoretic field. The first such alteration involved simply switching the polarity of the field in a regular pattern. Carle et al. (1986) showed that periodic reversals of polarity would induce the molecules to make U-turns in the gel. Even at very large sizes, this turning would permit separation of molecules. In the length of time the molecules were reversed was about one-third the time they were oriented forward, for example, three seconds forward and one second back, molecules as large as 2,000,000 bp could be resolved in a standard agarose gel in a few hours. The first practical demonstration of this method, called Field Inversion Gel Electrophoresis (FIGE), was to completely resolve intact yeast chromosomes. Since then, a variety of methods, collectively termed pulsed-field gel electrophoresis, have been developed.

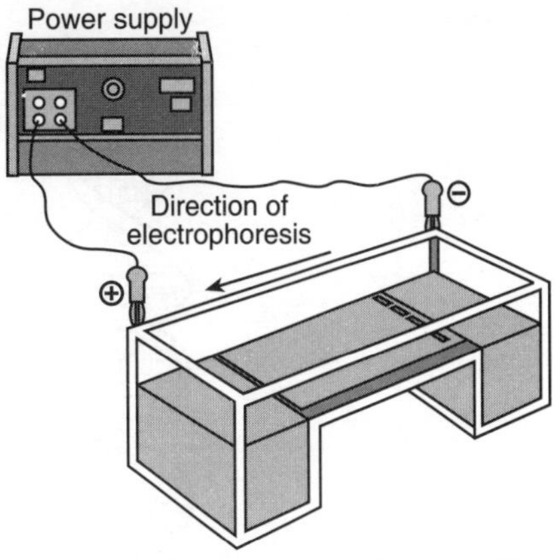

Fig. 8.9: A typical horizontal gel electrophoresis system

Many laboratories routinely use 0.8% gels, which are suitable for separating DNA molecules in the range 0.5-10 kb. Since agarose gels separate DNA according to size, the M_r of a DNA fragment may be determined from its electrophoretic mobility by running a number of standard DNA markers of known M_r on the same gel. This is most conveniently achieved by running a sample of bacteriophage λ DNA (49 kb) that has been cleaved with a restriction enzyme such as *Eco*RI. Since the base sequence of λ DNA is known, and the cleavage sites for *Eco*RI are known, this generates fragments of accurately known size.

DNA gels are invariably run as horizontal, submarine or submerged gels; so named because such a gel is totally immersed in buffer. Agarose, dissolved in gel buffer by boiling, is poured onto a glass or plastic plate, surrounded by a wall of adhesive tape or a plastic frame to provide a gel about 3 mm in depth. Loading wells are formed by placing a plastic well-forming template or comb in the poured gel solution, and removing this comb once the gel is set. The gel is placed in electrophoresis tank, covered with buffer, and samples loaded by directly injecting the sample into the wells. Samples are prepared by dissolving them in a buffer solution that contains sucrose, glycerol or Ficoll, which makes the solution dense and allows it to sink to the bottom of the well. A dye such as bromophenol blue is also included in the sample solvent; it makes it easier to see the sample that is being loaded and also acts as a marker of the electrophoresis front. No stacking gel is needed for the electrophoresis of DNA because the mobilities of DNA molecules are much greater in the well than in the gel, and therefore, all the molecules in the well pile up against the gel within a few minutes of the current being turned on, forming a tight band at the start of the run. General purpose gels are approximately 25 cm long and 12 cm wide, and are run at a voltage gradient of about 1.5 V cm^{-1} overnight. A higher voltage would cause excessive heating. For rapid analyses that do not need extensive separation of DNA molecules, it is common to use mini-gels that are less than 10 cm long. In this way information can be obtained in 2-3 h.

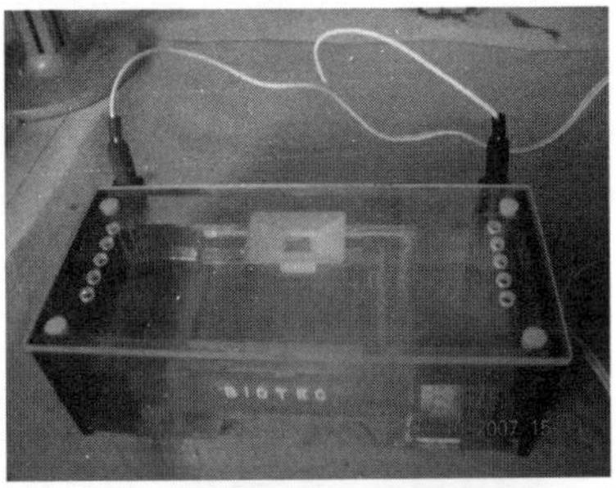

Fig. 8.10: Horizontal gel electrophoresis system (*Courtesy:* Dr D Dash, Deptt. of Biochemistry, IMS, BHU)

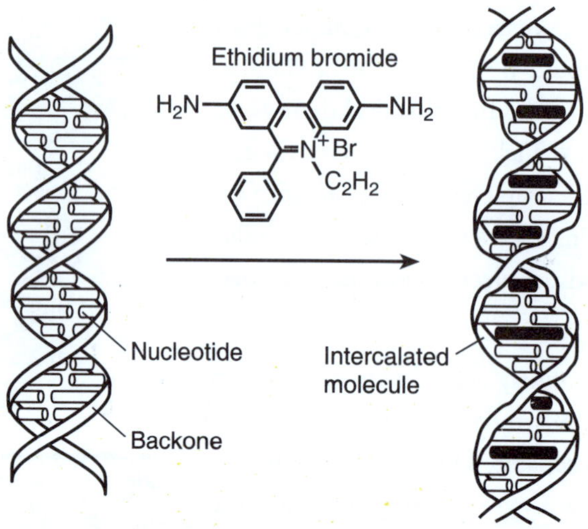

Fig. 8.11: Interaction of ethidium bromide with DNA molecule

Once the system has been run, the DNA in the gel needs to be stained and visualized. The reagent most widely used is the fluorescent dye ethidium bromide. The gel is rinsed gently in a solution of ethidium bromide (0.5 μg cm^{-1}) and then viewed under ultraviolet light (300 nm wavelength). Ethidium bromide is a cyclic planar molecule that binds between the stacked base-pairs of DNA (i.e., it intercalates). The ethidium bromide concentration, therefore, builds up at the site of DNA bands and under ultraviolet light the DNA bands fluoresce orange-red. As little as 10 ng of DNA can be visualized as a 1 cm wide band. It should be noted that extensive viewing of DNA with ultraviolet light can result in damage of the DNA by nicking and base-pair dimerisation. This is of no consequence if the gel is only to be viewed, but obviously viewing of the gel should be kept to a minimum if the DNA is to be recovered. It is essential to protect one's eye by wearing goggles when ultraviolet light is used. If viewing of gel under ultraviolet is carried out for long periods, a plastic mask that covers the whole face should be used to avoid 'sunburn'.

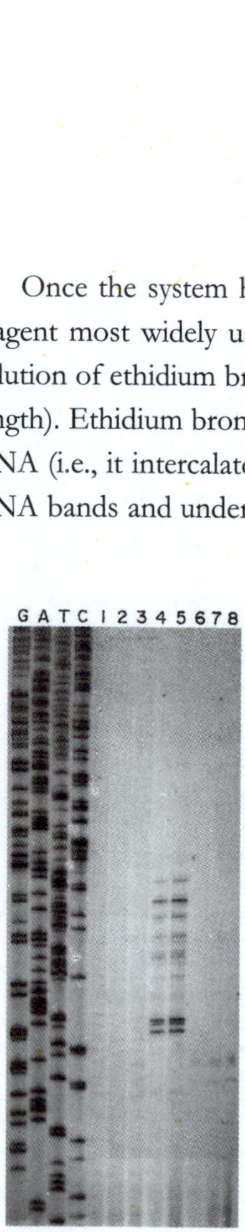

Fig. 8.12: DNA sequence gel

8.3.2 DNA sequencing gels

Although agarose gel electrophoresis of DNA is a 'workhorse' technique for the molecular biologists, a different form of electrophoresis has to be used when DNA sequences are to be determined. Whichever DNA sequencing method is used, the final analysis usually involves separating single-stranded DNA molecules shorter than about 1000 nt and differing in size by only 1 nt. To achieve this it is necessary to have a small-pored gel and so acrylamide gels are used instead of agarose. For example, 3.5% polyacrylamide gels are used to separate DNA in the range 80-1000 nt and 12% gels to resolve fragments of between 20-100 nt. If a wide range of sizes is being analysed it is often convenient to run a gradient gel, for example, from 3.5% to 7.5%. Sequencing gels are run the presence of denaturing agents, urea and formamide. Since it is necessary to separate DNA molecules that are very similar in size, DNA sequencing gels tend to be very long (100 cm) to maximize the separation achieved. A typical DNA sequencing gel is shown in Fig. 8.12.

As mentioned above, electrophoresis in agarose can be used as a preparative method for DNA. The DNA bands of interest can be cut out of the gel and the DNA recovered by: (a) electroelution, (b) macerating the gel piece in buffer, centrifuging and collecting the supernatant; or (c) if low melting point agarose is used, melting the gel piece and diluting with buffer. In each case, the DNA is finally recovered by precipitation of the supernatant with ethanol.

8.3.4 Pulsed-field gel electrophoresis

Manipulating and analyzing DNA are fundamental in the field of molecular biology. Indeed, separating complex mixtures of DNA into different sized fragments by electrophoresis was a well established technique by the early 1970s. Typically, DNA was isolated intact and then treated with restriction enzymes to generate pieces small enough to be resolved by electrophoresis in agarose or acrylamide. Although this procedure still forms the core of DNA separation and analysis in today's laboratories, the rules of the separation have changed. In 1984, Schwartz and Cantor described pulsed field gel electrophoresis (PFGE), introducing a new way to separate DNA. In particular, PFGE resolved extremely large DNA for the first time, raising the upper size limit of DNA separation in agarose from 30-50 kb to well over 10 Mb (10,000 kb). Now, instead of cloning a large number of small fragments of DNA, PFGE permits cloning and analysis of a smaller number of very large pieces of a genome.

Although the theory of pulsed field electrophoresis is a matter of debate, qualitative statements can be made about the movement of DNA in agarose gels during PFGE. During continuous field electrophoresis, DNA above 30-50 kb migrates with the same mobility regardless of size. This is seen in a gel as a single large diffuse band. If, however, the DNA is forced to change direction during electrophoresis, different sized fragments within this diffuse band begin to separate from each other.

With each reorientation of the electric field relative to the gel, smaller sized DNA will begin moving in the new direction more quickly than the larger DNA. Thus, the larger DNA lags behind, providing a separation from the smaller DNA. Currently, there are three models that attempt to describe the behaviour of DNA during PFGE, the biased reptation model (BRM), the chain model, and, most recently, the bag model.

Instrumentation

Although many types of PFGE instrumentation are available (Fig. 8.13), they generally fall into two categories. The simplest equipment is designed for field inversion gel electrophoresis (FIGE). FIGE

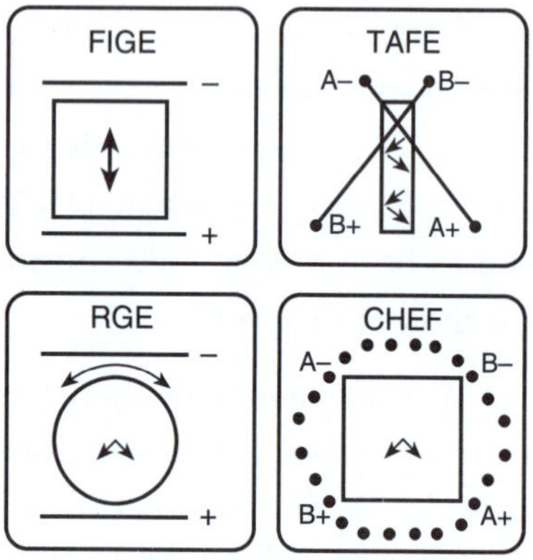

Fig. 8.13: Electrode configuration of commonly used pulsed field gel electroph,presis units

works by periodically inverting the polarity of the electrodes during electrophoresis. Because FIGE subjects DNA to a 180° reorientation, the DNA spends a certain amount of time moving backwards. Only an electrical field switching module is needed; any standard vertical or horizontal gel box that has temperature control can be used to run the gel. Although more complex in its approach, zero integrated field electrophoresis (ZIFE) also falls into this first category. Compared with simple FIGE, ZIFE is very slow. However, ZIFE is capable of resolving larger DNA and giving a larger useful portion of the gel.

The other category contains instruments that reorient the DNA at smaller oblique angle, generally between 96° and 120°. This causes DNA to always move forward in a zigzag pattern down the gel. For a similar size range under optimal conditions, these separations are faster, resolve a wider size range, and give a larger useful portion of the gel compared to their other counter parts.

Contour-clamped homogeneous electric field (CHEF); transverse alternating field electrophoresis (TAFE) and its relative ST/RIDEtm; and rotating gel electrophoresis (RGE) are all examples of commonly used transverse angle reorientation techniques for which instrumentation is available. In a further elaboration of the above procedures, Lai and Coworkers developed the programmable autonomously controlled electrophoresis (PACE) unit which allows complete control over reorientation angle, voltage, and switch time. In contrast with FIGE, these systems require both a special gel box with a specific electrode and gel geometry, and the associated electronic control for switching and programming the electrophoresis run.

Fig. 8.14: Increased separation of the 20-50 kb range with field inversion gel electrophoresis (FIGE)

Ideally, the DNA should separate in straight lanes to simplify lane-to-lane comparisons. The original pulsed-field systems used inhomogeneous electric fields that did not produce straight lanes, making interpretation of gels difficult. Again, the simplest approach to straight lanes is FIGE, which uses parallel electrodes to assure a homogeneous electric field. Although extremely useful for separating relatively small DNA, 4- 1,000 kb (Fig. 8.14), FIGE's reorientation angle of 180° results in a separation range most useful under 2,000 kb. Furthermore, like other PFGE techniques, FIGE has mobility inversions in which larger DNA can move ahead of smaller DNA during electrophoresis.

Ramping, where the reorientation pulse length is constantly increased during separation, will minimize inversions. This capability is included in most commercial instrumentation. Increasing both the separation range and the resolution of large DNA requires smaller reorientation angles, generally 96-140° with 120° most common. Smaller angles (e.g., 100°) increase the mobility of the DNA generally without seriously affecting resolution. The lower limit is approximately 96°. Below this, separation is seriously compromised.

TAFE and ST/RIDEtm use a complicated geometry between the electrodes and a vertically placed gel to get straight lanes. CHEF and RGE maintain a homogeneous electric field in combination with a horizontal gel. CHEF changes the direction of the electric field electronically to reorient the DNA by changing the polarity of an electrode array. With RGE the electric field is fixed; to move the DNA in a new direction, the gel simply rotates.

Rotating Gel Electrophoresis

RGE is one of the most recent commercial introductions of pulsed field equipment and combines variable angles with a homogeneous electric field. The electrodes are positioned along opposite sides of the buffer chamber with their polarity fixed. Briefly, the gel is cast on a circular running plate and then placed in the buffer chamber. The gel is coupled to a magnetic drive beneath the buffer chamber to eliminate the possibility of leakage that a direct connection might cause.

To force the migrating DNA to a new direction, the magnetic drive simply rotates the gel to the new angle. Because the reorientation angle of the DNA is determined by a straightforward mechanical coupling, RGE offers a lot of flexibility at a reduced cost. Voltage, angle, and pulse times are varied with the program stored into memory of the unit.

Along with the ability to separate large DNA came the need for new sample preparation and handling procedures. Large DNA (e.g., yeast chromosomes) is easily sheared and also difficult to pipette due to its high viscosity. The solution to this problem is to first embed the bacteria or yeast in agarose plugs and then treat the plugs with enzymes to digest away the cell wall and proteins, thus leaving the naked DNA undamaged in the agarose. The plugs then are cut to size, treated with restriction enzymes, if necessary, loaded in the sample well, and sealed into place with agarose.

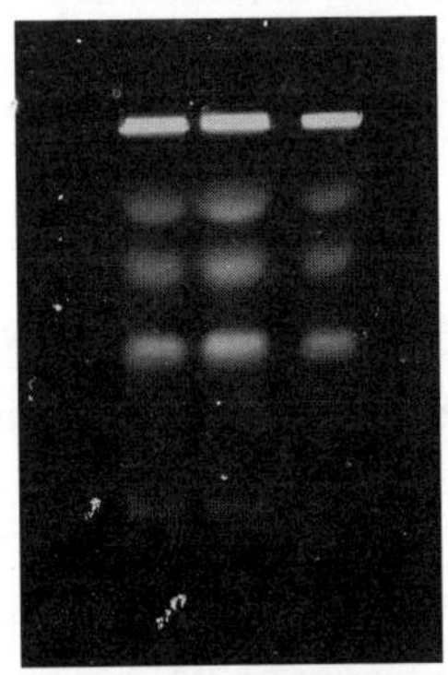

Fig. 8.15: Rotating gel electrophoresis (RGE) separation *Saccharomyces cerevisiae* chromosomes (245-2190 kb)

Several parameters act in concert during PFGE. These will be discussed briefly below as they relate to transverse field instruments such as RGE. The minimum amount of information needed to repeat a separation should include a short description of the pulsed field instrumentation used; applied voltage and field strength (e.g., 180 V at 5.3 V/cm); pulse length (e.g., 87 seconds); reorientation angle (e.g., 120°); the buffer (0.5X TBE); the agarose type and concentration; the buffer chamber temperature (e.g., 10°C); the type of standards; and, if possible, the amount of DNA loaded. Although the data listed above is necessary to faithfully reproduce a separation, the information supplied in publications is rarely this complete. Most PFGE systems separate DNA over a relatively small area, limiting the resolution of complex samples. RGE is an exception to this, with a useful separation distance up to 20 cm and a maximum gel size of 18 × 20 cm.

The field strength has a profound effect on pulsed field separations and is a compromise between separation time and resolution of a particular size class. Four to six volts/cm is generally required for resolving DNA up to 2000 kb (e.g., *S. cerevisiae* chromosomes) in a reasonable period of time (e.g., 1-2 days). However, these field strengths trap and immobilize even bigger DNA in the agarose matrix, and DNA > 3000 kb requires 2 V/cm or less for separation. Pulse time primarily changes the size range of separation. Longer pulse times lead to separation of larger DNA. For example, at 5.4 V/cm, the 1.6 Mb and 2.2 Mb chromosomes from *S. cerevisiae* separate as a single band with 90-second pulse length. Increasing the pulse length to 120 seconds resolves these into two bands. Any angle between 96° and 165° produces roughly equivalent separation. The smaller the angle, however, the faster the DNA mobility. And for separating extremely large DNA, 96° to 105° is almost a requirement to get a good separation in the shortest possible time.

The type of agarose also affects DNA separation, with the fastest mobilities and best resolution achieved in gels made of low electroendoosmosis (EEO) agarose. Although most standard electrophoresis grades of agarose are suitable for PFGE, agarose with minimal EEO will provide a faster separation. Several low EEO "pulsed field grades" are available, including Fast Lane and Gold, and Megarose.

The concentration of agarose affects both the resolution and mobility of the DNA. Higher concentrations of agarose yield sharper, but slower moving bands. And the typical concentrations used (0.8-1.2%) represent a trade off between speed and resolution. High percentages of low EEO agarose may improve resolution without sacrificing the speed of separation. Because DNA mobility also depends on the separation temperature, the temperature must be constant both during and between runs. Although higher temperatures increase DNA mobility, it does so at the expense of resolution.

8.3.5 Electrophoresis of RNA

Like that of DNA, electrophoresis of RNA is carried out in agarose gels; and the principle of the separation, based on the size, is the same. Often one requires a rapid method for checking the integrity of RNA immediately following extraction but before deciding whether to process it further. This can be achieved easily by electrophoresis in a 2% agarose gel in about 1 h. Ribosomal RNAs (18 S and 28 S) are clearly resolved and any degradation (seen as smear) or DNA contamination is seen easily. However, if greater resolution is required, a smaller-pored acrylamide gel is used to enhance resolution, for example, to resolve transfer RNAs (4 S) from 5 S ribosomal RNAs. This can be achieved on a 2.5-5% acrylamide gradient gel with an overnight run. Both these methods involve running native RNA. There will almost certainly be some secondary structure within the

RNA molecule owing to intramolecular hydrogen bonding. For this reason native RNA run on gels can be stained and visualised with ethidium bromide. However, if the study objective is to determine RNA size by gel electrophoresis, the full denaturation of the RNA is needed to prevent hydrogen bond formation within or even between polynucleotides that will otherwise affect the electrophoretic mobility. There are three denaturing agents (formaldehyde, glyoxal and methylmercuric hydroxide) that are compatible with both RNA and agarose. Either one of these may be incorporated into the agarose gel and electrophoresis buffer, and the sample is heat denatured in the presence of the denaturant prior to electrophoresis. After heat denaturation, each of these agents forms adducts with the amino groups of guanine and uracil, thereby hydrogen bond reformation at room temperature during electrophoresis. It is also necessary to run denaturing gels if the RNA is to be blotted and probed, to ensure that the base sequence is available to the probe. Denatured RNA stains only very weakly with ethidium bromide, so acrydine orange is commonly used to visualised RNA or denaturing gels. However, it should be noted that many workers will be using radiolabelled RNA and will, therefore, identify bands by autoradiography. An example of the electrophoresis of RNA is shown in Fig. 8.16.

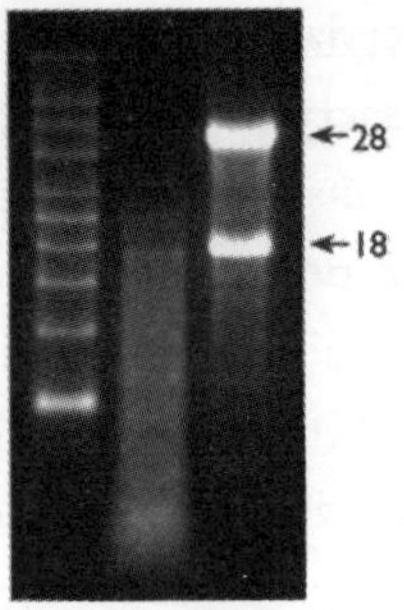

Fig. 8.16: Electrophoresis of RNA (*Courtesy:* Biochemistry and Molecular Biology Lab, Deptt. of Zoology, BHU)

8.4 POLYACRYLAMIDE GELS

Electrophoresis in acrylamide gels is frequently referred to as PAGE, being an abbreviation for Polyacrylamide gel electrophoresis.

Cross-linked polyacrylamide gels are formed from the polymerization of acrylamide monomer in the presence of smaller amounts of N-N-methylene-bisacrylamide (normally referred to as bisacrylamide) (Fig. 8.17). Note that bisacrylamide is basically two units of acrylamide molecules linked by a methylene group, and is used as a cross-linking agent.

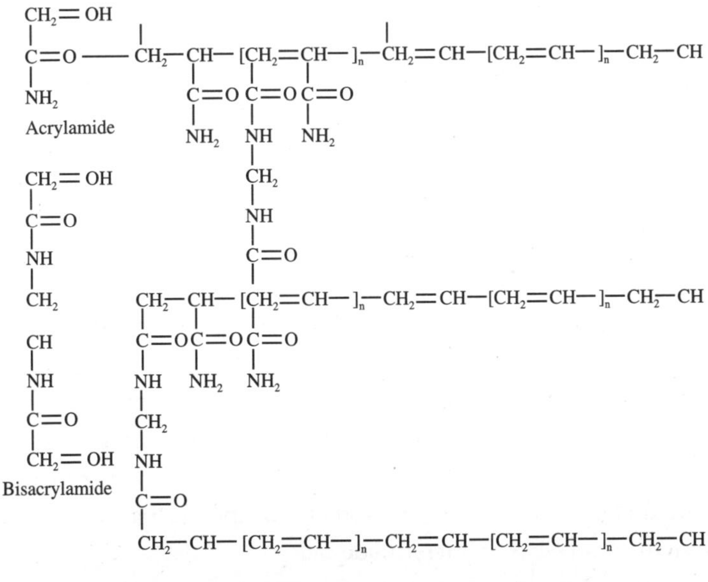

Fig. 8.17: Chemical structure of acrylamide, bisacrylamide, and polyacylamide gel

Acrylamide monomer is polymerized in a head-to-tail fashion into long chains and occasionally a bisacrylamide molecule is built into the growing chains, thus introducing a second site for chain extension. Proceeding in this way a cross-linked matrix of fairly well-defined structure is formed. the polymerization of acrylamide is an example of free radical catalysis, and is initiated by the addition of ammonium persulphate and the base N, N, N′, N′-tetramethylenediamine (TEMED). TEMED catalyses the decomoposition of persulphate ion to give a free radical (i.e., a molecule with an unpaired electron):

$$S_2O_8^{2-} + e^- \rightarrow SO_4^{2-} + SO_4^{-\bullet}$$

If this free radical is represented as $R^\bullet$ (where the dot represents an unpaired electron) and M as an acrylamide monomer molecule, then the polymerization can be represented as follow:

$$R^\bullet + M \rightarrow RM^\bullet$$

$$RM^\bullet + M \rightarrow RMM^\bullet$$

$$RMM^\bullet + M \rightarrow RMMM^\bullet, \text{ etc.}$$

Free radicals are highly reactive because of the presence of an unpaired electron that needs to be paired with another electron to stabilize the molecule. $R^\bullet$, therefore, reacts with M, forming a single bond by sharing its unpaired electron with one from the outer shell of the monomer molecule. In this way long chains of acrylamide are built up, being cross-linked by the introduction of the occasional bisacrylamide molecule into growing chain oxygen mops up free radicals and, therefore, all gel solutions are normally degassed (the solutions are briefly placed under vacuum to remove loosely dissolved air) prior to use. The degassing of the gel solution also serves a second purpose. The polymerization of acrylamide is an exothermic reaction (i.e., heat is liberated) and the warming up of the gel as it sets can liberate air bubbles that become trapped in the polymerized gel. The degassing step prevents this possibility.

Photopolymerisation is an alternative method that can be used to polymerize acrylamide gels. The ammonium persulphate and TEMED are replaced by riboflavin and when the gel is poured it is placed in front of a bright light for 2-3 h. Photodecomposition of riboflavin generates a free radical that initiates polymerization.

Acrylamide gels are defined in terms of the total percentage of acrylamide present, and the pore size in the gel can be varied by changing the concentrations of both the acrylamide and bisacrylamide. Acrylamide gels can be made with a content of between 3% and 30% acrylamide. Thus low percentage gels (e.g., 4%) have large pore sizes and are used, for example, in the electrophoresis of proteins, when free movement of the proteins by electrophoresis is required without any noticeable frictional effect and for another example, in flat-bed isoelectric focusing or the stacking gel system of an SDS-polyacrylamide gel.

Low percentage acrylamide gels are also used to separate DNA. Gels containing between 10% to 20% acrylamide are used in techniques such as SDS-gel electrophoresis, where the smaller pore size now introduces a sieving effect that contributes to the separation of proteins according to their size.

Proteins were originally separated on polyacrylamide gels that were polymerized in glass tubes, approximately 7 mm in diameter and about 10 cm in length. The tubes were easy to load and run, with minimum apparatus requirements. However, only one sample could be run per tube and, because conditions of separation could vary from tube to tube, comparison between different samples was not always accurate. The later introduction of vertical gel slabs allowed running of up to 20 samples under identical conditions in a single run. Vertical slabs are now used routinely for both analysis of proteins and for the separation of DNA fragments during DNA sequence analysis. Although some workers prepare their own acrylamide gels, others purchase commercially available ready-made gels for techniques such as SDS-PAGE, native gels and isoelectric focusing (IEF).

Electrophoresis of protein

8.5 SODIUM DODECYL SULPHATE-POLYACRYLAMIDE GEL ELECTROPHORESIS

SDS-polyacrylamide gel electrophoresis (SDS-PAGE) is the most widely used method for analyzing protein mixtures qualitatively. It is particularly useful for monitoring protein purification and, as the method is based on the separation of proteins according to size, it can also be used to determine the relative molecular mass of proteins. SDS (CH_3–$(CH_2)_{10}$–CH_2OSO_3–Na^+) is an anionic detergent. Samples to be run on SDS-PAGE are firstly boiled for 5 min in sample buffer containing β-mercaptoethanol and SDS. The mecaptoethnol reduces any disulphide bridges present that are

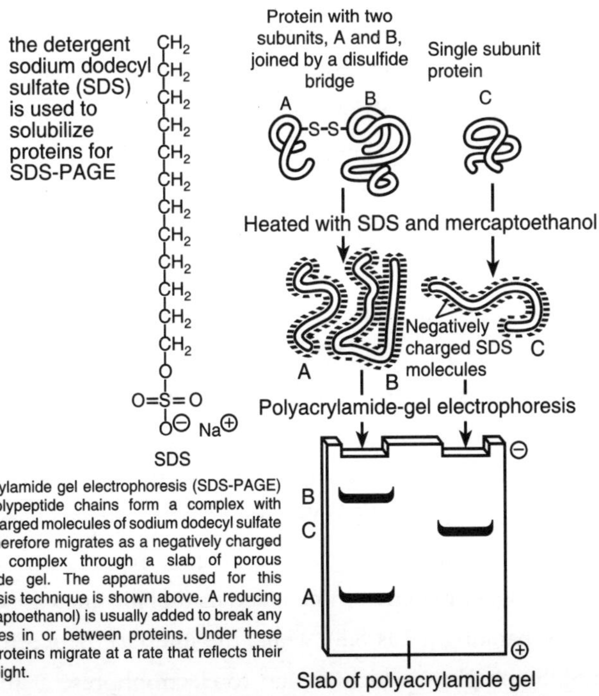

SDS Polyacrylamide gel electrophoresis (SDS-PAGE) Individual polypeptide chains form a complex with negatively charged molecules of sodium dodecyl sulfate (SDS) and therefore migrates as a negatively charged SDS-protein complex through a slab of porous polyacrylamide gel. The apparatus used for this electrophoresis technique is shown above. A reducing agent (inevcaptoethanol) is usually added to break any –S–S–linkages in or between proteins. Under these conditions, proteins migrate at a rate that reflects their molecular weight.

Fig. 8.18: Principle of SDS-PAGE

holding together the protein tertiary structure, and the SDS binds strongly to, and denatures, the protein. Each protein in the mixture is, therefore, fully denatured by this treatment and opens up into a rod-shaped structure with a series of negatively charged SDS molecules along the polypeptide chain. On average, one SDS molecule binds for every two amino acid residues. The original native charge on the molecule is, therefore, fully swamped by the negatively charged SDS molecules. The rod-like structure remains, as any rotation that tends to fold up the protein chain would result in repulsion between negative charges on different parts of the protein chain, returning the conformation back to rod-shape. The sample buffer also contains an ionisable tracking dye, usually bromophenol blue, that allows the electrophoretic run to be monitored, and sucrose or glycerol, which gives the sample solution density thus allowing the sample to settle easily through the electrophoresis buffer to bottom when injected into the loading well.

Before loading to gel the protein samples are lysed to denature them by boiling them in lysis buffer for 10 min. Lysis buffer contains buffer to maintain pH along with glycerol so as to make the protein heavy to properly settle in wells. Lysis buffer also have SDS to impart negative charge, DTT or β-mercaptoethanol to reduce the disulphide bonds in proteins.

Once the samples are all loaded, a current is passed through the gel. The samples to be separated are not in fact loaded directly into the main separating gel. When the main separating gel (normally about 5 cm long) has been poured between the glass plates and allowed to set, a shorter (approximately 0.5 cm) stacking gel is poured on top of the separating gel and it is into this gel that the wells are formed and the proteins loaded. The purpose of this stacking gel is to concentrate the protein sample into a sharp band before it enters the main separating gel. This is achieved by utilizing the differences in ionic strength and pH between the electrophoresis buffer and the stacking gel buffer and involves a phenomenon known as isotachophoresis. The stacking gel has a very large pore size (4% acrylamide), which allows the proteins to move freely and concentrate, or stack, under the effect of the electric field. The band-sharpening effect relies on the fact that negatively charged glycinate ions (in the electrophoresis buffer) have lower elctrophoretic mobility than do the protein–SDS complexes, which, in turn, have lower mobility than the chloride ions (Cl^-) of the loading buffer and the stacking gel buffer. When the current is switched on, all the ionic species have to migrate at the same speed as Cl^- only if they are in a region of high field strength. Field strength is inversely proportional to conductivity, which is proportional to concentration. The result is that the three species of interest adjust their concentration so that $[Cl^-]$ > [protein-SDS] > [glycinate]. There is only a small quantity of protein-SDS complexes, so they concentrate in a very tight band between glycinate and Cl^- boundaries. Once the gycinate reaches the separating gel it becomes more fully ionized in the higher pH environment and its mobility increases. (The pH of the stacking gel is 6.8, that of the separating gel is 8.8). Thus, the interface between glycinate and Cl^- leaves behind the protein-SDS complexs, which are left to electrophorese at their own rates. The negatively charged protein-SDS complexes now continue to move towards the anode, and as because they have the same charge per unit length, they travel into the separating gel under the applied electric field with the same mobility. However, as they pass through the separating gel the proteins separate, owing to the molecular sieving properties of the gel. Quite simply the smaller the protein the more easily it can pass through the pores of the gel, whereas large proteins are successively retarded by frictional resistance due to the sieving effect of the gels. Being a small molecule, the bromophenol blue dye is totally unretarded and, therefore, indicates the electrophoresis front. When the dye reaches the bottom of the gel, the current is turned off, and the gel is removed from between the glass plates and shaken in an appropriate stain solution and then washed in destain solution. The destain solution removes unbound background dye from the gel, leaving stained proteins visible as blue bands on a clear background. A typical minigel would take about 1 h to prepare and set, 40 min to run at 200 V and have a 1 h staining time with Coomassie Brilliant Blue. Upon destaining, strong protein bands would be seen in the gel within 10-20 min, but overnight destaining is needed to completely remove all background stain. Vertical slab gel are invariably run, since this allows up to 10n different sample to be loaded onto a single gel. A typical SDS-polyacrylamide gel is shown in Fig. 8.19.

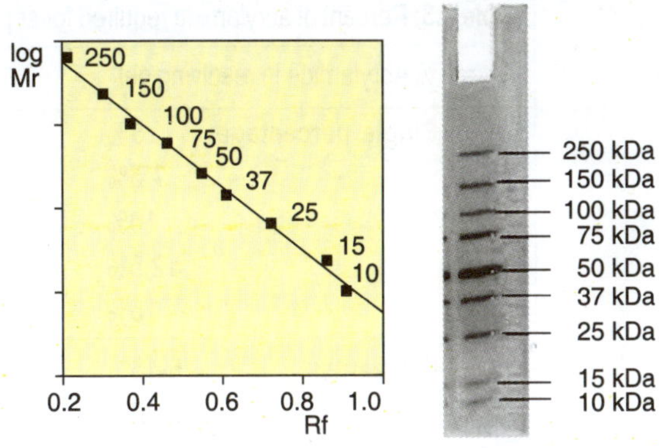

Fig. 8.19: Coomassie stained SDS-PAGE gel with separated starnds of molecular weight marker (right) and their relative mobility (left)

Typically, the separating gel used is a 15% polyacrylamide gel. This gives a gel a certain pore size in which proteins of relative molecular mass (M_r) 10,000 move through the gel relatively unhindered, whereas proteins of M_r 10,00,000 can only just enter the pores of this gel. Gels of 15% polyacrylamide are, therefore, useful for separating proteins in the range M_r 1,00,000 to 10,000. However, a protein of M_r 1,50,000, for example, would be unable to enter a 15% gel. In this case a larger-pored gel (e.g., a 10% or even 7.5% gel) would be used so that the protein could now enter the gel and be stained and identified. It is obvious, therefore, that the choice of gel to be used depends on the size of the proteins in the range of different percentage acrylamide gel is shown in Table 8.3. This shows, for example, that in a 10% polyacrylamide gel proteins greater than 200 kDa in mass cannot enter the gel, whereas proteins with relative molecular mass (M_r) in the range 200,000 to 15,000 will separate. Proteins of M_r 15,000 are too small to experience the sieving effect of the gel matrix, and all run together as a single band at the electrophoresis front.

The M_r of a protein can be determined by comparing its mobility with those of a number of standard proteins of known M_r that are run on a same gel. Plotting a graph of distance moved against log M_r for each of the standard proteins, a calibration curve can be constructed. The distance moved by the protein of unknown M_r is then measured, and then its log M_r and hence M_r can be determined from the calibration curve.

Table 8.3: Percent of acrylamide required for separation of different molecular weight proteins

% Acrylamide in resolving gel		Separation size range ($M_r \times 10^{-3}$)
Single percentage:	5%	36-200
	7.5%	24-200
	10%	14-200
	12.5%	14-100*
	15%	14-60*
Gradients:	5-15%	14-200
	5-20%	10-200
	10-20%	10-150

SDS-gel electrophoresis is often used after each step of a purification protocol to assess the purity or otherwise of the sample. A pure protein should give a single band on an SDS-polyacrylamide gel, unless the molecule is made up of two unequal subunits. In the latter case, two bands, corresponding to the two subunits, will be seen. Since only submicrogram amounts of the protein are needed for the gel, very little material is used in this form of purity assessment and at the same time a value for the relative molecular mass of the protein can be determined on the same gel run (as described above), with no more material being used.

8.6 NATIVE (BUFFER) GELS

While SDS-PAGE is the most frequently used gel system for studying proteins, the method is of no use if one is aiming to detect a particular protein (often an enzyme) on the basis of its biological activity, because the protein (enzyme) is denatured by SDS-PAGE procedure. In this case it is necessary to use non-denaturing conditions. In native or buffer gels, polyacrylamide gels are again used (normally a 7.5% gel) but the SDS is absent and the proteins are *not* denatured prior to loading. Since all the proteins in this sample being analyzed carry their native charge at the pH of the gel (normally pH 8.7), proteins separate according to their different electrophoretic mobilities *and* the sieving effects of the gel. It is, therefore, not possible to predict the behaviour of a given protein in a buffer gel but, because of the range of different charges and the sizes of the proteins in a given protein mixture, good resolution is achieved. The enzyme of interest can be identified by incubating the gel in an appropriate substrate solution such that a coloured product is produced at the site of the enzyme. An alternative method for enzyme detection is to include the substrate in an agarose gel that is poured over the acrylamide gel and allowed to set. Diffusion and interaction of enzyme and substrate between the two gels result in colour formation at the site of the enzyme. Often, duplicate samples will be run on the gel, the gel cut in half and one half stained for activity, the other for total protein. In this way the total protein content of the sample can be analysed and the particular band corresponding to the enzyme identified by reference to the activity stain gel.

8.7 GRADIENT GELS

This is again a polyacrylamide gel system, but instead of running a slab gel of uniform pore size throughout (e.g., a 15% gel) a gradient gel is formed, where the acrylamide concentration varies uniformly from typical 5% at the top of the gel to 25% acrylamide at the bottom of the gel. The gradient is formed via a gradient mixer and run down between the glass plates of a slab gel. The higher percentage acrylamide (e.g., 25%) is poured between the glass plates first and a continuous gradient of decreasing acrylamide concentration follows. Therefore, at the top of the gel there is a large pore size (5% acrylamide) but as the sample moves down the gel the acrylamide concentration slowly increases and the pore size correspondingly decreases. Gradient gels are normally run as SDS gels with a staking gel. There are two advantages to running gradient gels. First, a much greater range of protein M_r values can be separated than on a fixed-percentage gel. In a complex mixture, very low molecular weight proteins travel freely through the gel to begin with, and start to resolve when they reach the smaller pore sizes towards the lower part of the gel. Much larger proteins, on the other hand, can still enter the gel but start to separate immediately due to the sieving effect of the gel. The second advantage of the gradient gels is that proteins with very similar M_r values may be resolved, although they cannot otherwise be resolved in fixed percentage gels. As each protein moves through the gel the pore sizes become smaller until the protein reaches its pore size limit. The pore size in the gel is now too small to allow passage of the protein, and the protein sample stacks up at this point as a sharp band. A similar-sized protein, but with slightly lower M_r, will be able to travel a little further through the gel before reaching its pore size limit, at which point it will form a sharp band. These two proteins, of slightly different M_r values, therefore, separate as two close sharp bands.

8.8 CAPILLARY ELECTROPHORESIS (CE)

The process of electrophoresis is defined as 'the differential movement or migration of ions by attraction or repulsion in an electric field'. In practical terms, a positive (anode) and negative (cathode) electrode are placed in a solution containing ions. Then, when a voltage is applied across the electrodes, solute ions of different charge, i.e., anions (negative) and cations (positive), will move through the solution towards the electrode of opposite charge. Capillary electrophoresis, then, is the technique of performing electrophoresis in buffer-filled, narrow-bore capillaries, normally from 25 to 100 mm in internal diameter (ID).

Instrumentation

The instrumentation required for CE is remarkably simple in design, as Fig. 8.20 illustrates. The ends of a capillary are placed in separate buffer reservoirs, each containing an electrode connected to a high-voltage power supply capable of delivering up to 30 kV. The sample is injected onto the capillary by temporarily replacing one of the buffer reservoirs (normally at the anode) with a

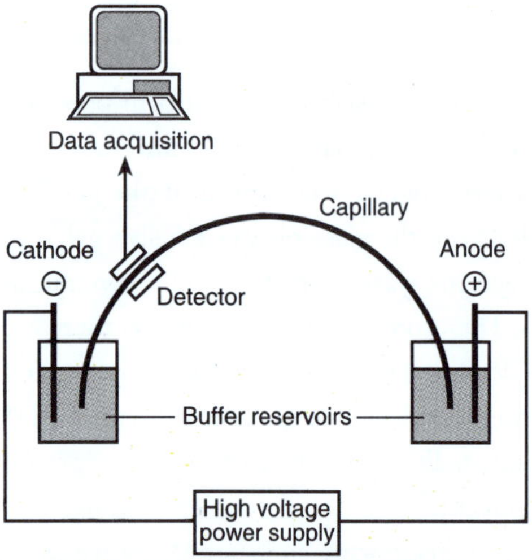

Fig. 8.20: A schematic representation of the arrangement of the main components of a typical CE instrument

sample reservoir and applying either an electric potential or external pressure for a few seconds. After replacing the buffer reservoir, an electric potential is applied across the capillary and the separation is performed. Optical (UV-visible or fluorometric) detection of separated analytes can be achieved directly through the capillary wall near the opposite end (normally near the cathode). CE is very suited to automation, and the arrangement of commercial CE instruments will seem familiar to those with knowledge of modern HPLC. Basic features of a CE instrument include an autosampler, a detection module, a high-voltage power supply, the capillary and, of course, a computer to control everything. So, if we consider that the power supply is equivalent to an HPLC pump and the capillary is equivalent to a column, the instrumentation is completely analogous. This is especially so as the software packages used to control most commercial CE instruments are based heavily on existing HPLC software.

Electroosmotic flow (EOF)

A vitally important feature of CE is the bulk flow of liquid through the capillary. This is called the electroosmotic flow and is caused as follows. An uncoated fused-silica capillary tube is typically used for CE. The surface of the inside of the tube has ionisable silanol groups, which are in contact with the buffer during CE. These silanol groups readily dissociate, giving the capillary wall a negative charge. Therefore, when the capillary is filled with buffer, the negatively charged capillary wall attracts positively charged ions from the buffer solution, creating an electrical double layer and a potential difference (zeta potential) close to the capillary wall, as described according to Stern's model in Fig. 8.21. Stern's model for an electrical double layer includes a rigid layer of adsorbed ions and a diffuse layer, in which ion diffusion may occur by thermal motion. The zeta potential is the potential at any given point in the double layer and decreases exponentially with increasing distance from the capillary wall surface.

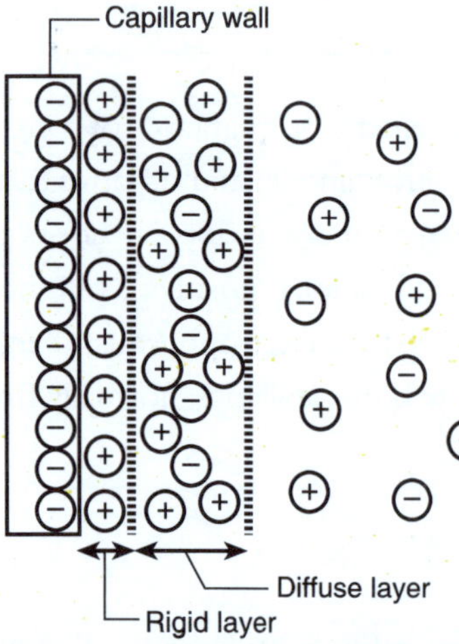

Fig. 8.21: Stern's model of the double-layer charge distribution at a negatively charged capillary wall leading to the generation of a zeta potential and EOF

When a voltage is applied across the capillary, cations in the diffuse layer are free to migrate towards the cathode, carrying the bulk solution with them. The result is a net flow in the direction of the cathode, with a velocity described by

$$v_{EOF} = \left(\frac{\varepsilon_0 \varepsilon \xi}{4\pi\eta}\right)E,$$

where ε_0 is the dielectric constant of a vacuum, ε is the dielectric constant of the buffer, ζ is the zeta potential, η is the viscosity of the buffer and E is the applied electric field. The terms enclosed in brackets equate to the mobility of the EOF (μ_{EOF}).

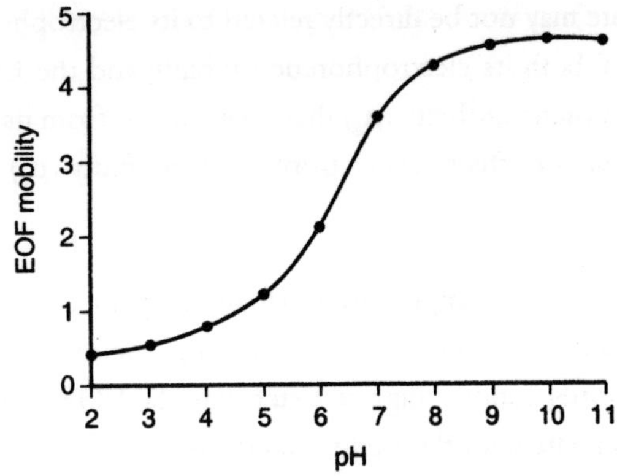

Fig. 8.22: The variation of EOF mobility with changing pH for a typical uncoated fused-silica capillary (simulated data)

The relationship between EOF mobility and EOF velocity is analogous to that between electrophoretic mobility and migration velocity. Indeed, the units for EOF mobility are the same as those for electrophoretic mobility.

Factors affecting EOF mobility

The main variables affecting EOF mobility are the dielectric constant and viscosity of the buffer and the size of the zeta potential. The use of buffer additives and/or other modifications of the buffer composition may influence the dielectric constant and viscosity of the buffer. Buffer viscosity will also depend on the temperature at which the CE separation is performed.

Zeta Potential

The zeta potential is proportional to the charge density on the capillary wall, which itself is pH dependent. Therefore, EOF mobility will vary according to the buffer pH, such that at high pH the EOF mobility will be significantly greater than at low pH. Fig. 8.22 depicts the variation of EOF mobility with pH for a typical fused-silica capillary. Above pH 9, silanols are completely ionised and the EOF mobility is at its greatest. Below pH 4, the ionisation of silanols is low and the EOF mobility is insignificant. The zeta potential will also depend upon the ionic strength of the buffer, because as ionic strength increases, the double layer will become compressed, which results in a decreased zeta potential and reduced EOF mobility.

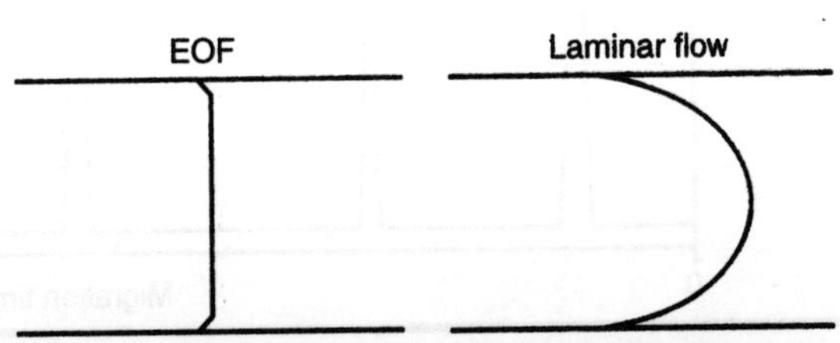

Fig. 8.23: Flow profiles of EOF and laminar flow

At pH>7, the EOF mobility is sufficient to ensure the net migration of most ions towards the cathode, regardless of their charge. Therefore, the observed migration velocity of a solute may not be directly related to its electrophoretic mobility. Instead, it is related to a combination of both its electrophoretic mobility and the EOF mobility. Therefore, a solute's apparent electrophoretic mobility (μ_a) that is calculated from its observed migration velocity, is the vector sum of its real (or effective) electrophoretic mobility (μ_e) and the EOF mobility (μ_{EOF}), i.e.,

$$\mu_a = \mu_e + \mu_{EOF}.$$

Since samples are normally introduced at the anode and EOF moves from the anode to the cathode, cations have positive μ_e , neutrals have zero μ_e and anions have negative μ_e. In other words, cations migrate faster than the EOF and anions migrate more slowly than the EOF. Neutrals migrate with the same velocity as the EOF.

Flow profile in CE

A further key feature of EOF is that it has flat flow profile, which is shown in figure 8.23, alongside the parabolic flow profile generated by an external pump, as used for HPLC. EOF has a flat profile because its driving force (i.e., charge on the capillary wall) is uniformly distributed along the capillary, which means that no pressure drops are encountered and the flow velocity is uniform across the capillary. This contrasts with pressure-driven flow, such as in HPLC, in which frictional forces at the column walls cause a pressure drop across the column, yielding a parabolic or laminar flow profile. The flat profile of EOF is important because it minimises zone broadening, leading to high separation efficiencies that allow separations on the basis of mobility differences as small as 0.05%.

The electropherogram

The data output from CE is presented in the form of an electropherogram, which is analogous to a chromatogram. An electropherogram is a plot of migration time vs. detector response. The detector response is usually concentration dependent, such as UV-visible absorbance or fluorescence. The appearance of a typical electropherogram is shown in Fig. 8.24 for the separation of a three component mixture of cationic, neutral and anionic solutes.

Fig. 8.24: A typical electropherogram showing the separation of a cation, a neutral and an anion

8.9 CELLULOSE ACETATE ELECTROPHORESIS

Although one of the older methods, cellulose acetate electrophoresis still has a number of applications, in particular it has retained a use in the clinical analysis of serum samples. Cellulose acetate has the advantage over paper, in that it is a much more homogeneous medium, with a uniform pore size, and does not absorb proteins in the way that paper does. There is, therefore, much less trialing of protein bands and resolution is better, although nothing like as good as that activated with polyacrylamide gels. The method is, however, far simpler to set up and run. Single samples are normally run on cellulose acetate strips (2.5 cm × 12 cm), although multiple samples are frequently run on wider sheets. The cellulose acetate is first wetted in electrophoresis buffer (pH 8.6 for serum samples) and the sample (1-2 mm^3) loaded as a 1 cm wide strip about one-third of the way along the strip. The ends of the strip make contact with the electrophoresis buffer tanks via a filter paper wick that overlaps the end of the cellulose acetate strip, and the electrophoresis conducted at 6-8 V cm^{-1} for about 3 h. Following electrophoresis, the strip is stained for protein, destained, and the bands visualised.

Enzymes can easily be detected, in samples electrophoresed on cellulose acetate, by using the zymogram technique. The cellulose strip is laid on a strip of filter paper soaked in buffer and substrate. After an enzyme appropriate incubation period, the strips are peeled apart and the paper zymogram treated accordingly to detect enzyme product; hence, it is possible to identify the position of the enzyme activity on the original strip. An alternative approach to detecting and semiquantifying any particular protein on a strip to treat the strip as the equivalent of a protein blot and to probe for the given protein using primary antibody and the enzyme-linked secondary antibody. Substrate colour development indicates the presence of a particular protein and the amount of colour developed in a given time is a semiquantitaive measure of the amount of protein. Thus, for example, large numbers of serum samples can be run on a wide sheet, the sheet is probed using antibodies, and elevated levels of a particular protein identified in certain samples by increased levels of colour development in these samples.

8.10 DETECTION, ESTIMATION AND RECOVERY OF PROTEIN GELS

The most commonly used general protein strain for detecting protein on gels is the sulphate trimethylamine dye Coomassie Brilliant Blue R-250 (CBB). Staining is usually carried out using 0.1% (w/v) CBB in methanol:water:glacial acetic acid (45:45:10, by vol.). This acid-methanol mixture acts as a denaturant to precipitate or fix the protein in the gel, which prevents the protein from being washed out whilst it is being stained. Staining of most gels is accomplished in about 2 h and destaining, usually overnight, is achieved by gentle agitation in the same acid methanol solution but in the absence of the dye. The Coomassie stain is highly sensitive; a very weakly staining band on a polyacrylamide gel would correspond to about 0.1 μg (100 ng) of proteins. The CBB stain is not used for staining cellulose acetate (or indeed protein blots) because it binds quite strongly to the paper. In this case, proteins are first denatured by brief immersion of the strip in 10% (v/v) trichloroacetic acid, and then immersed in a solution of a dye that dose not stain the support material; for example, Procion blue, Amino black of Procion S.

Although the Coomassie stain is highly sensitive, many workers require greater sensitivity such as that provided by silver staining. Silver stains are based either on techniques developed for histology or on methods based on the photographic process. In either case, silver ions (Ag^+) are reduced to metallic silver on the protein, where the silver is deposited to give a black or brown band. Silver stains can be used immediately after electrophoresis, or, alternatively, after staining with CBB. With the late approach, the major bands on the gel can be identified with CBB and then minor bands, not detected with CBB, resolved using the silver stain. The silver stain is at least 100 times more sensitive than CBB, detecting proteins down to 1ng amounts. Other stains with similar sensitivity include the fluorescent stains, Sypro Orange stains, with similar sensitivity include fluorescent stain Sypro Orange (30 ng) and Sypro Ruby (10 ng).

Glycoproteins have traditionally been detected on protein gels by use of the periodic acid-Shiff (PAS) stain. This allows components of a mixture of glycoproteins to be distinguished. However, the PAS stain is not very sensitive and often gives very weak, red-pink bands, difficult to observe on a gel. A far more sensitive method used nowadays is to blot the gel and use lectins to detect the glycoproteins. Lectins are protein molecules that detect carbohydrates, and different lectins have been found that have different specificities for different types of carbohydrate. For example, certain lectins recognize mannose, fucose or terminal glucosamine of the carbohydrate side-chains of glycoproteins. The sample to be analyzed is run on a number of tracks of SDS-polyacrylamide gel. Coloured bands appear at the point where the lectins bind if each blotted track is incubated with a different lectin, washed, incubated with a horseradish peroxidase-linked antibody to the lectin, and then peroxidase substrate added. In this way, by testing a protein sample against a series of lectins, it is possible to determine not only that a protein is a glycoprotein, but to obtain information about the type of glycosylation.

Quantitative analysis (i.e., measurement of the relative amounts of different protein sample) can be achieved by scanning densitometry. A number of commercial scanning densitometers are available, and work by passing the stained gel track over a beam of light (laser) and measuring the transmitted light. A graphic presentation of protein ones (peaks of absorbance) against migration distance is produced, and peak areas can be calculated to obtain quantitative data. However, such data must be interpreted with caution because there is only a limited range of protein concentrations over which there is a linear relationship between absorbance and concentration. Also, equal amounts of different proteins do not always stain equally with a given stain, so any data comparing the relative amounts of protein can only be semiquantitative. An alternative and much cheaper way of obtaining such data is cut out the stained bands of interest, elute the dye by shaking overnight in a known volume of 50% pyridine, and then to measure spectrophotometrically the amount of colour released. More recently gel documentation systems have been developed, which are replacing scanning densitometers. Such benchtop systems comprise a video imaging unit (computer linked) attached to a small 'darkroom' unit that is fitted with a choice of white or ultraviolet light (transilluminator). Gel images can be stored on the computer, enhanced accordingly and printed as required on a thermal printer, thus eliminating the need for wet developing in a purpose built darkroom, as in the case for traditional photography.

Although gel electrophoresis is used generally as an analytical tool, it can be utilized to separate proteins in a gel to achieve protein purification. Protein bands can be cut out of protein blots and sequence data obtained by placing the blot in a protein sequencer. Stained protein bands can be cut out of protein gels and the protein recovered by electrophoresis of the protein out of the gel piece (electroelution). A number of different designs of electroelution cells are commercially available, but perhaps the easiest method is to seal the gel piece in buffer in a dialysis sac and place the sac in buffer between two electrodes. Protein will electrophorese out of the gel piece towards the appropriate electrode but will be retained by the dialysis sac. After electroelution, the current is reversed for a few seconds to drive off any protein that has absorbed to the wall of the dialysis sac and then the protein solution within the sac is recovered.

8.11 ISOELECTRIC FOCUSING AND TWO-DIMENSIONAL GEL ELECTROPHORESIS

Two-dimensional electrophoresis (2-D electrophoresis) is a powerful and widely used method for the analysis of complex protein mixtures extracted from cells, tissues, or other biological samples. This technique sorts proteins according to two independent properties in two discrete steps: the first-dimension step, isoelectric focusing (IEF), separates proteins according to their isoelectric points (pI); the second-dimension step, SDS-polyacrylamide gel electrophoresis (SDS-PAGE), separates proteins according to their molecular weights (M_r, relative molecular weight). Each spot on the resulting two-dimensional array corresponds to a single protein species in the sample. Thousands of different proteins can thus be separated, and information such as the protein pI, the apparent molecular weight, and the amount of each protein are obtained.

Two-dimensional electrophoresis was first introduced by P. H. O'Farrell and J. Klose in 1975. In the original technique, the first-dimension separation was performed in carrier ampholyte-containing polyacrylamide gels cast in narrow tubes. A. Görg and colleagues developed the currently employed 2-D technique, where carrier ampholyte-generated pH gradients have been replaced with immobilized pH gradients and tube gels replaced with gels supported by a plastic backing.

A large and growing application of 2-D electrophoresis is "proteome analysis." The analysis involves the systematic separation, identification, and quantification of many proteins simultaneously from a single sample. Two-dimensional electrophoresis is used in this technique due to its unparalleled ability to separate thousands of proteins simultaneously. Two-dimensional electrophoresis is also unique in its ability to detect post- and co-translational modifications, which cannot be predicted from the genome sequence. Applications of 2-D electrophoresis include proteome analysis, cell differentiation, detection of disease markers, monitoring therapies, drug discovery, cancer research, purity checks, and microscale protein purification.

The 2-D process begins with sample preparation. Proper sample preparation is absolutely essential for a good 2-D result. The next step in the 2-D process is IPG (Isoelectric pH gradient) strip rehydration. IPG strips are provided dry and must be rehydrated with the appropriate additives prior to IEF (Immunoelectrophoresis). First-dimension IEF is performed on a flatbed system at very high voltages with active temperature control. Next, strip equilibration in SDS-containing buffer prepares the sample for the second-dimension separation. Following equilibration, the strip is placed on the second-dimension gel for SDS-PAGE. The final steps are visualization and analysis of the resultant two-dimensional array of spots. In summary, the experimental sequence for 2-D electrophoresis is:

1. Sample preparation
2. IPG strip rehydration
3. IEF
4. IPG strip equilibration
5. SDS-PAGE
6. Visualization
7. Analysis.

8.11.1. Sample preparation

Due to the great diversity of protein sample types and origins, only general guidelines for sample preparation are provided in this section. The optimal procedure must be determined empirically for each sample type. Ideally, the process will result in the complete solubilization, disaggregation, denaturation, and reduction of the proteins in the sample. When developing a sample preparation strategy, it is important to have a clear idea of what is desired in the final 2-D result. Is the goal to view as many proteins as possible, or is only a subset of the proteins in the sample of potential interest? Which is more important – complete sample representation, or a clear, reproducible pattern? Additional sample preparation steps can improve the quality of the final result, but each additional step can result in the selective loss of protein species. The trade-off between improved sample quality and complete protein representation must, therefore, be carefully considered.

In order to characterize specific proteins in a complex protein mixture, the proteins of interest must be completely soluble under electrophoresis conditions. Different treatments and conditions are required to solubilize different types of protein samples; some proteins are naturally found in complexes with membranes, nucleic acids, or other proteins, some proteins form various non-specific aggregates, and some proteins precipitate when removed from their normal environment. The effectiveness of solubilization depends on the choice of cell disruption method, protein concentration and dissolution method, choice of detergents, and composition of the sample solution. If any of these steps are not optimized for a particular sample, separations may be incomplete or distorted and information may be lost.

To fully analyze all intracellular proteins, the cells must be effectively disrupted. Choice of disruption method depends on whether the sample is from cells, solid tissue, or other biological material and whether the analysis is targeting all proteins or just a particular subcellular fraction. Both gentle and vigorous lysis methods are discussed below.

Table 8.4: Gentle lysis methods

Cell Disruption method	Application	General procedure
Osmotic lysis This very gentle method is well-suited for applications in which the lysate is to be subsequently fractionated into subcellular components.	Blood cells, tissue culture cells	Suspend cells in a hypoosmotic solution.
Freeze-thaw lysis Many types of cells can be lysed by subjecting them to one or more cycles of quick freezing and subsequent thawing.	Bacterial cells, tissue culture cells	Rapidly freeze cell suspension using liquid nitrogen, and then thaw. Repeat if necessary
Detergent lysis Detergents solubilize cellular membranes lysing cells and liberating their contents.	Tissue culture cells	Suspend cells in lysis solution containing detergent. Cells can often be lysed directly into sample solution or rehydration solution because these solutions always contain detergent. If an anionic detergent such as SDS is used for lysis, one of the following preparation steps is required to ensure that the SDS will not interfere with IEF: • Dilute the lysed sample into a solution containing an excess of non-ionic or zwitterionic detergent. OR, • Separate the SDS from the sample protein by acetone precipitation.
Enzymatic lysis Cells with cell-walls can be lysed gently following enzymatic removal of the cell-wall. This must be done with an enzyme specific for the type of cell to be lysed (e.g., lysozyme for bacterial cells, cellulase and pectinase or plant cells, lyticase for yeast cells).	Plant tissue, bacterial cells, fungal cells	Treat cells with enzyme in isoosmotic solution.

Table 8.5: Some vigorous lysis methods

Cell Disruption method	Application	General procedure
Sonication Ultrasonic waves generated by a sonicator lyse cells through shear forces. Complete shearing is obtained when maximal agitation is achieved, but care must be taken to minimize heating and foaming.	Cell suspensions	Sonicate cell suspension in short bursts to avoid heating. Cool on ice between bursts.

Contd.

Table 8.5: Some Vigorous lysis methods

Cell Disruption method	Application	General procedure
French pressure cell Cells are lysed by shear forces resulting orifice under high pressure.	Micro-organisms	Place cell suspension in chilled French pressure from forcing suspension through a small with cell walls cell. Apply pressure and collect extruded lysate.
Grinding Some cell types can be opened by hand grinding with a mortar and pestle.	Solid tissues, micro-organisms	Tissue or cells are normally frozen with liquid nitrogen and ground to a fine powder. Alumina (Al_2O_3) or sand may aid grinding.
Mechanical homogenization Many different devices can be used to mechanically homogenize tissues. Hand-held devices such as Dounce or Potter-Elvehjem homogenizers can be used to disrupt cell suspensions or relatively soft tissues. Blenders, or other motorized devices can be used for larger samples. Homogenization is rapid and poses little danger to proteins except by the proteases that may be liberated upon disruption.	Solid tissues	Chop tissue into small pieces, if necessary. Add chilled homogenization buffer (5-20 volume of tissue). Homogenize briefly. Clarify lysate by filtration and/or centrifugation.
Glass bead homogenization The abrasive action of the vortexed beads breaks cell-walls, liberating the cellular contents.	Cell suspensions, micro-organisms	Suspend cells in an equal volume of chilled lysis solution and place into a sturdy tube. Add 1-3 g of chilled glass beads per gram of wet cells. Vortex for 1 min and incubate cells on ice for 1 min. Repeat vortexing and chilling two to four times.

Proteases may be liberated upon cell disruption. Proteolysis greatly complicates analysis of the 2-D result, thus the protein sample should be protected from proteolysis during cell disruption and subsequent preparation. Proteases are less active at lower temperatures, so sample preparation at as low a temperature as possible is recommended. In addition, proteolysis can often be inhibited by preparing the sample in the presence of tris base, sodium carbonate, or basic carrier ampholyte mixtures. These approaches alone are often sufficient protection against proteolysis. However, some proteases may retain some activity even under these conditions. In these cases, protease inhibitors may be used. Individual protease inhibitors are only active against specific classes of proteases, so it is usually advisable to use a combination of protease inhibitors. Broadrange protease inhibitor "cocktails" are available from a number of commercial sources. Table 8.6 lists common protease inhibitors and the proteases they inhibit.

Table 8.6: List of common protease inhibitors and the proteases they inhibit

Protease inhibitor	Effective against	Limitations
PMSF (Phenylmethylsulphonyl fluoride) Most commonly used inhibitor. *Use at concentration up to* 1 mM.	PMSF is an irreversible inhibitor that inactivates: • serine proteases • some cysteine proteases.	PMSF rapidly becomes inactive in aqueous solutions: Prepare just prior to use. PMSF may be less effective in the presence of thiol reagents such as DTT or 2-mercaptoethanol. This limitation can be overcome by disrupting the sample into PMSF-containing solution lacking thiol reagents. Thiol reagents can be added at a later state. PMSF is very toxic.
AEBSF (Aminoethyl benzylsulphonyl fluoride or Pefabloc SC Serine Protease Inhibitor) *Use at concentrations up to* 4 mM.	AEBSF is similar to PMSF in its inhibitory activity, but is more soluble and less toxic.	AEBSF-induced modifications can potentially alter the pI of a protein.
1 mM EDTA or 1 mM EGTA *Generally used at* 1 mM.	These compounds inhibit metalloproteases by chelating free metal ions required for activity.	
Peptide protease inhibitors (e.g., leupeptin, pepstatin, aprotinin, bestatin) • reversible inhibitors • active in the presence of DTT • active at low concentrations under a variety of conditions. *Use at* 2-20 μg/ml.	Leupeptin inhibits many serine and cysteine proteases. Pepstatin inhibits aspartyl proteases (e.g., acidic proteases such as pepsin). Aprotinin inhibits many serine proteases. Bestatin inhibits aminopeptidases.	Peptide protease inhibitors are: • expensive • small peptides and thus may appear on the 2-D map, depending on the size range separated by the second-dimension gel. Pepstatin does not inhibit any proteases that are active at pH 9.
TLCK, TPCK (Tosyl lysine chloromethyl ketone, tosyl phenylalanine chloromethyl ketone). *Use at* 0.1-0.5 mM.	These similar compounds irreversibly inhibit many serine and cysteine proteases.	
Benzamidine *Use at* 1-3 mM.	Benzamidine inhibits serine proteases.	

If only a subset of the proteins in a tissue or cell type is of interest, prefractionation can be employed during sample preparation. If proteins from one particular subcellular compartment (e.g., nuclei, mitochondria, plasma membrane) are desired, the organelle of interest can be purified by differential centrifugation or other means prior to solubilization of proteins for 2-D electrophoresis. The sample can also be prefractionated by solubility under different extraction conditions prior to 2-D electrophoresis. Precipitation of the proteins in the sample and removal of interfering substances are optional steps. The decision to employ these steps depends on the nature of the sample and the experimental goal. Precipitation procedures, which are used both to concentrate the sample and to separate the proteins from potentially interfering substances, are described below. No precipitation technique is completely efficient and some proteins may not readily resuspend following precipitation. Thus, employing a precipitation step during sample preparation can alter the protein profile of a sample. Precipitation and resuspension should be avoided if the aim of a 2-D experiment is complete and accurate representation of all the proteins in a sample. If sample preparation requires precipitation, typically only one of the following precipitation techniques is employed:

1. **Ammonium sulphate precipitation:** ("Salting out") In the presence of high salt concentrations, proteins tend to aggregate and precipitate out of solution. Many potential contaminants (e.g., nucleic acids) will remain in solution.
2. **TCA precipitation:** TCA (trichloroacetic acid) is a very effective protein precipitant.
3. **Acetone precipitation:** This organic solvent is commonly used to precipitate proteins. Many organic-soluble contaminants (e.g., detergents, lipids) will remain in solution.
4. **Precipitation with TCA in acetone:** The combination of TCA and acetone is commonly used to precipitate proteins during sample preparation for 2-D electrophoresis and is more effective than either TCA or acetone alone.
5. **Precipitation with ammonium acetate in methanol following phenol extraction** This technique has proven useful with plant samples containing high levels of interfering substances.

Non-protein impurities in the sample can interfere with separation and subsequent visualization of the 2-D result, so sample preparation can include steps to rid the sample of these substances. Table 8.7 lists contaminants that affect 2-D results and techniques for their removal.

Table 8.7: List of contaminants and reasons for their removal

Contaminants	Reasons for removal
Salts, residual buffers, and other charged small molecules that carry over from sample preparation	Salts disturb the electrophoresis process and must be removed or maintained at as low a concentration as possible. Salts in the IPG strip result in high strip conductivity. Focusing of the proteins will not occur until the ions have moved to the ends of the strips, prolonging the

Contd.

Table 8.7: List of contaminants and reasons for their removal

Contaminants	Reasons for removal
	time required for IEF. Water movement can also result, causing one end of the strip to dry out and the other to swell. Salt in the IPG strip can result in large regions at either end of the IPG strip where proteins do not focus (seen as horizontal streaking or empty regions in the final result). If the sample is rehydrated into the IPG strip, the salt concentration in the rehydration solution should be lower than 10 mM. If the sample is applied in sample cups, salt concentrations of up to 50 mM in the sample may be tolerated; however, proteins may precipitate at the sample application point as they abruptly move into a lower salt environment.
Endogenous small ionic molecules, (nucleotides, metabolites, phospholipids, etc).	Endogenous small ionic molecules are present in any cell lysate. These substances are often negatively charged and can result in poor focusing towards the anode.
Phenolic compounds	Phenolic compounds are present in many plant tissues and can modify proteins through an enzyme-catalyzed oxidative reaction.
Insoluble material	Insoluble material in the sample can clog gel pores and result in poor focusing. Insoluble material is particularly problematic when the sample is applied using sample cups; it can prevent protein entry into the IPG strip.
Nucleic acids (DNA RNA)	Nucleic acids increase sample viscosity and cause background smears. High-molecular weight nucleic acids can clog gel pores. Nucleic acids can bind to proteins through electrostatic interactions, preventing focusing. If the separated sample proteins are visualized by silver staining, nucleic acids present in the gel will also stain, resulting in a background smear on the 2-D gel.
Polysaccharides	Polysaccharides can clog gel pores causing either precipitation or extended focusing times, resulting in horizontal streaking. Some polysaccharides contain negative charges and can complex with proteins by electrostatic interactions.
Lipids	Many proteins, particularly membrane proteins, are complexed with lipids. This reduces their solubility and can affect both the pI and the molecular weight. Lipids form complexes with detergents, reducing the effectiveness of the detergent as a protein-solubilizing agent. When extracts of lipid-rich tissues are centrifuged, there is often a lipid layer that can be difficult to remove.
Ionic detergent	Ionic detergent (usually SDS) is often used during protein extraction and solubilization, but can strongly interfere with IEF. SDS forms complexes with proteins, and the resulting negatively charged complex will not focus unless the SDS is removed or sequestered.

In general, it is advisable to keep sample preparation as simple as possible. A sample with low protein concentrations and a high salt concentration, for example, could be diluted normally and analyzed, or desalted, then concentrated by lyophilization, or precipitated with TCA and ice-cold acetone and re-solubilized with rehydration solution. The first option of simply diluting the sample with rehydration solution may be sufficient. If problems with protein concentration or interfering substances are otherwise insurmountable, then precipitation or removal steps may be necessary.

The composition of the sample solution is particularly critical for 2-D because solubilisation treatments for the first-dimension separation must not affect the protein pI, nor leave the sample in a highly conductive solution. In general, concentrated urea as well as one or more detergents are used. Sample solution composition is discussed herewith. These always include urea and one or more detergents. Complete denaturation ensures that each protein is present in only one configuration and that aggregation and intermolecular interaction are avoided. The lysis solution, which contains urea and the zwitterionic detergent CHAPS, has been found to be effective for solubilizing a wide range of samples. Reductant and IPG buffer are also frequently added to the sample solution to enhance sample solubility. IEF performed under denaturing conditions gives the highest resolution and the cleanest results. Urea, a neutral chaotrope, is used as the denaturant in the first-dimension of 2-D electrophoresis. Urea solubilizes and unfolds most proteins to their fully random conformation, with all ionizable groups exposed to solution. Recently, the use of thiourea in addition to urea has been found to further improve solubilization, particularly of membrane proteins. A non-ionic or zwitterionic detergent is always included in the sample solution to ensure complete sample solubilization and to prevent aggregation through hydrophobic interactions. Originally, either of two similar non-ionic detergents, NP-40 or Triton X-100, was used. When difficulties in achieving full sample solubilization are encountered, the anionic detergent SDS can be used as a solubilizing agent. SDS is a very effective protein solubilizer, but as because it is charged and forms complexes with proteins, it cannot be used as the sole detergent for solubilizing samples for 2-D electrophoresis. A widely used method for negating the interfering effect of SDS is dilution of the sample into a solution containing an excess of CHAPS, Triton X-100, or NP-40. The final concentration of SDS should be 0.25% or lower and the ratio of the excess detergent to SDS should be at least 8:1.Reducing agents are frequently included in the sample solution to break any disulphide bonds present and to maintain all proteins in their fully reduced state. The most commonly used reductant is dithiothreitol (DTT) at concentrations ranging from 20 to 100 mM. Dithioerythreitol (DTE) is similar to DTT and can also be used as a reducing agent. Originally, 2-mercaptoethanol was used as a reductant, but higher concentrations of the reductant are required and inherent impurities may result in artifacts. More recently, the non-thiol reductant tributyl phosphine (TBP), at a concentration of 2 mM, has been used as a reductant for 2-D samples. However, due to the limited solubility and instability of TBP in solution, a thiol reductant such as DTT should be used to maintain proteins in their reduced state through rehydration and first-dimension IEF, if TBP is employed as a reductant during sample preparation. Carrier ampholytes or IPG buffer (up to 2% (v/v)) can be included in the sample solution. They enhance protein solubility by minimizing protein aggregation due to charge-charge interactions. In some cases, buffers or bases (e.g., 40 mM Tris base) are added to the sample solution. This is done when basic conditions are required for full solubilization or to minimize proteolysis. However, introduction of such ionic compounds can result in first-dimension disturbances. Bases or buffers should be diluted to 5 mM or lower prior to loading the sample onto first-dimension IEF. A sample should remain in sample solution at room temperature for at least 30 min for full denaturation and solubilization prior to centrifugation and subsequent sample application. Heating of the sample in the presence of detergent can aid in solubilization, but should only be done prior to the addition of urea. Sonication helps speed up solubilization, particularly from material that is otherwise difficult to resuspend.

8.11.2. First-dimension Isoelectric Focusing (IEF)

A useful first-dimension separation requires selecting a first-dimension pH range appropriate for the sample, as well as a suitable sample application method. IEF is an electrophoretic method that separates proteins according to their isoelectric points (pI). Proteins are amphoteric molecules; they carry either positive, negative, or zero net charge, depending on the pH of their surroundings (Fig. 8.25). The net charge of a protein is the sum of all the negative and positive charges of its amino acid side-chains and amino- and carboxyl-termini. The isoelectric point (pI) is the specific pH at which the net charge of the protein is zero. Proteins are positively charged at pH values below their pI and negatively charged at pH values above their pI. If the net charge of a protein is plotted versus the pH of its environment, the resulting curve intersects the x-axis at the isoelectric point (Fig. 8.25). The presence of a pH gradient is critical to the IEF technique. In a pH gradient, under the influence of an electric field, a protein will move to the position in the gradient where its net charge is zero. A protein with a positive net charge will migrate toward the cathode, becoming progressively less positively charged as it moves through the pH gradient until it reaches its pI. A protein with a negative net charge will migrate toward the anode, becoming less negatively charged until it also reaches zero net charge. If a protein should diffuse away from its pI, it immediately gains charge and migrates back. This is the focusing effect of IEF, which concentrates proteins at their pIs and allows proteins to be separated on the basis of very small charge differences.

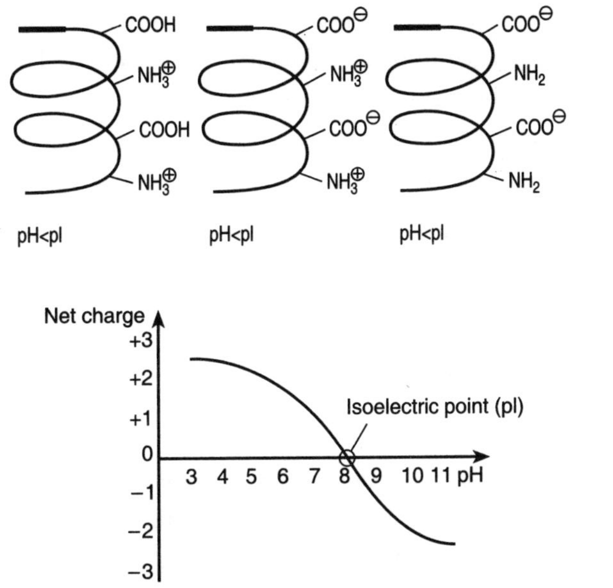

Fig. 8.25: Plot of the net charge of a protein versus the pH of its environment. The point of intersection of the curve at the x-axis represents the isoelectric point of the protein.

The resolution is determined by the slope of the pH gradient and the electric field strength. IEF is, therefore, performed at high voltages (typically in excess of 1 000 V). When the proteins have reached their final positions in the pH gradient, there is very little ionic movement in the system, resulting in a very low final current (typically below 1 mA). IEF of a given sample in a given electrophoresis system is generally performed for a constant number of volt-hours.

The original method for first-dimension IEF depended on carrier ampholyte-generated pH gradients in polyacrylamide gel rods in tubes. Carrier ampholytes are small, soluble, amphoteric molecules with a high buffering capacity near their pI. Commercial carrier ampholyte mixtures are comprised of hundreds of individual polymeric species with pIs spanning a specific pH range. When a voltage is applied across a carrier ampholyte mixture, the carrier ampholytes with the highest pI (and the most negative charge) move toward the anode and the carrier ampholytes with the lowest pI (and the most positive charge) move toward the cathode. The other carrier ampholytes align themselves between the extremes, according to their pIs, and buffer their environment to the corresponding pHs. The result is a continuous pH gradient. Although this basic method has been used in hundreds of 2-D electrophoresis studies, it has several limitations that have prevented its more widespread application:

- Carrier ampholytes are mixed polymers that are not well characterized and suffer from batch-to-batch manufacturing variations. These variations reduce the reproducibility of the first-dimension separation.
- Carrier ampholyte pH gradients are unstable and have a tendency to drift, usually toward the cathode, over time. Gradient drift adversely affects reproducibility by introducing a time variable. Gradient drift also causes a flattening of the pH gradient at each end, particularly above pH 9, rendering the 2-D technique less useful at pH extremes.
- The soft polyacrylamide tube gels have low mechanical stability. The gel rods may stretch or break, affecting reproducibility. Results are often dependent on the skill of the operator.

As a result of the limitations and problems with carrier ampholyte pH gradients, immobilized pH gradients were developed .An immobilized pH gradient (IPG) is created by covalently incorporating a gradient of acidic and basic buffering groups into a polyacrylamide gel at the time it is cast. The buffers are a set of well-characterized molecules, each with a single acidic or basic buffering group linked to an acrylamide monomer. The general structure of immobilized reagents is:

$$\mathrm{CH_2{=}CH{-}\underset{\underset{\displaystyle O}{\|}}{C}{-}NH{-}R}$$

R = weakly acidic or basic buffering group.

Immobilized pH gradients are formed using two solutions, one containing a relatively acidic mixture of acrylamide buffers and the other containing a relatively basic mixture. The concentrations of the various buffers in the two solutions define the range and shape of the pH gradient produced. Both solutions contain acrylamide monomers and catalysts. During polymerization, the acrylamide portion of the buffers copolymerizes with the acrylamide and bisacrylamide monomers to form a polyacrylamide gel. Fig. 8.26 is a graphic representation of the polyacrylamide matrix with attached buffering groups.

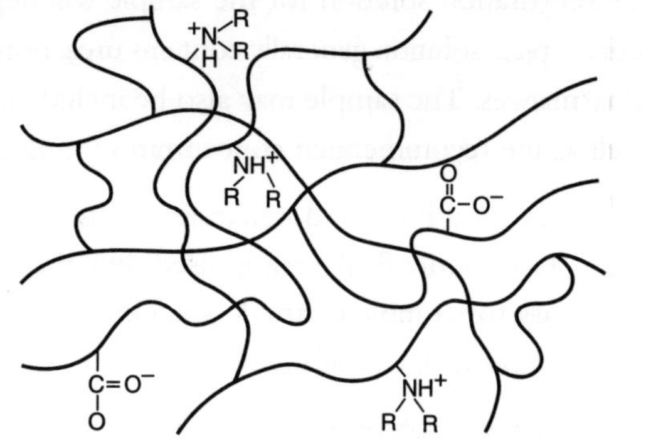

Fig. 8.26: Immobilized pH gradient polyacrylamide gel matrix showing attached buffering groups.

IPG strips must be rehydrated prior to IEF. The IPG strips are rehydrated in a solution containing the necessary additives and, optionally, the sample proteins. IEF is performed by gradually increasing the voltage across the IPG strips to at least 3500 V and maintaining this voltage for at least several thousand volt-hours. After IEF, the IPG strips are equilibrated in equilibration solution and applied onto flatbed or vertical SDS-polyacrylamide gels. When IPG strips are used for the first-dimension separation, the resultant 2-D maps are superior in terms of resolution and reproducibility. IPG strips are a marked improvement over tube gels with carrier ampholyte-generated pH gradients.

Ready-made IPG strips are commercially available. It is always advisable to choose shorter strips for fast screening or when the most abundant proteins are of interest and to use longer strips for maximal resolution and loading capacity.

Sample can be applied either by including it in the rehydration solution (rehydration loading) or by applying it directly to the rehydrated IPG strip via sample cups, sample wells, or paper bridge. Usually rehydration loading is preferable. Advantages to this mode of application include the following:

- Rehydration loading allows larger quantities of protein to be loaded and separated.
- Rehydration loading allows more dilute samples to be loaded.
- Because there is no discrete application point, this method eliminates the formation of precipitates at the application point that often occurs when loading with sample cups.
- The rehydration loading method is technically simpler, avoiding problems of leakage that can occur when using sample cups.
- There are, however, cases when one might prefer to load the sample following rehydration, immediately prior to IEF, e.g., if proteolysis or other protein modifications are a concern, overnight rehydration with sample may not be desired.

The IPG strips are rehydrated in the especially designed rehydration tray or cassette. Rehydration solution, which may or may not include the sample, is applied to the reservoir slots of the rehydration tray and then the IPG strips are soaked individually. The choice of the most appropriate rehydration solution for the sample will depend on its specific protein solubility requirements, but a typical solution generally contains urea, non-ionic or zwitterionic detergent, dithiothreitol (DTT), Pharmalytes. The sample may also be included. The role of each component is described below, as well as the recommended concentration range.

Urea solubilizes and denatures proteins, unfolding them to expose internal ionizable amino acids. Commonly 8 M urea is used, but the concentration can be increased to 9 M or 9.8 M, if necessary for complete sample solubilization. Thiourea, in addition to urea, can be used to further improve protein solubilization.

Detergent solubilizes hydrophobic proteins and minimizes protein aggregation. The detergent must have zero net charge—use only non-ionic and zwitterionic detergents. CHAPS, Triton X-100, or NP-40 in the range of 0.5 to 4% are most commonly used.

Reductant cleaves disulphide bonds to allow proteins to unfold completely. DTT or DTE, (20 to 100 mM) are commonly used. 2-Mercaptoethanol is not recommended, because higher concentrations are required, and impurities may result in artifacts. Tributyl phosphine (TBP) is not recommended as reductant for IEF due to its low solubility and poor stability in rehydration solution. Reductants should be added directly before use.

Pharmalyte (carrier ampholyte mixtures) improve separations, particularly with high sample loads. Carrier ampholyte mixtures enhance protein solubility and produce more uniform conductivity across the pH gradient without disturbing IEF or affecting the shape of the gradient.

Tracking dye (bromophenol blue) allows IEF progress to be monitored at the beginning of the protocol. If the tracking dye does not migrate toward the anode, no current is flowing. As isoelectric focusing proceeds, the bromophenol blue tracking dye migrates toward the anode. Note that the dye front leaves the IPG strip well before focusing is complete, so clearing of the dye is no indication that the sample is focused. If the dye does not migrate, no current is flowing.

Sample application by cup loading

If the sample was not applied by means of the rehydration solution, it can be applied using the sample cups, immediately prior to isoelectric focusing. When sample cups are used, the sample load limits are lower and more specific. Higher sample volumes and protein amounts can be applied with paper bridges, which are placed between the anodic or cathodic end of the IPG strip and the electrode strip. A large sample volume requires a large paper pad applied at the other side to absorb excess water.

In a typical IEF protocol, voltage is gradually increased to the final desired focusing voltage, which is held for up to several hours. With cup loading, a low initial voltage minimizes sample aggregation and a low initial voltage generally allows the parallel separation of samples with differing salt concentrations. The *main factors* determining the required voltage-hours (Vh) are the length of the IPG strips and the pH gradient used. Sample composition, rehydration solution composition, and sample application mode influence the required voltage-hours.

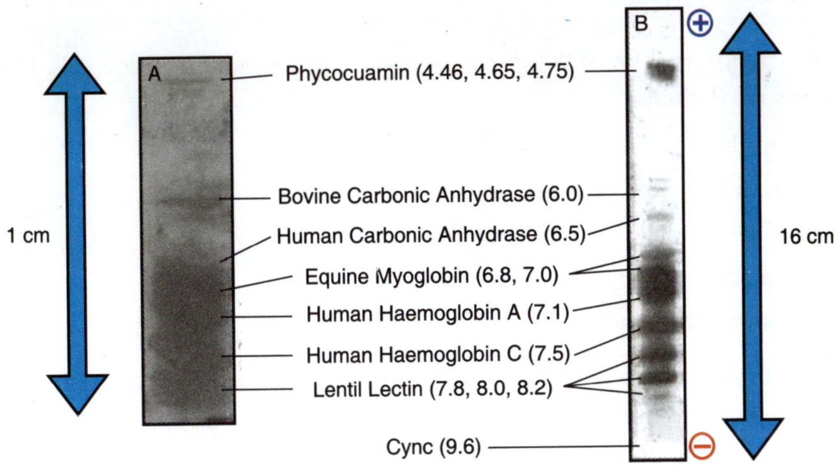

Fig. 8.27: Scanned pictures of IEF electrophoresis on (A) 1 cm^2 and (B) 16 cm^2 gels. Only one channel of the whole gel is shown.

Focusing for substantially longer than recommended will cause horizontal streaking and loss of proteins. This phenomenon is called "over-focusing". Therefore, focusing time should be reduced to the minimum necessary.

8.11.3. Second-dimension SDS-PAGE

After IEF, the second-dimension separation can be performed on various flatbed or vertical systems. SDS-PAGE consists of four steps: (1) preparing the second-dimension gel, (2) equilibrating the IPG strip(s) in SDS buffer, (3) placing the equilibrated IPG strip on the SDS gel, and (4) electrophoresis.

IPG strip equilibration

The equilibration step saturates the IPG strip with the SDS buffer system required for the second-dimension separation. The equilibration solution contains buffer, urea, glycerol, reductant, SDS, and dye. An additional equilibration step replaces the reductant with iodoacetamide. Equilibration is always performed immediately prior to the second-dimension run, never prior to storage of the IPG strips at –40 °C or lower.

Once equilibrated, place the IPG strips using forceps, gel-side down horizontally on vertical SDS-PAGE gel for electrophoresis.Electrophoresis is performed at constant current in two steps. During the initial migration and stacking period, the current is approximately half of the value required for the separation. Stop electrophoresis when the dye front is approximately 1 mm from the bottom of the gel. After electrophoresis, remove gels from their gel cassettes in preparation for staining or blotting.

8.11.4 Visualization of results

Most detection methods used for SDS gels can be applied to second-dimension gels. However, the following features are desired:

- High sensitivy
- Wide linear range for quantification
- Compatibility with mass spectrometry
- Low toxicity and environmentally safe
- Environmentally friendly.

Because none of the existing techniques can meet all these requirements, a 2-D electrophoresis laboratory may need to have more than one of the following methods in its repertoire:

Autoradiography and fluorography are the most sensitive detection methods (down to 200 fg protein). To employ these techniques, the sample must consist of protein radiolabelled in vivo using

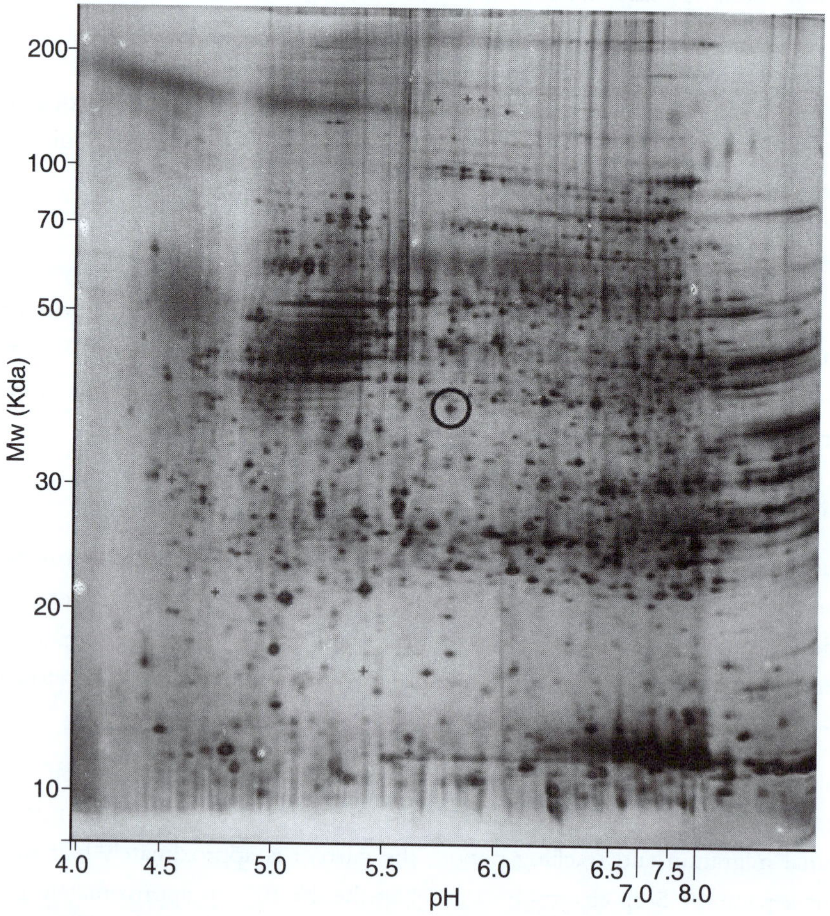

Fig. 8.28: A 2D gel (*Courtsey:* Dr D Dash; Deptt. of Biochemistry, IMS, BHU)

either ^{35}S, ^{14}C, ^{3}H or, in the case of phosphoproteins, ^{32}P. For autoradiographic detection, the gel is simply dried and exposed to X-ray film or—for quicker results and superior dynamic range of quantification—to a storage phosphor screen. Fluorography is a technique that provides extra sensitivity by impregnating the gel in a scintillant such as PPO (2,4-diphenyloxazole) prior to drying.

Silver staining is the most sensitive non-radioactive method (below 1 ng). Silver staining is a complex, multi-step process utilizing numerous reagents for which quality is critical. It is, therefore, often advantageous to purchase these reagents in the form of a dedicated kit, in which the reagents are quality assured specifically for the silver-staining application. By omitting glutardialdehyde from the sensitizer and formaldehyde from the silver nitrate solution the method becomes compatible with mass spectrometry analysis, however, at the expense of sensitivity.

Coomassie staining, although 50- to 100-fold less sensitive than silver staining, is a relatively simple method and more quantitative than silver staining. Coomassie blue is preferable when relative amounts of protein are to be determined by densitometry. Colloidal staining methods are recommended, because they show the highest sensitivity, down to 100 ng/protein spot.

Negative Zinc—Imidazole staining has a detection limit of approx. 15 ng protein/spot and is well compatible with mass spectrometry, but it is a poor quantification technique.

Fluorescent labelling and fluorescent staining with dyes have a sensitivity in-between colloidal Coomassie and Silver Staining. These techniques require fluorescence scanners, but they are compatible with mass spectrometry and show a wide dynamic range for quantification.

Apart from staining, Second-dimension gels can be blotted onto a nitrocellulose or PVDF membrane for immunochemical detection of specific proteins or chemical microsequencing.

8.11.5. Preserving the gels

The gels are optimally stored in sheet protectors after soaking them in 10% v/v glycerol for 30 min. Unbacked gels are shrunk back to their original sizes by soaking them in 30% (v/v) methanol or ethanol/4% glycerol until they match their original sizes. For autoradiography the gels are dried onto strong filter paper with a vacuum drier or in-between two sheets of wet cellophane sealed at ends.

8.11.6. Further analysis of protein spots

Picking the spots

Robotic systems are available that automatically picks selected protein spots from stained or destained gels using a pick list from the image analysis, and transfers them into microplates for further analysis.

Digestion of the proteins

The gel plugs are automatically digested in the computer controlled Digester; the supernatant peptides are mixed with MALDI matrix material and spotted onto MALDI slides using robotic spotter.

MALDI-ToF mass spectrometry

In the MALDI-ToF mass spectrometer, a laser beam is fired into the dried peptide-matrix spots for ionization of the peptides. After accurate determination of the peptide masses, databases are searched for identification of the original proteins.

8.12 MICROCHIP ELECTROPHORESIS

The further miniaturization of electrophoretic systems has led to the development of microchip electrophoresis, which has many advantages over conventional electrophoresis methods, allowing very high speed analyses at very low sample sizes. For example, microchip analysis can often be completed in tens of seconds whereas capillary electrophoresis (CE) can take 20 min and conventional gel electrophoresis at least 2 h. Using new detection systems, such as laser-induced fluorescence, picomole to attomole (10^{-18} moles) sensitivity can be achieved, which is at least two orders of magnitude greater than for conventional CE. Detection systems for molecules that do not fluoresce include electrochemical detectors, pulsed amperometric detection (PAD), and the sinusoidal voltometry. All these detection techniques offer high sensitivity and are ideally suited to micromachining and microfabrication technologies. Finally, the applied voltage required is only a few volts, and it eliminates the need for the high voltage used by CE.

The manufacturing process that produces microchips is called microfabrication. The process etches precise and reproducible capillary-like channels (typically, 50 μm wide and 10 μm deep; slightly smaller than a strand of human hair) on the surface of sheets of quartz, glass or plastic. A second sheet is then fused on top of the first sheet, turning the etched channels into closed microfluidic channels. The end of each channel connects to a reservoir through which fluids are introduced/ removed. Typically, the size of chips can be small as 2 cm^2. Basically the microchip provides an electrophoretic system similar to CE but with more flexibility.

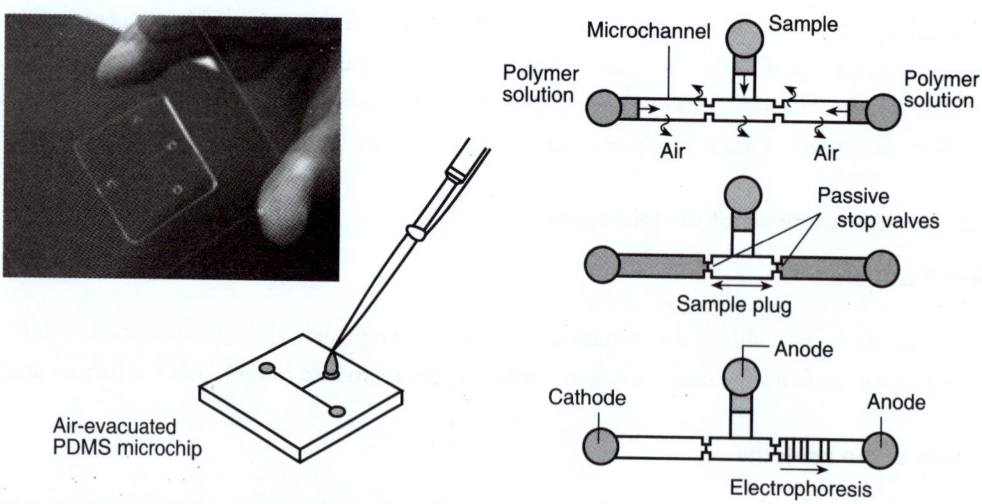

Fig. 8.29: Schematic diagram of microchip electrophoresis

Current developments of this technology are based on integrating functions other than just separation into the chip. For example, sample extraction, preconcentration of samples prior to separation, PCR amplification of DNA samples using infrared-mediated thermocycling for rapid on-chip amplification, and the extraction of separated molecules using microchamber-bound solid phases are all examples of where further functions have been built into a microchip electrophoresis system. An interface has also been developed for microchip electrophoresis-mass spectrometry (MCE-MS) where drugs have been separated by MCE and then identified by MS.

8.13 TIPS TO REMEMBER

Capillary Electrophoresis

This electrophoresis occurs in quartz capillary tubes with diameters of 50-100 μm, and lengths of 20-60 cm.

Isoelectric Focusing

Since proteins will typically have both basic and acidic groups, there will be a pH at which the charges balance and the net charge is zero. This is called the pI, or the isoelectric point. Therefore, in a gradient of pH were to exist between the anode and the cathode, a protein would then migrate until it reached the pH at which it is uncharged, and then would migrate no further. This is the principle of isoelectric focusing.

Gel Electrophoresis

native: mobility = (voltage)(charge)/(mass)

SDS-PAGE: minimizes contribution of charge

IEF: minimizes contribution of size.

Protein Detection Following Electrophoresis

General Proteins

- organic dyes (e.g., Coomassie blue)
- R-250 (slow)
- G-250 (fast)
- silver stain
- 10-100X more sensitive than CB
- fluorescence
- radioactivity.

Specific Proteins

- enzyme activity
- antibody/immunoblot.

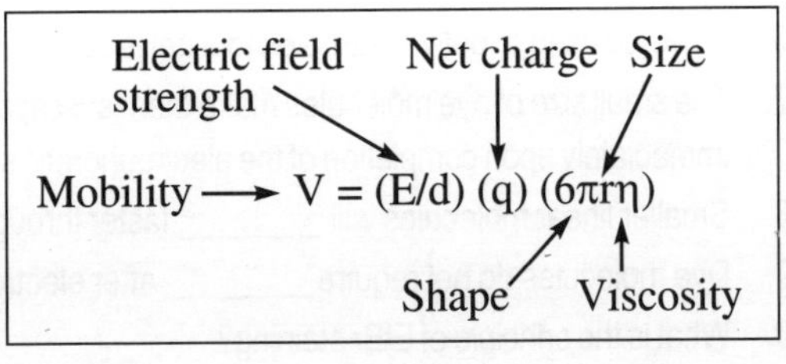

Voltage

- ↑ voltage = ↓ time, but ↑ heat
- resistance ↑ during electrophoresis (E = IR)

8.14 VIVA VOCE

Q.1. What is electrophoresis?

Q.2. What is the principle of electrophoresis?

Q.3. What are the factors which affect the mobility in electrophoresis?

Q.4. What is basis of separation in SDS –PAGE?

Q.5. The electrophoresis process is used to separate ________ by size, shape and charge.

Q.6. Because DNA is net negative in charge, it will migrate towards the ________ electrode.

Q.7. The type of electrophoresis unit commonly used for separating DNA is the ________ electrophoresis apparatus.

Q.8. The gel is made with ________, which is derived from seaweed.

Q.9. The gel consists of microscopic pores that act as a molecular ________.

Q.10. Electrophoresis units contain electrodes made of ________ which have superior electrical conductivity and permanency.

Q.11. By convention, the negative electrode (cathode) is colour-coded ___________.

Q.12. By convention, the positive electrode (anode) is colour-coded _______.

Q.13. The gel contains ______ for conductivity of electricity.

Q.14. The wells in the gel are formed by a template often called a ________.

Q.15. Electrical current is applied to the electrophoresis apparatus using a Direct Current (D.C.) ________ ________ (two words).

Q.16. What is the role of SDS in SDS-PAGE?

Q.17. Native PAGE is used for_______ .

Q.18. What is pulse field electrophoresis?

Q.19. What is the principle of immuno-electrophoresis?

Q.20. What is microchip electrophoresis?

Q.21. How acrylamide polymerises?

Q.22. What is 2-D gel electrophoresis and why is it used?

Q.23. By which ways one can stain and detect the protein resolved by electrophoresis?

Q.24. The percentage 0.8% is one of the many factors affecting mobility of molecules during electrophoresis, and is referred to as the gel ____________.

Q.25. Samples are prepared for electrophoresis by mixing them with components, such as sucrose or _______ that give the sample density.

Q.26. A device used to deliver samples to a gel is called a ___________.

Q.27. The small size of dye molecules make them susceptible to _________, and thus the gel results must be viewed immediately upon completion of the electrophoretic separation.

Q.28. Smaller linear molecules will ________ faster through the gel than larger ones.

Q.29. Dye molecules do not require ________ after electrophoresis, but DNAs require it for visualization.

Q.30. What is the principle of EtBr staining?

CHAPTER 9 Chromatographic Techniques

9.1 INTRODUCTION AND BASICS

Few methods of chemical analysis are truly specific to a particular analyte. It is often found that the analyte of interest must be separated from the myriad of individual compounds that may be present in a sample. Apart from providing the analytical scientist with methods of separation, chromatographic techniques can also provide methods of analysis.

Chromatography involves a sample (or sample extract) being dissolved in a *mobile phase* (which may be a gas, a liquid or a supercritical fluid). The mobile phase is then forced through an immobile, immiscible *stationary phase*. The phases are chosen such that components of the sample have differing solubilities in each phase. A component which is quite soluble in the stationary phase will take longer to travel through it than a component which is not very soluble in the stationary phase but very soluble in the mobile phase. As a result of these differences in mobilities, sample components will become separated from each other as they travel through the stationary phase.

Chromatography may be preparative or analytical. Preparative chromatography seeks to separate the components of a mixture for further use (and is thus a form of purification). Analytical chromatography normally operates with smaller amounts of material and seeks to measure the relative proportions of analytes in a mixture. The two are not mutually exclusive.

Before moving into the details of the topic it would be advisable that we should first get acclimatized with various technical terminologies of use in the topic.

- The analyte is the substance which is to be separated during chromatography.
- Analytical chromatography is used to determine the existence and possibly also the concentration of analyte(s) in a sample.
- A bonded phase is a stationary phase that is covalently bonded to the support particles or to the inside wall of the column tubing.

- A chromatogram is the visual output of the chromatograph. In the case of an optimal separation, different peaks or patterns on the chromatogram correspond to different components of the separated mixture.
- A chromatograph is an equipment that enables a sophisticated separation, e.g., gas chromatographic or liquid chromatographic separation.
- The effluent is the mobile phase leaving the column.
- An immobilized phase is a stationary phase which is immobilized on the support particles, or on the inner wall of the column tubing.
- The mobile phase is the phase which moves in a definite direction. It may be a liquid (LC and CEC), a gas (GC), or a supercritical fluid (supercritical-fluid chromatography, SFC).
- Preparative chromatography is used to nondestructively purify sufficient quantities of a substance for further use, rather than analysis.
- The retention time is the characteristic time it takes for a particular analyte to pass through the system (from the column inlet to the detector) under set conditions.
- The sample is the matter analysed in chromatography. It may consist of a single component or it may be a mixture of components. When the sample is treated in the course of an analysis, the phase or the phases containing the analytes of interest is/are referred to as the sample whereas everything out of interest separated from the sample before or in the course of the analysis is referred to as waste.
- The solute refers to the sample components in partition chromatography.
- The solvent refers to any substance capable of solubilizing other substance, especially the liquid mobile phase in LC.
- The stationary phase is the substance which is fixed in place for the chromatography procedure, e.g., the silica layer in thin layer chromatography.

9.1.1 Basic principles

Distribution of analytes between phases

The distribution of analytes between phases can often be described quite simply. An analyte is in equilibrium between the two phases;

$$\text{A mobile} \rightleftharpoons \text{A stationary}$$

The equilibrium constant, K, is termed the *partition coefficient*; defined as the molar concentration of analyte in the stationary phase divided by the molar concentration of the analyte in the mobile phase.

The time between sample injection and an analyte peak reaching a detector at the end of the column is termed the *retention time* (t_R). Each analyte in a sample will have a different retention time. The time taken for the mobile phase to pass through the column is called t_M.

A term called the *retention factor*, k′, is often used to describe the migration rate of an analyte on a column. You may also find it called the *capacity factor*. The retention factor for analyte A is defined as

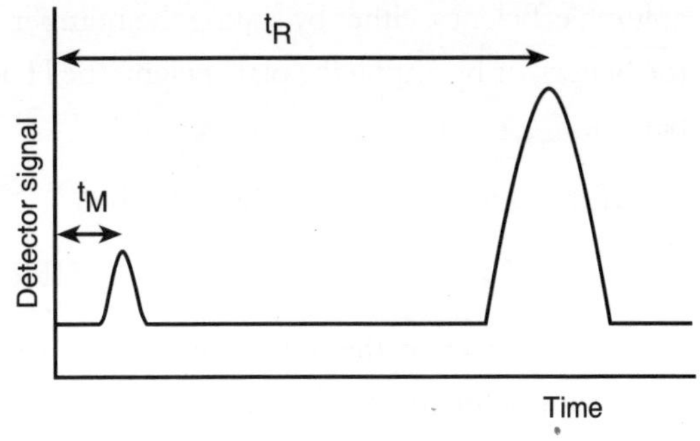

$$k'_A = (t_R - t_M) / t_M.$$

t_R and t_M are easily obtained from a chromatogram. When an analytes retention factor is less than one, elution is so fast that accurate determination of the retention time is very difficult. High retention factors (greater than 20) mean that elution takes a very long time. Ideally, the retention factor for an analyte is between one and five.

We define a quantity called the selectivity factor, α , which describes the separation of two species (A and B) on the column

$$\alpha = k'_B / k'_A.$$

When calculating the selectivity factor, species A elutes faster than species B. The selectivity factor is always greater than one.

Band broadening and column efficiency

To obtain optimal separations, sharp, symmetrical chromatographic peaks must be obtained. This means that band broadening must be limited. It is also beneficial to measure the efficiency of the column.

9.1.1.1 The Theoretical Plate Model of Chromatography

The plate model supposes that the chromatographic column contains a large number of separate layers, called *theoretical plates*. Separate equilibrations of the sample between the stationary and mobile phase occur in these "plates". The analyte moves down the column by transfer of equilibrated mobile phase from one plate to the next.

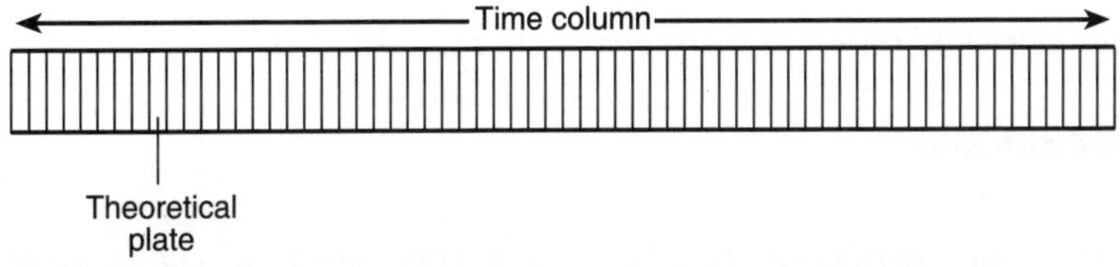

It is important to remember that the plates do not really exist; they are a figment of the imagination that helps us understand the processes at work in the column. They also serve as a way of measuring column efficiency, either by stating the number of theoretical plates in a column, N (the more plates the better), or by stating the plate height; the Height is equivalent to a theoretical plate (the smaller the better).

If the length of the column is L, then

$$\text{HETP} = L/N.$$

The number of theoretical plates that a real column possesses can be found by examining a chromatographic peak after elution.

$$N = \frac{5.55\,t_R^2}{w_{1/2}^2},$$

where $w_{1/2}$ is the peak width at half-height.

As can be seen from this equation, columns behave as if they have different numbers of plates for different solutes in a mixture.

9.1.1.2 The Rate Theory of Chromatography

A more realistic description of the processes at work inside a column takes account of the time taken for the solute to equilibrate between the stationary and mobile phases (unlike the plate model, which assumes that equilibration is infinitely fast). The resulting band shape of a chromatographic peak is, therefore, affected by the rate of elution. It is also affected by the different paths available to solute molecules as they travel between particles of stationary phase. If we consider the various mechanisms which contribute to band broadening, we arrive at the Van Deemter equation for plate height.

$$\text{HETP} = A + B/u + Cu,$$

where u is the average velocity of the mobile phase. A, B, and C are factors which contribute to band broadening.

(A) Eddy diffusion

The mobile phase moves through the column which is packed with stationary phase. Solute molecules will take different paths through the stationary phase at random. This will cause broadening of the solute band, because different paths are of different lengths.

(B) Longitudinal diffusion

The concentration of analyte is less at the edges of the band than at the centre. Analyte diffuses out from the centre to the edges. This causes band broadening. If the velocity of the mobile phase is high, then the analyte spends less time on the column, which decreases the effects of longitudinal diffusion.

(C) Resistance to mass transfer

The analyte takes a certain amount of time to equilibrate between the stationary and mobile phases. If the velocity of the mobile phase is high, and the analyte has a strong affinity for the stationary phase, then the analyte in the mobile phase will move ahead of the analyte in the stationary phase. The band of analyte is broadened. The higher the velocity of mobile phase, the worse the broadening becomes.

Van Deemter plots

A plot of plate height vs. average linear velocity of mobile phase.

Such plots are of considerable use in determining the optimum mobile phase flow rate.

A typical Van Deemter plot

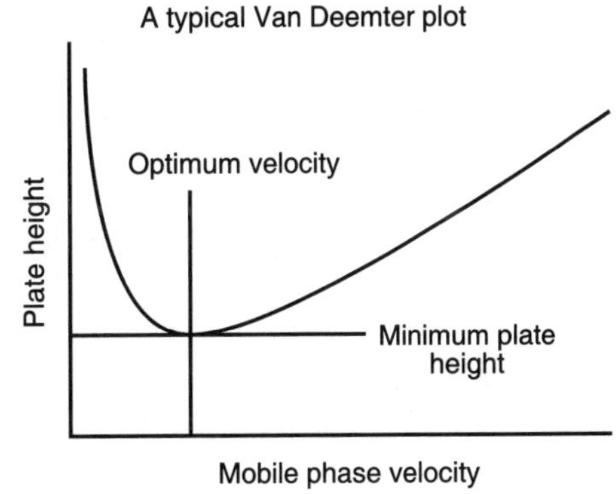

Resolution

Although the selectivity factor, α, describes the separation of band centres, it does not take into account peak widths. Another measure of how well species have been separated is provided by measurement of the *resolution*. The resolution of two species, A and B, is defined as

$$R = \frac{2\left[(t_R)_B - (t_R)_A\right]}{W_A + W_B}.$$

Baseline resolution is achieved when R = 1.5.

It is useful to relate the resolution to the number of plates in the column, the selectivity factor and the retention factors of the two solutes,

$$R = \frac{\sqrt{N}}{4}\left(\frac{\alpha - 1}{\alpha}\right)\left(\frac{1 + k'_B}{k'_B}\right).$$

To obtain high resolution, the three terms must be maximised. An increase in N, the number of theoretical plates, by lengthening the column leads to an increase in retention time and increased band broadening—which may not be desirable. Instead, to increase the number of plates, the height equivalent to a theoretical plate can be reduced by reducing the size of the stationary phase particles.

It is often found that by controlling the capacity factor, k′, separations can be greatly improved. This can be achieved by changing the temperature (in Gas Chromatography) or the composition of the mobile phase (in Liquid Chromatography).

The selectivity factor, α, can also be manipulated to improve separations. When α is close to unity, optimising k′ and increasing N is not sufficient to give good separation in a reasonable time. In these cases, k′ is optimised first, and then α is increased by one of the following procedures:

1. Changing mobile phase composition
2. Changing column temperature
3. Changing composition of stationary phase
4. Using special chemical effects (such as incorporating a species which complexes with one of the solutes into the stationary phase).

9.2 TYPES OF CHROMATOGRAPHY

There are different kinds of chromatographic techniques and these are classified according to the shape of bed, physical state of mobile phase, separation mechanisms. Apart form these there are certain modified forms of these chromatographic techniques involving different mechanisms and are hence categorized as modified or specialized chromatographic techniques. The detailed classification of these techniques is as follows:

1. Techniques by chromatographic bed shape
   - I. Column chromatography
   - II. Planer chromatography
     - (a) Paper chromatography
     - (b) Thin layer chromatography.
2. Techniques by physical state of mobile phase
   - I. Gas chromatography
   - II. Liquid chromatography
     - (a) High performance liquid chromatography (HPLC)
     - (b) Fast protein liquid chromatography
     - (c) Aqueous normal phase chromatography
   - III. Supercritical fluid chromatography.
3. Techniques by separation mechanism
   - I. Ion exchange chromatography
   - II. Size exclusion chromatography
   - III. Affinity chromatography.
4. Modified or special chromatographic techniques
   - I. Reversed phase chromatography
   - II. Two dimensional chromatography
   - III. Pyrolysis gas chromatography
   - IV. Counter current chromatography.

9.3 COLUMN CHROMATOGRAPHY

It is the preparative application of chromatography. It is used to obtain pure chemical compounds from a mixture of compounds on a scale from micrograms up to kilograms using large industrial columns.

The classical preparative chromatography column is a glass tube with a diameter from 5 to 50 mm and a height of 50 cm to 1 m with a tap at the bottom. Slurry is prepared of the eluent with the stationary phase powder and then carefully poured into the column. Care must be taken to avoid air bubbles. A solution of the organic material is pipetted on top of the stationary phase. This layer is usually topped with a small layer of sand or with cotton or glass wool to protect the shape of the organic layer from the velocity of newly added eluent. Eluent is slowly passed through the column to advance the organic material. Often a spherical eluent reservoir or an eluent-filled and stoppered separating funnel is put on top of the column.

Fig. 9.1: A chemist using column chromatographic apparatus (*Courtesy:* Deptt. of Biochemical Engineering, IT, BHU)

The individual components are retained by the stationary phase differently and separate from each other while they are running at different speeds through the column with the eluent. At the end of the column they elute one at a time. During the entire chromatography process the eluent is collected in a series of fractions. The composition of the eluent flow can be monitored and each fraction is analyzed for dissolved compounds, e.g., by analytical chromatography, UV absorption, or fluorescence. Coloured compounds (or fluorescent compounds with the aid of an UV lamp) can be seen through the glass wall as moving bands.

The *stationary phase* or *adsorbent* in column chromatography is a solid. The most common stationary phase for column chromatography is $-C_{18}H_{37}$, followed by alumina. Cellulose powder has often been used in the past. Also possible are ion exchange chromatography, reversed-phase chromatography (RP), affinity chromatography or expanded bed adsorption (EBA). The stationary phases are usually finely ground powders or gels and/or are microporous for an increased surface, though in EBA a fluidized bed is used.

The *mobile phase* or *eluent* is either a pure solvent or a mixture of different solvents. It is chosen so that the retention factor value of the compound of interest is roughly around 0.75 in order to minimize the time and the amount of eluent to run the chromatography. The eluent has also been chosen so that the different compounds can be separated effectively. The eluent is optimized in small scale pretests, often using thin layer chromatography (TLC) with the same stationary phase.

A faster flow rate of the eluent minimizes the time required to run a column and thereby minimizes diffusion, resulting in a better separation. A simple laboratory column runs by gravity flow. The flow rate of such a column can be increased by extending the fresh eluent filled column above the top of the stationary phase or decreased by the tap controls. Better flow rates can be achieved by using a pump or by using compressed gas (e.g., air, nitrogen, or argon) to push the solvent through the column (flash column chromatography).

Automated flash chromatography systems attempt to minimize human involvement in the purification process. Automated systems may include components normally found on HPLC systems (gradient pump, sample injection apparatus, UV detector) and a fraction collector to collect the eluent.The software controlling an automated system will coordinate the components and help the user to find the resulting purified material within the fraction collector. The software will also store results from the process for archival or later recall purposes.

9.4 PAPER CHROMATOGRAPHY

It is an analytical technique for separating and identifying mixtures that are or can be coloured, especially pigments. This can also be used in secondary or primary schools in ink experiments. This method has been largely replaced by thin layer chromatography, however it is still a powerful teaching tool. Two-way paper chromatography, also called two-dimensional chromatography, involves using two solvents and rotating the paper 90° in between. This is useful for separating complex mixtures of similar compounds, for example, amino acids.

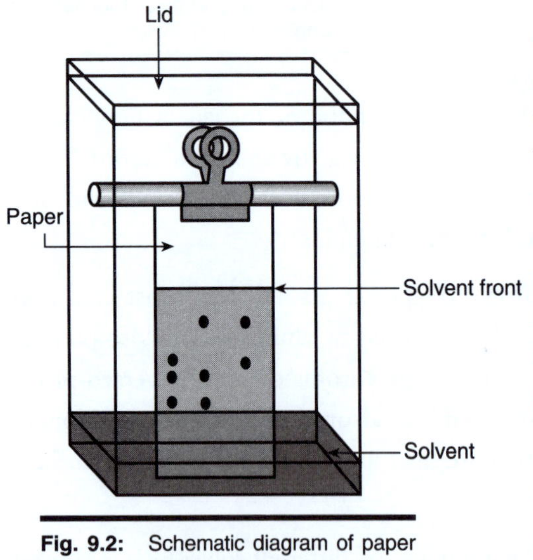

Fig. 9.2: Schematic diagram of paper chromatography

A small, ideally concentrated spot of solution that contains the sample is applied to a strip of chromatography paper about 1 cm from the base, usually using a capillary tube for maximum precision. This sample is absorbed onto the paper and may form interactions with it. Any substance that reacts or bonds with the paper cannot be measured using this technique. The paper is then dipped into a suitable solvent, such as ethanol or water, taking care that the spot is above the surface of the solvent, and placed in a sealed container.

The solvent moves up the paper by capillary action, which occurs as a result of the attraction of the solvent molecules to the paper and to one another. As the solvent rises through the paper it meets and dissolves the sample mixture, which will then travel up the paper with the solvent. Different compounds in the sample mixture travel at different rates due to differences in solubility in the solvent, and due to differences in their attraction to the fibers in the paper. Paper chromatography takes anywhere from several minutes to several hours.

In some cases, paper chromatography does not separate pigments completely; this occurs when two substances appear to have the same values in a particular solvent. In these cases, two-way chromatography is used to separate the multiple-pigment spots. The chromatogram is turned by ninety degrees, and placed in a different solvent in the same way as before; some spots separate in the presence of more than one pigment. As before, the value is calculated, and the two pigments are identified. The R_f value (retention factor) is the distance travelled by a particular component from the origin (where the sample was originally spotted) as a ratio to the distance travelled by the solvent front from the origin. R_f values for each substance will be unique, and can be used to identify components. A particular component will have the same R_f value if it is separated under identical conditions.

After development, the spots corresponding to different compounds may be located by their colour, ultraviolet light, ninhydrin (Triketohydrindane hydrate) or by treatment with iodine vapours. The final *chromatogram* can be compared with other known mixture chromatograms to identify sample mixture using the R_n value.

As in most other forms of chromatography, paper chromatography uses R_n values to help identify compounds. R_f values are calculated by dividing the distance the pigment travels up the paper by the distance the solvent travels (the solvent front). Because R_f values are standard for a given compound, known R_n values can be used to aid in the identification of an unknown substance in an experiment.

9.5 THIN LAYER CHROMATOGRAPHY

Thin-layer chromatography (TLC) is a chromatographic technique that is useful for separating organic compounds. It involves a *stationary phase* consisting of a thin layer of adsorbent material, usually silica gel, aluminium oxide, or cellulose immobilised onto a flat, inert carrier sheet. A *liquid phase* consisting of the solution to be separated dissolved in an appropriate solvent is drawn through the plate via capillary action, separating the experimental solution. When the solvent front reaches the other edge of the stationary phase, the plate is removed from the solvent reservoir. The separated spots are visualized with ultraviolet light or by placing the plate in iodine vapour. The different components in the mixture move up the plate at different rates due to differences in their partioning behavior between the mobile liquid phase and the stationary phase. It can be used to determine the pigments a plant contains, to detect pesticides or insecticides in food, in forensics to analyze the dye composition of fibers, or to identify compounds present in a given substance, among other uses. It is a quick, generic method for organic reaction monitoring.

TLC plates are made by mixing the adsorbent, such as silica gel, with a small amount of inert binder like calcium sulphate (gypsum) and water. This mixture is spread as thick slurry on an unreactive carrier sheet, usually glass, thick aluminum foil, or plastic, and the resultant plate is dried and *activated* by heating in an oven for thirty minutes at 110°C. The thickness of the adsorbent layer is typically around 0.1-0.25 mm for analytical purposes and around 1-2 mm for preparative TLC. Every type of chromatography contains a mobile phase and a stationary phase.

The process is similar to paper chromatography with the advantage of faster runs, better separations, and the choice between different stationary phases. Because of its simplicity and speed TLC is often used for monitoring chemical reactions and for the qualitative analysis of reaction products.

A small spot of solution containing the sample is applied to a plate, about one centimetre from the base. The plate is then dipped into a suitable solvent, such as ethanol or water, and placed in a sealed container. The solvent moves up the plate by capillary action and meets the sample mixture, which is dissolved and is carried up the plate by the solvent. Different compounds in the sample mixture travel at different rates due to differences in solubility in the solvent, and due to differences in their attraction to the stationary phase. Results also vary depending on the solvent used. For example, if the solvent were a 90:10 mixture of hexane to ethyl acetate, then the solvent would be mostly non-polar. This means that when analyzing the TLC, the non-polar parts will have moved further up the plate. The polar compounds, in contrast, will not have moved as much. The reverse is true when using a solvent that is more polar than non-polar (10:90 hexane to ethyl acetate). With these solvents, the polar compounds will move higher up the plate, while the non-polar compounds will not move as much.

Fig. 9.3: Spotting of TLC plate with a sample and then running the TLC plate in the solvent (*Courtsey:* Dr D Dash, Deptt. of Biochemistry, IMS, BHU)

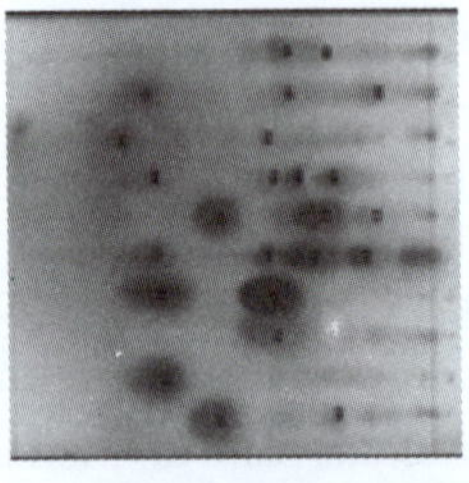

Fig. 9.4: Chromatogram of 10 essential oils coloured with vanillin reagent

The appropriate solvent in context of thin layer chromatography will be one which differs from the stationary phase material in polarity. If polar solvent is used to dissolve the sample and spot is applied over polar stationary phase of TLC, the sample spot will grow radially due to capillary action, which is not advisable as one spot may mix with the other. Hence, to restrict the radial growth of sample-spot, the solvent used for dissolving samples in order to apply them on plates should be as non-polar or semi-polar as possible when the stationary phase is polar, and vice versa.

As the chemicals being separated may be colourless, several methods exist to visualize the spots:

1. Often a small amount of a fluorescent compound, usually Manganese-activated Zinc Silicate, is added to the adsorbent that allows the visualization of spots under a blacklight(UV_{254}). The adsorbent layer will thus fluoresce light green by itself, but spots of analyte quench this fluorescence.
2. Iodine vapours are a general unspecific colour reagent
3. Specific colour reagents exist into which the TLC plate is dipped or which are sprayed onto the plate

Once visible, the R_f value of each spot can be determined by dividing the distance travelled by the product by the total distance travelled by the solvent (the solvent front). These values depend on the solvent used, and the type of TLC plate, and are not physical constants.

9.5.1 Applications

In organic chemistry, reactions are qualitatively monitored with TLC. Spots sampled with a capillary tube are placed on the plate: a spot of starting material, a spot from the reaction mixture, and a "co-spot" with both. A small (3 by 7 cm) TLC plate takes a couple of minutes to run. The analysis is qualitative, and it will show if starting material has disappeared, product has appeared, and how many products are generated. Unfortunately, TLCs from low-temperature reactions may give misleading results, because the sample is warmed to room temperature in the capillary. One such reaction is DIBALH reduction of ester to aldehyde.

9.6 GAS-LIQUID CHROMATOGRAPHY (GLC) OR SIMPLY GAS CHROMATOGRAPHY (GC)

It is a type of chromatography in which the mobile phase is a carrier gas, usually an inert gas such as helium or an unreactive gas such as nitrogen, and the stationary phase is a microscopic layer of liquid or polymer on an inert solid support, inside glass or metal tubing, called a column. The instrument used to perform gas chromatographic separations is called a *gas chromatograph* (also: *aerograph*, *gas separator*).

A gas chromatograph is a chemical analysis instrument for separating chemicals in a complex sample. A gas chromatograph uses a flow-through narrow tube known as the *column*, through which different chemical constituents of a sample pass in a gas stream (carrier gas,

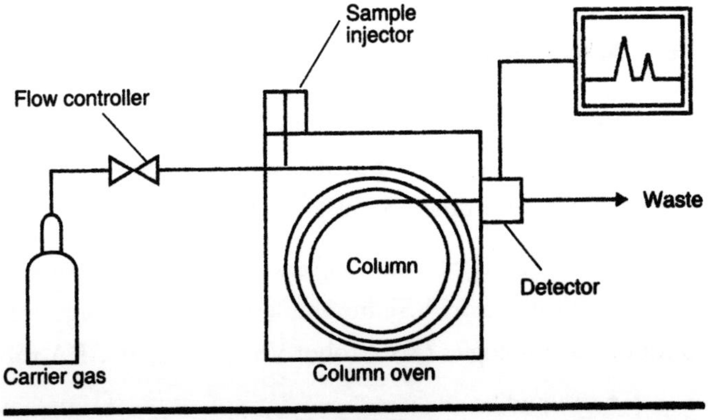

Fig. 9.5: Diagram of a gas chromatograph

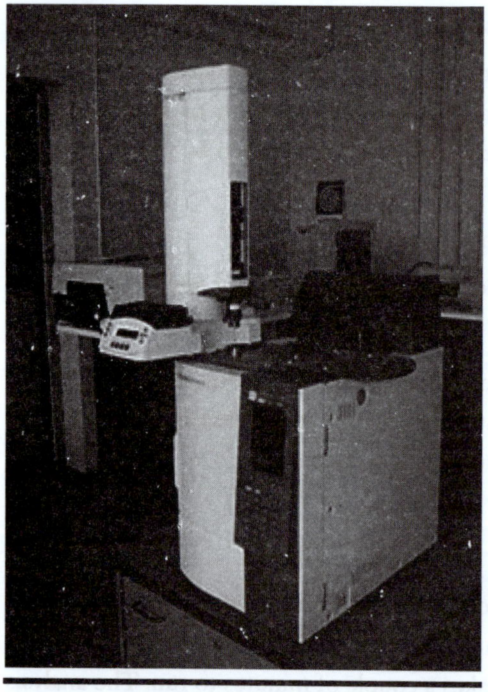

Fig. 9.6: A gas chromatograph with an headspace sampler (*Courtesy:* Deptt. of Pharmaceutics, IT, BHU)

mobile phase) at different rates depending on their various chemical and physical properties and their interaction with a specific column filling, called the *stationary phase.* As the chemicals exit the end of the column, they are detected and identified electronically. The function of the stationary phase in the column is to separate different components, causing each one to exit the column at a different time (*retention time*). Other parameters that can be used to alter the order or time of retention are the carrier gas flow rate, and the temperature.

In a GC analysis, a known volume of gaseous or liquid analyte is injected into the "entrance" (head) of the column, usually using a micro syringe (or, solid phase microextraction fibers, or a gas source switching system). As the carrier gas sweeps the analyte molecules through the column, this motion is inhibited by the adsorption of the analyte molecules either onto the column walls or onto packing materials in the column. The rate at which the molecules progress along the column depends on the strength of adsorption, which in turn depends on the type of molecule and on the stationary phase materials. Since each type of molecule has a different rate of progression, the various components of the analyte mixture are separated as they progress along the column and reach the end of the column at different times (retention time). A detector is used to monitor the outlet stream from the column; thus, the time at which each component reaches the outlet and the amount of that component can be determined. Generally, substances are identified (qualitatively) by the order in which they emerge (elute) from the column and by the retention time of the analyte in the column

9.6.1 Autosamplers

The autosampler provides the means to introduce automatically a sample into the inlets. Manual insertion of the sample is possible but very rare nowadays. Automatic insertion provides better reproducibility and time-optimization.

Different kinds of autosamplers exist. Autosamplers can be classified in relation to sample capacity (auto-injectors VS autosamplers, where auto-injectors can work a small number of samples), to robotic technologies (XYZ robot VS rotating/SCARA-robot - the most common), or to analysis:

- Liquid
- Static head-space by syringe technology

- Dynamic head-space by transfer-line technology
- Solid phase micro extraction (SPME).

9.6.2 Inlets

The column inlet (or *injector*) provides the means to introduce a sample into a continuous flow of carrier gas. The inlet is a piece of hardware attached to the *column head.*

Common inlet types are:

1. S/SL (Split/Splitless) injector: A sample is introduced into a heated small chamber via a syringe through a *septum*, the heat facilitates volatilization of the sample and sample matrix. The carrier gas then either sweeps the entirety (*splitless* mode) or a portion (*split* mode) of the sample into the column. In split mode, a part of the sample/carrier gas mixture in the injection chamber is exhausted through the *split vent.*
2. On-column inlet: The sample is here introduced in its entirety without heat.
3. PTV injector: Temperature-programmed sample introduction was first described by Vogt in 1979. Originally Vogt developed the technique as a method for the introduction of large sample volumes (up to 250 μl) in capillary GC. Vogt introduced the sample into the liner at a controlled injection rate. The temperature of the liner was chosen slightly below the boiling point of the solvent. The low-boiling solvent was continuously evaporated and vented through the split line. Based on this technique, Poy developed the Programmed Temperature Vaporising Injector PTV. By introducing the sample at a low initial liner temperature many of the disadvantages of the classic hot injection techniques could be circumvented.
4. Gas source inlet or gas switching valve: Gaseous samples in collection bottles are connected to what is most commonly a six-port switching valve. The carrier gas flow is not interrupted while a sample can be expanded into a previously evacuated sample loop. Upon switching, the contents of the sample loop are inserted into the carrier gas stream.
5. P/T (Purge-and-Trap) system: An inert gas is bubbled through an aqueous sample causing insoluble volatile chemicals to be purged from the matrix. The volatiles are 'trapped' on an absorbent column (known as a trap or concentrator) at ambient temperature. The trap is then heated and the volatiles are directed into the carrier gas stream. Samples requiring preconcentration or purification can be introduced via such a system, usually hooked up to the S/SL port.
6. SPME (solid phase microextraction) offers a convenient, low-cost alternative to P/T systems with the versatility of a syringe and simple use of the S/SL port.

9.6.3 Columns

Two types of columns are used in GC:

1. Packed columns are 1.5-10 m in length and have an internal diameter of 2-4 mm. The tubing is usually made of stainless steel or glass and contains a *packing* of finely divided,

inert, solid support material (e.g., diatomaceous earth) that is coated with a liquid or solid stationary phase. The nature of the coating material determines what type of materials will be most strongly adsorbed. Thus numerous columns are available that are designed to separate specific types of compounds.

2. Capillary columns have a very small internal diameter, on the order of a few tenths of millimetres, and lengths between 25-60 metres are common. The inner column walls are coated with the active materials (WCOT columns), some columns are quasi-solid filled with many parallel micropores (PLOT columns). Most capillary columns are made of fused-silica with a polyimide outer coating. These columns are flexible, so a very long column can be wound into a small coil.
3. New developments are sought where stationary phase incompatibilities lead to geometric solutions of parallel columns within one column. Among these new developments are:
   (i) Internally heated *microFAST* columns, where two columns, an internal heating wire and a temperature sensor are combined within a common column sheath (microFAST);
   (ii) *Micropacked columns* (1/16″ OD) are column-in-column packed columns where the outer column space has a packing different from the inner column space, thus providing the separation behaviour of two columns in one. They can be easily fit to inlets and detectors of a capillary column instrument.

The temperature-dependence of molecular adsorption and of the rate of progression along the column necessitates a careful control of the column temperature to within a few tenths of a degree for precise work. Reducing the temperature produces the greatest level of separation, but can result in very long elution times. For some cases, temperature is ramped either continuously or in steps to provide the desired separation. This is referred to as a temperature program. Electronic pressure control can also be used to modify flow rate during the analysis, aiding in faster run times while keeping acceptable levels of separation.

The choice of carrier gas (*mobile phase*) is important, with hydrogen being the most efficient and providing the best separation. However, helium has a larger range of flow rates that are comparable to hydrogen in efficiency, with the added advantage that helium is non-flammable, and works with a greater number of detectors. Therefore, helium is the most common carrier gas used.

9.6.4 Detectors

A number of detectors are used in gas chromatography. The most common are the flame ionization detector (FID) and the thermal conductivity detector (TCD). Both are sensitive to a wide range of components, and both work over a wide range of concentrations. While TCDs are essentially universal and can be used to detect any component other than the carrier gas (as long as their thermal conductivities are different than that of the carrier gas, at detector temperature), FIDs are sensitive primarily to hydrocarbons, and are more sensitive to them than TCD. However, an FID cannot detect water. Both detectors are also quite robust. Since TCD is non-destructive, it can be operated in series before an FID (destructive), thus providing complementary detection of the same eluents.

Other detectors are sensitive only to specific types of substances, or work well only in narrower ranges of concentrations. They include:

- discharge ionization detector (DID)
- electron capture detector (ECD)
- flame photometric detector (FPD)
- Hall electrolytic conductivity detector (ElCD)
- helium ionization detector (HID)
- nitrogen phosphorus detector (NPD)
- mass selective detector (MSD)
- photo-ionization detector (PID)
- pulsed discharge ionization detector (PDD)

Some gas chromatographs are connected to a mass spectrometer which acts as the detector. The combination is known as GC-MS. Some GC-MS are connected to an Nuclear Magnetic Resonance Spectrometer which acts as a back-up detector. This combination is known as GC-MS-NMR. Some GC-MS-NMR are connected to an infrared spectra which acts as a back-up detector. This combination is known as GC-MS-NMR-IR. It must, however, be stressed that this is very rare as most analysis needed can be concluded via purely GC-MS.

9.6.5 Methods

The method is the collection of conditions in which the GC operates for a given analysis. Method development is the process of determining what conditions are adequate and/or ideal for the analysis required.

Conditions which can be varied to accommodate a required analysis include inlet temperature, detector temperature, column temperature and temperature program, carrier gas and carrier gas flow rates, the column's stationary phase, diameter and length, inlet type and flow rates, sample size and injection technique. Depending on the detector(s) (see below) installed on the GC, there may be a number of detector conditions that can also be varied. Some GCs also include valves which can change the route of sample and carrier flow, and the timing of the turning of these valves can be important to method development.

9.6.6 Carrier gas selection and flow rates

Typical carrier gases include helium, nitrogen, argon, hydrogen and air. Which gas to use is usually determined by the detector being used, for example, a DID requires helium as the carrier gas. When analyzing gas samples, however, the carrier is sometimes selected based on the sample's matrix, for example, when analyzing a mixture in argon, an argon carrier is preferred, because the argon in the sample does not show upon the chromatogram. Safety and availability can also influence carrier selection, for example, hydrogen is flammable, and high-purity helium can be difficult to obtain in some areas of the world.

The purity of the carrier gas is also frequently determined by the detector, though the level of sensitivity needed can also play a significant role. Typically, purities of 99.995% or higher are used. Trade names for typical purities include "Zero Grade", "Ultra-High Purity (UHP) Grade", "4.5 Grade" and "5.0 Grade".

The carrier gas flow rate affects the analysis in the same way that temperature does (see above). The higher the flow rate the faster the analysis, but the lower the separation between analytes. Selecting the flow rate is, therefore, the same compromise between the level of separation and length of analysis as selecting the column temperature.

With GCs made before 1990s, carrier flow rate was controlled indirectly by controlling the carrier inlet pressure, or "column head pressure". The actual flow rate was measured at the outlet of the column or the detector with an electronic flow meter, or a bubble flow meter, and could be an involved, time consuming, and frustrating process. The pressure setting was not able to be varied during the run, and thus the flow was essentially constant during the analysis.

Many modern GCs, however, electronically measure the flow rate, and electronically control the carrier gas pressure to set the flow rate. Consequently, carrier pressures and flow rates can be adjusted during the run, creating pressure/flow programs similar to temperature programs.

9.6.7 Inlet types and flow rates

The choice of inlet type and injection technique depends on if the sample is in liquid, gas, adsorbed, or solid form, and on whether a solvent matrix is present that has to be vaporized. Dissolved samples can be introduced directly onto the column via a COC injector, if the conditions are well known; if a solvent matrix has to be vaporized and partially removed, a S/SL injector is used (most common injection technique); gaseous samples (e.g., air cylinders) are usually injected using a gas switching valve system; adsorbed samples (e.g., on adsorbent tubes) are introduced using either an external (on-line or off-line) desorption apparatus such as a purge-and-trap system, or are desorbed in the S/SL injector (SPME applications).

Sample size and injection technique

The real chromatographic analysis starts with the introduction of the sample onto the column. The development of capillary gas chromatography resulted in many practical problems with the injection technique. The technique of on-column injection, often used with packed columns, is usually not possible with capillary columns. The injection system, in the capillary gas chromatograph, should fulfill the following two requirements:

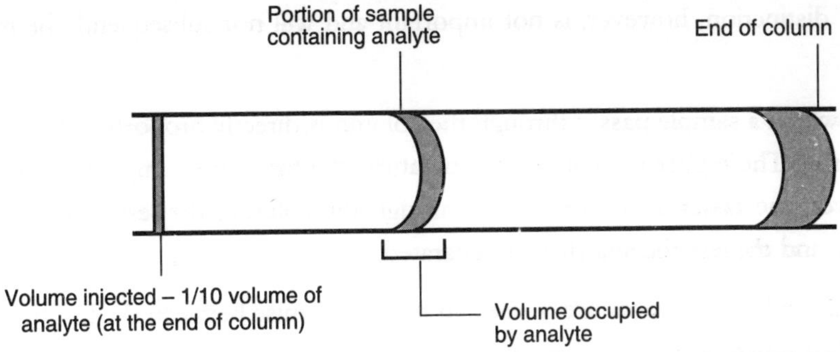

Fig. 9.7: The rule of ten in gas chromatography

1. The amount injected should not overload the column.
2. The width of the injected plug should be small compared to the spreading due to the chromatographic process. Failure to comply with this requirement will reduce the separation capability of the column. As a general rule, the volume injected, V_{inj}, and the volume of the detecor cell, V_{det}, should be about 1/10 of the volume occupied by the portion of sample containing the molecules of interest (analytes) when they exit the column.

Some general requirements, which a good injection technique should fulfill, are:

(i) It should be possible to obtain the column's optimum separation efficiency.

(ii) It should allow accurate and reproducible injections of small amounts of representative samples.

(iii) It should induce no change in sample composition. It should not exhibit discrimination based on differences in boiling point, polarity, concentration or thermal/catalytic stability.

(iv) It should be applicable for trace analysis as well as for undiluted samples.

Fig. 9.8: A gas chromatography oven, open to show a capillary column

9.6.8 Column selection

The column in a GC is contained in an oven, the temperature of which is precisely controlled electronically. (When discussing the "temperature of the column", an analyst is technically referring to the temperature of the column oven. The distinction, however, is not important and will not subsequently be made in this chapter.)

The rate at which a sample passes through the column is directly proportional to the temperature of the column. The higher the column temperature, the faster the sample moves through the column. However, the faster a sample moves through the column, the less it interacts with the stationary phase, and the less the analytes are separated.

In general, the column temperature is selected to compromise between the length of the analysis and the level of separation.

A method which holds the column at the same temperature for the entire analysis is called "isothermal". Most methods, however, increase the column temperature during the analysis, the initial temperature, rate of temperature increase (the temperature "ramp") and final temperature is called the "temperature program". A temperature program allows analytes that elute early in the analysis to separate adequately, while shortening the time it takes for late-eluting analytes to pass through the column.

9.6.9 Data reduction and analysis

Qualitative analysis

Generally chromatographic data is presented as a graph of detector response (y-axis) against retention time (x-axis). This provides a spectrum of peaks for a sample representing the analytes present in a sample eluting from the column at different times. Retention time can be used to identify analytes if the method conditions are constant. Also, the pattern of peaks will be constant for a sample under constant conditions and can identify complex mixtures of analytes. In most modern applications, however, the GC is connected to a mass spectrometer or similar detector that is capable of identifying the analytes represented by the peaks.

Quantitive analysis

The area under a peak is proportional to the amount of analyte present. By calculating the area of the peak using the mathematical function of integration, the concentration of an analyte in the original sample can be determined. Concentration can be calculated using a calibration curve created by finding the response for a series of concentrations of analyte, or by determining the response factor of an analyte. The response factor is the expected ratio of an analyte to an internal standard and is calculated by findng the response of a known amount of analyte and a constant amount of internal standard (a chemical added to the sample at a constant concentration, with a distinct retention time to the analyte).

In most modern GC-MS systems, computer software is used to draw and integrate peaks, and match MS spectra to library spectra.

9.6.9 Application

In general, substances that vaporize below 300°C (and therefore are stable up to that temperature) can be measured quantitatively. The samples are also required to be salt-free; they should not contain ions. Very minute amounts of a substance can be measured, but it is often required that the sample must be measured in comparison to a sample containing the pure, suspected substance.

Various temperature programs can be used to make the readings more meaningful; for example, to differentiate between substances that behave similarly during the GC process.

Professionals working with GC analyze the content of a chemical product; for example, in assuring the quality of products in the chemical industry; or measuring toxic substances in soil, air or water. GC is very accurate if used properly and can measure picomoles of a substance in a 1 ml liquid sample, or parts-per-billion concentrations in gaseous samples.

In practical courses at colleges, students sometimes get acquainted to the GC by studying the contents of Lavender oil or measuring the ethylene that is secreted by *Nicotiana benthamiana* plants after artificially injuring their leaves. These GC analyses hydrocarbons (C2-C40+). In a typical experiment, a packed column is used to separate the light gases, which are then detected with a TCD. The hydrocarbons are separated using a capillary column and detected with an FID. A complication with light gas analyses that include H_2 is that He, which is the most common and most sensitive inert carrier (sensitivity is proportional to molecular mass) has an almost identical thermal conductivity to hydrogen (it is the difference in thermal conductivity between two separate filaments in a Wheatstone Bridge type arrangement that shows when a component has been eluted). For this reason, dual TCD instruments are used with a separate channel for hydrogen that uses nitrogen as a carrier. Argon is often used when analysing gas phase chemistry reactions such as F-T synthesis so that a single carrier gas can be used rather than 2 separate ones. The sensitivity is less but this is a trade-off for simplicity in the gas supply.

9.6.10 Limitations associated with gas chromatography

In fact, a GC analysis takes much more time; sometimes a single sample must be run more than an hour according to the chosen program; and even more time is needed to "heat out" the column; so it is free from the first sample and can be used for the next. Equally, several runs are needed to confirm the results of a study—a GC analysis of a single sample may simply yield a result per chance.

Also, GC does not positively identify most samples; and not all substances in a sample will necessarily be detected. All a GC truly tells is at which relative time a component eluted from the column and that the detector was sensitive to it. To make results meaningful, analysts need to know which components at which concentrations are to be expected; and even then a small amount of a substance can hide itself behind a substance having both a higher concentration and the same relative elution time. Last but not least it is often needed to check the results of the sample against a GC analysis of a reference sample containing only the suspected substance.

A GC-MS can remove much of this ambiguity, since the mass spectrometer will identify the component's molecular weight. But this still takes time and skill to do properly.

Similarly, most GC analyses are not push-button operations. According to the substances one expects to find the operating program must be carefully chosen.

A push-button operation can exist for running similar samples repeatedly, such as in a chemical production environment or for comparing 20 samples from the same experiment to calculate the mean content of the same substance. However, for the kind of investigative work this is clearly not the case.

9.7 HIGH PERFORMANCE LIQUID CHROMATOGRAPHY

High-performance liquid chromatography (HPLC) is a form of column chromatography used frequently in biochemistry and analytical chemistry. It is also sometimes referred to as high-pressure liquid chromatography. HPLC is used to separate components of a mixture by using a variety of chemical interactions between the substance being analyzed (analyte) and the chromatography column.

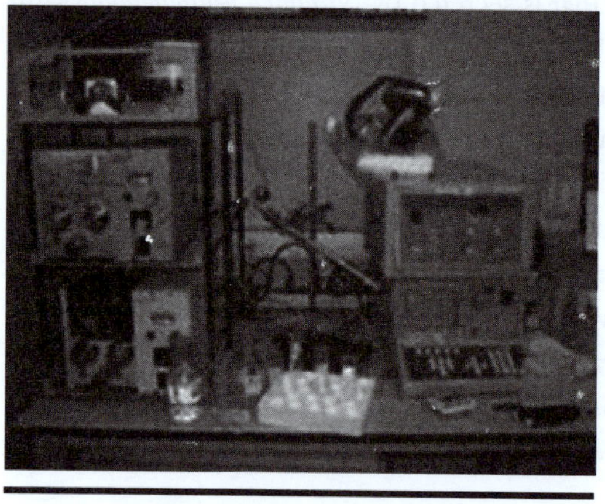

Fig. 9.9: A HPLC. From left to right: A pumping device generating a gradient of two different solvents, a steel enforced column and an apparatus for measuring the absorbance. (*Courtesy:* Deptt. of Phamacology, IMS, BHU)

In *isocratic HPLC,* the analyte is forced through a column of the stationary phase (usually a tube packed with small round particles with a certain surface chemistry) by pumping a liquid (mobile phase) at high pressure through the column. The sample to be analyzed is introduced in a small volume to the stream of mobile phase and is retarded by specific chemical or physical interactions with the stationary phase as it traverses the length of the column. The amount of retardation depends on the nature of the analyte, stationary phase and mobile phase composition. The time at which a specific analyte elutes (comes out of the end of the column) is called the retention time and is considered a reasonably unique identifying characteristic of a given analyte. The use of pressure increases the linear velocity (speed) giving the components less time to diffuse within the column, leading to improved resolution in the resulting chromatogram. Common solvents used include any miscible combinations of water or various organic liquids (the most common are methanol and acetonitrile). Water may contain buffers or salts to assist in the separation of the analyte components, or compounds such as Trifluoroacetic acid which acts as an ion pairing agent.

A further refinement to HPLC has been to vary the mobile phase composition during the analysis, this is known as *gradient elution.* A normal gradient for reverse phase chromatography might start at 5% methanol and progress linearly to 50% methanol over 25 minutes, depending on how hydrophobic the analyte is. The gradient separates the analyte mixtures as a function of the affinity of the analyte for the current mobile phase composition relative to the stationary phase. This partitioning process is similar to that which occurs during a liquid-liquid extraction but is continuous, not step-wise. In this example, using a water/methanol gradient, the more hydrophobic components will elute (come off the column) under conditions of relatively high methanol; whereas the more hydrophilic compounds will elute under conditions of relatively low methanol. The choice of solvents, additives and gradient, depends on the nature of the stationary phase and the analyte. Often a series of tests are performed on the analyte and a number of generic runs may be processed in order to find the optimum HPLC method for the analyte – the method which gives the best separation of peaks.

9.7.1 Types of HPLC

1. Normal phase chromatography

Normal phase HPLC (NP-HPLC) was the first kind of HPLC chemistry used, and separates analytes based on polarity. This method uses a polar stationary phase and a non-polar mobile phase, and is used when the analyte of interest is fairly polar in nature. The polar analyte associates with and is retained by the polar stationary phase. Adsorption strengths increase with increase in analyte polarity, and the interaction between the polar analyte and the polar stationary phase (relative to the mobile phase) increases the elution time. The interaction strength not only depends on the functional groups in the analyte molecule, but also on steric factors, and structural isomers are often resolved from one another. Use of more polar solvents in the mobile phase will decrease the retention time of the analytes while more hydrophobic solvents tend to increase retention times. Particularly polar solvents in a mixture tend to deactivate the column by occupying the stationary phase surface. This is somewhat particular to normal phase because it is most purely an adsorptive mechanism (the interactions are with a hard surface rather than a soft layer on a surface).

NP-HPLC had fallen out of favour in the 1970s with the development of reversed-phase HPLC because of a lack of reproducibility of retention times as water or protic organic solvents changed the hydration state of the silica or alumina chromatographic media. Recently it has become useful again with the development of HILIC bonded phases which utilize a partition mechanism which provides reproducibility.

2. Reversed phase chromatography

Reversed phase HPLC (RP-HPLC) consists of a non-polar stationary phase and a moderately polar mobile phase. One common stationary phase is a silica which has been treated with RMe_2SiCl, where R is a straight chain alkyl group such as $C_{18}H_{37}$ or C_8H_{17}. The retention time is, therefore, longer for molecules which are more non-polar in nature, allowing polar molecules to elute more readily. Retention time is increased by the addition of polar solvent to the mobile phase and decreased by the addition of more hydrophobic solvent. Reversed phase chromatography is so commonly used that it is not uncommon for it to be incorrectly referred to as "HPLC" without further specification.

RP-HPLC operates on the principle of hydrophobic interactions which result from repulsive forces between a relatively polar solvent, the relatively non-polar analyte, and the non-polar stationary phase. The driving force in the binding of the analyte to the stationary phase is the decrease in the area of the non-polar segment of the analyte molecule exposed to the solvent. This hydrophobic effect is dominated by the decrease in free energy from entropy associated with the minimization of the ordered molecule-polar solvent interface. The hydrophobic effect is decreased by adding more non-polar solvent into the mobile phase. This shifts the partition coefficient such that the analyte spends some portion of time moving down the column in the mobile phase, eventually eluting from the column.

The characteristics of the analyte molecule play an important role in its retention characteristics. In general, an analyte with a longer alkyl chain length results in a longer retention time because it increases the molecule's hydrophobicity. Very large molecules, however, can result in incomplete interaction between the large analyte surface and the alkyl chain. Retention time increases with hydrophobic surface area which is roughly inversely proportional to solute size. Branched chain compounds elute more rapidly than their corresponding isomers because the overall surface area is decreased.

Apart from mobile phase hydrophobicity, other mobile phase modifiers can affect analyte retention. For example, the addition of inorganic salts causes a linear increase in the surface tension of aqueous solutions, and because the entropy of the analyte-solvent interface is controlled by surface tension, the addition of salts tends to increase the retention time. Another important component is pH since this can change the hydrophobicity of the analyte. For this reason most methods use a buffering agent, such as sodium phosphate, to control the pH. An organic acid such as formic acid or most commonly trifluoroacetic acid is often added to the mobile phase. These serve multiple purposes: they control pH, neutralize the charge on any residual exposed silica on the stationary phase and act as ion pairing agents to neutralize charge on the analyte. The effect varies depending on use but generally improves the chromatography.

Reversed phase columns are quite difficult to damage compared with normal silica columns; however, many reverse phase columns consist of alkyl derivatized silica particles and should never be used with aqueous bases as these will destroy the underlying silica backbone. They can be used with aqueous acid but the column should not be exposed to the acid for too long, as it can corrode the metal parts of the HPLC equipment. The metal content of HPLC columns must be kept low if the best possible ability to separate substances is to be retained. A good test for the metal content of a column is to inject a sample which is a mixture of 2,2′- and 4,4′- bipyridine. Because the 2,2′-bipyridine can chelate the metal it is normal that when a metal ion is present on the surface of the silica the shape of the peak for the 2,2′-bipyridine will be distorted, tailing will be seen on this distorted peak.

3. Size exclusion chromatography

Size exclusion chromatography (SEC) is a chromatographic method in which particles are separated based on their size, or in more technical terms, their hydrodynamic volume. It is usually applied to large molecules or macromolecular complexes such as proteins and industrial polymers. When an aqueous solution is used to transport the sample through the column, the technique is known as gel filtration chromatography. The name gel permeation chromatography is used when an organic solvent is used as a mobile phase. The main application of gel filtration chromatography is the fractionation of proteins and other water-soluble polymers, while gel permeation chromatography is used to analyze the molecular weight distribution of organic-soluble polymers. Either technique should not be confused with gel electrophoresis, where an electric field is used to "pull" or "push" molecules through the gel depending on their electrical charges.

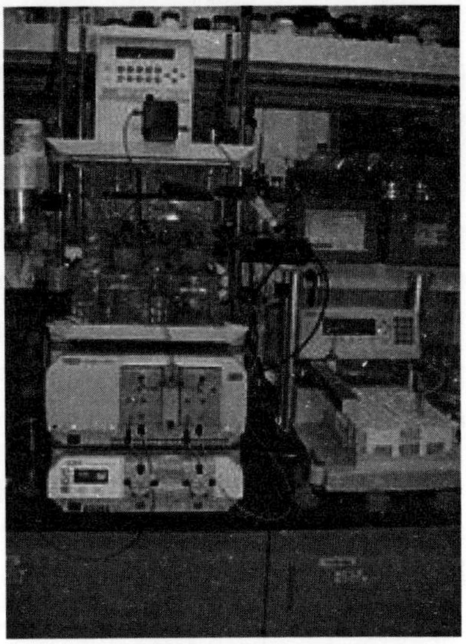

Fig. 9.10: Equipment for running size exclusion chromatography. The buffer is pumped through the column (right) by a computer controlled device.

SEC is a widely used technique for the purification and analysis of synthetic and biological polymers, such as proteins, polysaccharides and nucleic acids. Biologists and biochemists typically use a gel medium—usually polyacrylamide, dextran or agarose—and filter under low pressure. Polymer chemists typically use either a silica or cross-linked polystyrene medium under a higher pressure. These media are known as the stationary phase.

The advantage of this method is that the various solutions can be applied without interfering with the filtration process, while preserving the biological activity of the particles to be separated. The technique is generally combined with others that further separate molecules by other characteristics, such as acidity, basicity, charge, and affinity for certain compounds.

The underlying principle of SEC is that particles of different sizes will elute (filter) through a stationary phase at different rates. This results in the separation of a solution of particles based on size, provided that all the particles are loaded simultaneously or near simultaneously, particles of the same size should elute together.

This is usually achieved with an apparatus called a column, which consists of a hollow tube tightly packed with extremely small porous polymer beads designed to have pores of different sizes. These pores may be depressions on the surface or channels through the bead. As the solution travels down the column some particles enter into the pores. Larger particles cannot enter into as many pores. The larger the particles, the less overall volume to traverse over the length of the column, and the faster the elution.

The filtered solution that is collected at the end is known as the eluent. The void volume consists of any particles too large to enter the medium, and the solvent volume is known as the column volume.

In real life situations, particles in solution do not have a constant, fixed size, resulting in the probability that a particle which would otherwise be hampered by a pore may pass right by it. Also, the stationary phase particles are not ideally defined; both particles and pores may vary in size. Elution curves, therefore, resemble gaussian distributions. The stationary phase may also interact in undesirable ways with a particle and influence retention times, though great care is taken by column manufacturers to use stationary phases which are inert and minimize this issue.

Like other forms of chromatography, increasing the column length will tighten the resolution, and increasing the column diameter increases the capacity of the column. Proper column packing is important to maximize resolution: an overpacked column can collapse the pores in the beads, resulting in a loss of resolution. An underpacked column can reduce the relative surface area of the stationary phase accessible to smaller species, resulting in those species spending less time trapped in pores.

In simple manual columns the eluent is collected in constant volumes, known as fractions. The more similar the particles are in size, the more likely they will be in the same fraction and not detected separately. More advanced columns overcome this problem by constantly monitoring the eluent.

The collected fractions are often examined by spectroscopic techniques to determine the concentration of the particles eluted. Three common spectroscopy detection techniques are refractive index (RI), evaporative light scattering (ELS), and ultraviolet (UV). When eluting spectroscopically similar species (such as during biological purification) other techniques may be necessary to identify the contents of each fraction. The elution volume decreases roughly linearly with the logarithm of the molecular hydrodynamic volume (often assumed to be proportional to molecular weight). Columns are often calibrated using 4-5 standard samples (e.g., folded proteins of known molecular weight) to determine the void volume and the slope of the logarithmic dependence. This calibration may need to be repeated under different solution conditions.

Applications

1. **Proteomics:** SEC is generally considered a low resolution chromatography as it does not discern similar species very well, and is, therefore, often reserved for the final "polishing" step of a purification. The technique can determine the quaternary structure of purified proteins which have slow exchange times, since it can be carried out under native solution conditions, preserving macromolecular interactions. SEC can also assay protein tertiary structure as it measures the hydrodynamic volume (not molecular weight), allowing folded and unfolded versions of the same protein to be distinguished. For example, the apparent hydrodynamic radius of a typical protein domain might be 14Å and 36Å

for the folded and unfolded forms respectively. SEC allows the separation of these two forms as the folded form will elute much later due to its smaller size. Alternatively, folded and unfolded versions of the same metalloproteins can be separated according to their different isoelectric points by using quantitative preparative native continuous polyacrylamide gel electrophoresis (QPNC-PAGE).

2. **Polymer synthesis:** SEC can be used as a measure of both the size and the polydispersity of a synthesised polymer; that one is able to find distribution of sizes of polymer molecules. If standards of a known size are run previously, then a calibration curve can be created to determine the sizes of polymer molecules of interest. Alternatively, techniques such as light scattering and/or viscometry can be used online with SEC to yield absolute molecular weights that do not rely on calibration with standards of known molecular weight. Due to the difference in size of two polymers with identical molecular weights, the absolute determination methods are generally more desirable. A typical SEC system can quickly (in about half an hour) give polymer chemists information on the size and polydispersity of the sample.

4. Ion exchange chromatography

Ion-exchange chromatography (or *ion chromatography*) is a process that allows the separation of ions and polar molecules based on the charge properties of the molecules. It can be used for almost any kind of charged molecule including large proteins, small nucleotides and amino acids, with the experimental solution to be separated collectively known as the analyte. It is often used as a first step in protein purification.

Ion exchange chromatography retains analyte molecules based on coulombic (ionic) interactions. The stationary phase surface displays ionic functional groups that interact with analyte ions of opposite charge. This type of chromatography is further subdivided into cation exchange chromatography and anion exchange chromatography:

- Cation exchange chromatography retains positively charged cations because the stationary phase displays a negatively charged functional group such as a phosphoric acid.
- Anion exchange chromatography retains negatively charged anions using positively charged functional group such as a quaternary ammonium cation.

Proteins have numerous functional groups that can have both positive and negative charges. Ion exchange chromatography separates proteins according to their net charge, which is dependent on the composition of the mobile phase. By adjusting the pH or the ionic concentration of the mobile phase, various protein molecules can be separated. For example, if a protein has a net positive charge at pH 7, then it will bind to a column of negatively-charged beads, whereas a negatively charged protein would not. By changing the pH so that the net charge on the protein is negative, it too will be eluted.

Fig. 9.11: Ion exchange chromatography set-up

Elution by changing the ionic strength of the mobile phase is a more subtle effect – it works as ion from the mobile phase and will interact with the immobilized ion in preference over those on the stationary phase. This "shields" the stationary phase from the protein, (and vice versa) and allows the protein to elute.

A sample is introduced, either manually or with an autosampler, into a sample loop of known volume. A buffered aqueous solution known as the mobile phase carries the sample from the loop onto a column that contains some form of stationary phase material. This is typically a resin or gel matrix consisting of agarose or cellulose beads with covalently bonded charged functional groups. The target analytes (anions or cations) are retained on the stationary phase but can be eluted by increasing the concentration of a similarly charged species that will displace the analyte ions from the stationary phase. For example, in cation exchange chromatography, the positively charged analyte could be displaced by the addition of positively charged sodium ions. The analytes of interest must then be detected by some means, typically by conductivity or UV/Visible light absorbance.In order to control an IC system, a Chromatography Data System (CDS) is usually needed. In addition to IC systems, some of these CDSs can also control Gas Chromatography (GC) and HPLC systems.

5. Bioaffinity chromatography

This chromatographic process relies on the property of biologically active substances to form stable, specific, and reversible complexes. The formation of these complexes involves the participation of common molecular forces such as the van der Waals interaction, electrostatic interaction, dipole-dipole interaction, hydrophobic interaction, and the hydrogen bond. An efficient, biospecific bond is formed by a simultaneous and concerted action of several of these forces in the complementary binding sites.

9.7.2 Parameters

There are different parameters upon which the separation by HPLC relies. Some of the important parameters are discussed below.

1. Internal diameter

The internal diameter (ID) of an HPLC column is a critical aspect that determines quantity of analyte that can be loaded onto the column and also influences sensitivity. Larger columns are usually seen in industrial applications such as the purification of a drug product for later use. Low ID columns have improved sensitivity and lower solvent consumption at the expense of loading capacity.

- Larger ID columns (over 10 mm) are used to purify usable amounts of material because of their large loading capacity.

- Analytical scale columns (4.6 mm) have been the most common type of columns, though smaller columns are rapidly gaining popularity. They are used in traditional quantitative analysis of samples and often use a UV-Vis absorbance detector.
- Narrow-bore columns (1-2 mm) are used for applications when more sensitivity is desired either with special UV-Vis detectors, fluorescence detection or with other detection methods like liquid chromatography-mass spectrometry
- Capillary columns (under 0.3 mm) which are used almost exclusively with alternative detection means such as mass spectrometry. They are usually made from fused silica capillaries, rather than the stainless steel tubing that larger columns employ.

2. Particle size

Most traditional HPLC is performed with the stationary phase attached to the outside of small spherical silica particles (very small beads). These particles come in a variety of sizes with 5 μm beads being the most common. Smaller particles generally provide more surface area and better separations, but the pressure required for optimum linear velocity increases by the inverse of the particle diameter squared. This means that changing to particles that are half as big in the same size of column will double the performance, but increase the required pressure by a factor of four. Larger particles are more often used in non-HPLC applications such as solid-phase extraction.

3. Pore size

Many stationary phases are porous to provide greater surface area. Small pores provide greater surface area while larger pore size has better kinetics especially for larger analytes. For example, a protein which is only slightly smaller than a pore might enter the pore but not easily leave once inside.

4. Pump pressure

Pumps vary in pressure capacity, but their performance is measured on their ability to yield a consistent and reproducible flow rate. Pressure may reach as high as 6000 lbf/in^2 (~40 MPa, or about 400 atmospheres). Modern HPLC systems have been improved to work at much higher pressures, and therefore, be able to use much smaller particle sizes in the columns (< 2 micrometres). These "Ultra High Performance Liquid Chromatography" systems or UHPLCs can work at up to 15,000 lbf/in^2 (~100 MPa or about 1000 atmospheres). Note that the term "UPLC", sometimes found instead is a trademark of Waters Corporation and not the name for the technique in general.

9.8 SUPERCRITICAL FLUID CHROMATOGRAPHY (SFC)

It is a robust and easy-to-use form of normal phase chromatography ideally suited to the analysis and purification of low to moderate molecular weight, thermally labile molecules. It is especially suited to the separation of chiral compounds. Similar to high performance liquid chromatography

(HPLC), SFC typically utilizes carbon dioxide as the mobile phase; therefore, the entire chromatographic flow path must be pressurized.

In addition, SFC metering pumps require that the pump head be kept cold in order to maintain the carbon dioxide in a supercritical state, where it can be effectively metered at some specified flow rate. The chemist sets mobile phase flow rate, composition, and column temperature. In addition, SFC provides an additional control parameter, pressure, which the chemist similarly sets through the keyboard. From an operational standpoint, SFC is as simple and robust as HPLC.

Any molecule that will dissolve in methanol or a less polar solvent is an ideal candidate for SFC. SFC can even separate polar solutes. Many strong bases that are difficult to separate by other techniques separate rapidly and efficiently with good peak shapes.

9.9 AFFINITY CHROMATOGRAPHY

It is a chromatographic method of separating biochemical mixtures, based on a highly specific biologic interaction such as that between antigen and antibody, enzyme and substrate, or receptor

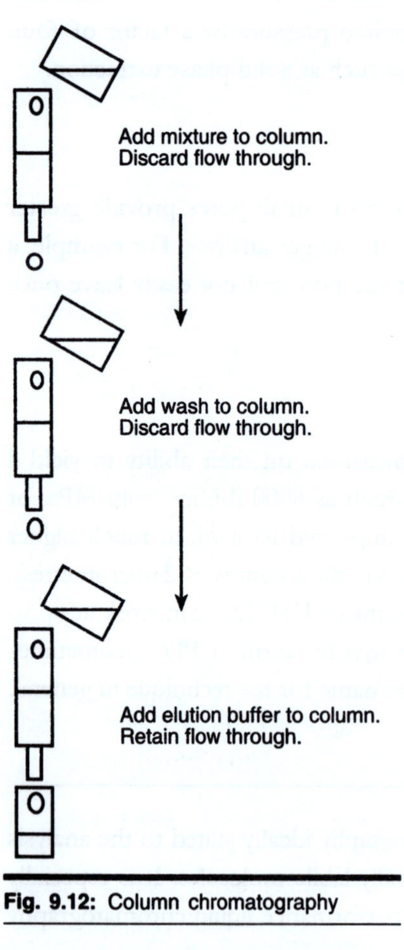

Fig. 9.12: Column chromatography

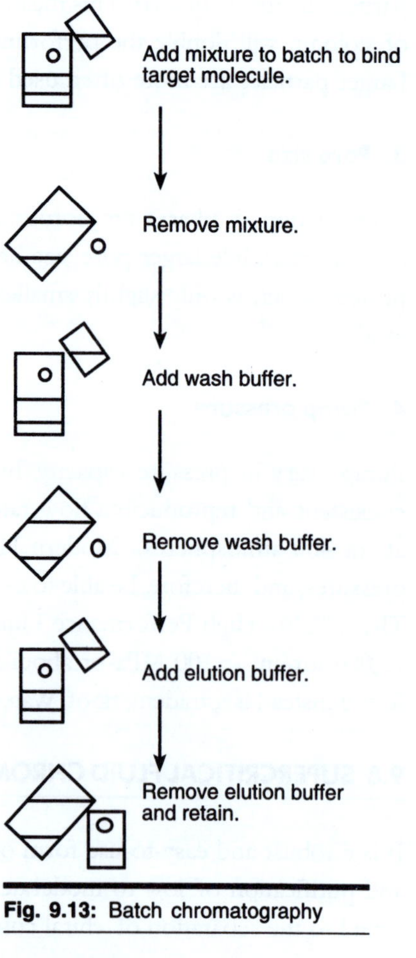

Fig. 9.13: Batch chromatography

and ligand. Affinity chromatography combines the size fractionation capability of gel permeation chromatography with the ability to design a stationary phase that reversibly binds to a known subset of molecules.

Affinity chromatography can be used to:

- Purify and concentrate a molecule from a mixture into a buffering solution.
- Reduce the amount of a molecule in a mixture.
- Discern what biological compounds bind to a particular molecule, such as drugs.

Usually the starting point is an undefined heterogeneous group of molecules in solution, such as a cell lysate, growth medium or blood serum. The molecule of interest will have a well known and defined property which can be exploited during the affinity purification process. The process itself can be thought of as an entrapment, with the target molecule becoming trapped on a solid or stationary phase or medium. The other molecules in solution will not become trapped as they do not possess this property. The solid medium can then be removed from the mixture, washed and the target molecule released from the entrapment in a process known as elution.

Binding to the solid phase may be achieved by column chromatography, whereby the solid medium is packed onto a chromatography column, the initial mixture run through the column to allow binding, a wash buffer run through the column and the elution buffer subsequently applied to the column and collected. These steps are usually done at ambient pressure (as opposed to HPLC or FPLC).

Alternatively binding may be achieved using a batch treatment, by adding the initial mixture to the solid phase in a vessel, mixing, separating the solid phase (by centrifugation, for example), removing the liquid phase, washing, re-centrifuging, adding the elution buffer, re-centrifuging and removing the eluate.

Sometimes a hybrid method is employed, the binding is done by the batch method, then the solid phase with the target molecule bound is packed onto a column and washing and elution are done on the column.

A third method, expanded bed adsorption, which combines the advantages of the two methods mentioned above, has also been developed. The solid phase particles are placed in a column where liquid phase is pumped in from the bottom and exits at the top. The gravity of the particles ensures that the solid phase does not exit the column with the liquid phase.

9.9.1 Application

Affinity chromatography can be used in a number of applications, including nucleic acid purification, protein purification from cell free extracts and antibody purification from blood serum.

1. Antibody affinity

Another use of the procedure is the affinity purification of antibodies from blood serum. If serum is known to contain antibodies against a specific antigen (for example, if the serum comes from an organism immunized against the antigen concerned), then it can be used for the affinity purification of that antigen. For example, if an organism is immunized against a GST-fusion protein it will produce antibodies against the fusion-protein, and possibly antibodies against the GST tag as well. The protein can then be covalently coupled to a solid support such as agarose.

For thoroughness the GST protein and the GST-fusion protein can each be coupled separately. The serum is initially allowed to bind to the GST affinity matrix. This will remove antibodies against the GST part of the fusion protein. The serum is then separated from the solid support and allowed to bind to the GST-fusion protein matrix. This allows any antibodies that recognize the antigen to be captured on the solid support. Elution of the antibodies of interest is most often achieved using a low pH buffer such as glycine, pH 2.8. The eluate is collected into a neutral tris or phosphate buffer to neutralize the low pH elution buffer and halts any degradation of the antibody's activity. This is a nice example as affinity purification is used to purify the initial GST-fusion protein to remove the undesirable anti-GST antibodies from the serum and to purify the target antibody.

2. Immobilized metal ion affinity chromatography

Immobilized metal ion affinity chromatography (IMAC) is based on the specific coordinate covalent binding of amino acids, allowing proteins with an affinity for metal ions to be retained in a column containing immobilized metal ions, such as cobalt, nickel, copper, zinc, or iron ions. Many naturally occurring proteins do not have an affinity for metal ions; therefore, recombinant DNA techniques are used to introduce this property into a protein of interest. Methods used to elute the protein of interest include changing the pH, or adding a specific molecule, such as imidazole. Similarly, the plant protein concanavalin A is able to bind with glucose immobilized on a column matrix. The concanavalin molecule is separated from the glucose by increasing the glucose concentration, displacing the concanavalin from the immobilized glucose.

3. Recombinant proteins

Possibly the most common use of affinity chromatography is for the purification of recombinant proteins. Proteins with a known affinity are tagged in order to aid their purification. The protein may have been genetically modified so as to allow it to be selected for affinity binding, this is known as a fusion protein. Tags include His-tags and GST (glutathione-S-transferase) tags. His_6-tags have an affinity for nickel or cobalt ions which are coordinated with NTA for the purposes of solid medium entrapment. For elution, an excess amount of a compound able to act as a nickel ligand, such as imidazole, is used. GST has an affinity for glutathione — commercially available immobilized as glutathione agarose. For elution, excess glutathione is used to displace the tagged protein.

9.10 REVERSED-PHASE CHROMATOGRAPHY

Reverse-phase chromatography (RPC) includes any chromatographic method that uses a non-polar stationary phase. All of the mathematical and experimental considerations used in other chromatographic methods apply (i.e., separation resolution proportional to the column length and inversely proportional to the column width). Reversed-phase column chromatography is widely used in the pharmaceutical, chemical, and biochemical industry for separating molecules of small molecular weight. In more recent years RPC has been used to separate larger molecules.

Any inert non-polar substance that achieves sufficient packing can be used for reversed-phase chromatography. Common examples include larger hydrocarbons, diphenyl, and divinylbenzene.

Typically mobile phases are made of water or a polar organic compound such as acetonitrile or the lighter alcohols. Buffer solutions are used to control the protonation of functional groups. For example, a buffer of high pH will encourage elution of alcohols because they are charged in basic solutions and, therefore, prefer to be in the mobile phase. For the same reason, very acidic solutions encourage the elution of nitrogen containing molecules. Using multiple buffers allows for selective elution of a wide variety of chemicals.

9.11 FAST PROTEIN LIQUID CHROMATOGRAPHY

Fast Protein Liquid Chromatography, usually referred to as FPLC, is a form of column chromatography used to separate or purify proteins from complex mixtures. It is very commonly used in biochemistry and enzymology. Columns used with an FPLC can separate macromolecules based on size, charge distribution, hydrophobicity, or biorecognition (as with affinity chromatography). Typical columns used for protein purification include:

- Ion exchange chromatography, cation or anion — separates proteins based on surface-charges
- Gel filtration chromatography, also known as Gel permeation chromatography or desalting — separates proteins based on size.
- Reversed phase or hydrophobic interaction — separates based on hydrophobicity which depends heavily on surface area.
- Affinity chromatography — columns which purify based on ligand affinity, such as a His-tagged protein uses a nickel column.

9.12 COUNTER-CURRENT CHROMATOGRAPHY

CCC or Partition Chromatography is a category of liquid-liquid chromatography techniques. Chromatography in general is used to separate components of a mixture based on their differing affinities for *mobile* and *stationary phases* of a column. The components can then be analyzed separately by various sorts of detectors which may or may not be integrated into an apparatus. In liquid-liquid chromatography, both the mobile and stationary phases are liquid. In contrast, standard column chromatography uses a solid stationary phase and a liquid mobile phase, while gas chromatography uses a liquid stationary phase on a solid support and a gaseous mobile phase. By eliminating solid supports, permanent adsorption of the analyte onto the column is avoided, and a near 100% recovery of the analyte can be achieved. The instrument is also easily switched between various modes of operation simply by changing solvents. With liquid-liquid chromatography, researchers are not limited by the composition of the columns commercially available for their instrument. Nearly any pair of immiscible solutions can be used in liquid-liquid chromatography, and most instruments can be operated in standard or reverse-phase modes. Solvent costs are also generally cheaper than that HPLC, and the cost of purchasing and disposing of solid adsorbents is completely eliminated. Another advantage is that experiments conducted in the lab can easily be scaled to industrial volumes. When GC or HPLC is done with large volumes, resolution is lost due to issues with surfact-to-volume ratios and flow dynamics; this is avoided when both phases are liquid.

CCC can be thought of as occurring in three stages: mixing, settling, and separation (although they often occur continuously). Mixing of the phases is necessary so that the interface between them has a large area, and the analyte can move between the phases according to its partition coefficient. A partition coefficient is a ratio of the amount of analyte found in each of the solvents at equilibrium and is related to the analyte's affinity for one over the other. The mobile phase is mixing with them settling from the stationary phase throughout the column. The degree of stationary phase retention (inversely proportional to the amount of stationary phase loss or "bleed" in the course of a separation) is a crucial parameter. Higher quality instruments have greater stationary phase retention. The settling time is a property of the solvent system and the sample matrix, both of which greatly influence stationary phase retention.

There are certain modifications of CCC. Major of them are:

1. **Droplet Counter-current Chromatography (DCCC)**: Droplet CCC is the oldest form of CCC. It uses only gravity to move the mobile phase through the stationary phase. In descending mode, droplets of the more dense mobile phase and sample are allowed to fall through a column of the lighter stationary phase using only gravity. If a less dense mobile phase is used it will rise through the stationary phase, this is called ascending mode. The eluent from one column is transferred to another; the more columns that are used, the more theoretical plates can be achieved. The disadvantage of DCCC is that flow rates are low, and poor mixing is achieved for most binary solvent systems, which make this technique both time-consuming and inefficient.
2. **Centrifugal Partition Chromatography (CPC)**: CPC uses centrifugal force to speed separation and achieve greater flow rates than DCCC (which relies on gravity)(e.g., the Kromaton FCPC). The columns are cut into a rotor, oriented out from the middle, and

connected by channels. The rotor is filled with the stationary phase, and the mobile phase is pumped through it as the rotor spins. Early CPC instruments had the disadvantage of requiring a rotary seal that frequently needed to be replaced, but a new generation of instruments are now reliable. CPC can also be operated in either descending or ascending mode, where the direction is relative to the force generated by the rotor rather than gravity.

3. **High-Speed Counter-current Chromatography:** The modern era of CCC began with the development by Dr Yoichiro Ito of the planetary centrifuge and the many possible column geometries it can support. These clever devices make use of a little-known means of making non-rotating connections between the stator and the rotor of a centrifuge. (It is beyond the scope of this discussion to describe the method of accomplishing this. Any of the several books available on CCC discuss it thoroughly.)

Functionally, the high-speed CCC consists of a helical coil of inert tubing which rotates on its *planetary* axis and simultaneously rotates eccentrically about another *solar* axis. (These axes can be made to coincide, but the most common or *type J* CCC is discussed here.) The effect is to create zones of mixing and zones of settling which progress along the helical coil at dizzying speed. This produces a highly favorable environment for chromatography.

There are numerous potential variants upon this instrument design. The most significant of these is the toroidal CCC (e.g., the Pharma-Tech TCC-1000). This instrument does not employ planetary motion. In some respects it is very alike CPC, but retains the advantage of not needing rotary seals. It also employs a capillary tube instead of the larger-diameter tubes employed in the helices of the other CCC models. This capillary passage makes the mixing of two phases very thorough, despite the lack of shaking or other mixing forces. This instrument provides rapid analytical scale separations, which can nonetheless be scaled up to either of the larger-scale CCC instruments.

9.12.1 Modes of Operation

CCC can be operated in different modes depending on the type of separtion required and sample to be separated. These different operation modes of CCC are as under:

- Head to tail: The denser phase is pumped through as the mobile phase. Derived from terminology for Archimedean screw force.
- Tail to head: The less dense phase is used as the mobile phase.
- Dual Mode: The mobile and stationary phases are reversed part way through the run.
- Gradient Mode: The concentration of one or more components in the mobile phase is varied throughout the run to achieve optimal resolution across a wider range of polarities.

For example, a methanol-water gradient may be employed using pure heptane as the stationary phase. This is not possible with all binary systems due to excessive loss of stationary phase.

- Elution Extrusion Mode (EECCC): The mobile phase is extruded after a certain point by switching the phase being pumped into the system. For example, during the elution portion of a separation using an EtOAc-water system running head to tail, the aqueous mobile phase is being pumped into the system. In order to switch to extrusion mode, organic phase is pumped into the system. This can be accomplished either with a valve on the inlet of single pump, or ideally with an integrated system of two or three pumps, each dedicated either to a single phase of a binary mixture, or to an intermediate wash solvent. This also allows for good resolution of compounds with high mobile-phase affinities. It requires only one column volume of solvent and leaves the column full of fresh stationary phase.
- pH Zone Refining: Acidic and basic solvents are used to elute analytes based on their pKa.

9.13 TWO-DIMENSIONAL CHROMATOGRAPHY

By using an additional physico-chemical (Chemical classification) criterion for separation of the mixture of analytes (sample), the resolution and quality of chromatographic separation can be increased. As a result, higher specificity concerning the separational capability of the chromatographic technique is obtained, allowing separation and preparation or analysis of compounds indistinguishable by one-dimensional chromatography.

In Gas-Phase Chromatography, two-dimensional separation is achieved by coupling a second, short column to the first long column. Coupling is achieved by different techniques, for example, shock-freezing the elutes in order of elution from the first column at fixed time-intervals, and then reheating them in order of elution, releasing them into the second column. The time of traversal through the second column needs to be shorter than the time remaining until the next sample is reheated to prevent compound build-up and to fully exploit the separational capability.

9.14 PYROLYSIS GAS CHROMATOGRAPHY

Pyrolysis-chromatography is a potent analytical tool able to thermally crack (fragment) essentially non-volatile molecules into fragments suitable for chromatographic analysis. The technique enables a reproducible and characteristic "fingerprint" to be generated of a non-volatile sample. The technique can be applied to such varied tasks as bacterial strain differentiation and forensic characterisation of paints, polymers and fibre cross-matching.

9.15 TIPS TO REMEMBER

Type of Chromatography	Applications in the Real World	Why and What is it
Liquid Chromatography	testing water samples to look for pollution	Used to analyze metal ions and organic compounds in solutions. It uses liquids which may incorporate hydrophilic, insoluble molecules.
Gas Chromatography	detecting bombs in airports, identifying and quantifying such drugs as alcohol, being used in forensics to compare fibres found on a victim	Used to analyze volatile gases. Helium is used to move the gaseous mixture through a column of absorbent material.
Thin-Layer Chromatography	detecting pesticide or insecticide residues in food, also used in forensics to analyze the dye composition of fibres	Uses an absorbent material on flat glass plates. This is a simple and rapid method to check the purity of the organic compound.
Paper Chromatography	separating amino acids and anions, RNA fingerprinting, separating and testing histamines, antibiotics	The most common type of chromatography. The paper is the stationary phase. This uses capillary action to pull the solutes up through the paper and separate the solutes.

Adsorption Chromatography

Adsorption chromatography is probably one of the oldest types of chromatography around. It utilizes a mobile liquid or gaseous phase that is adsorbed onto the surface of a stationary solid phase. The equilibration between the mobile and stationary phases accounts for the separation of different solutes.

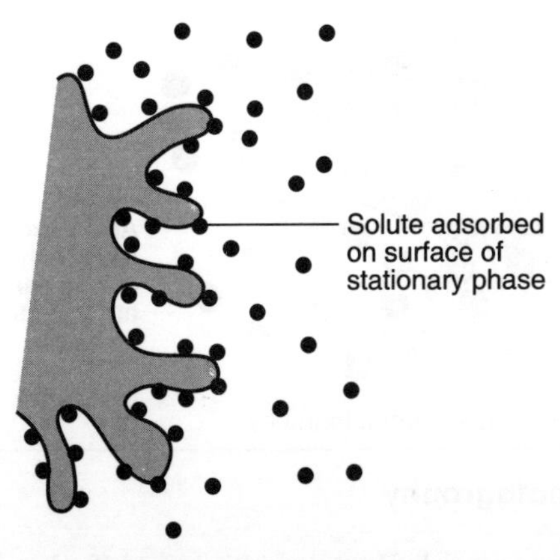

Adsorption chromatography

Partition Chromatography

This form of chromatography is based on a thin film formed on the surface of a solid support by a liquid stationary phase. Solute equilibrates between the mobile phase and the stationary liquid.

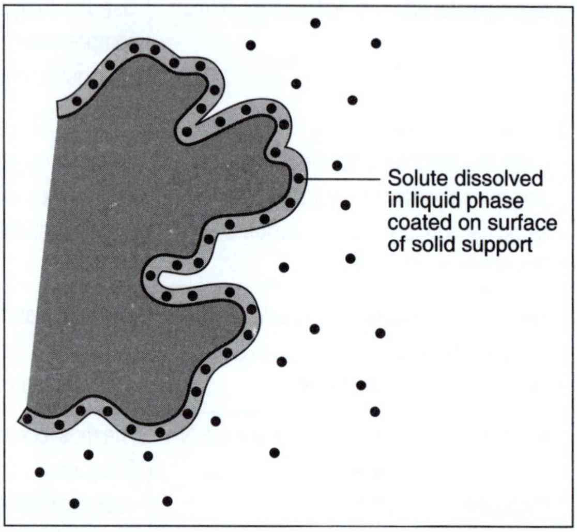

Ion Exchange Chromatography

In this type of chromatography, the use of a resin (the stationary solid phase) is used to covalently attach anions or cations onto it. Solute ions of the opposite charge in the mobile liquid phase are attracted to the resin by electrostatic forces.

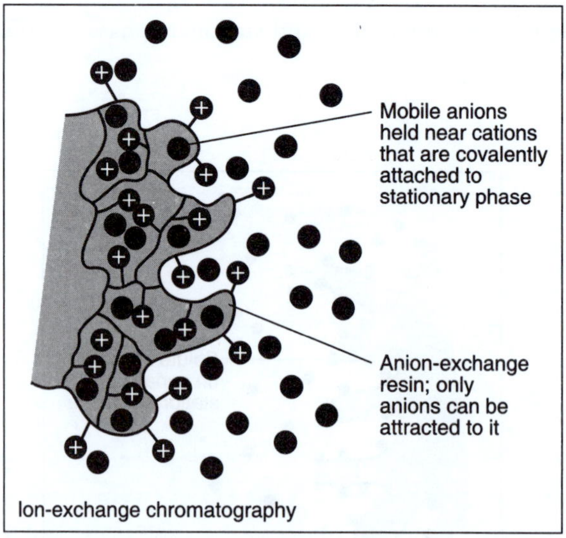

Ion-exchange chromatography

Molecular Exclusion Chromatography

Also known as gel permeation or gel filtration, this type of chromatography lacks an attractive interaction between the stationary phase and solute. The liquid or gaseous phase passes through a porous gel which separates the molecules according to its size. The pores are normally small and exclude the larger solute molecules, but allow smaller molecules to enter the gel, causing them to flow through a larger volume. This causes the larger molecules to pass through the column at a faster rate than the smaller ones.

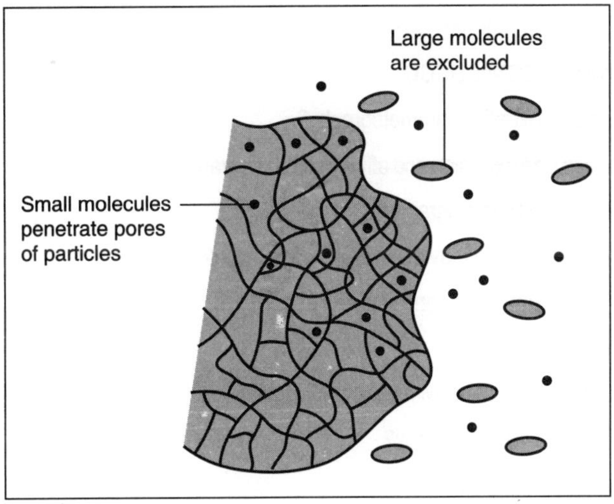

Affinity Chromatography

This is the most selective type of chromatography employed. It utilizes the specific interaction between one kind of solute molecule and a second molecule that is immobilized on a stationary phase. For example, the immobilized molecule may be an antibody to some specific protein. When solute containing a mixture of proteins are passed by this molecule, only the specific protein is reacted to this antibody, binding it to the stationary phase. This protein is later extracted by changing the ionic strength or pH.

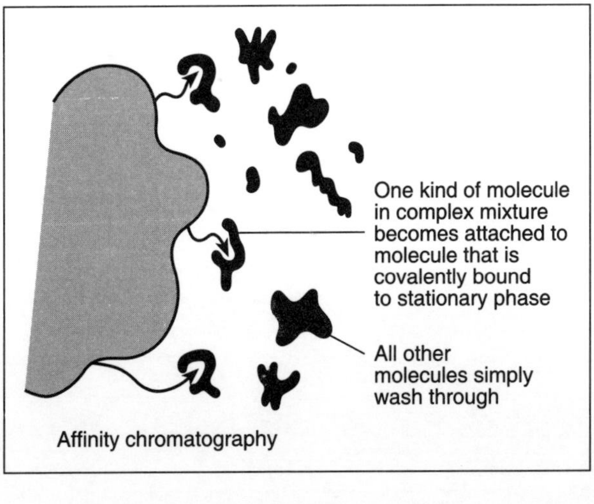

Affinity chromatography

9.16 VIVA VOCE

Q.1. What is chromatography?

Q.2. What are different types of chromatography?

Q.3. How can you separate a positively charged peptide from a mixture of proteins?

Q.4. Explain TLC.

Q.5. Differentiate between CCC and DCCC.

Q.6. What are the uses of gel filtration chromatography?

Q.7. Differentiate between size exclusion and affinity chromatography.

Q.8. What is two dimensional chromatography?

Q.9. What are the parameters which may affect HPLC separation?

Q.10. What is bioaffinity chromatography and where is it used?

Q.11. What is GC-MS?

Q.12. What are the limitations associated with gas chromatography?

Q.13. What is reverse phase chromatography?

Q.14. What are different types of HPLC?

Q.15. Name two cation exchangers.

Q.16. Name two anion exchangers.

Q.17. What is retention factor?

Q.18. Differentiate between solute and solvent.

Chapter 10

Cell Culture-I, Basics of Animal Cell Culture

10.1 WHAT IS CELL AND TISSUE CULTURE?

Tissue Culture is the general term for the removal of cells, tissues, or organs from an animal or plant and their subsequent placement into an artificial environment conducive to growth. This environment usually consists of a suitable glass or plastic culture vessel containing a liquid or semisolid medium that supplies the nutrients essential for survival and growth. The culture of whole organs or intact organ fragments with the intent of studying their continued function or development is called Organ Culture. When the cells are removed from the organ fragments prior to, or during cultivation, thus disrupting their normal relationships with neighboring cells, it is called *Cell Culture.*

Generally three types of culture are being done. They are animal cell culture, plant cell culture and microbial culture.

10.2 EQUIPMENT FOR CELL CULTURE

These include a tissue culture hood, incubators, autoclave and microscopes.

1. Cell culture hoods

This is the central piece as it is the place where all the cell handing is carried out. These hoods are designed not only to protect the culture from operator but also to protect operator from the cultures. They are generally referred to as laminar flow hoods as they generate a smooth, uninterrupted streamlined flow (laminar flow) of sterile air which has been filtered through a high efficiency particulate air (HEPA) filter. There are two types of laminar flow hoods classified as vertical or horizontal. The horizontal hood allows the air to flow directly towards operator and thus are generally used for non-infectious work like media preparation, etc. The vertical hoods also called as biological safety cabinets are best for working with hazardous organisms , since air within the hood is filtered before it passes in the surrounding environment.

Fig. 10.1: Cell culture hood

Currently, there are three different classes of hoods used, which provide various levels of protection to culture, to operator or to both.

Class-I hoods: These hoods along with class-II have screen at the front which acts as barrier between culture and operator but yet allows access to culture through a opening from bottom of the screen. This provides too much turbulence to air from outside and also provides good protection for the operator. These hoods are suitable for use with low risk organisms and when operator protection only is required.

Class-II hoods: These are most commonly used hoods. Unlike class-I, air drawn from outside is passed through the grill in front of the work area and filtered through the HEPA filter at the top of the hood before streaming down over the tissue culture. This mechanism not only protects the operator but also ensures the air over the culture is largely sterile. The hoods are adequate for animal cell culture , which involves low to moderate toxic or infectious agents, but are not recommended for use with high risk pathogens.

Class-III hoods: These are required when highest level of operator and product protection is required. These hoods are completely sealed, providing two gloves pockets through which operator can work with material inside the cabinet. Thus the operator is completely shielded making class-III hoods suitable to use with high risk involving cultures.

Safety aspects: All hoods must maintain culture-free and clean state at all times as too much culture may affect air flow, and contamination may introduce infections. Thus only those dedicated items must be placed inside the hoods that are required, and clean all the work places before and after use with 70% industrial methylated sprit (IMS) which acts against bacteria and fungal spores by dehydrating and fixing cells, thus preventing contamination of cultures.

Some cabinets may be equipped with a short wave ultraviolet light that can be used to irradiate the interior of the hood and to kill micro-organisms. When present, switch on the UV light 15 min before work to sterilize the work place. As UV can cause adverse effect to skin and eyes, precautions must be taken at all times that operator is not in direct contact with UV.

Additionally always turn on the hood at least 10 min prior to start of work to allow the flow of air to stabilize.

2. CO_2 incubators

Water jacketed incubators are required to facilitate optimum cell growth under strictly maintained and regulated conditions, normally requiring a constant temperature of 37°C and an atmosphere of 5-10% CO_2 plus air. The purpose of CO_2 is to ensure that culture medium is maintained at required physiological pH (usually 7.2-7.4). This is achieved by the supply of CO_2 by glass cylinder into the incubator through a wall that is triggered to draw in CO_2 whenever the level falls below the set level. The CO_2 that enters dissolves in the culture medium containg bicarbonate. The latter reacts with H^+ generated from cellular metabolism to form carbonic acid, which is in equilibrium with water and CO_2, thereby maintaining the pH in the medium at approximately 7.2.

Fig. 10.2: A CO_2 incubator

These incubators are generally humidified by a tray of sterile water at the bottom deck. The evaporation of water creates highly humidified atmosphere which helps to prevent evaporation of medium from culture.

An alternative to humidified incubators is the dry non-gassed unit that is not humidified and relies on the use of alternative buffering systems such as 4(2-hydroxyethyl1)-1-piperazine-ethanesulphonic acid (Hepes) or morpholinopropane sulphonic acid (mops) for maintaining a balanced pH within the culture medium. The advantage of this is that it can rule out the chances of infection by water tray. The disadvantage, however, is that culture medium will evaporate fast thereby stressing the cells.

3. Microscopes

Inverted phase contrast microscopes are usually employed for visualizing cells in culture medium. They are expensive but easy to operate with light source located above and the objective lenses below the stage on which the cells are placed. Visualisation of cells by microscopy provides useful information about morphology and state of the cells. Early signs of stress can easily be identified and thus precautionary measures can be taken.

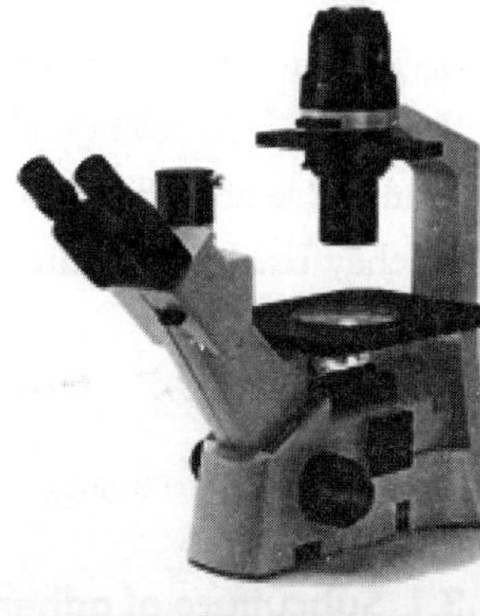

Fig. 10.3: Inverted Microscope for tissue culture

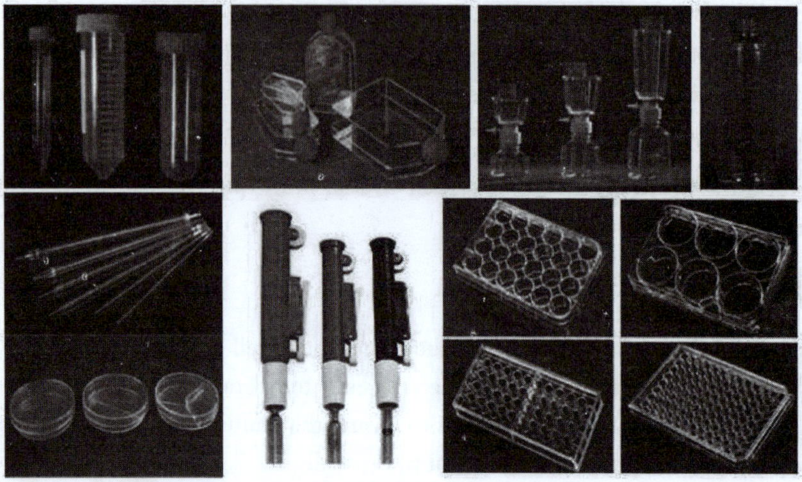

Fig. 10.4: Various general equipments required for tissue culture

4. Other general equipments

These include centrifuges to spin down the cells, water baths for thawing frozen samples and warming media to 37°C before starting the work, and refrigerators for storing media, etc. Some cells need to attach on a surface to grow, for this purpose non-toxic polystyrene plastics containing biologically inert surface on which cells can grow are available. Various kinds of plastics are available for this purpose which include Petri dishes, multiwell plates, and screw cap flasks.

10.3 SUBCULTURING OF CELLS

It is a process by which cells are harvested, diluted in fresh growth medium and replaced in a new culture flask to promote further growth. This process is known as passaging, which is essential if the cells are to be maintained in a healthy and viable state, otherwise they may die after a certain period in continuous culture. The reason for this is that adherent cells grow in continuous layer that eventually occupy the whole surface of the culture dish and at this point they are said to be confluent. Once confluent, the cells stop dividing and go into a resting state where they stop growing and eventually die. Thus to keep cells viable and to facilitate efficient transformation, they must be subcultured before they reach full contact inhibition. Ideally, cells should be harvested just before they reach confluent state.

Cells can be harvested and subcultured using one of several techniques. The precise method used is dependent to a large extent on whether the cells are adherent in suspension.

10.3.1 Subculture of adherent cells

Adherent cells can be harvested either mechanically, usually a rubber spatula (rubber policeman) or

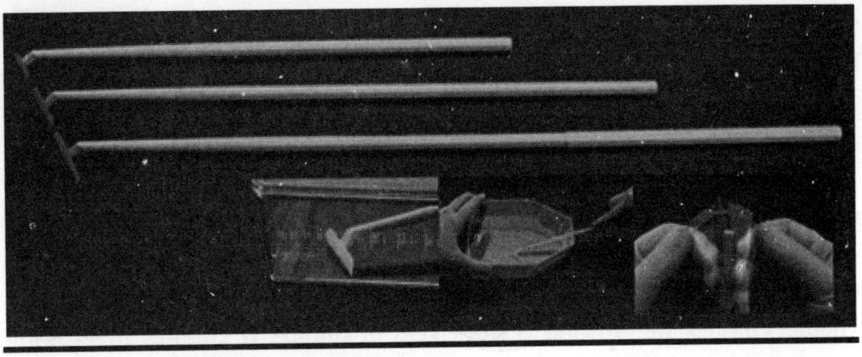

Fig. 10.5: Rubber spatula for scrapping of cells

enzymatically using proteolytic enzymes. Cells in suspension are simply diluted in fresh medium by taking a given volume of cell suspension and adding an equal volume of medium.

Harvesting of cells mechanically

This method is simple and easy. This involves gentle scrapping of cell from growth surface into the culture medium using a rubber spatula. This method is not suitable for all cell types as the scrapping may result in membrane damage and significant cell death.

Harvesting of cells using proteolytic enzymes

Several different proteolytic enzymes can be exploited including trypsin, proteolytic enzymes that destroys proteinaceous connections between cells and the surface of the flask in which they grow. As a result harvesting of cells using this enzyme results in the release of single cells, which is ideal for subculturing as each cell will then divide and grow, thus enhancing the propogation of cultures.

Trypsin is commonly used with EDTA, which enhances the action of enzyme. EDTA alone can also be effective in detaching adherent cells as it chelates the Ca^{2+} required by some adhesion molecules that facilitate cell-cell or cell-matrix interactions. Although EDTA alone is much gentler on the cells than trypsin, some cell types may adhere strongly to plastic, requiring trypsin to detach.

The standard procedure for detaching adherent cells using trypsin and EDTA involves making a working solution of 0.1% trypsin plus 0.02% EDTA in Ca^{2+}/Mg^{2+} free phosphate-buffered saline. The growth medium is aspirated from confluent cultures and washed at least twice with a serum-free medium such as Ca^{2+}/Mg^{2+} free PBS to remove traces of serum that may inactivate the trypsin. The trypsin-EDTA is then added to cell monolayer and swirled around for a few seconds. Excess trypsin-EDTA is aspirated, leaving just enough to form a thin film over the monolayer. The flask is then incubated at 37°C for 2-5 min but monitored under an inverted light microscope when cells are beginning to round up and detach. This is to ensure that cells are not overexposed to trypsin and cell surface may get damaged. It is, therefore, important that proteolysis reaction is appropriately terminated by adding complete medium having serum to deactivate the enzyme. The suspension of cells is then subjected to centrifuge and the pellet is resuspended in known volume of complete medium to give a required density of cells. All of the above procedure is carried out in a sterile tissue culture cabinet.

Subculture of cells in suspension

For cells in suspension it is important initially to examine an aliquot of cells under a microscope to establish whether cultures are growing as single cells or clumps. If cultures are growing as a single cell, an aliquot is counted and then reseeded at desired seeding density in a new flask by simply diluting the cell suspension with fresh medium, provided the original medium is not spent. Cells that grow in clumps should first be centrifuged and resuspended in fresh medium as single cell using a glass Pasteur or fine-bore pipette.

10.4 CELL QUANTIFICATION

It is essential that when cells are sub-cultured they are reseeded in appropriate density so as to facilitate optimum growth. If cells are seeded at lower seeding density they may take longer time to reach confluency and some may expire before getting this point. On the other hand, if seeded in high density they will reach confluency very quickly, resulting in irreproducible experimental results. This is because trypsin may digest surface proteins, including receptors for drugs and these may take time to renew. Therefore, appropriate seeding density is a must which requires quantification of the cells.

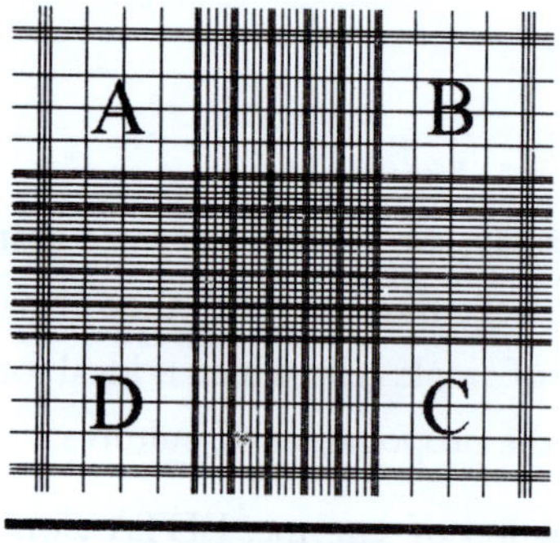

Fig. 10.6: Haemocytometer for counting of cells

It involves the use of haemocytometer, which is a thickened glass slide with small chambers of grids. The chamber has a fixed volume and is etched into nine large squares, of which the large corner square contain 16 small square each; each large square measures 1 mm × 1mm and is 0.1 mm deep. Thus with a coverslip in place, each square represents a volume of 0.1 mm^3. Knowing this, cell concentration can be determined and expressed per cubic centimetres.

Another alternative method available for cell quantification is use of electronic coulter counter. This is an automated method of counting and measuring size of microscopic particles. The instrument consists of glass probe with an electrode connected to an oscilloscope. The probe has a small aperture of fixed diameter near its bottom end. When immersed in solution of cell suspension, cells are flushed through the aperture causing the brief increase in resistance owing to a partial interruption of current flow. This will result in spikes being recorded on oscilloscope and each spike is counted as a cell. The disadvantage this technique had is, it cannot distinguish between live and dead cells.

Also indirectly cells can be counted by determining total cell protein and using protein versus cell number standard curve to determine cell number in test samples.

Again, the DNA content of cells may be used as an indicator of cell number, since the DNA content of diploid cells is usually constant. However, the DNA content of cells may change during the cell cycle and, therefore, does not give an accurate estimate of cell number.

10.5 ANIMAL CELL CULTURE

Although animal cell culture was first successfully undertaken by Ross Harrison in 1907, it was not until the late 1940s to early 1950s that several developments occurred that made cell culture widely available as a tool for scientists. First, there was the development of antibiotics that made it easier to avoid many of the contamination problems that plagued earlier cell culture attempts. Second was the development of the techniques, such as the use of trypsin to remove cells from culture vessels, necessary to obtain continuously growing cell lines (such as HeLa cells). Third, using these cell lines, scientists were able to develop standardized, chemically defined culture media that made it far easier to grow cells. These three areas combined to allow many more scientists to use cell, tissue and organ culture in their research. During the 1960s and 1970s, commercialization of this technology had further impact on cell culture that continues to this day. Companies, such as Corning, began to develop and sell disposable plastic and glass cell culture products and improve filtration products and materials, liquid and powdered tissue culture media, and laminar flow hoods. The overall result of these and other continuing technological developments has been a widespread increase in the number of laboratories and industries using cell culture today.

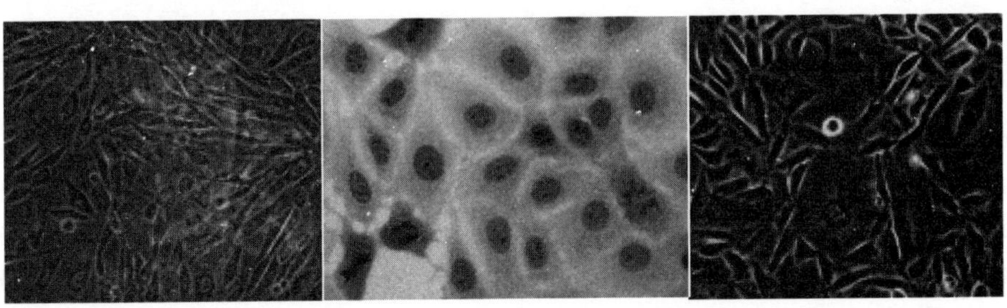

Fig. 10.7: A. Fibroblast-like 3T3 cells, B. Epithelial-like cell line (Cl-9), C. CHO-K1 cells

10.6 HOW ARE CELL CULTURES OBTAINED FOR CULTURE?

10.6.1 Primary Culture

When cells are surgically removed from an organism and placed into a suitable culture environment, they will attach, divide and grow. This is called a Primary Culture. There are two basic methods for doing this. First, for Explant Cultures, small pieces of tissue are attached to a glass or treated in

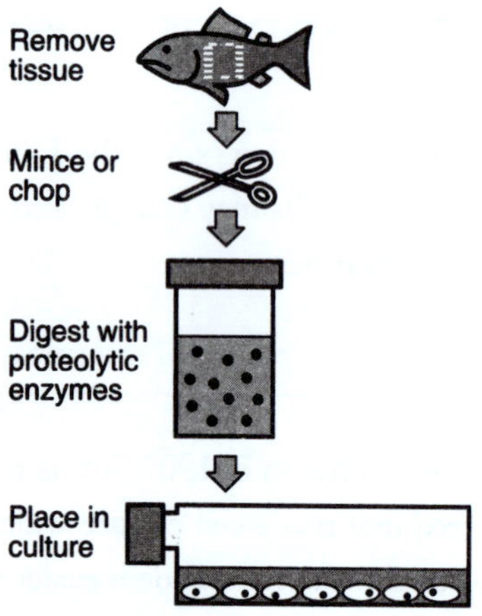

Fig. 10.8: Obtaining primary cultures by enzymatic dissociation

plastic culture vessel and bathed in culture medium. After a few days, individual cells will move from the tissue explant onto the culture vessel surface or substrate where they will begin to divide and grow. The second, more widely used method speeds up this process by adding digesting (proteolytic) enzymes, such as trypsin or collagenase, to the tissue fragments to dissolve the cement holding the cells together. This creates a suspension of single cells that are then placed into culture vessels containing culture medium and allows to grow and divide. This method is called Enzymatic Dissociation.

10.6.2 Subculturing

When the cells in the primary culture vessel have grown and filled up all of the available culture substrate, they must be subcultured to give them room for continued growth. This is usually done by removing them as gently as possible from the substrate with enzymes. These are similar to the enzymes used in obtaining the primary culture and are used to break the protein bonds attaching the cells to the substrate. Some cell lines can be harvested by gently scraping the cells off the bottom of the culture vessel. Once released, the cell suspension can then be subdivided and placed into new culture vessels. Once a surplus of cells is available, they can be treated with suitable cryoprotective agents, such as dimethylsulphoxide (DMSO) or glycerol, carefully frozen and then stored at cryogenic temperatures (below –130°C) until they are needed.

10.6.3 Buying and Borrowing

An alternative to establishing cultures by primary culture is to buy established cell cultures from organizations such as the American Type Culture Collection (ATCC; www.atcc.org) or the Coriell Institute for Medical Research (arginine.umdnj.edu). These two nonprofit organizations provide high quality cell lines that are carefully tested to ensure the authenticity of the cells. More frequently, researchers will obtain (borrow) cell lines from other laboratories. While this practice is widespread, it has one major drawback. There is a high probability that the cells obtained in this manner will not be healthy, useful cultures. This is usually due to previous mix-ups or contamination with other cell lines, or the result of contamination with micro-organisms such as mycoplasmas, bacteria, fungi or yeast.

10.7 CHARACTERISTICS OF CULTURED CELLS

Once in culture, cells exhibit a wide range of behaviours, characteristics and shapes. Some of the more common ones are described below.

10.7.1 Cell Culture Systems

Two basic culture systems are used for growing cells. These are based primarily upon the ability of the cells to either grow attached to a glass or treated plastic substrate (Monolayer Culture Sytems) or floating free in the culture medium (Suspension Culture Systems).

Monolayer cultures are usually grown in tissue culture treated dishes, T-flasks, roller bottles, or multiple well plates, the choice being based on the number of cells needed, the nature of the culture environment, cost and personal preference. Suspension cultures are usually grown either:

1. In magnetically rotated spinner flasks or shaken Erlenmeyer flasks where the cells are kept actively suspended in the medium;
2. In stationary culture vessels such as T-flasks and bottles where, although the cells are not kept agitated, they are unable to attach firmly to the substrate. Many cell lines, especially those derived from normal tissues, are considered to be Anchorage-Dependent, that is, they can only grow when attached to a suitable substrate. Some cell lines that are no longer considered normal (frequently designated as Transformed Cells) are frequently able to grow either attached to a substrate or floating free in suspension; they are Anchorage-Independent. In addition, some normal cells, such as those found in the blood, do not normally attach to substrates and always grow in suspension.

10.7.2 Types of Cells

Cultured cells are usually described based on their morphology (shape and appearance) or their functional characteristics. There are three basic morphologies:

1. Epithelial-like: cells that are attached to a substrate and appear flattened and polygonal in shape.
2. Lymphoblast-like: cells that do not attach normally to a substrate but remain in suspension with a spherical shape.
3. Fibroblast-like: cells that are attached to a substrate and appear elongated and bipolar, frequently forming swirls in heavy cultures. It is important to remember that the culture conditions play an important role in determining shape and that many cell cultures are capable of exhibiting multiple morphologies.

Using cell fusion techniques, it is also possible to obtain hybrid cells by fusing cells from two different parents. These may exhibit characteristics of either parent or both parents. This technique was used in 1975 to create cells capable of producing custom tailored monoclonal antibodies. These hybrid cells (called Hybridomas) are formed by fusing two different but related cells. The first is a spleen-derived lymphocyte that is capable of producing the desired antibody. The second is rapidly dividing myeloma cell machinery for making antibodies but is not programmed to produce any antibody. The resulting hybridomas can produce large quantities of the desired antibody. These antibodies, called Monoclonal Antibodies due to their purity, have many important clinical, diagnostic, and industrial applications with a yearly value of well over a billion dollars.

10.7.3 Functional Characteristics

The characteristics of cultured cells result from both their origin (liver, heart, etc.) and how well they adapt to the culture conditions. Biochemical markers can be used to determine if cells are still carrying on specialized functions that they performed in vivo (e.g., liver cells secreting albumin). Morphological or ultrastructural markers can also be examined (e.g., beating heart cells). Frequently, these characteristics are either lost or changed as a result of being placed in an artificial environment. Some cell lines will eventually stop dividing and show signs of aging. These lines are called Finite. Other lines are, or become immortal; these can continue to divide indefinitely and are called Continuous cell lines. When a "normal" finite cell line becomes immortal, it has undergone a fundamental irreversible change or "transformation". This can occur spontaneously or be brought about intentionally using drugs, radiation or viruses. Transformed Cells are usually easier and faster growing, may often have extra or abnormal chromosomes and frequently can be grown in suspension. Cells that have the normal number of chromosomes are called Diploid cells; those that have other than the normal number are Aneuploid. If the cells form tumours when they are injected into animals, they are considered to be *neoplastically transformed.*

10.8 PROBLEMS FACED BY CULTURED CELLS

10.8.1 Avoiding Contamination

Cell culture contamination is of two main types: chemical and biological. Chemical contamination is the most difficult to detect since it is caused by agents, such as endotoxins, plasticizers, metal ions or traces of chemical disinfectants, that are invisible. Biological contaminants in the form of fast growing yeast, bacteria and fungi usually have visible effects on the culture (changes in medium turbidity or pH) and thus are easier to detect (especially if antibiotics are omitted from the culture medium). However, two other forms of biological contamination, mycoplasmas and viruses, are not easy to detect visually and usually require special detection methods.

There are two major requirements to avoiding contamination. First, proper training in the use of good aseptic technique on the part of the cell culturist. Second, properly designed, maintained and sterilized equipment, plasticware, glassware, and media. The careful and selective (limited) use of antibiotics designed for use in tissue culture can also help avoid culture loss due to biological contamination.

10.8.2 Finding a "Happy" Environment

To cell culturists, a "happy" environment is one that does more than just allow cells to survive in culture. Usually, it means an environment that, at the very least, allows cells to increase in number by undergoing cell division (mitosis). Even better, when conditions are just right, some cultured cells will express their "happiness" with their environment by carrying out important in vivo physiological or biochemical functions, such as muscle contraction or the secretion of hormones and enzymes. To provide this environment, it is important to provide the cells with the appropriate temperature, a good substrate for attachment, and the proper culture medium.

Temperature is usually set at the same point as the body temperature of the host from which the cells were obtained. With cold-blooded vertebrates, a temperature range of 18-25°C is suitable; most mammalian cells require 36-37°C. This temperature range is usually maintained by use of carefully calibrated, and frequently checked, incubators.

Anchorage-dependent cells also require a good substrate for attachment and growth.

Glass and specially treated plastics (to make the normally hydrophobic plastic surface hydrophilic or wettable) are the most commonly used substrates. However, attachment factors, such as collagen, gelatin, fibronectin and laminin, can be used as substrate coatings to improve growth and function of normal cells derived from brain, blood vessels, kidney, liver, skin, etc. Often normal anchorage-dependent cells will also function better if they are grown on a permeable or porous surface. This allows them to polarize (have a top and bottom through which things can enter and leave the cell) as they do in the body. Transwell inserts are corning vessels with membrane-based permeable supports that allow these cells to develop polarity and acquire the ability to exhibit special functions such as transport. Many specialized cells can only be truly "happy" (function normally) when grown on a porous substrate in serum-free medium with the appropriate mixture of growth and attachment factors.

Cells can also be grown in suspension on beads made from glass, plastic, polyacrylamide and cross-linked dextran molecules. This technique has been used to enable anchorage-dependent cells to be grown in suspension culture systems and is increasingly important for the manufacture of cell-based biologicals. The culture medium is the most important and complex factor to control in making cells "happy". Besides meeting the basic nutritional requirement of the cells, the culture medium should also have any necessary growth factors, regulate the pH and osmolality, and provide essential gases (O_2 and CO_2). The 'food' portion of the culture medium consists of amino acids, vitamins, minerals, and carbohydrates. These allow the cells to build new proteins and other components essential for growth and function as well as providing the energy necessary for metabolism.

The growth factors and hormones help regulate and control the cells' growth rate and functional characteristics. Instead of being added directly to the medium, they are often added in an undefined manner by adding 5 to 20% of various animal sera to the medium. Unfortunately, the types and concentration of these factors in serum vary considerably from batch to batch. This often results in problems controlling growth and function. When growing normal functional cells, sera are often replaced by specific growth factors.

The medium also controls the pH range of the culture and buffers the cells from abrupt changes in pH. Usually a CO_2-bicarbonate based buffer or an organic buffer, such as HEPES, is used to help keep the medium pH in a range from 7.0 to 7.4 depending on the type of cell being cultured. When using a CO_2-bicarbonate buffer, it is necessary to regulate the amount of CO_2 dissolved in the medium.

This is usually done using an incubator with CO_2 controls set to provide an atmosphere with between 2% and 10% CO_2 (for Earle's salt-based buffers). However, some media use a CO_2-bicarbonate buffer (for Hanks' salt-based buffers) that requires no additional CO_2, but it must be used in a sealed vessel (not dishes or plates).

Finally, the osmolality (osmotic pressure) of the culture medium is important since it helps regulate the flow of substances in and out of the cell. It is controlled by the addition or subtraction of salt in the culture medium. Evaporation of culture media from open culture vessels (dishes, etc.) will rapidly increase the osmolality resulting in stressed, damaged or dead cells. For open (not sealed) culture systems, incubators with high humidity levels to reduce evaporation are essential.

10.9 HOW TO DECIDE IF CULTURED CELLS ARE "HAPPY"

Evaluating the general health or "happiness" of a culture is usually based on four important cell characteristics: morphology, growth rate, plating efficiency and expression of special functions. These same characteristics are also widely used in evaluating experimental results.

The Morphology or cell shape is the easiest to determine but is often the least useful. While changes in morphology are frequently observed in cultures, it is often difficult to relate these observations to the condition that caused them. It is also a very difficult characteristic to quantify or to measure precisely. Often, the first sign that something is wrong with a culture occurs when the cells are microscopically examined and poor or unusual patterns of cell attachment or growth are observed. When problems are suspected, staining the culture vessels with crystal violet or other simple histological stains may show growth patterns indicating a problem.

Cell counting and other methods for estimating cell number, on the other hand, allow the determination of the growth rate, which is sensitive to major changes in the culture environment. This allows the design of experiments to determine which set of conditions (culture media, substrate, serum, plasticware) is better for the cells, i.e., the conditions producing the best growth rate. These same or similar techniques can also be used to measure cell survival or death and are often used for in vitro cytotoxicity assays. Plating Efficiency is a testing method where small numbers of cells (20 to 200) are placed in a culture vessel and the number of colonies they form is measured.

The percentage of cells forming colonies is a measure of survival, while the colony size is a measure of growth rate. This testing method is similar in application to growth rate analysis but is more sensitive to small variations in culture conditions. The final characteristic, the expression of specialized functions, is usually the most difficult to observe and measure. Usually biochemical or immunological assays and tests are used. While cultured cells may grow very well in suboptimal conditions, highly specialized functions usually require near perfect culture conditions and are often quickly lost when cells are placed in culture.

10.10 APPLICATIONS OF ANIMAL CELL CULTURE

Cell culture has become one of the major tools used in cell and molecular biology. Some of the important areas where cell culture is currently playing a major role are briefly described below:

1. **Model Systems:** Cell cultures provide a good model system for studying (1) basic cell biology and biochemistry, (2) the interactions between disease-causing agents and cells, (3) the effects of drugs on cells, (4) the process and triggers for aging, and (5) nutritional studies.

2. **Toxicity Testing:** Cultured cells are widely used alone or in conjunction with animal tests to study the effects of new drugs, cosmetics and chemicals on survival and growth in a wide variety of cell types. Especially important are liver- and kidney-derived cell cultures.
3. **Cancer Research:** Since both normal cells and cancer cells can be grown in culture, the basic differences between them can be closely studied. In addition, it is possible, by the use of chemicals, viruses and radiation, to convert normal cultured cells to cancer causing cells. Thus, the mechanisms that cause the change can be studied. Cultured cancer cells also serve as a test system to determine suitable drugs and methods for selectively destroying types of cancer.
4. **Virology:** One of the earliest and major uses of cell culture is the replication of viruses in cell cultures (in place of animals) for use in vaccine production. Cell cultures are also widely used in the clinical detection and isolation of viruses, as well as basic research into how they grow and infect organisms.
5. **Cell-Based Manufacturing:** While cultured cells can be used to produce many important products, three areas are generating the most interest. The first is the large-scale production of viruses for use in vaccine production.These includes vaccines for polio, rabies, chicken pox, hepatitis B and measles.

   The second is the large-scale production of cells that have been genetically engineered to produce proteins that have medicinal or commercial value. These include monoclonal antibodies, insulin, hormones, etc. The third is the use of cells as replacement tissues and organs. Artificial skin for use in treating burns and ulcers is the first commercially available product. However, testing is underway on artificial organs such as pancreas, liver and kidney. A potential supply of replacement cells and tissues may come out of work currently being done with both embryonic and adult stem cells. These are cells that have the potential to differentiate into a variety of different cell types. It is hoped that learning how to control the development of these cells may offer new treatment approaches for a wide variety of medical conditions.
6. **Genetic Counselling:** Amniocentesis, a diagnostic technique that enables doctors to remove and culture fetal cells from pregnant women, has given doctors an important tool for the early diagnosis of fetal disorders. These cells can then be examined for abnormalities in their chromosomes and genes using karyotyping, chromosome painting and other molecular techniques.
7. **Genetic Engineering:** The ability to transfect or reprogram cultured cells with new genetic material (DNA and genes) has provided a major tool to molecular biologists wishing to study the cellular effects of the expression of theses genes (new proteins). These techniques can also be used to produce these new proteins in large quantity in cultured cells for further study. Insect cells are widely used as miniature cells factories to express substantial quantities of proteins that they manufacture after being infected with genetically engineered baculoviruses.
8. **Gene Therapy:** The ability to genetically engineer cells has also led to their use for gene therapy. Cells can be removed from a patient lacking a functional gene and the missing or

damaged gene can then be replaced. The cells can be grown for a while in culture and then replaced into the patient. An alternative approach is to place the missing gene into a viral vector and then "infect" the patient with the virus in the hope that the missing gene will then be expressed in the patient's cells.

9. **Drug Screening and Development:** Cell-based assays have become increasingly important for the pharmaceutical industry, not just for cytotoxicity testing but also for high throughput screening of compounds that may have potential use as drugs. Originally, these cell culture tests were done in 96 well plates, but increasing use is now being made of 384 and 1536 well plates.

10.11 TIPS TO REMEMBER

Basic environmental requirements for "Happy" Cells:

- Controlled temperature
- Good substrate for cell attachment
- Appropriate medium and incubator that maintains the correct pH and osmolality

Cell Culture Systems

Two basic culture systems are used for growing cells. These are based primarily upon the ability of the cells to either grow being attached to a glass or treated plastic substrate (**Monolayer Culture Sytems**)or floating free in the culture medium (**Suspension Culture Systems**).

Primary Culture

When cells are surgically removed from an organism and placed into a suitable culture environment, they will attach, divide and grow.

10.12 VIVA VOCE

Q.1. What do you understand by cell culture?
Q.2. What are the types of cell culture hoods?
Q.3. What is the importance of CO_2 incubator?
Q.4. What do you understand by subculturing the cells?
Q.5. Explain the different harvesting techniques employed in cell culture.
Q.6. What are the sources of primary culture?
Q.7. What are the precautionary measures one should take for maintaining a cell line?
Q.8. What are the criteria for "Happy cell"?
Q.9. Differentiate between animal and plant cell culture.
Q.10. What are the characters of cultured cells?
Q.11. List the major problems occurring during cell culture. Also suggest the ways by which they can be overcome.
Q.12. List the applications of cell culture.

CHAPTER 11

Cell Culture-II, Plant and Microbial Cell Culture

PLANT TISSUE CULTURE

Most methods of plant transformation applied to crops require that a whole plant is regenerated from isolated plant cell or tissue which has been genetically transformed. This regeneration is conducted *in vitro* so that the environment and growth medium can be manipulated to ensure a high frequency of regeneration. In addition to a high frequency of regeneration, the regenerable cells must be accessible to gene transfer by whatever technique is chosen. The primary aim is, therefore, to produce, as easily and as quickly as possible, a large number of regenerable cells that are accessible to gene transfer. The subsequent regeneration step is often the most difficult step in plant transformation studies. However, it is important to remember that a high frequency of regeneration does not necessarily correlate with high transformation efficiency.

This chapter will consider some basic issues concerned with plant tissue culture *in vitro*, particularly as applied to plant transformation. It will also look at the basic culture types used for plant transformation and cover some of the techniques that can be used to regenerate whole transformed plants from transformed cells or tissue.

Practically any plant transformation experiment relies at some point on tissue culture. There are some exceptions to this generalisation, but the ability to regenerate plants from isolated cells or tissues *in vitro* underpins most plant transformation systems.

11.1 PLASTICITY AND TOTIPOTENCY

Two concepts, plasticity and totipotency, are central to understanding plant cell culture and regeneration.

Plants, due to their sessile nature and long life span, have developed a greater ability to endure extreme conditions and predation than animals have. Many of the processes involved in plant growth and development adapt to environmental conditions. This plasticity allows plants to alter their metabolism, growth and development to best suit their environment. Particularly important aspects of this adaptation, as far as plant tissue culture and regeneration are concerned, are the abilities to initiate cell division from almost any tissue of the plant and to regenerate lost organs or undergo different developmental pathways in response to particular stimuli. When plant cells and tissues are cultured *in vitro* they generally exhibit a very high degree of plasticity, which allows one type of tissue or organ to be initiated from another type. In this way, whole plants can be subsequently regenerated.

This regeneration of whole organisms depends upon the concept that all plant cells can, given the correct stimuli, express the total genetic potential of the parent plant. This maintenance of genetic potential is called 'totipotency'. Plant cell culture and regeneration do, in fact, provide the most compelling evidence for totipotency.

In practical terms though, identifying the culture conditions and stimuli required to manifest this totipotency can be extremely difficult and it is still a largely empirical process.

11.2 THE CULTURE ENVIRONMENT

When cultured *in vitro*, all the needs, both chemical (see Table11.1) and physical, of the plant cells have to met by the culture vessel, the growth medium and the external environment (light, temperature, etc.). The growth medium has to supply all the essential mineral ions required for growth and development. In many cases (as the biosynthetic capability of cells cultured *in vitro* may not replicate that of the parent plant), it must also supply additional organic supplements such as amino acids and vitamins. Many plant cell cultures, as they are not photosynthetic, also require the addition of a fixed carbon source in the form of a sugar (most often sucrose). One other vital component that must also be supplied is water, the principal biological solvent. Physical factors, such as temperature, pH, the gaseous environment, light (quality and duration) and osmotic pressure, also have to be maintained within acceptable limits.

11.3 PLANT CELL CULTURE MEDIA

Culture media used for the in vitro cultivation of plant cells are composed of three basic components:

1. Essential elements, or mineral ions, supplied as a complex mixture of salts;
2. An organic supplement supplying vitamins and/or amino acids; and
3. A source of fixed carbon; usually supplied as the sugar sucrose.

Table 11.1: Some of the elements important for plant nutrition and their physiological function

Element	Function
Nitrogen	Component of proteins, nucleic acids and some coenzymes element required in greatest amount
Potassium	Regulates osmotic potential, principal inorganic cation
Calcium	Cell wall synthesis, membrane function, cell signalling
Magnesium	Enzyme cofactor, component of chlorophyll
Phosphorus	Component of nucleic acids, energy transfer, component of intermediates in respiration and photosynthesis
Sulphur	Component of some amino acids (methionine, cysteine) and some cofactors
Chlorine	Required for photosynthesis
Iron	Electron transfer as a component of cytochromes
Manganese	Enzyme cofactor
Cobalt	Component of some vitamins
Copper	Enzyme cofactor, electron-transfer reactions
Zinc	Enzyme cofactor, chlorophyll biosynthesis
Molybdenum	Enzyme cofactor, component of nitrate reductase

These elements have to be supplied by the culture medium in order to support the growth of healthy cultures *in vitro.*

For practical purposes, the essential elements are further divided into the following categories:

1. Macroelements (or macronutrients);
2. Microelements (or micronutrients); and
3. An iron source.

Complete plant cell culture medium is usually made by combining several different components, as outlined in Table 11.2. The table represents one of the most common plant culture media called Murashige and Skoog.

Many other commonly used plant culture media (such as Gamborg's B5 and Schenk and Hildebrandt (SH) medium) are similar in composition to MS medium and can be thought of as 'high-salt' media. MS is an extremely widely used medium and forms the basis for many other media formulations.

Table 11.2: Composition of a typical Murashige and Skoog plant culture medium

Essential element	Concentration in stock solution (mg/l)	Concentration in medium (mg/l)
Macroelements[b]		
NH_4NO_3	33 000	1 650
KNO_3	38 000	1 900
$CaCl_2 \cdot 2H_2O$	8 800	440
$MgSO_4 \cdot 7H_2O$	7 400	370
KH_2PO_4	3 400	170
Microelements[c]		
KI	166	0.83
H_3BO_3	1 240	6.2
$MnSO_4 \cdot 4H_2O$	4 460	22.3
$ZnSO_4 \cdot 7H_2O$	1 720	8.6
$Na_2MoO_4 \cdot 2H_2O$	50	0.25
$CuSO_4 \cdot 5H_2O$	5	0.025
$CoCl_2 \cdot 6H_2O$	5	0.025
Iron source[c]		
$FeSO_4 \cdot 7H_2O$	5 560	27.8
$Na_2EDTA \cdot 2H_2O$	7 460	37.3
Organic supplement[c]		
Myoinositol	20 000	100
Nicotinic acid	100	0.5
Pyridoxine-HCl	100	0.5
Thiamine-HCl	100	0.5
Glycine	400	2
Carbon source[d]		
Sucrose	Added as solid	30 000

11.3.1 Media components

It is useful to briefly consider some of the individual components of the stock solutions.

1. Macroelements

As is implied by the name, the stock solution supplies those elements required in large amounts for plant growth and development. Nitrogen, phosphorus, potassium, magnesium, calcium and sulphur (and carbon, which is added separately) are usually regarded as macroelements. These elements usually comprise at least 0.1% of the dry weight of plants.

Nitrogen is most commonly supplied as a mixture of nitrate ions (from the KNO_3) and ammonium ions (from the NH_4NO_3). Theoretically, there is an advantage in supplying nitrogen in the form of ammonium ions, as nitrogen must be in the reduced form to be incorporated into macromolecules. Nitrate ions, therefore, need to be reduced before incorporation. However, at high concentrations, ammonium ions can be toxic to plant cell cultures and uptake of ammonium ions from the medium causes acidification of the medium. In order to use ammonium ions as the sole nitrogen source, the medium needs to be buffered. High concentrations of ammonium ions can also cause culture problems by increasing the frequency of vitrification (the culture appears pale and 'glassy' and is usually unsuitable for further culture). Using a mixture of nitrate and ammonium ions has the advantage of weakly buffering the medium as the uptake of nitrate ions causes OH^- ions to be excreted.

Phosphorus is usually supplied as the phosphate ion of ammonium, sodium or potassium salts. High concentrations of phosphate can lead to the precipitation of medium elements as insoluble phosphates.

2. Microelements

These elements are required in trace amounts for plant growth and development, and have many and diverse roles. Manganese, iodine, copper, cobalt, boron, molybdenum, iron and zinc usually comprise the microelements, although other elements such as nickel and aluminium are frequently found in some formulations.

Iron is usually added as iron sulphate, although iron citrate can also be used. Ethylenediamine tetra-acetic acid (EDTA) is usually used in conjunction with the iron sulphate. The EDTA complexes with the iron so as to allow the slow and continuous release of iron into the medium. Uncomplexed iron can precipitate out of the medium as ferric oxide.

3. Organic supplements

Only two vitamins, thiamine (vitamin B_1) and myoinositol (considered a B vitamin) are considered essential for the culture of plant cells in vitro. However, other vitamins are often added to plant cell culture media for historical reasons.

Amino acids are also commonly included in the organic supplement. The most frequently used is glycine (arginine, asparagine, aspartic acid, alanine, glutamic acid, glutamine and proline are also used), but in many cases its inclusion is not essential. Amino acids provide a source of reduced nitrogen and, like ammonium ions, uptake causes acidification of the medium. Casein hydrolysate can be used as a relatively cheap source of a mix of amino acids.

4. Carbon source

Sucrose is cheap, easily available, readily assimilated and relatively stable and is, therefore, the most commonly used carbon source. Other carbohydrates (such as glucose, maltose, galactose and sorbitol) can also be used, and in specialised circumstances may prove superior to sucrose.

5. Gelling agents

Media for plant cell culture in vitro can be used in either liquid or 'solid' forms, depending on the type of culture being grown. For any culture types that require the plant cells or tissues to be grown on the surface of the medium, it must be solidified (more correctly termed 'gelled'). Agar, produced from seaweed, is the most common type of gelling agent, and is ideal for routine applications. However, because it is a natural product, the agar quality can vary from supplier to supplier and from batch to batch. For more demanding applications, a range of purer (and in some cases, considerably more expensive) gelling agents are available. Purified agar or agarose can be used, as can a variety of gellan gums.

These components, then, are the basic 'chemical' necessities for plant cell culture media. However, other additions are made in order to manipulate the pattern of growth and development of the plant cell culture.

11.4 PLANT GROWTH REGULATORS IN PLANT TISSUE CULTURE

We have already briefly considered the concepts of plasticity and totipotency. The essential point as far as plant cell culture is concerned is that, due to this plasticity and totipotency, specific media manipulations can be used to direct the development of plant cells in culture.

Plant growth regulators are the critical media components in determining the developmental pathway of the plant cells. The plant growth regulators used most commonly are plant hormones or their synthetic analogues. Generalisations about plant growth regulators and their use in plant cell culture media have been developed from initial observations made in the 1950s. There is, however, some considerable difficulty in predicting the effects of plant growth regulators: this is because of the great differences in culture response between species, cultivars and even plants of the same cultivar grown under different conditions.

However, some principles do hold true and have become the paradigm on which most plant tissue culture regimes are based. Auxins and cytokinins are the most widely used plant growth regulators in plant tissue culture and are usually used together, the ratio of the auxin to the cytokinin determining the type of culture established or regenerated (see Fig. 11.1). A high auxin to cytokinin ratio generally favours root formation, whereas a high cytokinin to auxin ratio favours shoot formation. An intermediate ratio favours callus production.

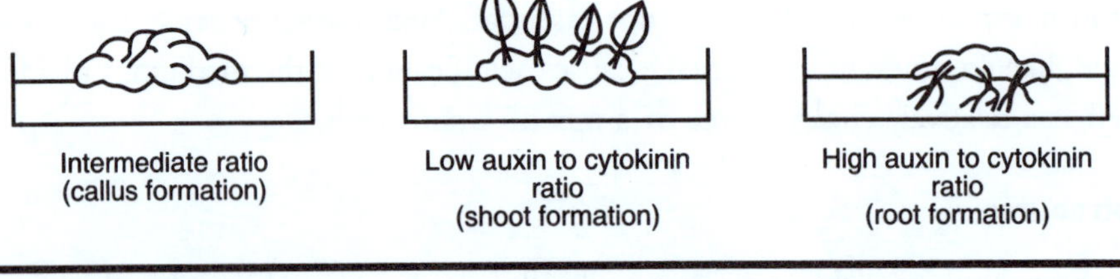

Fig. 11.1: The effect of different ratios of auxin to cytokinin on the growth and morphogenesis of callus. High auxin to cytokinin ratios promote root development, low ratios promote shoot development. Intermediate ratios promote continued growth of the callus without differentiation.

11.5 CULTURE TYPES

Cultures are generally initiated from sterile pieces of a whole plant. These pieces are termed 'explants', and may consist of pieces of organs, such as leaves or roots, or may be specific cell types, such as pollen or endosperm. Many features of the explant are known to affect the efficiency of culture initiation.

Generally, younger, more rapidly growing tissue (or tissue at an early stage of development) is most effective. Several different culture types most commonly used in plant transformation studies will now be examined in more detail.

11.5.1 Callus

Explants, when cultured on the appropriate medium, usually with both an auxin and a cytokinin, can give rise to an unorganized, growing and dividing mass of cells. It is thought that any plant tissue can be used as an explant, if the correct conditions are found. In culture, this proliferation can be maintained more or less indefinitely, provided that the callus is subcultured on to fresh medium periodically. During callus formation there is some degree of dedifferentiation (i.e., the changes that occur during development and specialization are, to some extent, reversed), both in morphology (callus is usually composed of unspecialised parenchyma cells) and metabolism. One major consequence of this dedifferentiation is that most plant cultures lose the ability to photosynthesise. This has important consequences for the culture of callus tissue, as the metabolic profile will probably not match that of the donor plant. This necessitates the addition of other components—such as vitamins and, most importantly, a carbon source to the culture medium, in addition to the usual mineral nutrients.

Callus culture is often performed in the dark (the lack of photosynthetic capability being no drawback) as light can encourage differentiation of the callus.

During long-term culture, the culture may lose the requirement for auxin and/or cytokinin. This process, known as 'habituation', is common in callus cultures from some plant species (such as sugar beet). Callus cultures are extremely important in plant biotechnology. Manipulation of the auxin to cytokinin ratio in the medium can lead to the development of shoots, roots or somatic embryos from which whole plants can subsequently be produced. Callus cultures can also be used to initiate cell suspensions, which are used in a variety of ways in plant transformation studies.

11.5.2 Cell-suspension cultures

Callus cultures, broadly speaking, fall into one of two categories: compact or friable. In compact callus the cells are densely aggregated, whereas in friable callus the cells are only loosely associated with each other and the callus becomes soft and breaks apart easily. Friable callus provides the inoculum to form cell-suspension cultures. Explants from some plant species or particular cell types tend not to form friable callus, making cell-suspension initiation a difficult task. The friability of callus can sometimes be improved by manipulating the medium components or by repeated subculturing. The friability of the callus can also sometimes be improved by culturing it on 'semi-solid' medium (medium with a low concentration of gelling agent). When friable callus is placed into a liquid medium (usually the same composition as the solid medium used for the callus culture) and then agitated, single cells and/or small clumps of cells are released into the medium. Under the correct conditions, these released cells continue to grow and divide, eventually producing a cell-suspension culture. A relatively large inoculum should be used when initiating cell suspensions so that the released cell numbers build up quickly. The inoculum should not be too large though, as toxic products released from damaged or stressed cells can build up to lethal levels. Large cell clumps can be removed during subculture of the cell suspension.

Cell suspensions can be maintained relatively simply as batch cultures in conical flasks. They are continually cultured by repeated subculturing into fresh medium. This results in dilution of the suspension and the initiation of another batch growth cycle. The degree of dilution during subculture should be determined empirically for each culture. Too great a degree of dilution will result in a greatly extended lag period or, in extreme cases, death of the transferred cells.

After subculture, the cells divide and the biomass of the culture increases in a characteristic fashion, until nutrients in the medium are exhausted and/or toxic by-products build up to inhibitory levels—this is called the 'stationary phase'. If cells are left in the stationary phase for too long, they will die and the culture will be lost. Therefore, cells should be transferred as they enter the stationary phase. It is, therefore, important that the batch growth-cycle parameters are determined for each cell-suspension culture.

11.5.3 Protoplasts

Protoplasts are plant cells with the cell wall removed. Protoplasts are most commonly isolated from either leaf mesophyll cells or cell suspensions, although other sources can be used to advantage. Two general approaches to removing the cell wall (a difficult task without damaging the protoplast) can be taken—mechanical or enzymatic isolation. Mechanical isolation, although possible, often results in low yields, poor quality and poor performance in culture due to substances released from damaged cells.

Enzymatic isolation is usually carried out in a simple salt solution with a high osmoticum, plus the cell wall degrading enzymes. It is usual to use a mix of both cellulase and pectinase enzymes, which must be of high quality and purity.

Protoplasts are fragile and easily damaged, and therefore, must be cultured carefully. Liquid medium is not agitated and a high osmotic potential is maintained, at least in the initial stages. The liquid medium must be shallow enough to allow aeration in the absence of agitation. Protoplasts can be plated out on to solid medium and callus produced. Whole plants can be regenerated by organogenesis or somatic embryogenesis from this callus. Protoplasts are ideal targets for transformation by a variety of means.

11.5.4 Root cultures

Root cultures can be established in vitro from explants of the root tip of either primary or lateral roots and can be cultured on fairly simple media. The growth of roots *in vitro* is potentially unlimited, as roots are indeterminate organs. Although the establishment of root cultures was one of the first achievements of modern plant tissue culture, they are not widely used in plant transformation studies.

11.5.5 Shoot tip and meristem culture

The tips of shoots (which contain the shoot apical meristem) can be cultured *in vitro*, producing clumps of shoots from either axillary or adventitious buds. This method can be used for clonal propagation. Shoot meristem cultures are potential alternatives to the more commonly used methods for cereal regeneration as they are less genotype-dependent and more efficient (seedlings can be used as donor material).

11.5.6 Embryo culture

Embryos can be used as explants to generate callus cultures or somatic embryos.

Both immature and mature embryos can be used as explants. Immature, embryo-derived embryogenic callus is the most popular method of monocot plant regeneration.

11.5.7 Microspore culture

Haploid tissue can be cultured *in vitro* by using pollen or anthers as an explant. Pollen contains the male gametophyte, which is termed the 'microspore'. Both callus and embryos can be produced from pollen. Two main approaches can be taken to produce in vitro cultures from haploid tissue. The first method depends on using the anther as the explant. Anthers (somatic tissue that surrounds and contains the pollen) can be cultured on solid medium (agar should not be used to solidify the medium as it contains inhibitory substances). Pollen-derived embryos are subsequently produced via dehiscence of the mature anthers. The dehiscence of the anther depends both on its isolation at the correct stage and on the correct culture conditions. In some species, the reliance on natural dehiscence can be circumvented by cutting the wall of the anther, although this does, of course, take a considerable amount of time. Anthers can also be cultured in liquid medium, and pollen released from the anthers can be induced to form embryos, although the efficiency of plant regeneration is often very low. Immature pollen can also be extracted from developing anthers and cultured directly, although this is a very time-consuming process. Both methods have advantages and disadvantages. Some beneficial effects to the culture are observed when anthers are used as the explant material.

There is, however, the danger that some of the embryos produced from anther culture will originate from the somatic anther tissue rather than the haploid microspore cells. If isolated pollen is used there is no danger of mixed embryo formation, but the efficiency is low and the process is time-consuming. In microspore culture, the condition of the donor plant is of critical importance, as is the timing of isolation. Pretreatments, such as a cold treatment, are often found to increase the efficiency. These pretreatments can be applied before culture, or, in some species, after placing the anthers in culture. Plant species can be divided into two groups, depending on whether they require the addition of plant growth regulators to the medium for pollen/anther culture; those that do also often require organic supplements, e.g., amino acids. Many of the cereals (rice, wheat, barley and maize) require medium supplemented with plant growth regulators for pollen/anther culture. Regeneration from microspore explants can be obtained by direct embryogenesis, or via a callus stage and subsequent embryogenesis. Haploid tissue cultures can also be initiated from the female gametophyte (the ovule). In some cases, this is a more efficient method than using pollen or anthers. The ploidy of the plants obtained from haploid cultures may not be haploid. This can be a consequence of chromosome doubling during the culture period. Chromosome doubling (which often has to be induced by treatment with chemicals such as colchicine) may be an advantage, as in many cases haploid plants are not the desired outcome of regeneration from haploid tissues.

Such plants are often referred to as 'di-haploids', because they contain two copies of the same haploid genome.

11.6 PLANT REGENERATION

Having looked at the main types of plant culture that can be established in vitro, we can now look at how whole plants can be regenerated from these cultures.

In broad terms, two methods of plant regeneration are widely used in plant transformation studies, i.e., somatic embryogenesis and organogenesis.

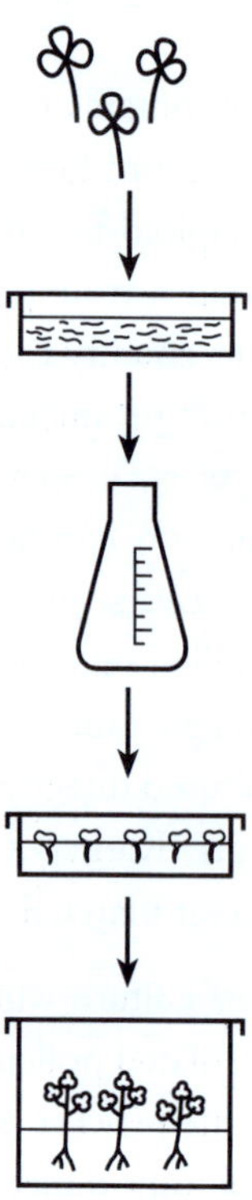

Fig. 11.2: Direct somatic embryogenesis in alfalfa. (1) Explants are removed from plants grown in vitro. (2) Explants are placed in liquid medium for embryo induction. (3) Embryos develop to the globular stage in liquid medium supplemented with maltose and polyethylene glycol. (4) Embryos mature on gelled medium containing abscisic acid (ABA). (5) Embryos develop into plants on solid medium.

11.6.1 Somatic embryogenesis

In somatic (asexual) embryogenesis, embryo-like structures, which can develop into whole plants in a way analogous to zygotic embryos, are formed from somatic tissues (Fig. 11.2). These somatic embryos can be produced either directly or indirectly. In direct somatic embryogenesis, the embryo is formed directly from a cell or small group of cells without the production of an intervening callus. Though common from some tissues (usually reproductive tissues such as the nucellus, styles or pollen), direct somatic embryogenesis is generally rare in comparison with indirect somatic embryogenesis. An example of direct somatic embryogenesis is explained in figure. In indirect somatic embryogenesis, callus is first produced from the explant. Embryos can then be produced from the callus tissue or from a cell suspension produced from that callus. Somatic embryo-genesis from carrot is the classical example of indirect somatic embryogenesis.

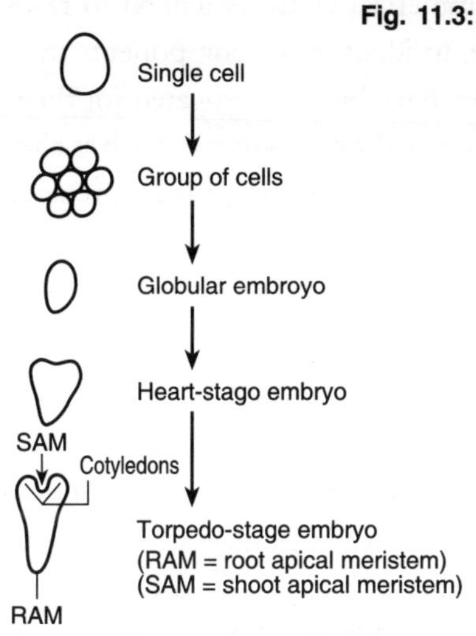

Fig. 11.3: A schematic representation of the sequential stages of somatic embryo development. Somatic embryos may develop from single cells or from a small group of cells. Repeated cell divisions lead to the production of a group of cells that develop into an organised structure known as a 'globular-stage embryo'. Further development results in heart and torpedo-stage embryos, from which plants can be regenerated. Zygotic embryos undergo a fundamentally similar development through the globular (which is formed after the 16-cell stage), heart and torpedo-stages. Polarity is established early in embryo development. Signs of tissue differentiation become apparent at the globular stage and apical meristems are apparent in heart-stage embryos.

Somatic embryogenesis usually proceeds in two distinct stages (Fig. 11.3). In the initial stage (embryo initiation), a high concentration of 2,4-D is used. In the second stage (embryo production) embryos are produced in a medium with no or very low levels of 2,4-D.

11.6.1.1 Cereal regeneration via somatic embryogenesis from immature or mature embryos

The principal method adopted for the tissue culture and regeneration of a wide range of cereal species is somatic embryogenesis, using cultures initiated from immature zygotic embryos. Embryogenic callus is normally initiated by placing the immature embryo on to a medium containing 2,4-D. Shoot regeneration is initiated by placing the embryogenic callus on a medium with BAP (with or without 2,4-D). These shoots can be subsequently rooted. The medium used for induction of embryogenic callus is usually a modified MS (for Triticeae) or N6 (for rice and maize). Maltose is often used as the carbon source in preference to sucrose, and additional organic supplements (such as specific amino acids, yeast extract and/or casein hydrolysate) are common. The isolation and culture of immature embryos is, however, a labour-intensive and relatively expensive procedure. An additional problem is the small target size if the immature embryos are to used for biolistic transformation. Alternatives are, therefore, being sought. One alternative is to use mature embryos (or seeds) as the explant to initiate embryogenic callus. This approach has been successfully applied to several cereal species such as rice and oats. The culture techniques and media used for culture establishment and regeneration from mature embryo/seed-derived cultures are fundamentally the same as those used for cultures initiated from immature embryos.

11.6.1.2 The importance of genotype

The major influence on tissue-culture response appears to be genetic, with culture requirements varying between species and cultivars. Model genotypes that responded well to culture *in vitro* were initially used in plant transformation studies. However, most of the model genotypes used were not elite, commercial cultivars. The commercial cultivars tended to respond poorly to culture *in vitro*. One of the main aims is, therefore, to identify the components that make up a widely applicable, optimal culture regime. Many factors have been investigated for their ability to improve the culture response from elite cultivars, including media components (such as alternative carbon sources, macro- and microelement concentrations and composition), media preparation method and donor plant condition and growth condition.

11.6.2 Organogenesis

Somatic embryogenesis relies on plant regeneration through a process analogous to zygotic embryo germination. Organogenesis relies on the production of organs, either directly from an explant or from a callus culture. There are three methods of plant regeneration via organogenesis. The first two methods depend on adventitious organs arising either from a callus culture or directly from an explant (Fig. 11.4). Alternatively, axillary bud formation and growth can also be used to regenerate whole plants from some types of tissue culture. Organogenesis relies on the inherent plasticity of plant tissues, and is regulated by altering the components of the medium. In particular, it is the auxin to cytokinin ratio of the medium that determines which developmental pathway the regenerating tissue will take.

It is usual to induce shoot formation by increasing the cytokinin to auxin ratio of the culture medium. These shoots can then be rooted relatively simply.

11.7 INTEGRATION OF PLANT TISSUE CULTURE INTO PLANT TRANSFORMATION PROTOCOLS

Various methods of plant regeneration are available to the plant biotechnologist. Some plant species may be amenable to regeneration by a variety of methods, but some may only be regenerated by one method. Although there are various methods that can be used to transform plants, but here we will be briefly considering the interaction of plant regeneration methodology and transformation methodology only (Fig. 11.4). Not all plant tissue is suited to every plant transformation method, and not all plant species can be regenerated by every method. There is, therefore, a need to find both a suitable plant tissue culture/regeneration regime and a compatible plant transformation methodology.

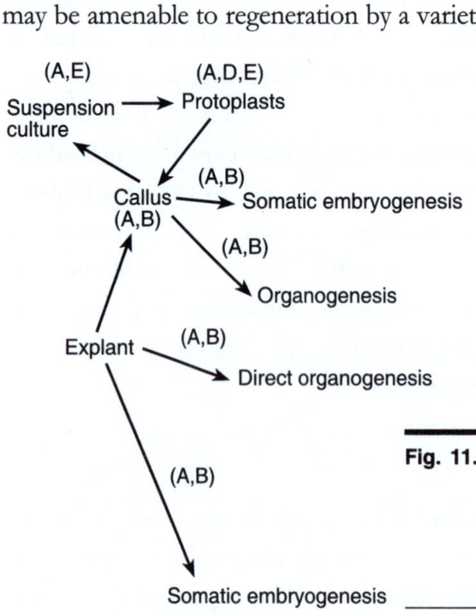

Fig. 11.4: A simplified scheme for the integration of plant tissue culture into plant transformation protocols

Tissue culture and plant regeneration are an integral part of most plant transformation strategies, and can often prove to be the most challenging aspect of a plant transformation protocol. Key to success in integrating plant tissue culture into plant transformation strategies is the realisation that a quick (to avoid too many deleterious effects from somaclonal variation) and efficient regeneration system must be developed. However, this system must also allow high transformation efficiencies from whichever transformation technique is adopted. Not all regeneration protocols are compatible with all transformation techniques. Some crops may be amenable to a variety of regeneration and transformation strategies, others may currently only be amenable to one particular protocol. Advances are being made all the time, so it is impossible to say that a particular crop will never be regenerated by a particular protocol. However, some protocols, at least at the moment, are clearly more efficient than others. Regeneration from immature embryo-derived somatic embryos is, for example, the favoured method for regenerating monocot species. This chapter has only scratched the surface of the problems and potentials of plant tissue culture. General rules are sometimes difficult to make because of the variability of response to particular protocols of different plant species or even cultivars. However, plant tissue culture has been successfully integrated into plant transformation strategies and the list of plant species that can be routinely transformed continues to grow

11.8 APPLICATIONS OF PLANT TISSUE CULTURE

Plant tissue culture is used widely in plant science; it also has a number of commercial applications. Applications include:

- Micropropagation can be used to produce large numbers of identical individuals.
- Micropropagation is widely used in forestry and in floriculture. Micropropagation can also be used to conserve rare or endangered plant species.
- A plant breeder may use tissue culture to screen cells rather than plants for advantageous characters, e.g., herbicide resistance/tolerance.
- Large-scale growth of plant cells in liquid culture inside bioreactors as a source of secondary products, like recombinant proteins used as biopharmaceuticals.
- To cross distantly related species by protoplast fusion and regeneration of the novel hybrid.
- To cross-pollinate distantly related species and then tissue culture the resulting embryo which would otherwise normally die (Embryo Rescue).
- For production of doubled monoploid plants from haploid cultures to achieve homozygous lines more rapidly in breeding programmes, usually by treatment with colchicine which causes doubling of the chromosome number.
- As a tissue for transformation, followed by either short-term testing of genetic constructs or regeneration of transgenic plants.
- Certain techniques such as meristem tip culture may be employed that can be used to produce clean plant material from virused stock, such as potatoes and many species of soft fruit.

11.9 MICROBIAL CELL CULTURE

Micro-organism grown in pure culture is simpler to use than animal and plant cells culture. The composition of the medium largely depends upon latter's nutritional classification as chemolithotroph, chemo-organotroph, photolithotroph, or photoorganotroph and whether or not it is a wild type or mutant cell line. Typical minimal media for a fully biosynthetically competent chemo-organotroph would contain Na^{2+}, K^+, Ca^{2+}, Mg^{2+}, NH_4^{2+}, or NO_3^-, Cl^-, HPO_2^{2-} and SO_4^{2-} and a simple carbon source such as glucose.

Microbial culture can be performed in either closed or open system. In closed system finite amount of medium is supplied and growth continues until either one factor in the medium becomes limiting or a toxic by product of growth accumulates. In open system of so-called continous culture, outflow from the system and cells is maintained by addition of equal volume of fresh medium.

Closed systems also called as batch culture have disadvantage that they usually give rise to unbalanced growth with heterogeneous population at any given point in the life cycle. If fresh medium is subsequently added increasing the total culture volume they are called as fed-batch culture. The objective of fed-batch culture is to maintain important nutrient in a narrow concentration range. The specific advantage of continuous culture over batch culture is that former provides growth in steady state conditions in which there is tight coupling between cell division and biosynthesis. It follows that if rates of cell division and biosynthesis are constant and equal, the mean composition of cells in culture does not vary with time, and growth is said to be balanced.

Microbial cultures have widespread industrial uses for large scale production of alcohol, amiono acids, coenzymes, growth regulators, organic acids, polysaccharides, solvents, sterols, surfactants, vitamins and much more. Microbial degradation of waste products especially agriculture and food industries waste is an important application of microbial culture. Biomass production as single cell protein (SCP) following growth of micro-organisms on different substrate is a further important use. Micro-organisms also have value in replacing animals in preliminary toxicological testing of xenobiotics.

11.10 TIPS TO REMEMBER

Totipotency

The regeneration of whole organisms (plant) depends upon the concept that all plant cells can, given the correct stimuli, express the total genetic potential of the parent plant. This maintenance of genetic potential is called 'totipotency'.

Classes of plant growth regulators

There are five main classes of plant growth regulator used in plant cell culture,namely:

1. auxins;

2. cytokinins;
3. gibberellins;
4. abscisic acid;
5. ethylene.

Two methods of plant regeneration are widely used in plant transformation studies, i.e., somatic embryogenesis and organogenesis.

11.11 VIVA VOCE

Q.1. Why plant tissue or cell culture is necessary?

Q.2. What are the major components of plant tissue culture media? And also list their sources.

Q.3. What is plant regeneration and what are the ways by which it is achieved?

Q.4. What are the different types of plant tissue or cell cultures?

Q.5. What are plant growth regulators?

Q.6. How plant growth regulators are important for plant tissue culture?

Q.7. What are the differences between protoplast and callus culture?

Q.8. Name some common plant tissue culture media.

Q.9. What is plant transformation and how is it different from plant regeneration?

Q.10. Explain explant material.

Q.11. How microbial culture differs from plant and animal culture?

Q.12. List the main components of typical microbial culture.

Q.13. What are the applications of microbial culture?

CHAPTER 12 Spectroscopic Techniques

12.1 PRINCIPLES OF SPECTROSCOPY

Spectroscopy is the study of the interaction of electromagnetic radiation with matter. When matter is energized (excited) by the application of thermal, electrical, nuclear or radiant energy, electromagnetic radiation is often emitted as the matter relaxes back to its original (ground) state. The spectrum of radiation emitted by a substance that has absorbed energy is called an emission spectrum and the science is appropriately called emission spectroscopy.

Another approach often used to study the interaction of electromagnetic radiation with matter is one whereby a continuous range of radiation (e.g., white light) is allowed to fall on a substance; then the frequencies absorbed by the substance are examined. The resulting spectrum from the substance contains the original range of radiation with dark spaces that correspond to missing, or absorbed frequencies. This type of spectrum is called an absorption spectrum. In spectroscopy the emitted or absorbed radiation is usually analyzed, i.e., separated into the various frequency components and the intensity is measured by means of an instrument called spectrometer.

The resultant spectrum is mainly a graph of intensity of emitted or absorbed radiation versus wavelength or frequency. There are in general three types of spectra: continuous, line, and band. The sun and heated solids produce continuous spectra in which the emitted radiation contains all frequencies within a region of the electromagnetic spectrum. A rainbow and light from a light bulb are examples of continuous spectra. Line spectra are produced by excited atoms in the gas phase and contain only certain frequencies, all other frequencies being absent. Each chemical element of the periodic chart has a unique and, therefore, characteristic line spectrum. Band spectra are produced by excited molecules emitting radiation in groups of closely spaced lines that merge to form bands.

These categories of emission and absorption spectra contain tremendous amounts of useful information about the structure and composition of matter. Spectroscopy is a powerful and sensi-

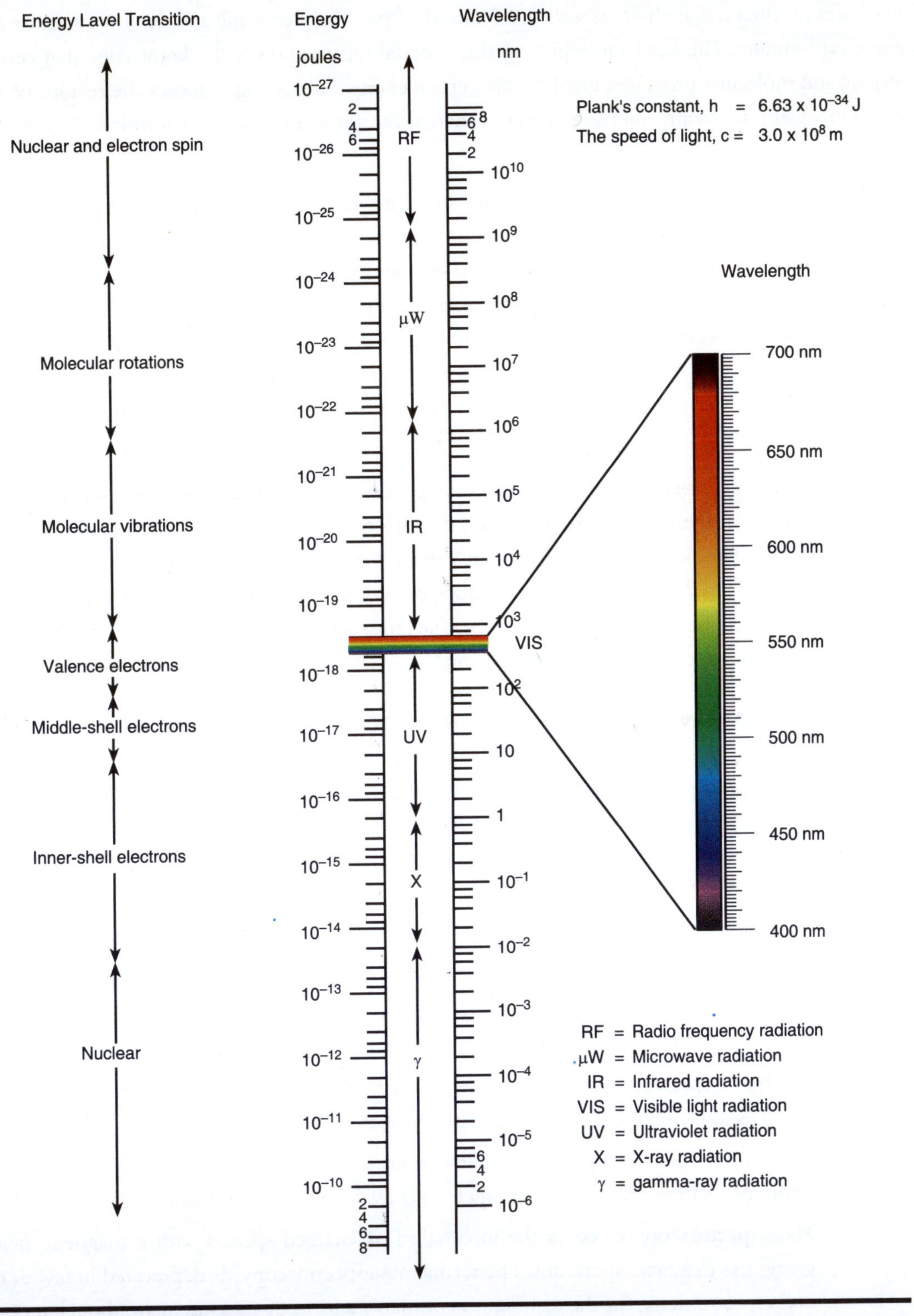

Fig. 12.1: The electromagnetic radiation ruler

tive form of chemical analysis, as well as a method of probing electronic and nuclear structure and chemical bonding. The key to interpreting this spectral information is the knowledge that certain atomic and molecular processes involve only certain energy ranges. Fig. 1 shows the regions of the electromagnetic spectrum and the associated energy transitions that occur in atomic and molecular processes.

Much of the scientific knowledge of the structure of the universe, from stars to atoms, is derived from interpretations of the interaction of radiation with matter. One example of the power of these techniques is the determination of the composition, the velocities, and the evolutionary dynamics of stars. The source of the incredible amount of energy produced by the sun is nuclear fusion reactions going on within the hot interior (temperature 40 × 106 K). Two fusion cycles, the carbon cycle and the proton cycle, convert hydrogen nuclei into helium nuclei via heavier nuclei, such as carbon 12 and nitrogen 14. The enormous radiation of energy from the hot core seethes outwards by convection.

This radiation consists of the entire electromagnetic spectrum as a continuous spectrum. Towards the surface of the sun (the photosphere), the different elements all absorb at their characteristic frequencies. The radiation that shoots into space toward the earth is a continuous emission spectrum with about 22,000 dark absorption lines present in it (Fraunhofer lines), of which about 70% have been identified. These absorption lines, i.e., missing frequencies, prove that more than 60 terrestrial elements are certainly present in the sun.

12.2 CLASSIFICATION OF SPECTROSCOPIC METHODS

Different spectroscopic techniques have been classified mainly on two parameters, first what type of radiation is to be measured or by what measurement procedure is employed.

1. Nature of radiation measured

This category of spectroscopy depends on the physical quantity measured. Normally, the quantity that is measured is an amount or intensity of something.

- Electromagnetic spectroscopy involves interactions with electromagnetic radiation, or light. Ultraviolet-visible spectroscopy is an example.
- Electronic spectroscopy involves interactions with electron beams. Auger spectroscopy involves inducing the Auger effect with an electron beam.
- Mechanical spectroscopy involves interactions with macroscopic vibrations, such as phonons. An example is acoustic spectroscopy, involving sound waves.
- Mass spectroscopy involves the interaction of charged species with a magnetic field, giving rise to a mass spectrum. The term "mass spectroscopy" is deprecated in favour of mass spectrometry, for the technique is primarily a form of measurement, though it does produce a spectrum for observation.

2. Measurement process

Most spectroscopic methods are differentiated as either atomic or molecular based on whether or not they apply to atoms or molecules. Along with that distinction, they can be classified on the nature of their interaction:

- Absorption spectroscopy uses the range of the electromagnetic spectra in which a substance absorbs. This includes atomic absorption spectroscopy and various molecular techniques, such as infrared spectroscopy in that region and nuclear magnetic resonance (NMR) spectroscopy in the radio region.

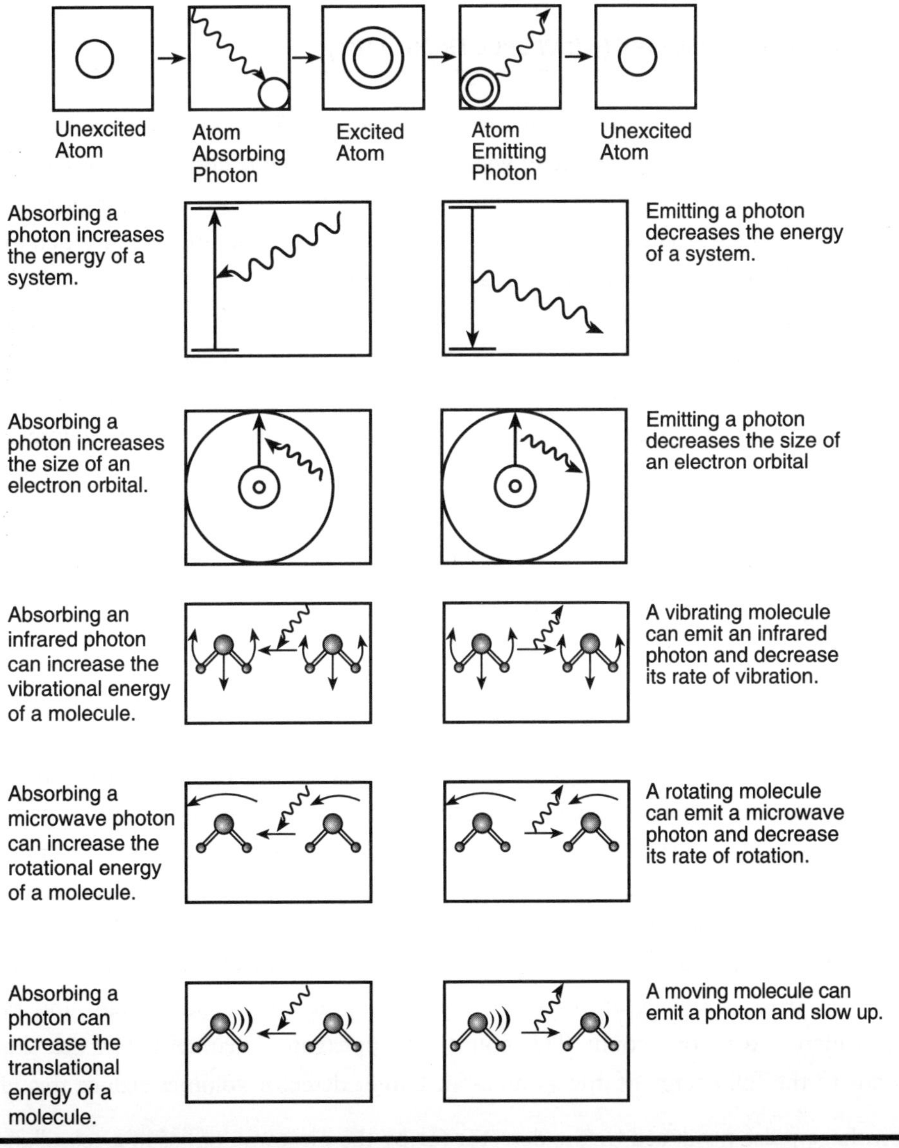

Fig. 12.2: Fundamental processes during spectroscopy

- Emission spectroscopy uses the range of electromagnetic spectra in which a substance radiates (emits). The substance first must absorb energy. This energy can be from a variety of sources, which determines the name of the subsequent emission, like luminescence. Molecular luminescence techniques include spectrofluorimetry.
- Scattering spectroscopy measures the amount of light that a substance scatters at certain wavelengths, incident angles, and polarisation angles. The scattering process is much faster than the absorption/emission process. One of the most useful applications of light scattering spectroscopy is Raman spectroscopy.

Now we will analyze details of different spectroscopic methods:

12.3 GAMMA SPECTROSCOPY (γ-RAY SPECTROSCOPY)

Gmma spectroscopy is a radionuclidic measurement method. While a Geiger counter determines only the count rate, a gamma spectrometer will determine the energy and the count rate of gamma-rays emitted by radioactive substances. Gamma spectroscopy is an extremely important method. Most radioactive sources produce gamma-rays of various energies and intensities. When these emissions are collected and analyzed with a gamma spectroscopy system, a gamma energy spectrum can be produced. A detailed analysis of this spectrum is typically used to determine the identity and quantity of gamma emitters present in the source. The gamma spectrum is characteristic of the gamma emitting nuclides contained in the source, just as in optical spectroscopy, the optical spectrum is characteristic of the atoms and molecules contained in the probe. The equipment used in gamma spectroscopy includes an energy sensitive radiation detector, a pulse sorter (multichannel analyzer), and associated amplifiers and data readout devices. The most common detectors include sodium iodide (NaI) scintillation counter and high purity germanium detectors.

Instrumentation

A gamma spectroscopy system consists of a detector, electronic system to collect and process the signals produced by the detector, and a computer with processing software to generate the spectrum, and display and store it for analysis. Gamma spectroscopy detectors are passive materials that wait for a gamma interaction to occur in the detector volume. The most important interaction mechanisms are the photoelectric effect, the Compton effect, or pair production. The photoelectric effect is preferred, as it absorbs all of the energy of the incident gamma-ray. Full energy absorption is also possible when a series of these interaction mechanisms take place within the detector volume. When a gamma-ray undergoes a Compton interaction or Pair Production, and a portion of the energy escapes from the detector volume without being absorbed, the background rate in the spectrum is increased by one count. This count will appear in a channel below the channel that corresponds to the full energy of the gamma-ray. Larger detector volumes reduce this effect.

The voltage pulse produced by the detector (or by the photomultiplier in a scintillation detector) is shaped by a multichannel analyzer (MCA). The multichannel analyzer takes the very small voltage signal produced by the detector, reshapes it into a Gaussian or trapezoidal shape, and converts it into a digital signal. In some systems, the analog to digital conversion is performed before the peak is reshaped. The analog to digital converter (ADC) also sorts the pulses by their height. ADCs have specific numbers of "bins" to sort the pulses into; these are the "channels" in the spectrum. The number of channels can be changed in most modern gamma spectroscopy system by changing a software or hardware setting. The number of bins is a power of two. Common values include 512, 1024, 2048, 4096, 8192, or 16384 channels. The choice of number of channels depends on the resolution of the system and the energy range being studied. The MCA output is sent to a computer which stores, displays, and analyzes the data. A variety of software packages are available from several manufacturers, and generally include spectrum analysis tools such as energy calibration, peak area and net area calculation, and resolution calculation. Other components, such as rate meters and peak position stabilizers, may also be included.

Gamma spectroscopy systems are selected to take advantage of several performance characteristics. Two of the most important include detector resolution and detector efficiency.

1. Detector Resolution

Gamma rays detected in a spectroscopic system produce peaks in the spectrum. These peaks can also be called lines, by analogy to optical spectroscopy. The width of the peaks is determined by the resolution of the detector, a very important characteristic of gamma spectroscopic detectors. Resolution is analogous to Resolving power in optical spectroscopy. High resolution enables the spectroscopist to separate two gamma lines which are close to each other. Gamma spectroscopy systems are designed and adjusted to produce symmetrical peaks of the best possible resolution. The peak shape is usually a Gaussian distribution. In most spectra, the horizontal position of the peak is determined by the gamma-ray's energy, and the area of the peak is determined by the intensity of the gamma-ray and the efficiency of the detector. The most common figure used to express detector resolution is Full Width at Half Maximum (FWHM). This is the width of the gamma-ray peak at half of the highest point on the peak distribution. Resolution figures should be given with reference to specified gamma-ray energies. Resolution can be expressed in absolute terms (eV or keV), or relative terms. For example, a NaI detector may have a FWHM of 9.15 keV at 122 keV, and 82.75 keV at 662 keV. These are resolution values expressed in absolute terms. To express the resolution in relative terms, the FWHM in eV or keV are divided by the energy of the gamma-ray and multiplied by 100. In these examples, the resolution is 7.5% at 122 keV, and 12.5% at 662 keV. A Germanium detector might give resolution of 560 eV at 122 keV, a relative resolution of 0.46%.

2. Detector Efficiency

Not all gamma-rays that are emitted by the source and pass through the detector will produce a count in the system. The probability that an emitted gamma-ray will interact with the detector and produce a count is the efficiency of the detector. High efficiency detectors produce spectra in less time than low efficiency detectors. In general, larger detectors have higher efficiency than smaller detectors, although the shielding properties of the detector material are also an important factor. Detector efficiency is measured by taking a spectrum from a source of known activity, and comparing the count rates in each peak to the count rates expected from the known intensities of each gamma-ray. Efficiency, like resolution, can also be expressed in absolute or relative terms. The same unit are used, percentages, so the spectroscopist must take care to determine which kind of efficiency is being given for the detector. Absolute efficiency values give the probability that a gamma-ray of a specified energy passing through the detector will interact with the crystal and be detected. Relative efficiency values are often used for Germanium detectors, and compare the efficiency of the detector at 1332 keV to that of a 3" by 3" NaI detector (1.2E-3 cps/Bq). Relative efficiency values greater than 100% can, therefore, be encountered when working with very large Germanium detectors. The energy of the gamma-rays being detected is an important factor in the efficiency of the detector. By plotting the efficiency at various energies, an efficiency curve can be obtained. This curve can then be used to determine the efficiency of the detector at energies different from those used to obtain the curve.

Scintillation Detectors

Scintillation detectors use crystals that emit light when gamma-rays interact with the atoms in the crystals. The intensity of the light produced is proportional to the energy deposited in the crystal by the gamma-ray. The mechanism is similar to that in a Thermoluminescent Dosimeter. The detectors are joined to photomultipliers that convert the light into electrons and amplify the electrical signal provided by the electrons. Because the photomultipliers are also sensitive to ambient light, scintillators are packaged in light-tight coverings. Common scintillators include Thallium-doped Sodium Iodide (NaI(Tl)), often simply called Sodium Iodide (NaI) detectors, and Bismuth germanate (BGO) detectors. Scintillation detectors also have other uses, such as alpha- and beta-detectors. In details these detectors have been discussed in chapter entitled radioisotopic techniques.

(a) **Sodium Iodide {NaI(Tl)} Detectors:** Thallium-doped Sodium iodide has two principal advantages; it can be produced in large crystals, giving good efficiency, and it produces intense bursts of light compared to other spectroscopic scintillators. It is also convenient to use, making it popular for field work such as identification of unknown materials for law enforcement purposes. An example of a NaI spectrum is the gamma spectrum of the isotope ^{137}Cs shown in the illustration. ^{137}Cs emits a single gamma line of 662 keV. It should be noted that the 662 keV line shown is actually produced by ^{137}Ba$_m$, the decay product of ^{137}Cs, which is in secular equilibrium with ^{137}Cs.

The spectrum was measured using a NaI-crystal on a photomultiplier, an amplifier and a multichannel analyzer and plotted on an x-y-plotter. The figure shows the number of counts (within the measuring period) versus channel number. The spectrum shows the following peaks (from left to right):

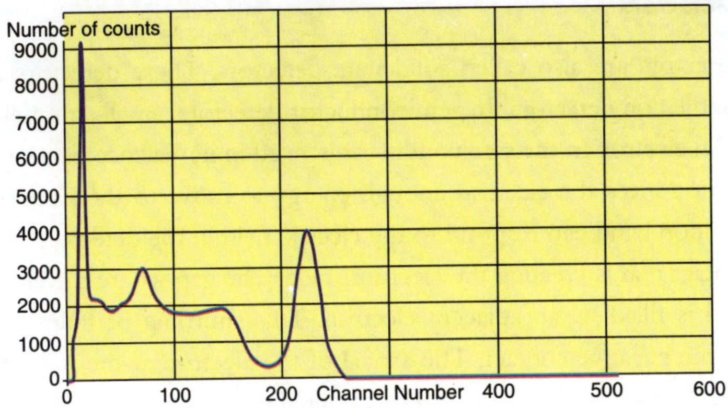

Fig. 12.3: The spectrum using NaI-crystal

- low energy x radiation (due to internal conversion of the gamma-ray)
- a backscatter peak at the low energy end of the Compton distribution
- a photopeak (Full Energy Peak) at an energy of 662 keV

The Compton distribution is a continuous distribution which goes up to channel 150 in this figure. It is due to primary gamma-rays undergoing Compton effect within the crystal. Depending on the scattering angle, the Compton electrons have different energies and hence produce pulses of different heights. If many gamma-rays are present in a spectrum, Compton distributions are a disturbing nuisance. In order to reduce them, one can use an anticoincidence shield. This is especially useful for small Ge(Li) detectors.

The next figure shows another example: the gamma spectrum of the isotope ^{60}Co with two gamma-rays with 1.17 MeV and 1.33 MeV, respectively, again measured by a NaI counter. The two gamma lines can be seen well separated; the rise to the left of channel 200 probably indicates a strong background that has not been subtracted. At channel 150, one can see a backscatter peak (like in the figure above). The multichannel spectrum was plotted by means of an x-y plotter. Sodium Iodide systems, like all scintillator systems, are sensitive to temperature changes. Changes in the operating temperature caused by changes in environmental temperature will shift the spectrum on the horizontal axis. Peak shifts of tens of channels or more are commonly observed. Spectrum stabilizers are used to prevent such shifts. Because of their poor resolution, NaI detectors are not suitable for the identification of complicated mixtures of gamma-ray producing materials. Such analyses require higher resolution.

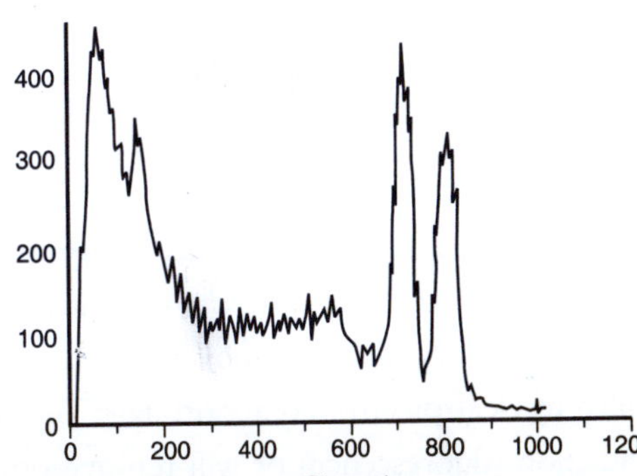

Fig. 12.4: The gamma spectrum of the isotope ^{60}Co

Semiconductor Detectors

Semiconductor detectors are also called solid-state detectors. These detectors are fundamentally different from Scintillation detectors. In semiconductor detectors, an electric field is applied to the detector volume. An electron in the semiconductor is fixed in its Valence band in the crystal until a gamma-ray interaction gives the electron enough energy to move to the Conduction band. Electrons in the conduction band can respond to the electric field in the detector, and therefore, move to the positive contact that is creating the electrical field. The gap where the electron used to be is called a "hole" and is filled by an adjacent electron. This shuffling of holes effectively moves a positive charge to the negative contact. The arrival of the electron at the positive contact and the hole at the negative contact produces the electrical signal that is sent to the preamplifier, the MCA, and on through the system for analysis. The movement of electrons and holes in a solid state detector is very similar to the movement of ions within the sensitive volume of gas-filled detectors such as ionization chambers. Common semiconductor detectors include Germanium, Cadmium Telluride, and Cadmium Zinc Telluride. Germanium Detectors produce much higher energy resolution than NaI, as shown in the discussion of resolution above. Germanium detectors produce the highest resolution commonly available today.

Calibration, background

If the spectrometer is used to identify a sample of unknown composition, its energy scale must be calibrated first. This is done using the peaks of a known source (like ^{137}Cs or ^{60}Co shown above). Since the channel number is proportional to energy, the channel scale can then be converted to an energy scale. If the size of the detector crystal is known, one can also perform an intensity calibration, so that not only the energies but also the intensities of an unknown source (or the amount of a certain isotope in the source) can be determined. Because some radioactivity is present everywhere, one must also determine the "background", i.e., the spectrum when no source is present. The background must then be subtracted from the actual measurement. By using lead absorbers around the apparatus, one can reduce the background.

12.4 X-RAY SPECTROSCOPY

X-ray spectroscopy is a gathering name for several spectroscopic techniques for determining the electronic structure of materials by using X-ray excitation. When X-rays of sufficient frequency (energy) interact with a substance, inner shell electrons in the atom are excited to outer empty orbitals, or they may be removed completely, ionizing the atom. The inner shell "hole" will then be filled by electrons from outer orbitals. The energy available in this de-excitation process is emitted as radiation (fluorescence) or will remove other less-bound electrons from the atom (Auger effect). The absorption or emission frequencies (energies) are characteristic of the specific atom. In addition, for a specific atom small frequency (energy) variations occur which are characteristic of the chemical bonding. With a suitable apparatus, these characteristic X-ray frequencies or Auger electron energies can be measured. X-ray absorption and emission spectroscopy is used in chemistry and material sciences to determine elemental composition and chemical bonding.

X-ray crystallography is a scattering process; crystalline materials scatter X-rays at well-defined angles. If the wavelength of the incident X-rays is known, this allows calculation of the distances between planes of atoms within the crystal. The intensities of the scattered X-rays give information about the atomic positions and allow the arrangement of the atoms within the crystal structure to be calculated.

X-ray emission spectroscopy

Karl Manne Georg Siegbahn from Uppsala, Sweden (Nobel Prize 1924), who painstakingly produced numerous diamond-ruled glass diffraction gratings for his spectrometers, was one of the pioneers in developing X-ray emission spectroscopy (also called X-ray fluorescence spectroscopy). He measured the X-ray wavelengths of many elements to high precision, using high-energy electrons as excitation source. Intense and wavelength-tunable X-rays are now typically generated with synchrotrons. In a material, the X-rays may suffer an energy loss compared to the incoming beam. This energy loss of the re-emerging beam reflects an internal excitation of the atomic system, an X-ray analogue to the well-known Raman spectroscopy that is widely used in the optical region. In the X-ray region there is sufficient energy to probe changes in the electronic state (transitions between orbitals; this is in contrast with the optical region, where the energy loss is often due to changes in the state of the rotational or vibrational degrees of freedom). For instance, in the ultra soft X-ray region (below about 1 keV), crystal field excitations give rise to the energy loss.

We may think of the photon-in-photon-out process as a scattering event. When the X-ray energy corresponds to the binding energy of a core level electron this scattering process is resonantly enhanced by many orders of magnitude. This type of X-ray emission spectroscopy is often referred to as resonant inelastic X-ray scattering (RIXS). Due to the wide separation of orbital energies of the core levels, it is possible to select a certain atom of interest. The small spatial extent of core level orbitals forces the RIXS process to reflect the electronic structure in close vicinity of the chosen atom. Thus RIXS experiments give valuable information about the local electronic structure of complex systems, and theoretical calculations are relatively simple to perform.

Instrumentation

There exist several efficient designs for analyzing an X-ray emission spectrum in the ultra soft X-ray region. The figure of merit for such instruments is the spectral throughput, i.e., the product of detected intensity and spectral resolving power. Usually, it is possible to change the parameters within a certain range while keeping their product constant.

Grating spectrometers

Typically, the X-rays emerging from a sample must pass a source-defining slit, then optical elements (mirrors and/or gratings) disperse them by diffraction according to their wavelength and, finally, a detector is placed at their focal points.

Spherical grating mounts

Henry Augustus Rowland (1848-1901) devised an instrument that allowed the use of a single optical element that combines diffraction and focusing: a spherical grating. Reflectivity of X-rays is low regardless of the used material and, therefore, grazing incidence upon the grating is necessary. X-ray beams impinging on a smooth surface at a few degrees glancing angle of incidence undergo external total reflection which is taken advantage of to enhance the instrumental efficiency substantially. Denote by R the radius of a spherical grating. Imagine a circle with half the radius R tangent to the centre of the grating surface. This small circle is called the Rowland circle. If the entrance slit is anywhere on this circle, then a beam passing the slit and striking the grating will be split into a specularly reflected beam, and beams of all diffraction orders, that come into focus at certain points on the same circle.

Plane grating mounts

Similar as in optical spectrometers, a plane grating spectrometer first needs optics that turns the divergent rays emitted by the X-ray source into a parallel beam. This may be achieved by using a parabolic mirror. The parallel rays emerging from this mirror strike a plane grating (with constant groove distance) at the same angle and are diffracted according to their wavelength. A second parabolic mirror then collects the diffracted rays at a certain angle and creates an image on a detector. A spectrum within a certain wavelength range can be recorded simultaneously by using a 2-dimensional position sensitive detector such as a microchannel photomultiplier plate or an X-ray sensitive CCD chip (film plates are also possible to use).

Interferometers

Instead of using the concept of multiple beam interference that gratings produce, one may simply let two rays interfere. By recording the intensity of two such co-linearly at some fixed point and changing their relative phase one obtains an intensity spectrum as a function of path length difference. One can show that this is equivalent to a Fourier transformed spectrum as a function of frequency. The highest recordable frequency of such a spectrum is dependent on the minimum step size chosen in the scan and the frequency resolution (i.e., how well a certain wave can be defined in terms of its frequency) depends on the maximum path length difference achieved. The latter feature allows a much more compact design for achieving higher resolution than for a grating spectrometer because X-ray wavelengths are small compared to attainable path length differences. Other important type of X-ray spectroscopic techniques include X-ray absorption spectroscopy and X-ray magnetic circular dichroism.

12.5 ULTRAVIOLET-VISIBLE SPECTROSCOPY

Ultraviolet-visible spectroscopy or ultraviolet-visible spectrophotometry (UV/VIS) involves the spectroscopy of photons and spectrophotometry. It uses light in the visible and adjacent to ultraviolet (UV) and near infrared (NIR) ranges. In this region of energy space, molecules undergo electronic transitions.

All atoms absorb in the UV region because photons are energetic enough to excite outer electrons. If the frequency is high enough, photoionisation takes place. UV spectroscopy is also used in quantifying protein and DNA concentration as well as the ratio of protein to DNA concentration in a solution. Several amino acids usually found in protein, such as tryptophan, absorb light in the 280 nm range and DNA absorbs light in the 260 nm range. For this reason, the ratio of 260/280 nm absorbance is a good general indicator of the relative purity of a solution in terms of these two macromolecules. Reasonable estimates of protein or DNA concentration can also be made this way using Beer's law.

UV/Vis spectroscopy is routinely used in the quantitative determination of solutions of transition metal ions and highly conjugated organic compounds. Solutions of transition metal ions can be coloured (i.e., absorb visible light) because electrons within the metal atoms can be excited from one electronic state to another. The colour of metal ion solutions is strongly affected by the presence of other species, such as certain anions or ligands. For instance, the colour of a dilute solution of copper sulphate is very light blue; adding ammonia intensifies the colour and changes the wavelength of maximum absorption (λ max). Organic compounds, especially those with a high degree of conjugation, also absorb light in the UV or visible regions of the electromagnetic spectrum. The solvents for these determinations are often water for water soluble compounds, or ethanol for organic-soluble compounds. (Organic solvents may have significant UV absorption; not all solvents are suitable for use in UV spectroscopy. Ethanol absorbs very weakly at most wavelengths). While charge transfer complexes also give rise to colours, the colours are often too intense to be used for quantitative measurement. The Beer-Lambert law states that the absorbance of a solution is directly proportional to the solution's concentration. Thus UV/Vis spectroscopy can be used to determine the concentration of a solution. It is necessary to know how quickly the absorbance changes with concentration. This can be taken from references (tables of molar extinction coefficients), or more accurately, determined from a calibration curve. A UV/Vis spectrophotometer may be used as a detector for HPLC. The presence of an analyte gives a response which can be assumed to be proportional to the concentration. For accurate results, the instrument's response to the analyte in the unknown should be compared with the response to a standard; this is very similar to the use of calibration curves. The response (e.g., peak height) for a particular concentration is known as the response factor.

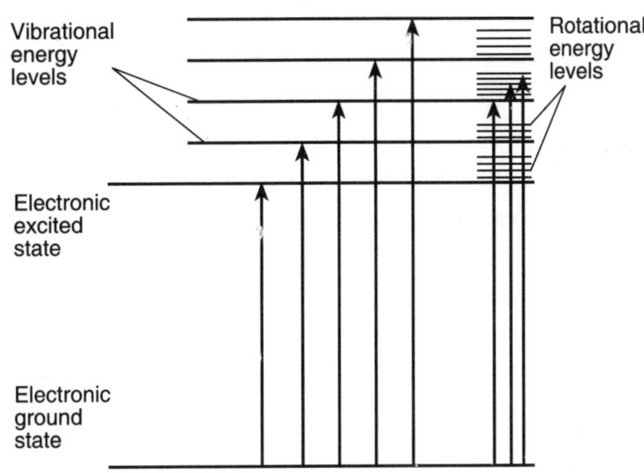

Fig. 12.5: Different energy levels obtained by an atoms

Beer-Lambert law

The Beer-Lambert law (also called the Beer-Lambert-Bouguer law or simply Beer's law) is the linear relationship between absorbance and concentration of an absorber of electromagnetic radiation. The general Beer-Lambert law is usually written as:

$$A = a_\lambda \times b \times c,$$

where A is the measured absorbance, a_λ is a wavelength-dependent absorptivity coefficient, b is the path length, and c is the analyte concentration. When working in concentration units of molarity, the Beer-Lambert law is written as:

$$A = \varepsilon_\lambda \times b \times c,$$

where ε_λ is the wavelength-dependent molar absorptivity coefficient with units of M^{-1} cm^{-1}. The λ subscript is often dropped with the understanding that a value for ε is for a specific wavelength. If multiple species that absorb light at a given wavelength are present in a sample, the total absorbance at that wavelength is the sum due to all absorbers:

$$A = (\varepsilon_1 \times b \times c_1) + (\varepsilon_2 \times b \times c_2) + \cdots,$$

where the subscripts refer to the molar absorptivity and concentration of the different absorbing species that are present.

Theory

Experimental measurements are usually made in terms of transmittance (T), which is defined as:

$$T = \frac{P}{P_0},$$

where P is the power of light after it passes through the sample and P_0 is the initial light power. The relation between A and T is:

$$A = -\log(T) = -\log\left(\frac{P}{P_0}\right).$$

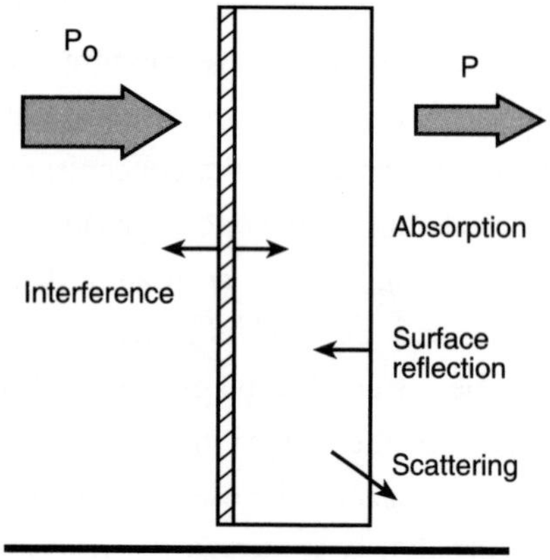

Fig. 12.6: Absorption of light

The figure shows the case of absorption of light through an optical filter and includes other processes that decrease the transmittance such as surface reflectance and scattering.

Modern absorption instruments can usually display the data as either transmittance, %-transmittance, or absorbance. An unknown concentration of an analyte can be determined by measuring the amount of light that a sample absorbs and applying Beer's law. If the absorptivity coefficient is not known, the unknown concentration can be determined using a working curve of absorbance versus concentration derived from standards. In analytical applications we often want to measure the concentration of an analyte independent of the effects of reflection, solvent absorption, or other interferences. The figure to the right shows the two transmittance measurements that are necessary to use absorption to determine the concentration of an analyte in solution. The top diagram is for solvent only and the bottom is for an absorbing sample in the same solvent. In this example, P_s is the source light power that is incident on a sample, P is the measured light power after passing through the analyte, solvent, and sample holder, and P_o is the measured light power after passing through only the solvent and sample holder. The measured transmittance in this case is attributed to only the analyte.

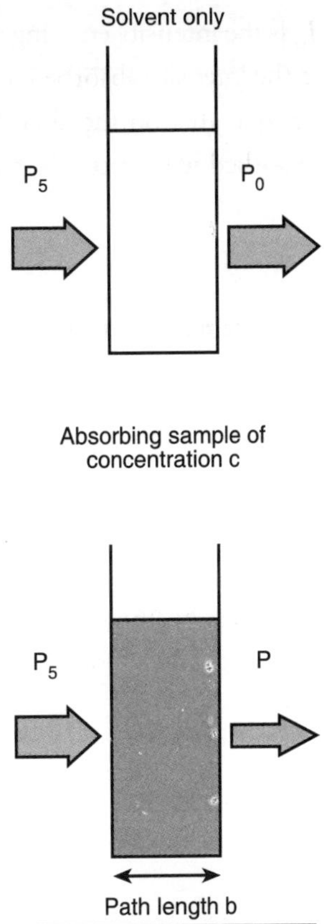

Fig. 12.7: Absorption of light by a sample

Depending on the type of instrument, the reference measurement (top diagram) might be made simultaneously with the sample measurement (bottom diagram) or a reference measurement might be saved on computer to generate the full spectrum.

Derivation of the Beer-Lambert law

The Beer-Lambert law can be derived from an approximation for the absorption coefficient for a molecule by approximating the molecule by an opaque disk whose cross-sectional area, σ, represents the effective area seen by a photon of frequency w. If the frequency of the light is far from resonance, the area is approximately 0, and if w is close to resonance the area is a maximum. Taking an infinitesimal slab, dz, of sample:

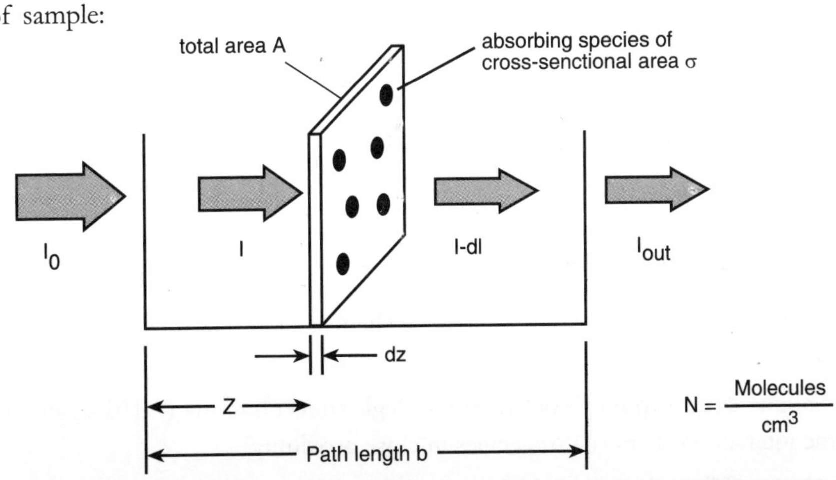

Fig. 12.8: Details of absorption

I_o is the intensity entering the sample at z=0, I_z is the intensity entering the infinitesimal slab at z, dI is the intensity absorbed in the slab, and I is the intensity of light leaving the sample. Then, the total opaque area on the slab due to the absorbers is σ * N * A * dz. Then, the fraction of photons absorbed will be σ * N * A * dz / A; so,

$$\frac{dI}{I_z} = -\sigma N dz.$$

Integration of this equation from z = 0 to z = b gives:

$$\text{In}(I) - \text{In}(I_o) = -\sigma Nb$$

or

$$\text{In}\left(\frac{I}{I_o}\right) = -\sigma Nb$$

Since N (molecules/cm^3) * (1 mole / 6.023×10^{23} molecules) * 1000 cm^3 / liter = c (moles/liter) and 2.303 * log(x) = ln(x), then

$$-\log\left(\frac{I}{I_o}\right) = -\sigma\left(\frac{6.023\times10^{20}}{2.303}\right)cb$$

or

$$-\log\left(\frac{I}{I_o}\right) = A = \varepsilon cb$$

where ε = σ * (6.023×10^{20}/2.303) = σ * 2.61×10^{20}.

Typical cross sections and molar absorptivities are:

	σ (cm^2)	ε (M^{-1} cm^{-1})
absorption-atoms	10^{-12}	3×10^8
absorption-molecules	10^{-16}	3×10^4
absorption-infrared	10^{-19}	3×10
Raman scattering	10^{-29}	3×10^{-9}

Limitations of the Beer-Lambert law

The linearity of the Beer-Lambert law is limited by chemical and instrumental factors. Causes of nonlinearity include:

- deviations in absorptivity coefficients at high concentrations (>0.01M) due to electrostatic interactions between molecules in close proximity
- scattering of light due to particulates in the sample
- fluoresecence or phosphorescence of the sample

- changes in refractive index at high analyte concentration
- shifts in chemical equilibria as a function of concentration
- non-monochromatic radiation, deviations can be minimized by using a relatively flat part of the absorption spectrum such as the maximum of an absorption band
- stray light.

Ultraviolet-visible spectrophotometer

The UV-Vis spectral range is approximately 190 to 900 nm, as defined by the working range of typical commercial UV-Vis spectrophotometers. The short-wavelength limit for simple UV-Vis spectrometers is the absorption of ultraviolet wavelengths less than 180 nm by atmospheric gases. Purging a spectrometer with nitrogen gas extends this limit to 175 nm. Working beyond 175 nm requires a vacuum spectrometer and a suitable UV light source. The long-wavelength limit is usually determined by the wavelength response of the detector in the spectrometer. High-end commercial UV-Vis spectrophotometers extend the measurable spectral range into the NIR region as far as 3300 nm.

The light source is usually a deuterium discharge lamp for UV measurements and a tungsten-halogen lamp for visible and NIR measurements. The instruments automatically swap lamps when scanning between the UV and visible regions. The wavelengths of these continuous light sources are typically dispersed by a holographic grating in a single or double monochromator or spectrograph. The spectral bandpass is then determined by the monochromator slit width or by the array-element width in array-detector spectrometers. Spectrometer designs and optical components are optimized to reject stray light, which is one of the limiting factors in quantitative absorbance measurements. The detector in single-detector instruments is a photodiode, phototube, or photomultiplier tube (PMT). UV-Vis-NIR spectrometers utilize a combination of a PMT and a Peltier-cooled PbS IR detector. The light beam is redirected automatically to the appropriate detector when scanning between the visible and NIR regions. The diffraction grating and instrument parameters such as slit width can also change.

Most commercial UV-Vis absorption spectrometers use one of three overall optical designs: a fixed or scanning spectrometer with a single light beam and sample holder, a scanning spectrometer with dual light beams and dual sample holders for simultaneous measurement of P and P_0, or a non-scanning spectrometer with an array detector for simultaneous measurement of multiple wavelengths. In single-beam and dual-beam spectrometers, the light from a lamp is dispersed before reaching the sample cell. In an array-detector instrument, all wavelengths pass through the sample and the dispersing element is between the sample and the array detector.

Instrument Designs

- Single-beam UV-Vis spectrophotometer
- Dual-beam UV-Vis spectrophotometer
- Array detector UV-Vis spectrophotometer

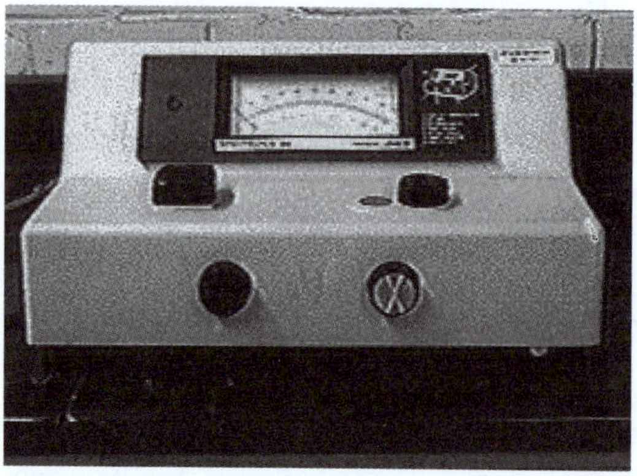

Fig. 12.9: Single beam UV-Vis Spectrophotometer (*Courtesy:* Dr D Dash, Deptt. of Biochemistry, IMS, BHU)

Single-Beam UV-Vis Spectrophotometer

Single-Beam spectrophotometers are often sufficient for making quantitative absorption measurements in the UV-Vis spectral region. The concentration of an analyte in solution can be determined by measuring the absorbance at a single wavelength and applying the Beer-Lambert Law.

Single-beam spectrophotometers can utilize a fixed wavelength light source or a continuous source. The simplest instruments use a single-wavelength light source, such as a light-emitting diode (LED), a sample container, and a photodiode detector. Instruments with a continuous source have a dispersing element and aperture or slit to select a single wavelength before the light passes through the sample cell (see schematic below). In either type of single-beam instrument, the instrument is calibrated with a reference cell containing only solvent to determine the P_o value necessary for an absorbance measurement.

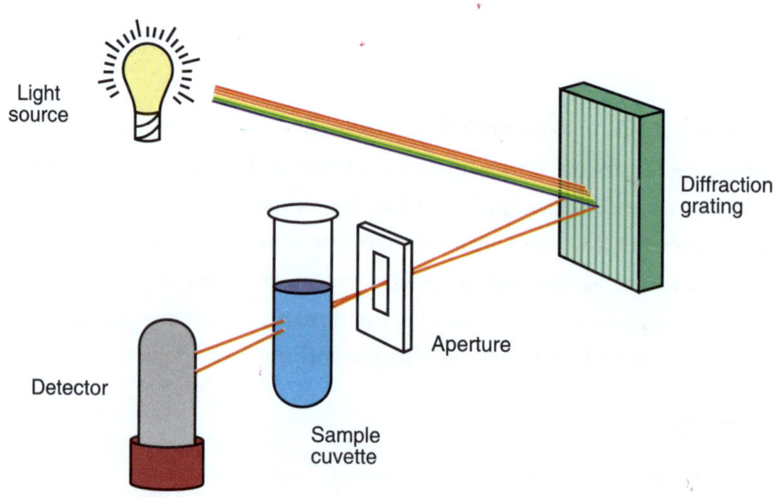

Fig. 12.10: Schematic of a wavelength-selectable, single-beam UV-Vis spectrophotometer

Dual-Beam UV-Vis Spectrophotometer

Fig. 12.11: Dual-Beam UV-Vis Spectrophotometer (*Courtesy:* Dr D Dash, Deptt. of Biochemistry, IMS, BHU)

In single-beam UV-Vis absorption spectroscopy, obtaining a spectrum requires manually measuring the transmittance of the sample and solvent at each wavelength. The double-beam design greatly simplifies this process by measuring the transmittance of the sample and solvent simultaneously.

The detection electronics can then manipulate the measurements to give the absorbance. The dual-beam design greatly simplifies this process by simultaneously measuring P and P_0 of the sample and reference cells, respectively. Most spectrometers use a mirrored rotating chopper wheel to alternately direct the light beam through the sample and reference cells. The detection electronics or software program can then manipulate the P and P_0 values as the wavelength scans to produce the spectrum of absorbance or transmittance as a function of wavelength.

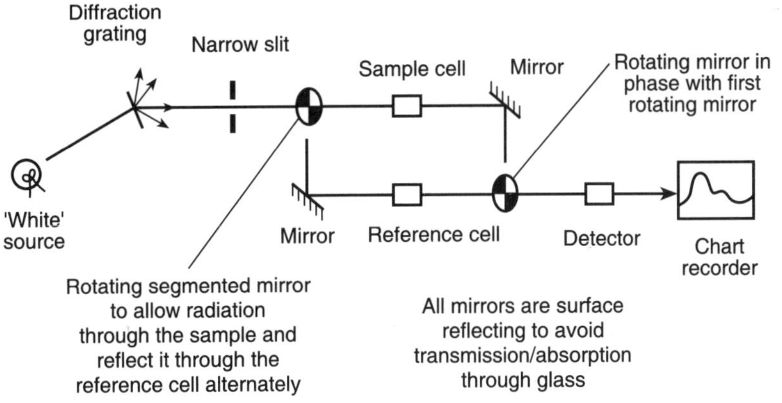

Fig. 12.12: Schematic of a dual-beam UV-Vis spectrophotometer

Array-Detector Spectrophotometer

Array-detector spectrophotometers allow rapid recording of absorption spectra. Dispersing the source light after it passes through a sample allows the use of an array detector to simultaneously record the transmitted light power at multiple wavelengths. There are a large number of applications where absorbance spectra must be recorded very quickly. Some examples include HPLC detection, process monitoring, and measurement of reaction kinetics. These spectrometers use photodiode arrays (PDAs) or charge-coupled devices (CCDs) as the detector. The spectral range of these array detectors is typically 200 to 1000 nm. The light source is a continuum source such as a tungsten lamp. All wavelengths pass through the sample. The light is dispersed by a diffraction grating after the sample and the separated wavelengths fall on different pixels of the array detector. The resolution depends on the grating, spectrometer design, and pixel size, and is usually fixed for a given instrument. Besides allowing rapid spectral recording, these instruments are relatively small and robust. Portable spectrometers have been developed that use optical fibers to deliver light to and from a sample. These instruments use only a single light beam, so a reference spectrum is recorded and stored in memory to produce transmittance or absorbance spectra after recording the sample spectrum.

Ultraviolet-visible spectrum

An ultraviolet-visible spectrum is essentially a graph of light absorbance versus wavelength in a range of ultraviolet or visible regions. Such a spectrum can often be produced directly by a more sophisticated spectrophotometer, or the data can be collected one wavelength at a time by simpler instruments. Wavelength is often represented by the symbol λ. Similarly, for a given substance, a standard graph of the extinction coefficient (ε) vs. wavelength (λ) may be made or used if one is already available. Such a standard graph would be effectively "concentration-corrected" and thus independent of concentration. For the given substance, the wavelength at which maximum absorption in the spectrum occurs is called λ_{max}, pronounced "Lambda-max". The Woodward-Fieser rules are a set of empirical observations which can be used to predict λ_{max}, the wavelength of the most intense UV/Vis absorption, for conjugated organic compounds such as dienes and ketones. This spectrum can be used qualitatively to identify components in a sample as each component has their own unique absorbance spectrum (like a fingerprint).

12.6 FLUORESCENCE SPECTROSCOPY

Fluorescence spectroscopy or fluorometry or spectrofluorimetry is a type of electromagnetic spectroscopy which analyzes fluorescence from a sample. It involves using a beam of light, usually ultraviolet light, that excites the electrons in molecules of certain compounds and causes them to emit light of a lower energy; typically, but not necessarily, visible light. A complementary technique is absorption spectroscopy. Devices that measure fluorescence are called fluorometers or fluorimeters.

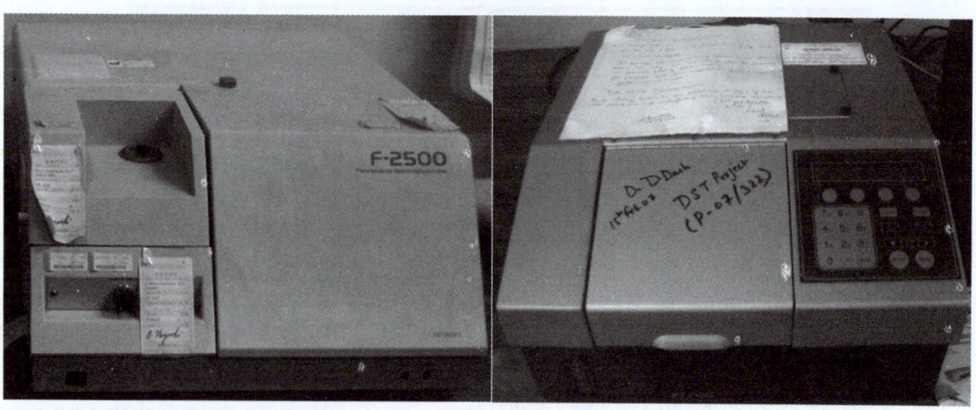

Fig. 12.13: Spectrofluorimeter and Fluorescence plate reader (*Courtesy:* Dr D Dash, Deptt. of Biochemistry, IMS, BHU)

Molecules have various states referred to as energy levels. Fluorescence spectroscopy is primarily concerned with electronic states and vibrational states. Generally, the species being examined will have a ground electronic state (a low energy state) of interest, and an excited electronic state of higher energy. Each of these electronic states has various vibrational states. Photons of light are small "packets" of energy, each with an energy proportional to its frequency; photons of high frequency light have higher energy than those of low frequency light. These can be absorbed by molecules, with the molecule gaining the energy of the photon, or emitted by molecules, with the photon carrying some of the energy of the molecule away. In fluorescence spectroscopy, the species is first excited, by absorbing a photon of light, from its ground electronic state to one of the various vibrational states in the excited electronic state. Collisions with other molecules cause the excited molecule to lose vibrational energy until it reaches the lowest vibrational state of the excited electronic state. The molecule then drops down to one of the various vibrational levels of the ground electronic state again, emitting a photon in the process. As molecules may drop down into any of the vibrational levels of this ground state, the photons will have different energies, and thus frequencies. Therefore, by analysing the different frequencies of light emitted in fluorescent spectroscopy, the structure of these different vibrational levels can be determined. Typically, the different frequencies of fluorescent light emitted by a sample, when the excited light is held at a constant wavelength are measured. This is called an emission spectrum. An excitation spectrum is measured by recording the sum of the fluorescent light emitted at all frequencies as function of the frequency of the monochromatic incident light.

Instrumentation

Two general types of instruments exist:

1. Filter fluorometers use filters to isolate the incident light and fluorescent light.

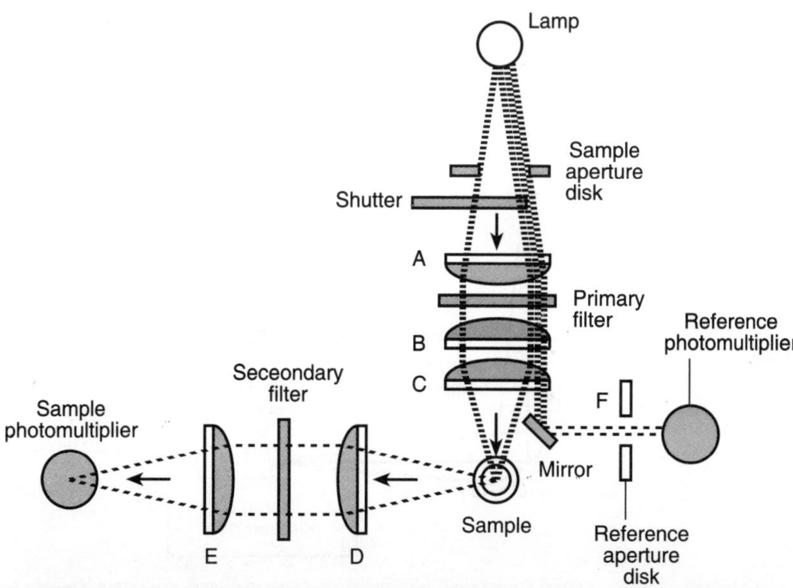

Fig. 12.14: Schematic diagram of filter flourimeter

2. Spectrofluorometers use diffraction grating monochromators (two) to isolate the incident light and fluorescent light.

Both types utilize the following scheme: The light from an excitation source passes through a filter or monochromator, and passes through the sample. Here some of it probably is absorbed, making some of the molecules in the sample fluoresce. A part of the fluorescent light is then focused on a filter or monochromator, which often is placed at a 90° angle to the excitation light. The light is then detected by a detection device. Various light sources may be used as excitation sources, including lasers, photodiodes, and lamps; xenon arcs and mercury vapour lamps in particular. A laser only emits light of high irradiance at a very narrow wavelength interval, typically under 0.01 nm, which makes an excitation monochromator or filter unnecessary. The downside is that the wavelength of a laser cannot be changed much. A mercury vapour lamp is a line lamp, meaning it emits light near peak wavelengths contrary to the Xe arc, which have a continuous emission spectrum with nearly constant intensity in the range from 300-800 nm and have a sufficient irradiance for measurements down to just above 200 nm. Filters and/or monochromators may be used in fluorimeters. A monochromator transmits light of an adjustable wavelength with an adjustable tolerance. The most common type of monochromator utilizes diffraction grating, that is, collimated light enters a grating and exits with a different angle depending on the wavelength. The monochromator can then select which wavelengths to transmit. For allowing anisotropy measure-

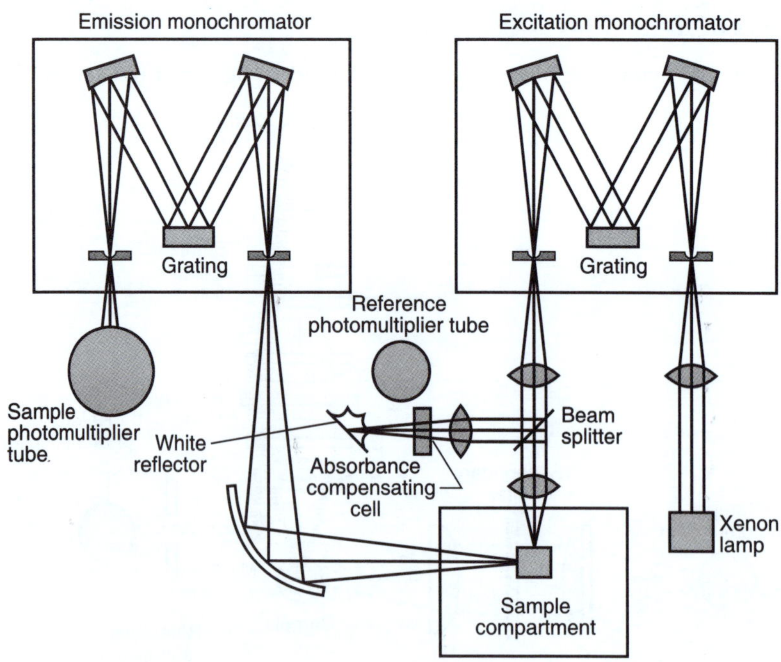

Fig. 12.15: Schematic diagram of spectroflurometer

ments the addition of two polarization filters are necessary: One after the excitation monochromator or filter, and one before the emission monochromator or filter. As mentioned before, the fluorescence is most often measured at a 90° angle relative to the excitation light. This geometry is used instead of placing the sensor at the line of the excitation light at a 180° angle in order to avoid interference of the transmitted excitation light. No monochromator is perfect and it will transmit some stray light, that is, light with other wavelengths than the targeted. An ideal monochromator would only transmit light in the specified range and have a high wavelength-independent transmission. When measuring at a 90° angle, only the light scattered by the sample causes stray light. This results in a better signal-to-noise ratio, and lowers the detection limit by approximately a factor 10000, when compared to the 180° geometry. Furthermore, the fluorescence can also be measured from the front, which is often done for turbid samples.

The detector can either be single-channeled or multichanneled. The single-channeled detector can only detect the intensity of one wavelength at a time, while the multichanneled one detects the intensity at all wavelengths simultaneously, making the emission monochromator or filter unnecessary. The different types of detectors have both advantages and disadvantages. The most versatile fluorimeters with dual monochromators and a continuous excitation light source can record both an excitation spectrum and a fluorescence spectrum. When measuring fluorescence spectra, the wavelength of the excitation light is kept constant, preferably at a wavelength of high absorption, and the emission monochromator scans the spectrum. For measuring excitation spectra, the wavelength passing though the emission filter or monochromator is kept constant and the excitation monochromator is scanning. The excitation spectrum generally is identical to the absorption spectrum as the fluorescence intensity is proportional to the absorption.

Analysis of data

At low concentrations the fluorescence intensity will generally be proportional to the concentration of the fluorophore. Contrary to UV/Vis spectroscopy 'standard', device-independent spectra are not easily attained, however. Several factors influence and distort the spectra or represents factors that need to be corrected to attain 'true' spectra. The different types of distortions will here be classified as being either instrumentally or sample related. Firstly, the distortion arising from the instrument is discussed. As a start, the light source intensity and wavelength characteristics varies over time during each experiment and between each experiment. Furthermore, no lamp has a constant intensity at all frequencies. To correct this, a beam splitter can be applied after the excitation monochromator or filter to direct a portion of the light to a reference detector. Additionally, the transmission efficiency of monochromators and filters has to be taken into account. These may also change over time. The transmission efficiency of the monochromator also varies depending on wavelength. This is the reason that an optional reference detector should be placed after the excitation monochromator or filter. The percentage of the fluorescence picked up by the detector is also dependent upon the system. Furthermore, the detector quantum efficiency, that is, the percentage of photons detected, varies between different detectors, with wavelength and with time, as the detector inevitably deteriorates. Correction of all these instrumental factors for getting a 'standard' spectrum is a tedious process, which is only applied in practice when it is strictly necessary. This is the case when measuring the quantum yield or when finding the wavelength with the highest emission intensity for instance. As mentioned earlier, distortions arise from the sample as well. Therefore, some aspects of the sample must be taken into account too. Firstly, photodecomposition may decrease the intensity of fluorescence over time. Scattering of light must also be taken into account. The most significant types of scattering in this context are Rayleigh and Raman scattering. Light scattered by Rayleigh scattering has the same wavelength as the incident light, whereas in Raman scattering the scattered light changes wavelength usually to longer wavelengths. Raman scattering is the result of a virtual electronic state induced by the excitation light. From this virtual state, the molecules may relax back to a vibrational level other than the vibrational ground state. In fluorescence spectra, it is always seen at a constant wave number difference relative to the excitation wave number, e.g., the peak appears at a wave number 3600 cm^{-1} lower than the excitation light in water. Other aspects to consider are the inner filter effects. These include reabsorption. Reabsorption happens because another molecule or part of a macromolecule absorbs at the wavelengths at which the fluorophore emits radiation. If this is the case, some or all of the photons emitted by the fluorophore may be absorbed again. Another inner filter effect occurs because of high concentrations of absorbing molecules, including the fluorophore. The result is that the intensity of the excitation light is not constant throughout the solution—only a small percentage of the excitation light reaches the fluorophores that are visible for the detection system. The inner filter effects change the spectrum and intensity of the emitted light and they must, therefore, be considered when analysing the emission spectrum of fluorescent light.

Applications

It is used in, among others, biochemical, medical, and chemical research fields, for analyzing organic compounds. There has been report of its use in differentiating malign skin tumours from benign: a kind of tumour cells.

12.7 CIRCULAR DICHROISM SPECTROSCOPY

Circular dichroism (CD) spectroscopy is a type of absorption spectroscopy that can provide information on the structures of many types of biological macromolecules. Circular Dichroism is the difference between the absorption of left and right handed circularly-polarised light and is measured as a function of wavelength. CD is measured as a quantity called mean residue ellipticity, whose units are degrees-cm^2/dmol. Chiral or asymmetric molecules produce a CD spectrum because they absorb left and right handed polarised light to different extents and thus are considered to be "optically active". Biological macromolecules such as proteins, polysaccharides and DNA are composed of optically active elements and because they can adopt different types of three-dimensional structures, each type of molecule produces distinct CD spectra.

The wavelengths of light that are most useful for examining the structures of proteins and DNA are in the ultraviolet (UV) or vacuum ultraviolet (VUV) ranges (from 160 to 300 nm) because these are the regions of the electronic transitions of the peptide backbone and side chains in proteins and the purine and pyrimidine bases in DNA. Typically the range of data collected on a commercially available CD instrument will be between 190 nm and 300 nm.

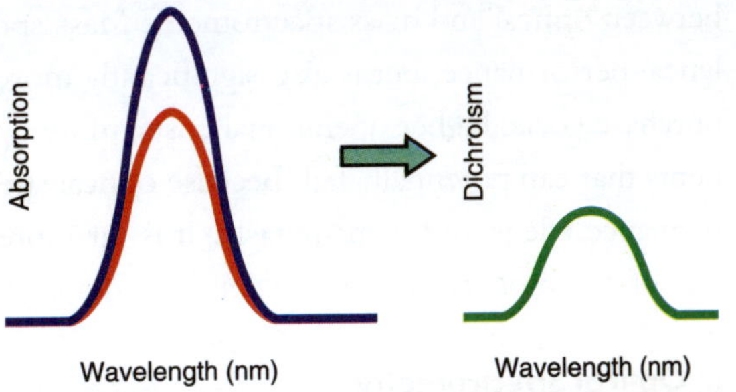

Fig. 12.16: Absorption and CD spectra

Different types of protein secondary structures (helices, sheets, turns and coils) give rise to different CD spectra.

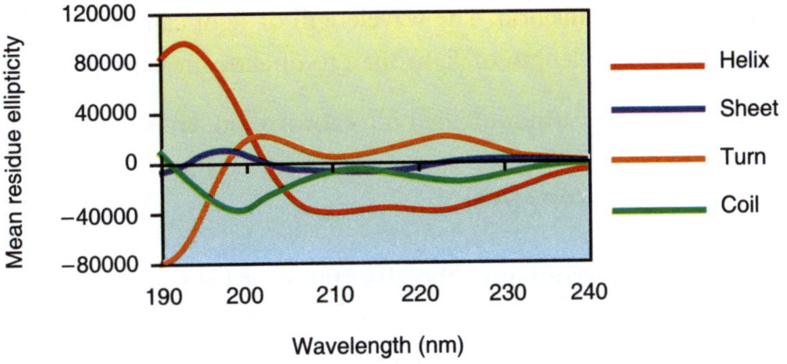

Fig. 12.17: Spectra of different secondary protein structure

Since the spectrum of a protein is directly related to its secondary structure content, the spectrum will be a linear combination of each of these "reference" spectra, weighted by the fraction of the type of secondary structure. Therefore, it is possible to mathematically extract the secondary structure information for an unknown protein from its CD spectrum.

Some of the advantages of CD spectroscopy are that it requires only very small amounts of material (100 micrograms or less) and measurements can be done very quickly (in 30 minutes or less). CD spectroscopy has been used to monitor: (1) secondary structure, (2) conformational changes, (3) environmental effects, (4) protein folding and denaturation, and (5) dynamics.

12.8 ATOMIC SPECTROSCOPY

Atomic spectroscopy is the determination of elemental composition by its electromagnetic or mass spectrum. Atomic spectroscopy is closely related to other forms of spectroscopy. It can be divided by atomization source or by the type of spectroscopy used. In the latter case, the main division is between optical and mass spectrometry. Mass spectrometry generally gives significantly better analytical performance, but is also significantly more complex. This complexity translates into higher purchase costs, higher operational costs, more operator training, and a greater number of components that can potentially fail. Because optical spectroscopy is generally less expensive and has performance adequate for many tasks, it is far more common. Atomic absorption spectrometers are one of the most commonly sold and used analytical devices.

I. Optical Spectrometry

Electrons exist in energy levels within an atom. These levels have well defined energies and electrons moving between them must absorb or emit energy equal to the difference between them. In optical spectroscopy, the energy absorbed to move an electron to a more energetic level and/or the energy emitted as the electron moves to a less energetic energy level is in the form of a photon (a particle of light). Because this energy is well-defined, an atom's identity (i.e., what element it is) can be identified by the energy of this transition. The wavelength of light can be related to its energy. It is usually easier to measure the wavelength of light than to directly measure its energy.

Optical spectroscopy can be further divided into absorption, emission, and fluorescence.

1. Atomic-Absorption Spectroscopy (AA)

Atomic-absorption (AA) spectroscopy uses the absorption of light to measure the concentration of gas-phase atoms. Since samples are usually liquids or solids, the analyte atoms or ions must be vaporized in a flame or graphite furnace. The atoms absorb ultraviolet or visible light and make transitions to higher electronic energy levels. The analyte concentration is determined from the amount of absorption. Applying the Beer-Lambert law directly in AA spectroscopy is difficult due to variations in the atomization efficiency from the sample matrix, and non-uniformity of concentration and path length of analyte atoms (in graphite furnace AA). Concentration measurements are usually determined from a working curve after calibrating the instrument with standards of known concentration.

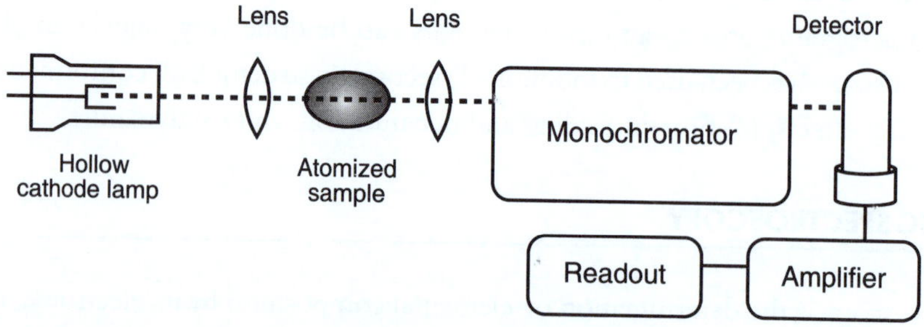

Fig. 12.18: Schematic of an atomic-absorption experiment

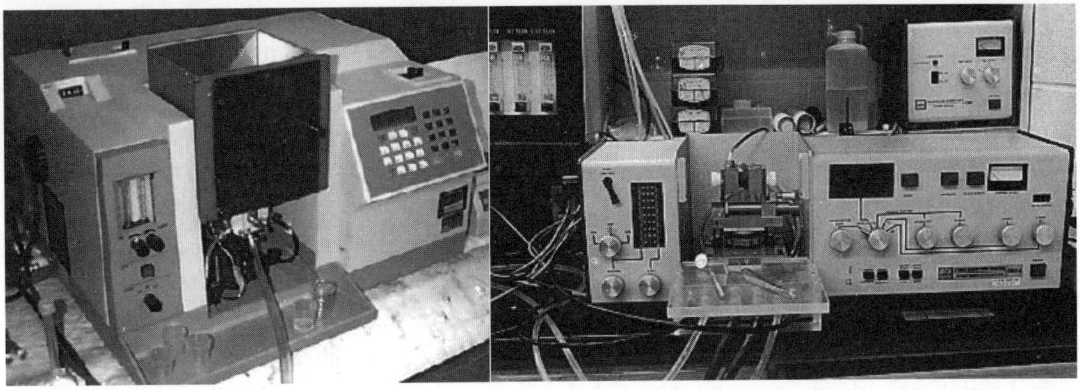

A B

Fig. 12.19: A. Flame atomic-absorption spectrometer, B. Graphite-furnace atomic-absorption spectrometer

Instrumentation

The light source is usually a hollow-cathode lamp of the element that is being measured. Lasers are also used in research instruments. Since lasers are intense enough to excite atoms to higher energy levels, they allow AA and atomic fluorescence measurements in a single instrument. The disadvantage of these narrow-band light sources is that only one element is measurable at a time.

AA spectroscopy requires that the analyte atoms be in the gas phase. Ions or atoms in a sample must undergo desolvation and vaporization in a high-temperature source such as a flame or graphite furnace. Flame AA can only analyze solutions, while graphite furnace AA can accept solutions, slurries, or solid samples. Flame AA uses a slot type burner to increase the path length, and, therefore, to increase the total absorbance (see Beer-Lambert law). Sample solutions are usually aspirated with the gas flow into a nebulizing/mixing chamber to form small droplets before entering the flame. The graphite furnace has several advantages over a flame. It is a much more efficient atomizer than a flame and it can directly accept very small absolute quantities of sample. It also provides a reducing environment for easily oxidized elements. Samples are placed directly in the graphite furnace and the furnace is electrically heated in several steps to dry the sample, ash organic matter, and vaporize the analyte atoms. AA spectrometers use monochromators and detectors for UV and visible light. The main purpose of the monochromator is to isolate the absorption line from background light due to interferences. Simple dedicated AA instruments often replace the monochromator with a bandpass interference filter. Photomultiplier tubes are the most common detectors for AA spectroscopy.

2. Atomic Emission Spectroscopy (AES, OES)

Atomic emission spectroscopy (AES or OES [optical emission spectroscopy]) uses quantitative measurement of the optical emission from excited atoms to determine analyte concentration. Analyte atoms in solution are aspirated into the excitation region where they are desolvated, vaporized, and

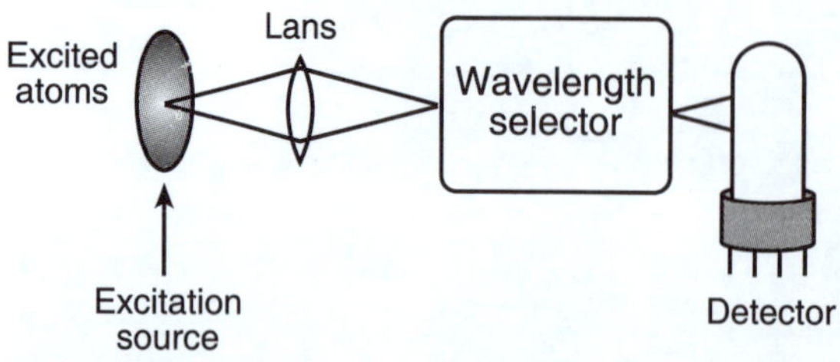

Fig. 12.20: Schematic of an AES experiment

atomized by a flame, discharge, or plasma. These high-temperature atomization sources provide sufficient energy to promote the atoms into high energy levels. The atoms decay back to lower levels by emitting light. Since the transitions are between distinct atomic energy levels, the emission lines in the spectra are narrow. The spectra of samples containing many elements can be very congested, and spectral separation of nearby atomic transitions requires a high-resolution spectrometer. Since all atoms in a sample are excited simultaneously, they can be detected simultaneously using a polychromator with multiple detectors. This ability to simultaneously measure multiple elements is a major advantage of AES compared to atomic-absorption (AA) spectroscopy.

Instrumentation

As in AA spectroscopy, the sample must be converted to free atoms, usually in a high-temperature excitation source. Liquid samples are nebulized and carried into the excitation source by a flowing gas. Solid samples can be introduced into the source by slurry or by laser ablation of the solid sample in a gas stream. Solids can also be directly vaporized and excited by a spark between electrodes or by a laser pulse. The excitation source must desolvate, atomize, and excite the analyte atoms. The various excitation sources used are:

- Direct-current plasma (DCP)
- Flame
- Inductively-coupled plasma (ICP)
- Laser-induced breakdown (LIBS)
- Laser-induced plasma
- Microwave-induced plasma (MIP)
- Spark or arc.

Since the atomic emission lines are very narrow, a high-resolution polychromator is needed to selectively monitor each emission line.

Fig. 12.21: Inductively-coupled plasma atomic emission spectrometer

3. Atomic-Fluorescence Spectroscopy (AFS)

Atomic fluorescence is the optical emission from gas-phase atoms that have been excited to higher energy levels by absorption of electromagnetic radiation. The main advantage of fluorescence detection compared to absorption measurements is the greater sensitivity achievable because the fluorescence signal has a very low background. The resonant excitation provides selective excitation of the analyte to avoid interferences. AFS is useful to study the electronic structure of atoms and to make quantitative measurements. Analytical applications include flames and plasmas diagnostics, and enhanced sensitivity in atomic analysis.

Instrumentation

Analysis of solutions or solids requires that the analyte atoms be desolvated, vaporized, and atomized at a relatively low temperature in a heat pipe, flame, or graphite furnace. A hollow-cathode lamp or laser provides the resonant excitation to promote the atoms to higher energy levels. The atomic fluorescence is dispersed and detected by monochromators and photomultiplier tubes, similar to atomic-emission spectroscopy instrumentation.

II. Atomic Mass Spectrometry

Atomic mass spectrometry is similar to other types of mass spectrometry in that it consists of an ion source, a mass analyzer, and a detector. Atoms' identities are determined by their mass-to-charge ratio (via the mass analyzer) and their concentrations are determined by the number of ions detected. Although considerable research has gone into customizing mass spectrometers for atomic ion sources, it is the ion source that differs most from other forms of mass spectrometry. These ion sources must also atomize samples, or an atomization step must take place before ionization. Atomic ion sources are generally modifications of atomic optical spectroscopy atom sources.

Ion and atom sources

Sources can be adapted in many ways, but the lists below gives the general uses of a number of sources. Of these, flames are the most common due to their low cost and their simplicity. Although significantly less common, inductively-coupled plasmas, especially when used with mass spectrometers, are recognized for their outstanding analytical performance and their versatility. For all atomic spectroscopy, a sample must be vaporized and atomized. For atomic mass spectrometry, a sample must also be ionized. Vaporization, atomization, and ionization are often, but not always, accomplished with a single source. Alternatively, one source may be used to vaporize a sample while another is used to atomize (and possibly ionize). An example of this would be laser ablation inductively-coupled plasma atomic emission spectrometry, where a laser is used to vaporize a solid sample and an inductively-coupled plasma is used to atomize the vapour. With the exception of flames and graphite furnaces, which are most commonly used for atomic absorption spectroscopy, most sources are used primarily for atomic emission spectroscopy.

Liquid-sampling sources include

- Flames and sparks (atom source)
- Inductively-coupled plasma (atom and ion source)
- Graphite furnace (atom source)
- Microwave plasma (atom and ion source)
- Direct-current plasma (atom and ion source)

Solid-sampling sources include

- Lasers (atom and vapour source)
- Glow discharge (atom and ion source)
- Arc (atom and ion source)
- Spark (atom and ion source)
- Graphite furnace (atom and vapour source)

Gas-sampling sources include

- Flame (atom source)
- Inductively-coupled plasma (atom and ion source)
- Microwave plasma (atom and ion source)
- Direct-current plasma (atom and ion source)
- Glow discharge (atom and ion source)

12.9 INFRARED SPECTROSCOPY

The infrared region of the electromagnetic spectrum extends from 14,000 cm^{-1} to 10 cm^{-1}. The region of most interest for chemical analysis is the mid-infrared region (4,000 cm^{-1} to 400 cm^{-1}) which corresponds to changes in vibrational energies within molecules. The far infrared region (400 cm^{-1} to 10 cm^{-1}) is useful for molecules containing heavy atoms such as inorganic compounds but requires rather specialized experimental techniques. It is rarely, if ever, possible to identify an unknown compound by using IR spectroscopy alone. Its principal strengths are: (i) it is a quick and relatively cheap spectroscopic technique, (ii) it is useful for identifying certain functional groups in molecules and (iii) an IR spectrum of a given compound is unique and can, therefore, serve as a fingerprint for this compound.

Molecular Vibrations

All molecules vibrate, even at a temperature of absolute zero. In general, a polyatomic molecule with N atoms has 3Ñ6 distinct vibrations. Each of these vibrations has an associated set of quantum states and in IR spectroscopy the IR radiation induces a jump from the ground (lowest) to the first excited quantum state. Although approximate, each vibration in a molecule can be associated with motion in a particular group. A simple example is methanal, whose 6 vibrations involve the following motions:

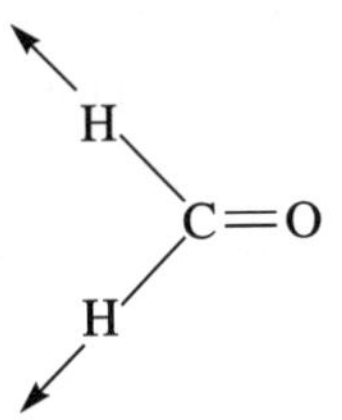

Symmetric C—H stretch

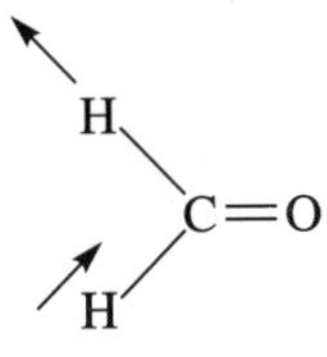

Asymmetric C—H stretch

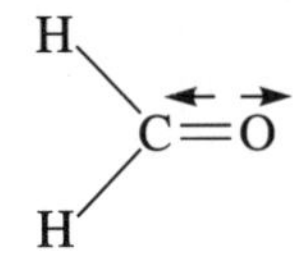

C=O stretch

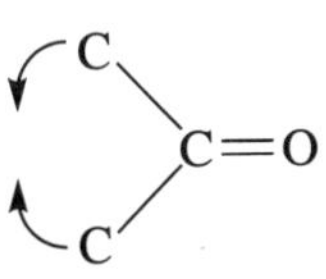

CH_2 bend

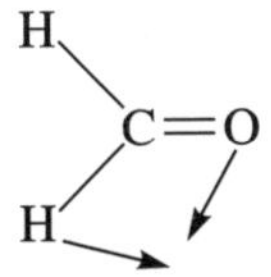

HCO bend

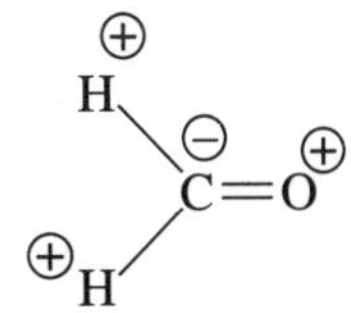

Out-of-plane bend (⊕ and ⊖ here signify directions of motion not atomic charges)

Clearly, even for this simple molecule, a description of the vibrations is quite complicated and this complexity rapidly increases as the size of the molecule increases. Fortunately, as we will see, it is almost never necessary to be able to picture all the molecular vibrations for IR spectroscopy to be useful.

Infrared Activity

Not all possible vibrations within a molecule will result in an absorption band in the infrared region. To be infrared active the vibration must result in a change of dipole moment during the vibration. This means that for homonuclear diatomic molecules such as Hydrogen (H_2), Nitrogen (N_2) and Oxygen (O_2), no infrared absorption is observed, as these molecules have zero dipole moment and stretching of the bonds will not produce one. For heteronuclear diatomic molecules such Carbon Monoxide (CO) and Hydrogen Chloride (HCl), which do possess a permanent dipole moment, infrared activity occurs because stretching of this bond leads to a change in dipole moment.

It is important to remember that it is not necessary for a compound to have a permanent dipole moment to be infrared active. In the case of Carbon Dioxide (CO_2) the molecule is linear and centrosymmetric and therefore does not have a permanent dipole moment. This means that the symmetric stretch will not be infrared active. However, in the case of the asymmetric stretch a dipole moment will be periodically produced and destroyed resulting in a changing dipole moment and, therefore, infrared activity.

The Fingerprint Region

The fact that there are many different vibrations even within relatively simple molecules means that the infrared spectrum of a compound usually contains a large number or peaks, many of which will be impossible to confidently assign to vibration of a particular group. Particularly notable is the complex pattern of peaks below 1500 cm^{-1} which are very difficult to assign. However, this complexity has an important advantage; in that it can serve as a fingerprint for a given compound. Consequently, by referring to known spectra, the region can be used to identify a compound.

Interpretation of Spectra

To obtain a more detailed interpretation of an IR spectrum it is necessary to refer to correlation charts and tables of infrared data. There are many different tables available for reference and a brief summary is given below for some of the main groups. When assigning peaks to specific groups in the infrared region it is usually the stretching vibrations which are most useful. Broadly speaking, these can be divided into four regions:

3700 - 2500 cm^{-1} Single bonds to hydrogen

2300 - 2000 cm^{-1} Triple bonds

1900 - 1500 cm^{-1} Double bonds

1400 - 650 cm^{-1} Single bonds (other than hydrogen)

It should also be noted that the region 1650 - 650 cm^{-1} contains peaks due to bending vibrations but it is rarely possible to assign a specific peak to a specific group.

Single Bonds to Hydrogen

Bond	Wave number/cm-1	Notes
C–H	3000 - 2850	Saturated alkanes, limited value as most organic compounds contain C–H
=C–H	3100 - 3000	Unsaturated alkene or aromatic
C–H	3300	Terminal Alkyne
O=C–H	2800 and 2700	Aldehyde, two weak peaks
O–H O–H (free)	3400 - 3000 ~ 3600	Alcohols and Phenols. If hydrogen bonding present peak will be broad 3000-2500 (e.g., carboxylic acids).
N–H	3450 - 3100	Amines: Primary – several peaks, Secondary – one peak, tertiary – no peaks

Double Bonds

Bond	Wave number/cm-1	Notes
C=O	1840 - 1800 & 1780 - 1740	Anhydrides
C=O	1815 - 1760	Acyl halides

Contd.

Double Bonds

Bond	Wave number/cm^{-1}	Notes
C=O	1750 - 1715	Esters
C=O	1740 - 1680	Aldehydes
C=O	1725 - 1665	Ketones
C=O	1720 - 1670	Carboxylic acids
C=O	1690 - 1630	Amides
C=C	1675 - 1600	Often weak
C=N	1690 - 1630	Often difficult to assign
N=O	1560 - 1510 & 1370 - 1330	Nitro compounds

Triple Bonds

Bond	Wave number/cm^{-1}	Notes
C≡C	2260 - 2120	Alkynes, bands are weak
C≡N	2260 - 2220	Nitriles

Single Bonds (not to Hydrogen)

Bond	Wave number/cm^{-1}	Notes
C–C	Variable	No diagnostic value
C–O, C–N	1400 - 1000	Difficult to assign
C–Cl	800 - 700	Difficult to interpret
C–Br, C–I	Below 650	Often out of range of instrumentation

Bending Vibrations

Bond	Wave number/cm^{-1}	Notes
R–N–H	1650 - 1500	Take care not to confuse N–H bend with the C=O stretch in amides
R–C–H	1480 - 700	Difficult to assign
R–C–H	1000 - 680	Unsaturated alkenes and aromatics

Sample Preparation

There are a variety of techniques for sample preparation dependent on the physical form of the sample to be analysed.

Solids

There are two main methods for sample preparation involving the use of Nujol mull or potassium bromide disks. However, there is also a third option of preparing a solution in a suitable solvent (not infrared active in the region of interest).

Nujol Mull

The sample is ground using an agate mortar and pestle to give a very fine powder. A small amount is then mixed with nujol to give a paste and several drops of this paste are then applied between two sodium chloride plates (these do not absorb infrared in the region of interest). The plates are then placed in the instrument sample holder ready for scanning.

Potassium Bromide disk

A very small amount of the solid (approximately $\tilde{1}2$ mg) is added to pure potassium bromide powder (approximately 200 mg) and ground up as fine as possible. This is then placed in a small die and put under pressure mechanically. The pressure is maintained for several minutes before removing the die and the KBr disk formed. The disk is then placed in a sample holder ready for scanning.

The success of this technique is dependent on the powder being ground as fine as possible to minimise infrared light scattering off the surface of the particles. It is also important that the sample be dry before preparation. KBr has no infrared absorption in the region 400-650 cm^{-1}.

Thin Films

The infrared spectrum of a thin film can be easily obtained by placing a sample in a suitable holder, such as a card with a slot cut for the sample window. This method is often used for checking the calibration of an instrument with a polystyrene sample as the bands produced by this material are accurately known.

Liquids

This is possibly the simplest and most common method of sample preparation. A drop of the sample is placed between two potassium bromide or sodium chloride circular plates to produce a thin capillary film. The plates are then placed in a holder ready for analysis.

Gases

To obtain an infrared spectrum of a gas requires the use of a cylindrical gas cell with windows at each end composed of an infrared inactive material such as KBr, NaCl or CaF_2. The cell usually has an inlet and an outlet ports with a tap to enable the cell to be easily filled with the gas to be analysed.

Example Spectra

1. **Alcohols**

n-Butanol $CH_3CH_2CH_2CH_2OH$

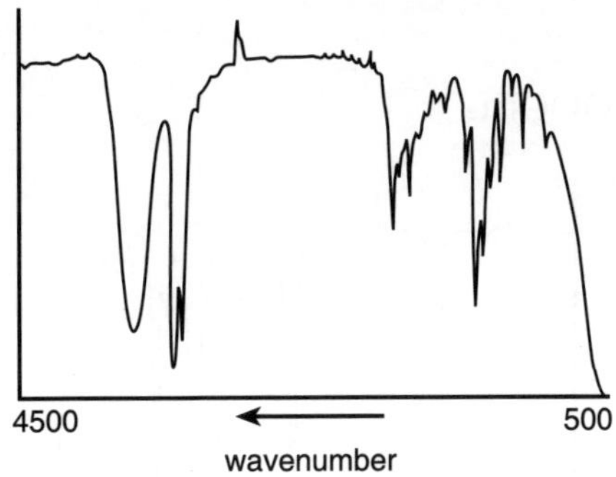

O-H Stretch
3330 cm^{-1}
Hydrogen Bonded

C-O Stretch
1070 cm^{-1}

Carboxylic Acids

Ethanoic Acid CH_3COOH

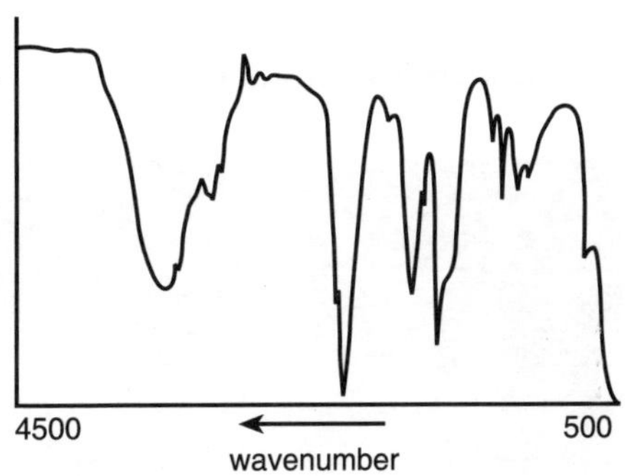

O-H Stretch
3050 cm^{-1}
Hydrogen Bonded

C=O Stretch
1715 cm^{-1}

C-O Stretch
1295 cm^{-1}

Aldehydes and Ketones

Butanone $CH_3COCH_2CH_3$

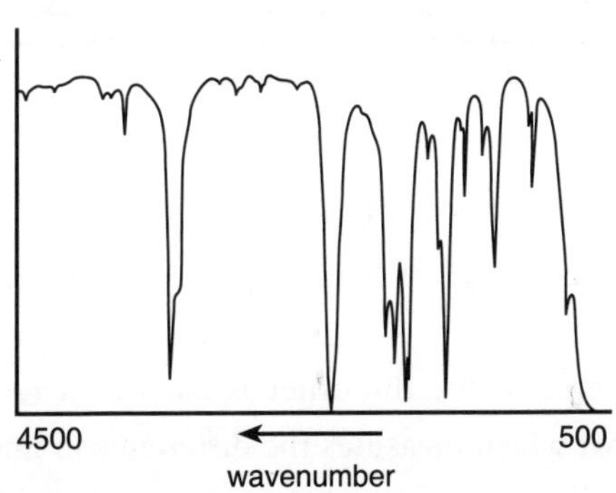

C=O Stretch
1715 cm^{-1}

Esters

Ethyl Ethanoate $CH_3COOC_2H_5$

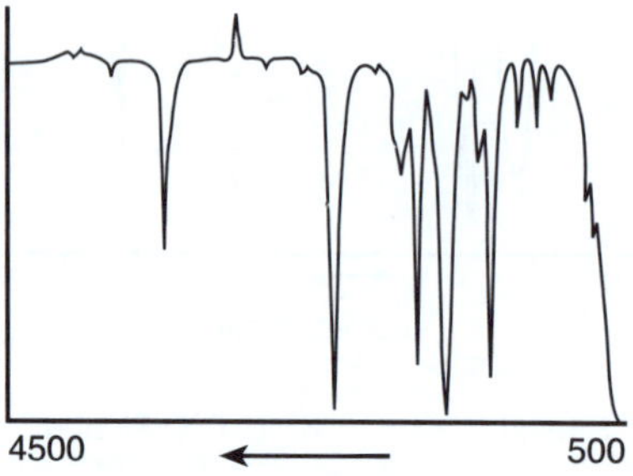

C-O Stretch $1710\ cm^{-1}$ C-O Stretch $1240\ cm^{-1}$ C-O Stretch $1050\ cm^{-1}$

Instrumentation

The Double Beam Infrared Spectrometer

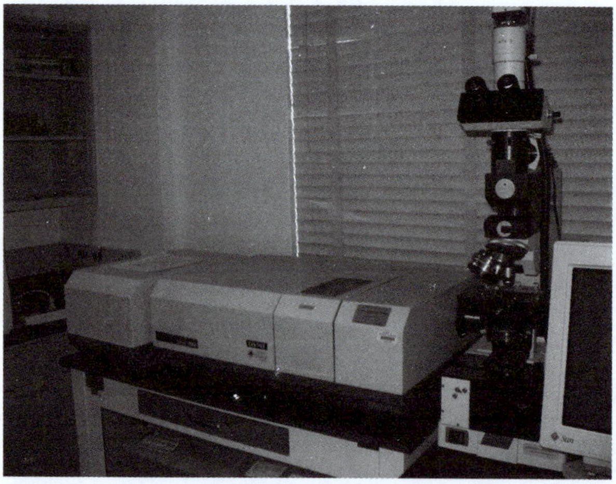

Fig. 12.22: Infrared spectrometer

This instrument uses a source of infrared radiation such as a nichrome wire or cooled rod of silicon carbide to produce a range of frequencies which are then separated into individual frequencies using a monochromator diffraction grating. The beam produced is then split into two and one passes through the sample whilst the other is used as a reference beam. The two beams then converge on the detector which measures the difference in intensity and then sends a proportional signal to the recorder.

The resulting plot is a measure of transmission against frequency which is usually plotted as wave number (cm^{-1}). Wave number can be related to the frequency by using the following equations:

$$\text{Wave number} = \frac{1}{\text{Wavelength}(\lambda)*}$$

*The units for wavenumber are cm^{-1}, therefore, wavelength should be converted to cm rather than metres for this calculation and wavelength is related to frequency as:

$$\text{Velocity of light (c)} = \frac{\text{Frequency}(\upsilon)}{\text{Wavelength}(\lambda)}.$$

Units

Wave number	cm^{-1}
Wavelength	m
Frequency	Hz
Velocity of light	ms^{-1}

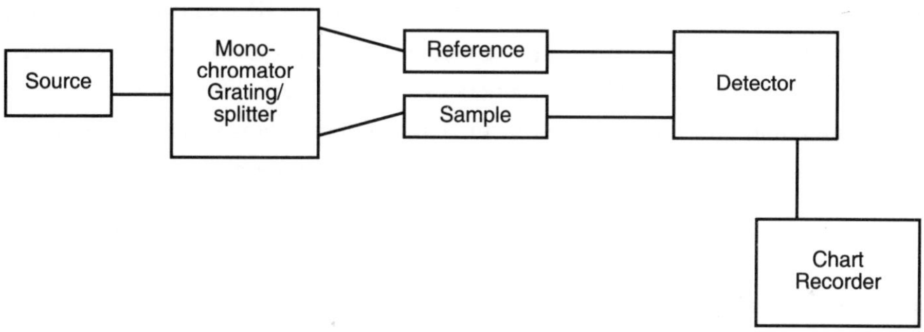

Fig. 12.23: Schematic Diagram of an Infrared Spectrometer

Fourier Transform Spectrometers

Fourier transform spectrometers have recently replaced dispersive instruments for most applications due to their superior speed and sensitivity. They have greatly extended the capabilities of infrared spectroscopy and have been applied to many areas that are very difficult or nearly impossible to analyze by dispersive instruments. Instead of viewing each component frequency sequentially, as in a dispersive IR spectrometer, all frequencies are examined simultaneously in Fourier transform infrared (FTIR) spectroscopy.

Instrumentation

There are three basic spectrometer components in an FT system: radiation source, interferometer, and detector. A simplified optical layout of a typical FTIR spectrometer is illustrated in the concerned figure below. The same types of radiation sources are used for both dispersive and Fourier

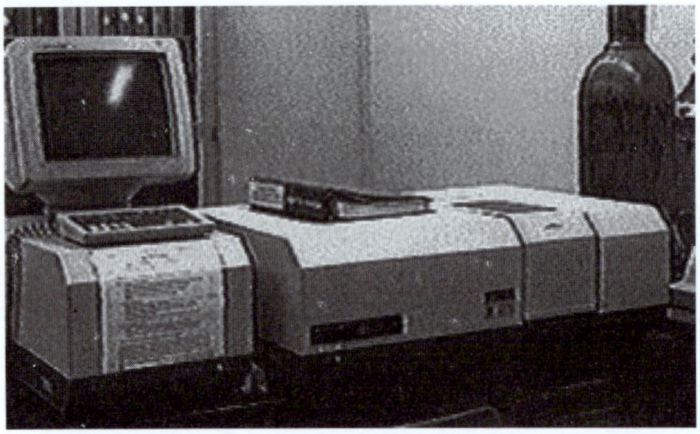

Fig. 12.24: FTIR absorption spectrometer

transforms spectrometers. However, the source is more often water-cooled in FTIR instruments to provide better power and stability.

In contrast, a completely different approach is taken in an FTIR spectrometer to differentiate and measure the absorption at component frequencies. The monochromator is replaced by an interferometer, which divides radiant beams, generates an optical path difference between the beams, then recombines them in order to produce repetitive interference signals measured as a function of optical path difference by a detector. As its name implies, the interferometer produces interference signals, which contain infrared spectral information generated after passing through a sample.

The most commonly used interferometer is a Michelson interferometer. It consists of three active components: a moving mirror, a fixed mirror, and a beamsplitter. The two mirrors are perpendicular to each other. The beamsplitter is a semireflecting device and is often made by depositing a thin film of germanium onto a flat KBr substrate. Radiation from the broadband IR source is collimated and directed into the interferometer, and impinges on the beamsplitter. At the

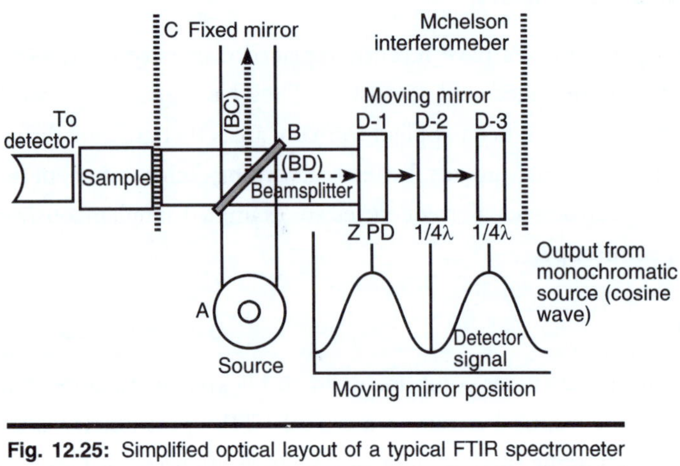

Fig. 12.25: Simplified optical layout of a typical FTIR spectrometer

beamsplitter, half the IR beam is transmitted to the fixed mirror and the remaining half is reflected to the moving mirror. After the divided beams are reflected from the two mirrors, they are recombined at the beamsplitter. Due to changes in the relative position of the moving mirror to the fixed mirror, an interference pattern is generated. The resulting beam then passes through the sample and is eventually focused on the detector.

For an easier explanation, the detector response for a single-frequency component from the IR source is first considered. This simulates an idealized situation where the source is monochromatic, such as a laser source. As previously described, differences in the optical paths between the two split beams are created by varying the relative position of moving mirror to the fixed mirror. If the two arms of the interferometer are of equal length, the two split beams travel through the exact same path length. The two beams are totally in phase with each other; thus, they interfere constructively and lead to a maximum in the detector response. This position of the moving mirror is called the point of zero path difference (ZPD). When the moving mirror travels in either direction by the distance $\lambda/4$, the optical path (beamsplitter-mirror-beamsplitter) is changed by 2 $(\lambda/4)$, or $\lambda/2$. The two beams are 180° out of phase with each other, and thus interfere destructively. As the moving mirror travels another $\lambda/4$, the optical path difference is now 2 $(\lambda/2)$, or λ. The two beams are again in phase with each other and result in another constructive interference. When the mirror is moved at a constant velocity, the intensity of radiation reaching the detector varies in a sinusoidal manner to produce the interferogram output shown in Fig. 12.25. The interferogram is the record of the interference signal. It is actually a time domain spectrum and records the detector response changes versus time within the mirror scan. If the sample happens to absorb at this frequency, the amplitude of the sinusoidal wave is reduced by an amount proportional to the amount of sample in the beam.

Extension of the same process to three component frequencies results in a more complex interferogram, which is the summation of three individual modulated waves. In contrast to this simple, symmetric interferogram, the interferogram produced with a broadband IR source displays extensive interference patterns. It is a complex summation of superimposed sinusoidal waves, each wave corresponding to a single frequency. When this IR beam is directed through the sample, the amplitudes of a set of waves are reduced by absorption if the frequency of this set of waves is the same as one of the characteristic frequencies of the sample. The interferogram contains information over the entire IR region to which the detector is responsive. A mathematical operation known as Fourier transformation converts the interferogram (a time domain spectrum displaying intensity versus time within the mirror scan) to the final IR spectrum, which is the familiar frequency domain spectrum showing intensity versus frequency. This also explains how the term Fourier transform infrared spectrometry is created.

The detector signal is sampled at small, precise intervals during the mirror scan. The sampling rate is controlled by an internal, independent reference, a modulated monochromatic beam from helium neon (HeNe) laser focused on a separate detector.

The two most popular detectors for a FTIR spectrometer are deuterated triglycine sulphate (DTGS) and mercury cadmium telluride (MCT). The response times of many detectors (for example, thermocouple and thermistor) used in dispersive IR instruments are too slow for the rapid scan times (1 sec or less) of the interferometer. The DTGS detector is a pyroelectric detector that delivers rapid responses because it measures the changes in temperature rather than the value of temperature. The MCT detector is a photon (or quantum) detector that depends on the quantum nature of radiation and also exhibits very fast responses. Whereas DTGS detectors operate at room temperature, MCT detectors must be maintained at liquid nitrogen temperature (77°K) to be effective. In general, the MCT detector is faster and more sensitive than the DTGS detector.

Spectrometer Design

The basic instrument design is quite simple. The IR radiation from a broadband source is first directed into an interferometer, where it is divided and then recombined after the split beams travel different optical paths to generate constructive and destructive interference. Next, the resulting beam passes through the sample compartment and reaches to the detector. Most benchtop FTIR spectrometers are single-beam instruments. Unlike double-beam grating spectrometers, single-beam FTIR does not obtain transmittance or absorbance IR spectra in real time.

FTIR Advantages

FTIR instruments have distinct advantages over dispersive spectrometers:

- Better speed and sensitivity (Felgett advantage). A complete spectrum can be obtained during a single scan of the moving mirror, while the detector observes all frequencies simultaneously.

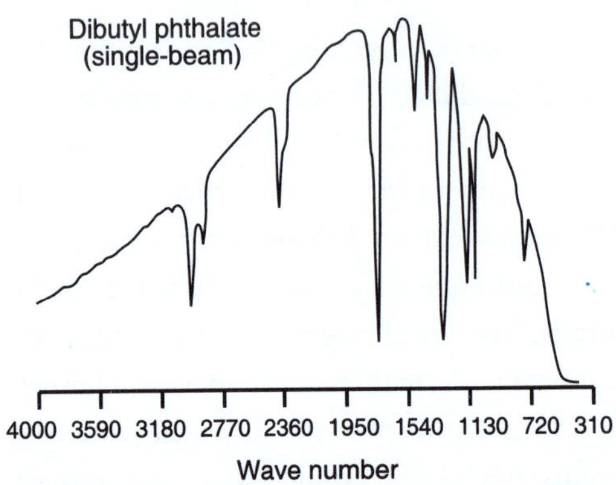

Fig. 12.26: A single-beam IR spectrum of dibutyl phthalate (a liquid sample)

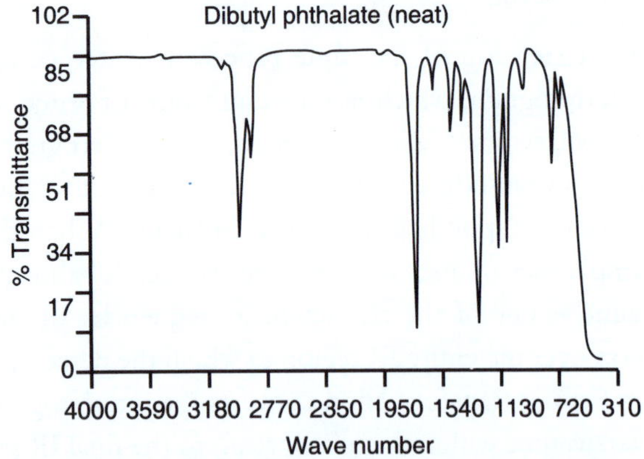

Fig. 12.27: The "double-beam" IR spectrum of dibutyl phthalate, produced by ratio of the corresponding single-beam sample spectrum against the single-beam background spectrum

An FTIR instrument can achieve the same signal-to-noise (S/N) ratio of a dispersive spectrometer in a fraction of the time (1 sec or less versus 10 to 15 min). The S/N ratio is proportional to the square root of the total number of measurements. Because multiple spectra can be readily collected in 1 min or less, sensitivity can be greatly improved by increasing S/N through coaddition of many repeated scans.

- *Increased optical throughput (Jaquinot advantage)*: Energy-wasting slits are not required in the interferometer because dispersion or filtering is not needed. Instead, a circular optical aperture is commonly used in FTIR systems. The beam area of an FT instrument is usually 75 to 100 times larger than the slit width of a dispersive spectrometer. Thus, more radiation energy is made available. This constitutes a major advantage for many samples or sampling techniques that are energy-limited.
- *Internal laser reference (Connes advantage)*: The use of a helium neon laser as the internal reference in many FTIR systems provides an automatic calibration in an accuracy of better than 0.01 cm^{-1}. This eliminates the need for external calibrations.
- *Simpler mechanical design*: There is only one moving part, the moving mirror, resulting in less wear and better reliability.
- *Elimination of stray light and emission contributions*: The interferometer in FTIR modulates all the frequencies. The unmodulated stray light and sample emissions (if any) are not detected.
- *Powerful data station*: Modern FTIR spectrometers are usually equipped with a powerful, computerized data system. It can perform a wide variety of data processing tasks such as Fourier transformation, interactive spectral subtraction, baseline correction, smoothing, integration, and library searching.

Although the spectra of many samples can be satisfactorily run on either FTIR or dispersive instruments, FTIR spectrometers are the preferred choice for samples that are energy-limited or when increased sensitivity is desired. A wide range of sampling accessories is available to take advantage of the capabilities of FTIR instruments.

Two-dimensional infrared spectroscopy

Two-dimensional infrared correlation spectroscopy analysis is the application of 2D correlation analysis on infrared spectra. By extending the spectral information of a perturbed sample, spectral analysis is simplified and resolution is enhanced. The 2D synchronous and 2D asynchronous spectra represent a graphical overview of the spectral changes due to a perturbation (such as a changing concentration or changing temperature) as well as the relationship between the spectral changes at two different wave numbers.

Nonlinear two-dimensional infrared spectroscopy

It is the infrared version of correlation spectroscopy. Nonlinear two-dimensional infrared spectroscopy is a technique that has come available with the development of femtosecond infrared laser pulses. In this experiment first a set of pump pulses are applied to the sample. This is followed by a waiting time, where the system is allowed to relax. The waiting time typically lasts from zero to several picoseconds and the duration can be controlled with a resolution of tens of femtoseconds. A probe pulse is then applied resulting in the emission of a signal from the sample. The nonlinear two-dimensional infrared spectrum is a two-dimensional correlation plot of the frequency ω_1 that was excited by the initial pump pulses and the frequency ω_3 excited by the probe pulse after the waiting time. This allows the observation of coupling between different vibrational modes. Because of its extremely high time resolution it can be used to follow changes in molecular configurations taking place on a picosecond timescale. It is still a largely unexplored technique and is becoming increasingly popular for fundamental research.

Like in two-dimensional nuclear magnetic resonance (2DNMR) spectroscopy this technique spreads the spectrum in two dimensions and allow for the observation of cross peaks that contain information on the coupling between different modes. In contrast to 2DNMR nonlinear two-dimensional infrared spectroscopy also involves the excitation to overtones. These excitations result in excited state absorption peaks located below the diagonal and cross peaks. In 2DNMR two distinct techniques, COSY and NOESY are frequently used. The cross peaks in the first are related to the scalar coupling, while in the latter they are related to the spin transfer between different nuclei. In nonlinear two-dimensional infrared spectroscopy analogs have been drawn to these 2DNMR techniques. Nonlinear two-dimensional infrared spectroscopy with zero waiting time corresponds to COSY and nonlinear two-dimensional infrared spectroscopy with finite waiting time allowing vibrational population transfer corresponds to NOESY. The COSY variant of nonlinear two-dimensional infrared spectroscopy has been used for determination of the secondary structure content proteins.

Uses and applications

Infrared spectroscopy is widely used in both research and industry as a simple and reliable technique for measurement, quality control and dynamic measurement. The instruments are now small, and can be transported, even for use in field trials. With increasing technology in computer filtering and manipulation of the results, samples in solution can now be measured accurately (water produces a broad absorbance across the range of interest, and thus renders the spectra unreadable without this computer treatment). Some machines will also automatically tell you what substance is being measured from a store of thousands of reference spectra held in storage.

By measuring at a specific frequency over time, changes in the character or quantity of a particular bond can be measured. This is especially useful in measuring the degree of polymerization in polymer manufacture. Modern research machines can take infrared measurements across the whole range of interest as frequently as 32 times a second. This can be done whilst simultaneous measurements are made using other techniques. This makes the observations of chemical reactions and processes quicker and more accurate.

Techniques have been developed to assess the quality of tea-leaves using infrared spectroscopy. This will mean that highly trained experts (also called 'noses') can be used more sparingly, at a significant cost saving. Infrared spectroscopy has been highly successful for applications in both organic and inorganic chemistry. Infrared spectroscopy has also been successfully utilized in the field of semiconductor microelectronics for example, infrared spectroscopy can be applied to semiconductors like silicon, gallium arsenide, gallium nitride, zinc selenide, amorphous silicon, silicon nitride, etc.

Near-Infrared Absorption Spectroscopy (NIR)

NIR spectroscopy is the measurement of the wavelength and intensity of the absorption of near-infrared light by a sample. Near-infrared light spans the 800 nm - 2.5 m (12500 - 4000 cm^{-1}) range and is energetic enough to excite overtones and combinations of molecular vibrations to higher energy levels. NIR spectroscopy is typically used for quantitative measurement of organic functional groups, especially O–H, N–H, and C=O. Detection limits are typically 0.1% and applications include pharmaceutical, agricultural, polymer, and clinical analysis.

The components and design of NIR instrumentation are similar to UV-Vis absorption spectrometers. The light source is usually a tungsten lamp and the detector is usually a PbS solid-state detector. Sample holders can be glass or quartz and typical solvents are CCl_4 and CS_2. The convenient instrumentation of NIR spectroscopy compared to IR spectroscopy makes it much more suitable for on-line monitoring and process control.

12.10 RAMAN SPECTROSCOPY

Raman spectroscopy is a spectroscopic technique used in condensed matter physics and chemistry to study vibrational, rotational, and other low-frequency modes in a system. It relies on inelastic scattering, or Raman scattering of monochromatic light, usually from a laser in the visible, near infrared, or near ultraviolet range. The laser light interacts with phonons or other excitations in the system, resulting in the energy of the laser photons being shifted up or down. The shift in energy gives information about the phonon modes in the system. Infrared spectroscopy yields similar, but complementary information. Typically, a sample is illuminated with a laser beam. Light from the illuminated spot is collected with a lens and sent through a monochromator. Wavelengths close to the laser line (due to elastic Rayleigh scattering) are filtered out and those in a certain spectral window away from the laser line are dispersed onto a detector. Spontaneous Raman scattering is typically very weak, and as a result the main difficulty of Raman spectroscopy is separating the weak inelastically scattered light from the intense Rayleigh scattered laser light. Raman spectrometers typically use holographic diffraction gratings and multiple dispersion stages to achieve a high degree of laser rejection. In the past, PMTs were the detectors of choice for dispersive Raman setups, which resulted in long acquisition times. However, the recent uses of CCD detectors have made dispersive Raman spectral acquisition much more rapid. Raman spectroscopy has a stimulated version, analogous to stimulated emission, called stimulated Raman scattering.

Basics

The Raman effect occurs when light impinges upon a molecule and interacts with the electron cloud of the bonds of that moiecule. The incident photon excites one of the electrons into a virtual state.

For the spontaneous Raman effect, the molecule will be excited from the ground state to a virtual energy state, and relax into a vibrational excited state, which generates Stokes Raman scattering. If the molecule was already in an elevated vibrational energy state, the Raman scattering is then called anti-Stokes Raman scattering. A molecular polarizability change or amount of deformation of the electron cloud, with respect to the vibrational coordinate is required for the molecule to exhibit the Raman effect. The amount of the polarizability change will determine the intensity, whereas the Raman shift is equal to the vibrational level that is involved.

History

Although the inelastic scattering of light was predicted by Smekal in 1923, it was 1928 before it was observed in practice. The Raman effect was named after one of its discoverers, the Indian scientist Sir C. V. Raman who observed the effect by means of sunlight (1928, together with K. S. Krishnan and independently by Grigory Landsberg and Leonid Mandelstam). Raman won the Nobel Prize in Physics in 1930 for this discovery accomplished using sunlight, a narrow band photographic filter to create monochromatic light and a "crossed" filter to block this monochromatic light. He found that light of changed frequency passed through the "crossed" filter. Subsequently the mercury arc became the principal light source, first with photographic detection and then with spectrophotometric detection. Currently lasers are used as light sources.

Applications

Raman spectroscopy is commonly used in chemistry, since vibrational information is very specific for the chemical bonds in molecules. It, therefore, provides a fingerprint by which the molecule can be identified. The fingerprint region of organic molecules is in the range 500-2000 cm^{-1}. Another way that the technique is used is to study changes in chemical bonding, e.g., when a substrate is added to an enzyme. Raman gas analyzers have many practical applications, for instance they are used in medicine for real-time monitoring of anaesthetic and respiratory gas mixtures during surgery. In solid state physics, spontaneous Raman spectroscopy is used, among other things, to characterize materials, measure temperature, and find the crystallographic orientation of a sample. As with single molecules, a given solid material has characteristic phonon modes that can help an experimenter identify it. In addition, Raman spectroscopy can be used to observe other low frequency excitations of the solid, such as plasmons, magnons, and superconducting gap excitations. The spontaneous Raman signal gives information on the population of a given phonon mode in the ratio between the Stokes (downshifted) intensity and anti-Stokes (upshifted) intensity. Raman scattering by an anisotropic crystal gives information on the crystal orientation. The polarization of the Raman scattered light with respect to the crystal and the polarization of the laser light can be used to find the orientation of the crystal, if the crystal structure (specifically, its point group) is known. Raman active fibres, such as aramid and carbon, have vibrational modes that show a shift in Raman frequency with applied stress. Polypropylene fibres also exhibit similar shifts. The radial breathing mode is a commonly used technique to evaluate the diameter of carbon nanotubes.

Spatially Offset Raman Spectroscopy (SORS), which is less sensitive to surface layers than conventional Raman, can be used to discover counterfeit drugs without opening their internal packaging, and for non-invasive monitoring of biological tissue. Raman spectroscopy can be used to investigate the chemical composition of historical documents such as the Book of Kells and contribute to knowledge of the social and economic conditions at the time the documents were produced. This is especially helpful because Raman spectroscopy offers a non-invasive way to determine the best course of preservation or conservation treatment for such materials.

Raman microspectroscopy

Raman spectroscopy offers several advantages for microscopic analysis. Since it is a scattering technique, specimens do not need to be fixed or sectioned. Raman spectra can be collected from a very small volume (< 1 μm in diameter); these spectra allow the identification of species present in that volume. Water does not interfere very strongly. Thus, Raman spectroscopy is suitable for the microscopic examination of minerals, materials such as polymers and ceramics, cells and proteins. A Raman microscope begins with a standard optical microscope, and adds an excitation laser, a monochromator, and a sensitive detector (such as a charge-coupled device (CCD) or photomultiplier tube (PMT)). FT-Raman has also been used with microscopes. In direct imaging, the whole field of view is examined for scattering over a small range of wave numbers (Raman shifts). For instance, a wave number characteristic for cholesterol could be used to record the distribution of cholesterol within a cell culture. The other approach is hyperspectral imaging or chemical imaging, in which thousands of Raman spectra are acquired from all over the field of view. The data can then be used to generate images showing the location and amount of different components. Taking the cell

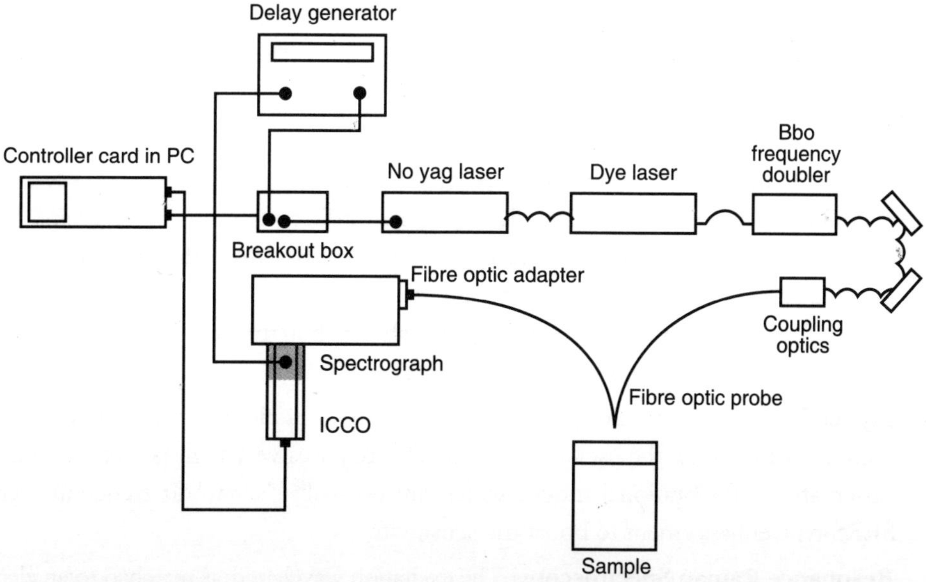

Fig. 12.28: Schematic diagram of Raman spectroscopy

culture example, a hyperspectral image could show the distribution of cholesterol, as well as proteins, nucleic acids, and fatty acids. Sophisticated signal- and image-processing techniques can be used to ignore the presence of water, culture media, buffers, and other interferences. Raman microscopy, and in particular confocal microscopy, has very high spatial resolution. For example, the lateral and depth resolutions were 250 nm and 1.7 μm, respectively, using a confocal Raman microspectrometer with the 632.8 nm line from a He-Ne laser with a pinhole of 100 μm diameter. Since the objective lenses of microscopes focus the laser beam to several micrometres in diameter, the resulting photon flux is much higher than achieved in conventional Raman set-ups. This has the added benefit of enhanced fluorescence quenching. However, the high photon flux can also cause sample degradation, and for this reason some setups require a thermally conducting substrate (which acts as a heat sink) in order to mitigate this process.

By using Raman microspectroscopy, in vivo time- and space-resolved Raman spectra of microscopic regions of samples can be measured. As a result, the fluorescence of water, media, and buffers can be removed. Consequently in vivo time- and space-resolved Raman spectroscopy is suitable to examine proteins, cells and organs. Raman microscopy for biological and medical specimens generally uses near-infrared (NIR) lasers (785 nm diodes and 1064 nm Nd:YAG are especially common). This reduces the risk of damaging the specimen by applying high power. However, the intensity of NIR Raman is low (owing to the ω^{-4} dependence of Raman scattering intensity), and most detectors required very long collection times. Recently, more sensitive detectors have become available, making the technique better suited to general use. Raman microscopy of inorganic specimens, such as rocks and ceramics and polymers, can use a broader range of excitation wavelengths.

Variations

Several variations of Raman spectroscopy have been developed. The usual purpose is to enhance the sensitivity (e.g., surface-enhanced Raman), to improve the spatial resolution (Raman microscopy), or to acquire very specific information (resonance Raman).

1. **Surface Enhanced Raman Spectroscopy (SERS):** Normally done in a silver or gold colloid or a substrate containing silver or gold. Surface plasmons of silver and gold are easily excited by the laser, and the resulting electric fields cause other nearby molecules to become Raman active. The result is amplification of the Raman signal (by up to 10^{11}). This effect was originally observed by Fleishman but the prevailing explanation was proposed by Van Duyne in 1977.
2. **Hyper Raman:** A non-linear effect in which the vibrational modes interact with the second harmonic of the excitation beam. This requires very high power, but allows the observation of vibrational modes which are normally "silent". It frequently relies on SERS-type enhancement to boost the sensitivity.
3. **Resonance Raman Spectroscopy:** The excitation wavelength is matched to an electronic transition of the molecule or crystal, so that vibrational modes associated with the excited

electronic state are greatly enhanced. This is useful for studying large molecules such as polypeptides, which might show hundreds of bands in "conventional" Raman spectra. It is also useful for associating normal modes with their observed frequency shifts.

4. **Spontaneous Raman Spectroscopy:** Used to study the temperature dependence of the Raman spectra of molecules.
5. **Optical Tweezers Raman Spectroscopy (OTRS):** Used to study individual particles, and even biochemical processes in single cells trapped by optical tweezers.
6. **Stimulated Raman Spectroscopy:** A two-colour pulse transfers the population from ground to a vibrationally excited state, if the difference in energy corresponds to an allowed Raman transition. Two-photon UV ionization, applied after the population transfer but before relaxation, allows the intra-molecular or inter-molecular Raman spectrum of a gas or molecular cluster (indeed, a given conformation of molecular cluster) to be collected. This is a useful molecular dynamics technique.
7. **Spatially Offset Raman Spectroscopy (SORS):** The Raman scatter is collected from regions laterally offset away from the excitation laser spot, leading to significantly lower contributions from the surface layer than with traditional Raman spectroscopy.
8. **Coherent Anti-Stokes Raman Spectroscopy (CARS):** Two laser beams are used to generate a coherent anti-Stokes frequency beam, which can be enhanced by resonance
9. **Resonance Enhanced Multi-Photon Ionization (REMPI)** is a technique applied to the spectroscopy of atoms and small molecules. In practice, a tunable laser can be used to access an excited intermediate state. The selection rules associated with a two-photon or other multi-photon photoabsorption are different from the selection rules for a single photon transition. The REMPI technique typically involves a resonant single or multiple photon absorption to an electronically excited intermediate state followed by another photon which ionizes the atom or molecule. The light intensity to achieve a typical multiphoton transition is generally significantly larger than the light intensity to achieve a single photon photoabsorption. Because of this, a subsequent photoabsorption is often very likely. An ion and a free electron will result if the photons have imparted enough energy to exceed the ionization threshold energy of the system. In many cases, REMPI provides spectroscopic information that can be unavailable to single photon spectroscopic methods, for example rotational structure in molecules is easily seen with this technique.

Rydberg states, Rydberg molecules

High photon intensity experiments can involve multiphoton processes with the absorption of integer multiples of the photon energy. In experiments that involve a multiphoton resonance, the intermediate is often a low-lying Rydberg state, and the final state is often an ion. The initial state of the system, photon energy, angular momentum and other selection rules can help in determining the nature of the intermediate state. This approach is exploited in Resonance Enhanced Multi-Photon Ionization Spectroscopy (REMPI). The technique is in wide use in both atomic and molecular spectroscopy. An advantage of the REMPI technique is that the ions can be detected with almost complete efficiency and even time resolved for their mass. It is also possible to gain additional information by performing experiments to look at the energy of the liberated photoelectron in these experiments. (Robert N. Compton formerly of O.R.N.L. and Philip M. Johnson pioneered the development of REMPI. The technique was named by Johnson.)

12.11 MASS SPECTROSCOPY

Mass Spectrometry is one of the most powerful analytical tools that the chemist has for the characterisation of individual compounds. The principles of the technique have been established for some time although several new methods of ionisation have been developed recently.

Basics

All mass spectrometers require the sample compound to be ionised and those ions to be transferred into the gas phase for analysis; this may be a simultaneous process or a distinct series of events. In conventional Electron Impact Ionisation [EI], the sample is vaporised and then ionised by collisions with a beam of energetic electrons. The ions produced are accelerated by an applied electric field and then they are separated according to their mass to charge ratio either as a continuous beam or a discrete pulse. After separation the ions arrive at a detector and produce a small electric current, which is amplified and recorded by a computer-controlled system.

The Spectrum

Most ions formed by conventional techniques are singly charged. The mass spectrum is produced by relating a scan of the analyser to the signals received at the detector. In the case of a magnetic sector instrument, the signal is measured as a function of magnetic field strength. The magnetic field strength can be related to the ion mass, thus yielding a plot of signal intensity versus mass the mass spectrum. From the spectrum it is often possible to determine the molecular mass of the compound, some information about its structure and, if in a mixture, its concentration.

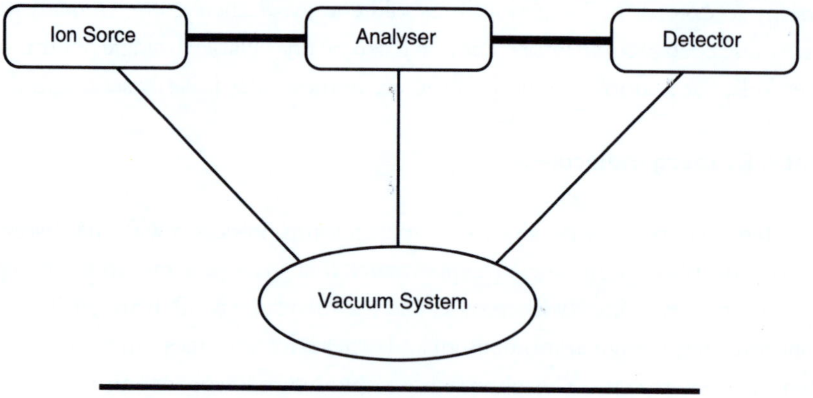

Fig. 12.29: A typical Mass Spectrometer layout

The Ion Source

Some instruments have a single means of sample introduction and ionisation, others may have many.

EI requires the sample to be vaporised and, therefore, solids are introduced on a heatable probe, liquids via a heated septum and bleed valve, and gases through a membrane or a needle valve system. The vapour is crossed by an energetic electron beam, which can ionise the sample giving molecular ions, as well as fragment ions, all with a positive charge.

FAB [fast atom bombardment] requires samples to be dissolved in a matrix such as glycerol and, therefore, the sample must be inserted on a probe system. The surface of this mixture is subjected to a beam of energetic xenon [or argon] atoms which sputter ions from the mixture. These may be positive or negative and each can be detected separately.

ESI [i.e., electrospray] is a technique that ionises directly from the liquid state and at atmospheric pressure. The idea is simple and relies on the solution being ionic. The ionic solution is pumped through a capillary at high electrical potential and the ions are rapidly separated from solvent molecules and their corresponding counter-ions. The ions then pass through a tiny nozzle connected to the vacuum system and enter the spectrometer. Again, ions formed in this manner may be positive or negative depending on the applied potential.

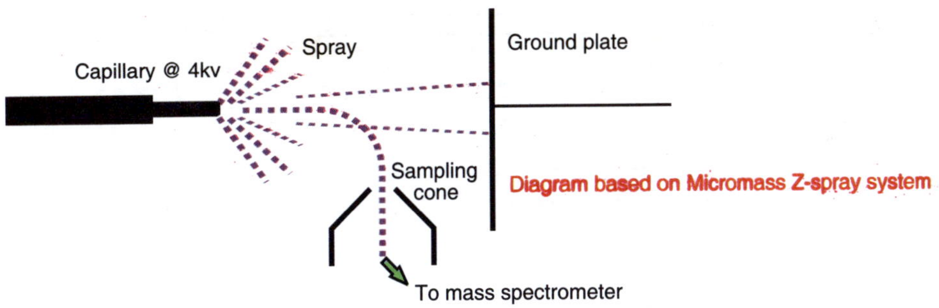

Fig. 12.30: Electrospray

MALDI [matrix assisted laser desorption ionisation] works on solid sample spots co deposited with a light-absorbing matrix, e.g., cyanohydroxycinnamic acid. By irradiating these sample spots in the matrix with high intensity laser radiation, both positive and negative ions, are produced.

The Analyser

Magnetic Sectors are the traditional means of separating ions of different masses using the principle that when a beam of ions passes through a magnetic field they are deflected through an arc whose radius is related to the ratio of mass to charge. By varying the field in a known manner, i.e., sweeping from high to low field with time, ions of different mass arrive at the detector sequentially and produce the mass spectrum.

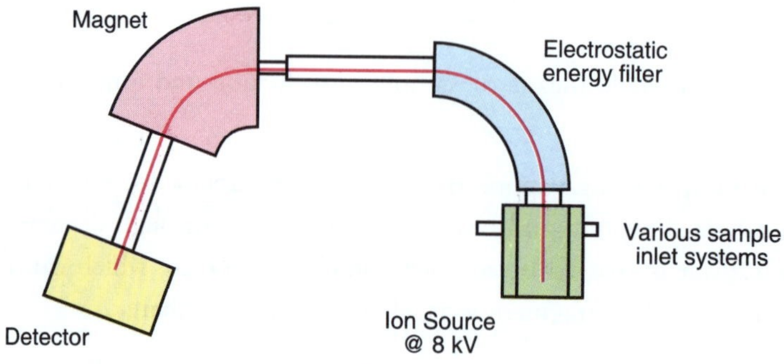

Fig. 12.31: The analyzer arrangement

Quadrupoles, often used on smaller instruments linked to chromatographic systems, use an oscillating electric field between opposite pairs of a set of 4 rods. The beam of ions passing between these rods is deflected by differing amounts depending on their mass/charge ratio. By varying the field strength or the frequency of oscillation, ions of different mass reach the detector to produce a mass spectrum.

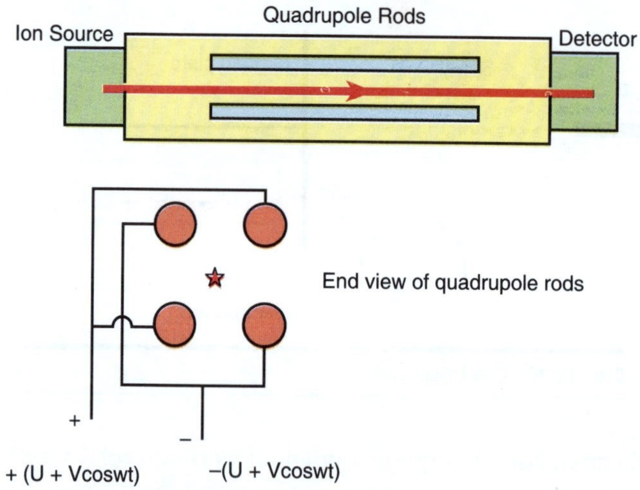

Fig. 12.32: The Quadrupoles arrangement

Time-of-Flight — Given the same amount of energy [e.g., 20 kV potential] ions of different mass will take different times to travel the same distance, the lightest being the fastest. This idea is put to use in time-of-flight instruments. In essence, these consist of an ion source/acceleration region, a drift tube along which the ions travel, and finally a detector which measures the ion signal at the end of the drift tube as a function of time. The short pulse time of laser ionisation using intense laser pulses is ideally suited to this form of analyser.

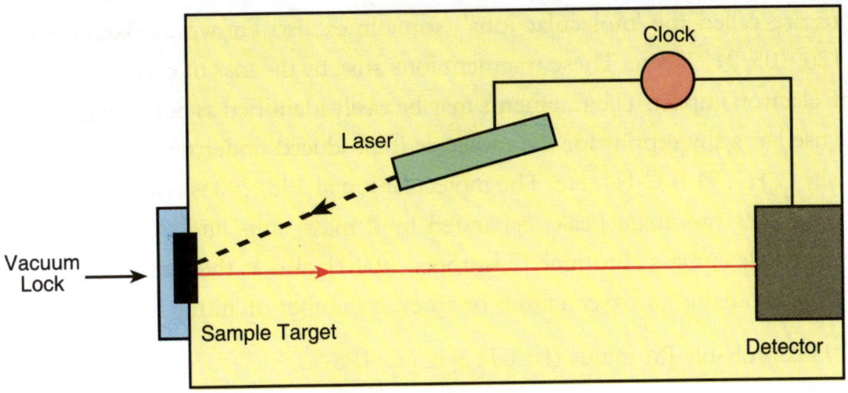

Fig. 12.33: time-of-flight instrument

Other types of analysers, e.g., ion traps, cyclotron resonance, are also used in mass spectrometry but are beyond the scope of these brief notes.

Detectors

These are usually electron multipliers which produce up to 106 electrons for every ion that hits the detectors. This produces a current which can be amplified and measured.

The Mass Spectrum

The example below illustrates a typical EI spectrum

[C_8H_7OBr - Ethanone,2-bromo-1-phenyl-]

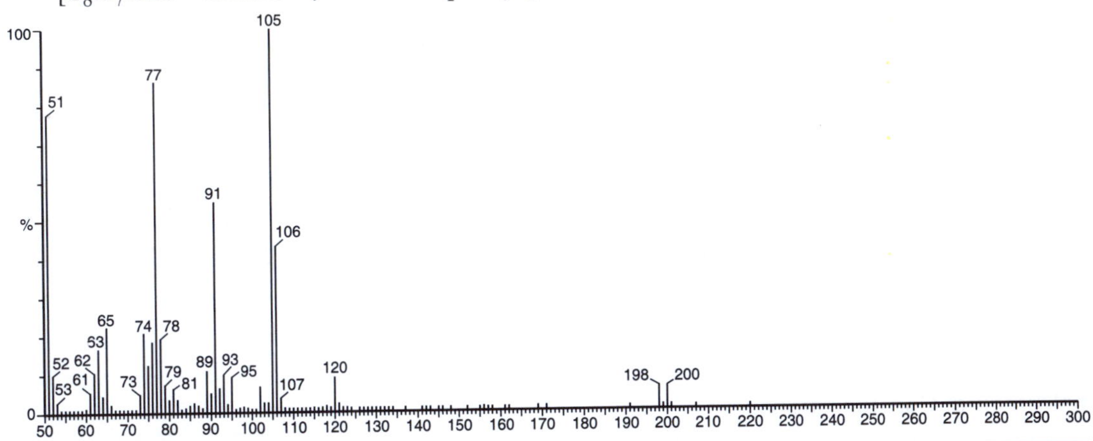

Fig. 12.34: Typical EI spectrum

Ions are shown as vertical lines of differing intensity along an x-axis of mass/charge. As all the charges here are '1+' the scale can be regarded as that of mass. Various features of the spectrum may be distinguished, e.g., the ions at highest mass [in this case a small cluster at 198/200]. These unfragmented ions are called the 'molecular ions' (sometimes also known as parent ions) with fragment ions at 120, 105, 91, 77, etc. These fragment ions arise by the loss of certain groups due to the high energy of electron impact. The fragments may be easily identified as belonging to a specific group and can be used as a fingerprint for the molecule if produced under standard conditions. A mass of 77 is usually $C_6H_5^+$, 91 is $C_7H_7^+$, etc. The molecular ion at 198/200 is also very informative. The fact that it comprises two main peaks separated by 2 mass units and of similar intensities indicates that the molecule contains bromine [2 isotopes 79/81]; also as the mass is even this indicates that the molecule contains an oxygen atom or an even number of nitrogen atoms.

In this case 105 is probably [M minus ($Br-CH_2$)]+, i.e., 198-93.

As a contrast the ESI spectrum below is of a protein.

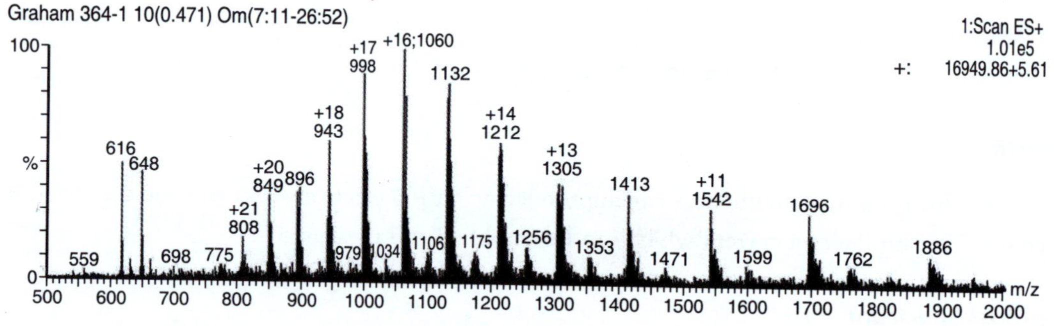

Fig. 12.35: ESI spectrum

The mass of this protein is 16951 but because of the multiple charges on the ions formed under electrospray conditions the various ion clusters appear as a pattern on the normal scale of 800-2000.

The separation of the various ions allows the allocation of the number of charges by application of a suitable algorithm; hence the numbers in this case show ions of +10 to +21; this in turn [assuming that the ions are formed by association with H^+] allows the calculation of the true molecular mass of the protein. Very few fragment ions are observed, although impurities may contribute to some of the extra signals

Instrumental Considerations

(a) **Resolution:** The resolution of any mass spectrometer determines its ability to distinguish between adjacent masses; for many applications it is only required to separate unit values of mass, e.g., 500 from 501, but where there is the possibility of mass differences of 0.01 or smaller the ability to increase the resolution is necessary. Quadrupoles generally only work at unit mass differences [the resolution defined as $R = m/(m_1-m_2)$, where m is the average of the ions and m_1-m_2 the mass difference, with peak separation showing a valley between the ions of 50% peak height]. Two-Sector instruments usually have the ability to adjust the resolution and can resolve peaks 500.00 and 500.01 to a valley of 10% peak height or better. Resolution in time-of-flight instruments is somewhere between these other two instrument types.

(b) **Mass accuracy:** The ability to measure masses to a high degree of accuracy is required for the determination of the elemental composition of a molecule; generally this is only achieved on two-sector instruments (and some of the more specialised techniques, e.g., Fourier Transform Ion Cyclotron Resonance). The accuracy depends on many factors, but crucial is the use of accurately known calibrant ion peaks.

(c) **Sensitivity:** Mass Spectrometers are very sensitive, often operating with very minute amounts of material, e.g., 1 femtogram (0.000000000000001g). However, different ionisation techniques have different efficiencies and so EI and MALDI are the most sensitive with ESI close and FAB the least sensitive.

Uses of Mass Spectrometry

Chemical Identification — by means of fragmentation and accurate mass.

Forensic Testing — drugs, explosives, etc. are often detected using mass spectrometry.

Environmental Monitoring — atmospheric pollutants, spillage incidents and various emissions can be effectively detected and mapped.

Pharmaceutical Testing — new drugs, etc., have to be thoroughly tested for side effects and harmful metabolites before release for use. Mass spectrometry is one of the main analytical techniques used.

Archaeological Studies — isotope ratios are measured by mass spectrometry and used to determine the age of artefacts and often the origin of materials.

Medical Research — Many illnesses produce particular chemical species in the body, or changes to normal components. These may be detected and monitored by suitable mass spectrometry techniques.

Biological Studies — The natural world produces many active chemical species that may be useful, their detection and formation pathways may often require mass spectrometry.

12.12 ELECTRON PARAMAGNETIC RESONANCE (ESR/EPR)

Electron paramagnetic resonance (EPR) or electron spin resonance (ESR) spectroscopy is a technique for studying chemical species that have one or more unpaired electrons, such as organic and inorganic free radicals or inorganic complexes possessing a transition metal ion. The basic physical concepts of EPR are analogous to those of nuclear magnetic resonance (NMR), but it is electron spins that are excited instead of spins of atomic nuclei. Since most stable molecules have all their electrons paired, the EPR technique is less widely used than NMR. However, this limitation to paramagnetic species also means that the EPR technique is one of great specificity, since ordinary chemical solvents and matrices do not give rise to EPR spectra.

Origin of an EPR signal

Every electron has a magnetic moment and spin quantum number s = 1/2, with magnetic components ms = +1/2 and ms = –1/2. In the presence of an external magnetic field with strength B_0, the electron's magnetic moment aligns itself either parallel (m_s = –1/2) or antiparallel (m_s = +1/2) to the field, each alignment having a specific energy. The parallel alignment corresponds to the lower energy state, and the separation between it and the upper state is $\Delta E = g_e \mu_B B_0$, where g_e is the electron's so-called g-factor. This equation implies that the splitting of the energy levels is directly proportional to the magnetic field's strength, as shown in the diagram below.

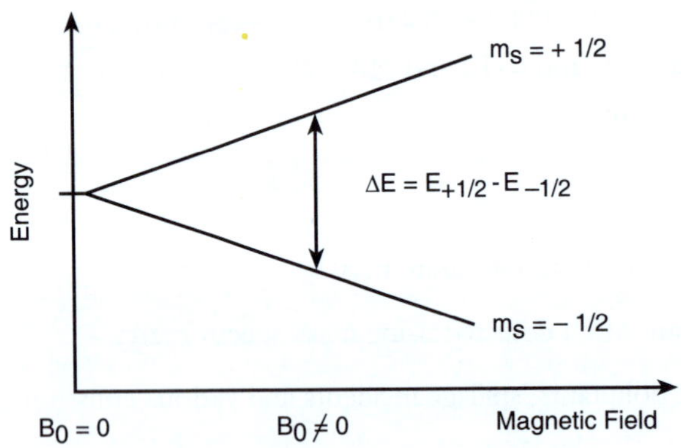

Fig. 12.36: Splitting of energy level in EPR signal

An unpaired electron can move between the two energy levels by either absorbing or emitting electromagnetic radiation of energy $\varepsilon = h\nu$ such that the resonance condition, $\varepsilon = \Delta E$, is obeyed. Substituting in $\varepsilon = h\nu$ and $\Delta E = g_e \mu_B B_0$ leads to the fundamental equation of EPR spectroscopy: $h\nu = g_e \mu_B B_0$. Experimentally, this equation permits a large combination of frequency and magnetic field values, but the great majority of EPR measurements are made with microwaves in the 9000 - 10000 MHz (9 - 10 GHz) region, with fields corresponding to about 3500 G (0.3500 T). In principle, EPR spectra can be generated by either varying the photon frequency incident on a sample while holding the magnetic field constant, or doing the reverse. In practice, it is usually the frequency which is kept fixed. A collection of paramagnetic centres, such as free radicals, is exposed to microwaves at a fixed frequency. By increasing an external magnetic field, the gap between the m_s = +1/2 and ms = –1/2 energy states is widened until it matches the energy of the microwaves, as represented by the double-arrow in the diagram above. At this point the unpaired electrons can move between their two spin states. Since there are typically more electrons in the lower state, due to the Maxwell-Boltzmann distribution, there is a net absorption of energy, and it is this absorption which is monitored and converted into a spectrum.

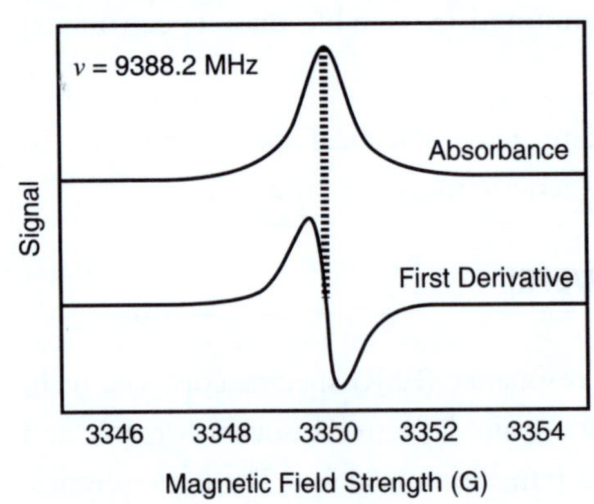

Fig. 12.37: The Maxwell-Boltzmann distribution

As an example of how $h\nu = g_e\mu_B B_0$ can be used, consider the case of a free electron, which has $g_e = 2.0023$, and the simulated spectrum shown at the right in two different forms. For the microwave frequency of 9388.2 MHz, the predicted resonance position is a magnetic field of about $B_0 = h\nu/g_e\mu_B = 0.3350$ tesla = 3350 gauss, as shown. Note that while two forms of the same spectrum are presented in the figure, most EPR spectra are recorded and published only as first derivatives. Because of electron-nuclear mass differences, the magnetic moment of an electron is substantially larger than the corresponding quantity for any nucleus, so that a much higher electromagnetic frequency is needed to bring about a spin resonance with an electron than with a nucleus, at identical magnetic field strengths. For example, for the field of 3350 G shown at the right, spin resonance occurs near 9388.2 MHz for an electron compared to only about 14.3 MHz for ^1H nuclei. (For NMR spectroscopy, the corresponding resonance equation is $h\nu = g_N\mu_N B_0$, where g_N and μ_N depend on the nucleus under study.)

Maxwell-Boltzmann distribution

In practice, EPR samples consist of collections of many paramagnetic species, and not single isolated paramagnetic centers. If the population of radicals is in thermodynamic equilibrium, its statistical distribution is described by the Maxwell-Boltzman equation

$$\frac{n_{upper}}{n_{lower}} = \exp\left(-\frac{E_{upper} - E_{lower}}{kT}\right) = \exp\left(-\frac{\Delta E}{kT}\right) = \exp\left(-\frac{\varepsilon}{kT}\right) = \exp\left(-\frac{h\nu}{kT}\right)$$

where n_{upper} is the number of paramagnetic centres occupying the upper energy state, k is the Boltzmann constant, and T is the temperature in kelvin. At 298 K, X-band microwave frequencies ($\nu \approx 9.75$ GHz) give $n_{upper} / n_{lower} \approx 0.998$, meaning that the upper energy level has a smaller population than the lower one. Therefore, transitions from the lower to the higher level are more probable than the reverse, which is why there is a net absorption of energy. The sensitivity of the EPR method (i.e., the minimum number of detectable spins, N_{min}) depends on the photon frequency ν according to

$$N_{min} = \frac{k_1 V}{Q_0 k_f \nu^2 P^{1/2}},$$

where k_1 is a constant, V is the sample's volume, Q_0 is the unloaded quality factor of the microwave cavity (sample chamber), k_f is the cavity filling coefficient, and P is the microwave power in the spectrometer cavity. With k_f and P being constants, $N_{min} \sim (Q_0\nu^2)^{-1}$, i.e., $N_{min} \sim \nu^{-\alpha}$, where $\alpha \approx 1.5$. In practice, α can change varying from 0.5 to 4.5 depending on spectrometer characteristics, resonance conditions, and sample size. In other words, the higher the spectrometer frequency the lower the detection limit (N_{min}), meaning greater sensitivity.

EPR spectral parameters

In real systems, electrons are normally not solitary, but are associated with one or more atoms. There are several important consequences of this:

1. An unpaired electron can gain or lose angular momentum, which can change the value of its g-factor, causing it to differ from g_e. This is especially significant for chemical systems with transition-metal ions.
2. If an atom with which an unpaired electron is associated has a non-zero nuclear spin, then its magnetic moment will affect the electron. This leads to the phenomenon of hyperfine coupling, analogous to J-coupling in NMR, splitting the EPR resonance signal into doublets, triplets and so forth.
3. Interactions of an unpaired electron with its environment influence the shape of an EPR spectral line. Line shapes can yield information about, for example, rates of chemical reactions.
4. The g-factor and hyperfine coupling in an atom or molecule may not be the same for all orientations of an unparied electron in an external magnetic field. This anisotropy depends upon the electronic structure of the atom or molecule (e.g., free radical) in question, and so can provide information about the atomic or molecular orbital containing the unpaired electron.

The g-factor

Knowledge of the g-factor can give information about a paramagnetic centre's electronic structure. An unpaired electron responds not only to a spectrometer's applied magnetic field B_0, but also to any local magnetic fields of atoms or molecules. The effective field B_{eff} experienced by an electron is thus written as

$$B_{eff} = B_0(1 - \sigma),$$

where σ includes the effects of local fields (σ can be positive or negative). Therefore, the $h\nu = g_e\mu_B B_{eff}$ resonance condition (above) is rewritten as follows:

$$h\nu = g_e\mu_B B_{eff} = g_e\mu_B B_0(1 - \sigma).$$

The quantity $g_e(1 - \sigma)$ is denoted by g and called simply the g-factor, so that the final resonance equation becomes

$$h\nu = g\mu_B B_0 .$$

This last equation is used to determine g in an EPR experiment by measuring the field and the frequency at which resonance occurs. If g does not equal g_e the implication is that the electron's spin magnetic moment is constant (approximately the Bohr magneton), then the electron must have gained or lost angular momentum through spin-orbit coupling. Because the mechanisms of spin-orbit coupling are well understood, the magnitude of the change gives information about the nature of the atomic or molecular orbital containing the unpaired electron.

Since the source of an EPR spectrum is a change in an electron's spin state, it might be thought that all EPR spectra would consist of a single line. However, the intereaction of an unpaired electron, by way of its magnetic moment, with nearby nuclear spins, results in additional allowed energy states and, in turn, multi-lined spectra. In such cases, the spacing between the EPR spectral lines indicates the degree of interaction between the unpaired electron and the perturbing nuclei. Two common mechanisms by which electrons and spectral line spacing and, in the simplest cases, is essentially the spacing itself.

Two common mechanisms by which electrons and nuclei interact are Fermi-contact interaction and dipolar interaction. The former applies largely to the case of isotropic interactions (independent of sample orientation in a magnetic field) and the latter to the case of anisotropic interactions (spectra dependent on sample orientation in a magnetic field). Spin polarization is a third mechanism for interactions between an unpaired electron and a nuclear spin, being especially important for π-electron organic radicals, such as the benzene radical anion. The symbols "a" or "A" are used for isotropic hyperfine coupling constants while "B" is usually employed for anisotropic hyperfine coupling constants.

In many cases like some explained below, the isotropic hyperfine splitting pattern for a radical freely tumbling in a solution (isotropic system) can be predicted.

- For a radical having M equivalent nuclei, each with a spin of I, the number of EPR lines expected is 2MI + 1. As an example, the methyl radical, CH_3, has three ^1H nuclei each with I = 1/2, and so the number of lines expected is 2MI + 1 = 2(3)(1/2) + 1 = 4, which is as observed.
- For a radical having M_1 equivalent nuclei, each with a spin of I_1, and a group of M_2 equivalent nuclei, each with a spin of I_2, the number of lines expected is $(2M_1I_1 + 1)(2M_2I_2 + 1)$. As an example, the methoxymethyl radical, $H_2C(OCH_3)$, has two equivalent ^1H nuclei each with I = 1/2 and three equivalent ^1H nuclei each with I = 1/2, and so the number of lines expected is $(2M_1I_1 + 1)(2M_2I_2 + 1)$ = [2(2)(1/2) + 1] [2(3)(1/2) + 1] = [3] [4] = 12, again as observed.
- The above can be extended to predict the number of lines for any number of nuclei.

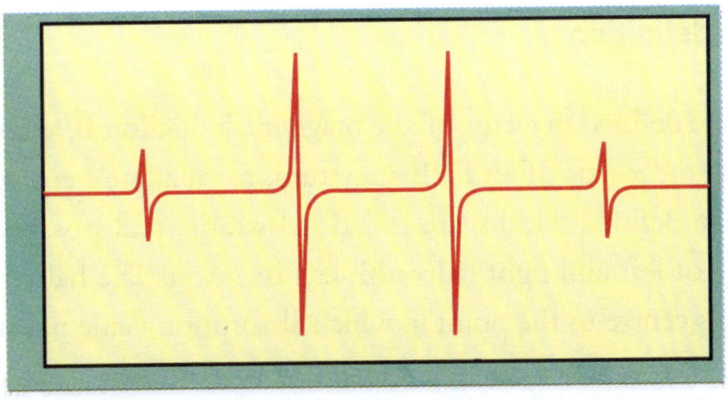

Fig. 12.38: Simulated EPR spectrum of the CH_3 radical

While it is easy to predict the number of lines a radical's EPR spectrum should show, the reverse problem, unravelling a complex multi-line EPR spectrum and assigning the various spacings to

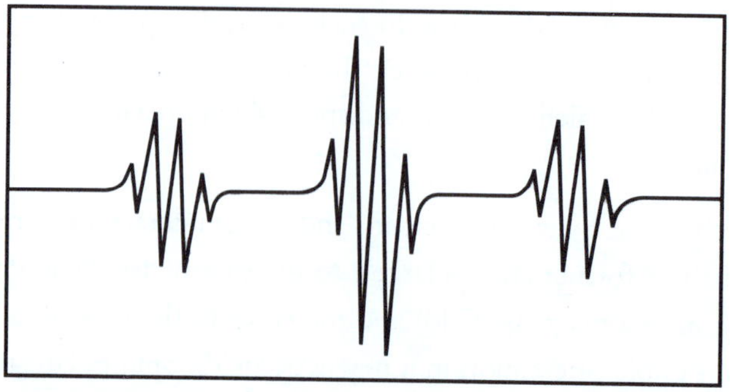

Fig. 12.39: Simulated EPR spectrum of the $H_2C(OCH_3)$ radical

specific nuclei, is more difficult. In the oft-encountered case of I = 1/2 nuclei (e.g., ^1H, ^{19}F, ^{31}P), the line intensities produced by a population of radicals, each possessing M equivalent nuclei, will follow Pascal's triangle. For example, the spectrum at the right shows that the three 1H nuclei of the CH_3 radical give rise to 2MI + 1 = 2(3)(1/2) + 1 = 4 lines with a 1:3:3:1 ratio. The line spacing gives a hyperfine coupling constant of a_H = 23 G for each of the three ^1H nuclei. Note again that the lines in this spectrum are first derivatives of absorptions.

As a second example, consider the methoxymethyl radical, $H_2C(OCH_3)$. The two equivalent methyl hydrogens will give an overall 1:2:1 EPR pattern, each component which is further split by the three methoxy hydrogens into a 1:3:3:1 pattern to give a total of 3 × 4 = 12 lines, a triplet of quartets. A simulation of the observed EPR spectrum is shown at the right, and agrees with the 12-line prediction and the expected line intensities. Note that the smaller coupling constant (smaller line spacing) is due to the three methoxy hydrogens, while the larger coupling constant (line spacing) is from the two hydrogens bonded directly to the carbon atom bearing the unpaired electron. It is often the case that coupling constants decrease in size with distance from a radical's unpaired electron, but there are some notable exceptions, such as the ethyl radical (CH_2CH_3).

Resonance linewidth definition

Resonance linewidths are defined in terms of the magnetic induction B, and its corresponding units, and are measured along the x-axis of an EPR spectrum, from a line's centre to a chosen reference point of the line. These defined widths are called halfwidths and possess some advantages; for asymmetric lines values of left and right halfwidth can be given. The halfwidth ΔB_h is the distance measured from the line's centre to the point in which absorption value has half of maximal absorption value in the centre of resonance line. First inclination width $\Delta B_{1/2}$ is a distance from centre of the line to the point of maximal absorption curve inclination. In practice, a full definition of linewidth is used. For symmetric lines, halfwidth $\Delta B_{1/2} = 2\Delta B_h$, and full inclination width $\Delta B_{max} = 2\Delta B_{1s}$.

The EPR waveband is stipulated by the frequency or wavelength of a spectrometer's microwave source (see Table). EPR experiments often are conducted at X and, less commonly, Q bands, mainly due to the ready availability of the necessary microwave components (which originally were developed for radar applications). A second reason for widespread X and Q band measurements is that electromagnets can reliably generate fields up to about 1 tesla. However, the low spectral resolution over g-factor at these wavebands limits the study of paramagnetic centres with comparatively low anisotropic magnetic parameters. Measurements at $\nu > 40$ GHz, in the millimetre wavelength region, offer the following advantages:

(a) EPR spectra are simplified due to the reduction of second-order effects at high fields.

(b) Increase in orientation selectivity and sensitivity in the investigation of disordered systems.

(c) The informativity and precision of pulse methods, e.g., ENDOR also increase at high magnetic fields.

(d) Accessibility of spin systems with larger zero-field splitting due to the larger microwave quantum energy $h\nu$.

(e) The higher spectral resolution over g-factor, which increases with irradiation frequency ν and external magnetic field B_0. This is used to investigate the structure, polarity, and dynamics of radical microenvironments in spin-modified organic and biological systems through the spin label and probe method. The figure shows how spectral resolution improves with increasing frequency.

(f) Saturation of paramagnetic centres occurs at a comparatively low microwave polarizing field B_1, due to the exponential dependence of the number of excited spins on the radiation frequency ν. This effect can be successfully used to study the relaxation and dynamics of paramagnetic centres as well as of superslow motion in the systems under study.

The cross-relaxation of paramagnetic centres decreases dramatically at high magnetic fields, making it easier to obtain more-precise and more-complete information about the system under study.

EPR applications

EPR is used in solid-state physics, for the identification and quantification of radicals in chemistry to identify reaction pathways, as well as in biology and medicine for tagging biological spin probes. Since radicals are very reactive, they do not normally occur in high concentrations in biological environments. With the help of specially designed nonreactive radical molecules that attach to specific sites in a biological cell, it is possible to obtain information on the environment of these so-called spin-label or spin-probe molecules.

12.13 NMR SPECTROSCOPY

Nuclear magnetic resonane spectroscopy most commonly known as NMR spectroscopy is the name given to the technique which exploits the magnetic properties of certain nuclei. This phenomenon and its origins are detailed in a separate section on Nuclear magnetic resonance. The most important applications for the organic chemist are proton NMR and carbon-13 NMR, although in principle NMR is applicable to any nucleus possessing spin. Many types of information can be obtained from this single phenomenon. Much like using infrared spectroscopy to identify functional groups, analysis of a 1D NMR spectrum tells the scientist what chemical environments are present, and (in most cases) their relative abundances within the sample. The impact of NMR spectroscopy on the natural sciences has been substantial. It can, amongst other things, be used to study mixtures of analytes, to understand dynamic effects such as change in temperature and reaction mechanisms, and is an invaluable tool in understanding protein and nucleic acid structure and function. It can be applied to a wide variety of samples, both in the solution and the solid state.

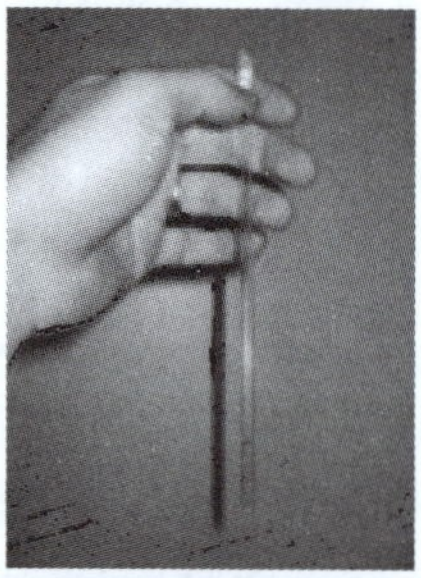

Fig. 12.40: The NMR sample is prepared in a thin-walled glass tube – an NMR tube

Basic NMR Techniques

When placed in a magnetic field, NMR active nuclei (like ^1H or ^{13}C) resonate at a specific frequency for each element or isotope, directly proportional to the strength of the magnetic field. For example, in a 21 tesla magnetic field, protons resonate at 900 MHz. It is common to refer to a 21 T magnet as a 900 MHz magnet but different nuclei resonate at a different frequency at this field strength.

Chemical Shift

Depending on the local chemical environment, different protons in a molecule resonate at slightly different frequencies. Since this frequency shift is proportional to the strength of the magnetic field, it is converted into a field-independent dimensionless value known as the chemical shift. The chemical shift is reported as a relative measure from some reference resonance frequency. (For the nuclei ^1H, ^{13}C, and ^{29}Si, TMS (tetramethylsilane) is commonly used as a reference.) This difference between the frequency of the signal and the frequency of the reference is divided by frequency of the reference signal to give the chemical shift. The frequency shifts are extremely small (say 100s of Hz) in comparison to the fundamental NMR frequency (typically 100s of MHz) and so are generally expressed as parts per million (ppm).

By understanding different chemical environments, the chemical shift can be used to obtain some structural information about the molecule in question to assign signals to an atom or a group of atoms. For example, in a proton spectrum for ethanol (CH_3CH_2OH) one would expect three specific signals at three specific chemical shifts. One for the CH_3 group, one for the CH_2 group and one for the OH. A typical CH_3 group has a shift around 1 ppm, the CH_2 attached to a OH has a shift of around 4 ppm and the OH has a shift around 2-3 ppm depending on the solvent used. Because the molecular motion at room temperature makes each of the three methyl protons average out during the course of the NMR experiment (which typically takes a few ms), the protons become degenerate and form a peak at the same chemical shift. Interestingly the shape and size of peaks are indicators of chemical structure too. In the example above–the proton spectrum of ethanol–the CH_3 peak would be three times as large as the OH. Similarly, the CH_2 peak would be twice the size of the OH peak but only 2/3 of the size of the CH_3 peak. Modern analysis software allows analysis of the size of peaks to understand how many protons give rise to the peak. This is known as integration–a mathematical process which calculates the area under a graph (essentially what a spectrum is). The analyst must integrate the peak and not measure its height because the peaks also have width, and thus its size is dependent on its area not on its height. However, it should be mentioned that the number of protons, or any other observed nucleus, is only proportional to the intensity, or the integral of the NMR signal, in the very simplest one-dimensional NMR experiments. In more elaborate experiments, for instance, experiments typically used to obtain carbon-13 NMR spectra, the integral of the signals depends on the relaxation rate of the nucleus, and its scalar and dipolar coupling constants. Very often these factors are poorly understood. Therefore, the integral of the NMR signal is very difficult to interpret in more complicated NMR experiments.

J-coupling

Some of the most useful information for structure determination in a one dimensional NMR spectrum comes from J-coupling or scalar coupling (a special case of spin-spin coupling) between NMR active nuclei. This coupling arises from the interaction of different spin states through the chemical bonds of a molecule and results in the splitting of NMR signals into recognizable patterns based on the pairing of spin states. As a result, this coupling can tell us information about the connectivity of atoms in a molecule. Coupling to n equivalent (spin ½) nuclei splits the signal into a n+1 multiplet with intensity ratios following Pascal's triangle as described on the right. Coupling to additional spins will lead to further splittings of each component of the multiplet, e.g., coupling to two different spin ½ nuclei with significantly different coupling constants will lead to a doublet of doublets (abbreviation: dd). Note that coupling between nuclei that are chemically equivalent (that is, have the same chemical shift) has no effect of the NMR spectra and couplings between nuclei that are distant (usually more than 3 bonds apart for protons in flexible molecules) are usually too small to cause observable splittings. Long-range couplings over more than three bonds can often be observed in cyclic and aromatic compounds, leading to more complex splitting patterns.

For example, in the proton spectrum for ethanol described above, the CH_3 group is split into a triplet with an intensity ratio of 1:2:1 by the two neighbouring CH_2 protons. Similarly, the CH_2 is split into a quartet with an intensity ratio of 1:3:3:1 by the three neighbouring CH_3 protons. In principle, the two CH_2 protons would also be split again into a doublet to form a doublet of quartets by the hydroxyl proton, but intermolecular exchange of the acidic hydroxyl proton often results in a loss of coupling information. Coupling to any spin ½ nuclei such as phosphorus-31 or fluorine-19 works in this fashion (although the magnitudes of the coupling constants may be very different). But the splitting patterns differ from those described above for nuclei with spin greater than ½ because the spin quantum number has more than two possible values. For instance, coupling to deuterium (a spin 1 nucleus) splits the signal into a 1:1:1 triplet because the spin 1 has three spin states. Similarly, a spin 3/2 nucleus splits a signal into a 1:1:1:1 quartet and so on. Coupling combined with the chemical shift (and the integration for protons) tells us not only about the chemical environment of the nuclei, but also the number of neighbouring NMR active nuclei within the molecule. In more complex spectra with multiple peaks at similar chemical shifts or in spectra of nuclei other than hydrogen, coupling is often the only way to distinguish different nuclei.

Second-order (or strong) coupling

The above description assumes that the coupling constant is small in comparison with the difference in NMR frequencies between the inequivalent spins. If the shift separation decreases (or the coupling strength increases), the multiplet intensity patterns are first distorted, and then become more complex and less easily analyzed (especially if more than two spins are involved). Intensification of some peaks in a multiplet is achieved at the expense of the remainder, which sometimes almost disappear in the background noise, although the integrated area under the peaks remains constant. In most high-field NMR, however, the distortions are usually modest and the characteristic distortions (roofing) can in fact help to identify related peaks. Second-order effects decrease as the frequency difference between multiplets increases, so that high-field (i.e., high-frequency) NMR spectra display less distortion than lower frequency spectra. Early spectra at 60 MHz were more prone to distortion than spectra from later machines typically operating at frequencies at 200 MHz or above.

Magnetic inequivalence

More subtle effects can occur if chemically equivalent spins (i.e., nuclei related by symmetry and so having the same NMR frequency) have different coupling relationships to external spins. Spins that are chemically equivalent but are not indistinguishable (based on their coupling relationships) are termed magnetically inequivalent. For example, the 4 H sites of 1,2-dichlorobenzene divide into two chemically equivalent pairs by symmetry, but an individual member of one of the pairs has different couplings to the spins making up the other pair. Magnetic inequivalence leads to highly complex spectra which can only be analyzed by computational modelling. Such effects are more common in NMR spectra of aromatic and other non-flexible systems, while rapid rotation about C-C bonds in flexible molecules ensures that the couplings between protons on adjacent carbons are identical, avoiding problems with magnetic inequivalence.

Correlation spectroscopy

Correlation spectroscopy is one of several types of two-dimensional nuclear magnetic resonance (NMR) spectroscopy. This type of NMR experiment is best known by its acronym, COSY. Other types of two-dimensional NMR include J-spectroscopy, exchange spectroscopy (EXSY), Nuclear Overhauser effect spectroscopy (NOESY), total correlation spectroscopy (TOCSY) and heteronuclear correlation experiments, such as HSQC, HMQC, and HMBC. Two-dimensional NMR spectra provide more information about a molecule than one-dimensional NMR spectra and are especially useful in determining the structure of a molecule, particularly for molecules that are too complicated to work with using one-dimensional NMR. The first two-dimensional experiment, COSY, was proposed by Jean Jeener, a professor at Université Libre de Bruxelles, in 1971. This experiment was later implemented by Walter P. Aue, Enrico Bartholdi and Richard R. Ernst, who published their work in 1976.

Solid-State nuclear magnetic resonance

A variety of physical circumstances does not allow molecules to be studied in solution, and at the same time not by other spectroscopic techniques to an atomic level, either. In solid-phase media, such as crystals, microcrystalline powders, gels, anisotropic solutions, etc., it is in particular the dipolar coupling and chemical shift anisotropy that become dominant to the behaviour of the nuclear spin systems. In conventional solution-state NMR spectroscopy, these additional interactions would lead to a significant broadening of spectral lines. A variety of techniques allows to establish high-resolution conditions, that can, at least for ^{13}C spectra, be comparable to solution-state NMR spectra. Two important concepts for high-resolution solid-state NMR spectroscopy are the limitation of possible molecular orientation by sample orientation, and the reduction of anisotropic nuclear magnetic interactions by sample spinning. Of the latter approach, fast spinning around the magic angle is a very prominent method, when the system comprises spin 1/2 nuclei. A number of intermediate techniques, with samples of partial alignment or reduced mobility, is currently being used in NMR spectroscopy. Applications in which solid-state NMR effects occur are often related to structure investigations on membrane proteins, protein fibrils or all kinds of polymers, and chemical analysis in inorganic chemistry, but also include "exotic" applications like the plant leaves and fuel cells.

NMR spectroscopy applied to proteins

Much of the recent innovation within NMR spectroscopy has been within the field of protein NMR, which has become a very important technique in structural biology. One common goal of these investigations is to obtain high resolution 3-dimensional structures of the protein, similar to what can be achieved by X-ray crystallography. In contrast to X-ray crystallography, NMR is primarily limited to relatively small proteins, usually smaller than 25 kDa, though technical advances allow ever larger structures to be solved. NMR spectroscopy is often the only way to obtain high resolution information on partially or wholly intrinsically unstructured proteins. Proteins are orders of magnitude larger than the small organic molecules discussed earlier in this article, but the same NMR theory applies. Because of the increased number of each element present in the molecule, the basic 1D spectra become crowded with overlapping signals to an extent where analysis is impossible. Therefore, multidimensional (2, 3 or 4D) experiments have been devised to deal with this problem. To facilitate these experiments, it is desirable to isotopically label the protein with ^{13}C and ^{15}N because the predominant naturally-occurring isotope ^{12}C is not NMR-active, whereas the nuclear quadrupole moment of the predominant naturally-occurring ^{14}N isotope prevents high resolution information to be obtained from this nitrogen isotope. The most important method used for structure determination of proteins utilizes NOE experiments to measure distances between pairs of atoms within the molecule. Subsequently, the obtained distances are used to generate a 3D structure of the molecule using a computer program.

12.14 LASER-INDUCED BREAKDOWN SPECTROSCOPY

Laser-induced breakdown spectroscopy (LIBS) is a type of atomic emission spectroscopy which utilises a highly energetic laser pulse as the excitation source. LIBS can analyse any matter regardless of its physical state, be it solid, liquid or gas. Even slurries, aerosols, gels, and more can be readily investigated. Because all elements emit light when excited to sufficiently high temperatures, LIBS can detect all elements, limited only by the power of the laser as well as the sensitivity and wavelength range of the spectrograph & detector. Operationally, LIBS is very similar to arc/spark emission spectroscopy. LIBS can often be referred to as its alternative name: laser-induced plasma spectroscopy (LIPS). Unfortunately the term LIPS has alternative meanings that are outside the field of analytical spectroscopy, therefore, the term LIBS is preferred. LIBS is technically very similar to a number of other laser-based analytical techniques, sharing much of the same hardware. These techniques are the vibrational spectroscopic technique of Raman spectroscopy, and the fluorescence spectroscopic technique of laser-induced fluorescence (LIF). In fact, devices are now being manufactured which combine these techniques in a single instrument, allowing the atomic, molecular and structural characterisation of a specimen as well as giving a deeper insight into physical properties.

Design

A typical LIBS system consists of a neodymium doped yttrium aluminium garnet (Nd:YAG) solid-state laser and a spectrometer with a wide spectral range and a high sensitivity, fast response rate, time gated detector. This is coupled to a computer which can rapidly process and interpret the acquired data. As such LIBS is one of the most experimentally simple spectroscopic analytical techniques, making it one of the cheapest to purchase and to operate. The Nd:YAG laser generates energy in the near infrared region of the electromagnetic spectrum, with a wavelength of 1064 nm. The pulse duration is in the region of 10 ns generating a power density which can exceed 1 GW·cm-2 at the focal point. Other lasers have been used for LIBS mainly Excimer (Excited dimer) type generating energy in the visible and ultraviolet regions. The spectrometer consists of either a monochromator (scanning) or a polychromator (non-scanning) and a photomultiplier or CCD detector respectively. The most common monochromator is the Czerny-Turner type whilst the most common polychromator is the Echelle type, even so the Czerny-Turner type can be (and is often) used to disperse the radiation onto CCD effectively making it a polychromator. The polychromator spectrometer is the type most commonly used in LIBS as it allows simultaneous acquisition of the entire wavelength range of interest.

The spectrometer collects electromagnetic radiation over the widest wavelength range possible, maximising the number of emission lines detected for each particular element. Spectrometer response is typically from 1100 nm (near infrared) to 170 nm (deep ultraviolet), the approximate response range of a CCD detector. All elements have emission lines within this wavelength range. The energy resolution of the spectrometer can also affect the quality of the LIBS measurement, since high resolution systems can separate spectral emission lines in close juxtaposition, reducing interference and increasing selectivity. This feature is particularly important in specimens which have a complex matrix, containing a large number of different elements. Accompanying the spectrometer and detector is a delay generator which accurately gates the detector's response time, allowing temporal resolution of the spectrum. LIBS operates by focusing the laser onto a small area at the surface of the specimen, when the laser is discharged it ablates a very small amount of material, in the range of nanogram to picogram which instantaneously generates a plasma plume with temperatures of about 10,000-20,000 K. At these temperatures the ablated material dissociates (breaks down) into excited ionic and atomic species. During this time the plasma emits a continuum of radiation which does not contain any useful information about the species present, but within a very small timeframe the plasma expands at supersonic velocities and cools. At this point the characteristic atomic emission lines of the elements can be observed. The delay between the emission of continuum radiation and characteristic radiation is in the order of 10 μs, this is why it is necessary to temporally gate the detector.

Advantages

Because such a small amount of material is consumed during the LIBS process the technique is considered essentially non-destructive or minimally-destructive, and with an average power density of less than one watt radiated onto the specimen there is almost no specimen heating surrounding the ablation site. Due to the nature of this technique sample preparation is typically minimised to homogenisation or is often unnecessary where heterogeneity is to be investigated or where a specimen is known to be sufficiently homogeneous, this reduces the possibility of contamination during chemical preparation steps. One of the major advantages of the LIBS technique is its ability to depth profile a specimen by repeatedly discharging the laser in the same position, effectively going deeper into the specimen with each shot. This can also be applied to the removal of surface contamination, where the laser is discharged a number of times prior to the analysing shot. LIBS is also a very rapid technique giving results within seconds, making it particularly useful for high volume analyses or on-line industrial monitoring. LIBS is an entirely optical technique, therefore it requires only optical access to the specimen. This is of major significance as fibre optics can be employed for remote analyses. And being an optical technique it is non-invasive, non-contact and can even be used as a stand-off analytical technique when coupled to appropriate telescopic apparatus. These attributes have significance for use in areas from hazardous environments to space exploration. Additionally LIBS systems can easily be coupled to an optical microscope for micro-sampling adding a new dimension of analytical flexibility. The use of specialised optics or a mechanically positioned specimen stage can be used raster the laser over the surface of the specimen allowing spatially resolved chemical analysis and the creation of 'elemental maps'. This is very significant as chemical imaging is becoming more important in all branches of science and technology. Portable LIBS systems are more sensitive, faster and can detect a wider range of elements (particularly the light elements) than competing techniques such as portable X-ray fluorescence. And LIBS does not utilise ionizing radiation to excite the sample, which is both penetrating and potentially carcinogenic.

Disadvantages

LIBS, like all other analytical techniques is not without limitations. LIBS is subject to the matrix effect which can be minimised by good specimen preparation and the use of accurate calibration standards, it is also subject to variation in the laser spark and resultant plasma which often limits reproducibility. The accuracy of LIBS measurements is typically better than 10% and precision is often better than 5%. The detection limits for LIBS vary from one element to the next depending on the specimen type and the experimental apparatus used. Even so detection limits of 1 to 30 ppm by mass are not uncommon, but can range from >100 ppm to <1 ppm.

Recent interest in LIBS has focused on the miniaturization of the components and the development of compact, low power, portable systems. This direction has been pushed along by interest from groups such as NASA, ESA as well as the military. The Mars Science Laboratory mission plans to bring combined Raman/LIBS onto Mars. Recent developments in LIBS have seen the introduction of double-pulsed laser systems. For double-pulse LIBS one distinguishes between orthogonal and perpendicular configuration. In perpendicular configuration the laser is fired twice on the same spot on the specimen with a pulse separation in the order of one to a couple of tens of microseconds. Depending on pulse separation, the second pulse is more or less absorbed by the plasma plume caused by the previous pulse, resulting in a reheating of the laser plasma leading to signal enhancement. In orthogonal configuration a laser pulse is fired parallel to the sample surface either before or after the perpendicular pulse hits the specimen. The laser plasma ignited in the surrounding medium above the surface by a first pulse causes (by its shock wave) an area of reduced pressure above the specimen into which the actual plasma from the sample can expand. This has similar positive effects on sensitivity like LIBS performed at reduced pressures. If the orthogonal laser pulse is delayed with respect to the perpendicular one, the effects are similar as in the perpendicular configuration. Both double-pulse LIBS as well as LIBS at reduced pressures are aimed at increasing the sensitivity of LIBS and the reduction of errors caused by the differential volatility of elements (e.g., Hydrogen as an impurity in solids). It also significantly reduces the matrix effects. Double-pulsed systems are also proving useful in conducting analysis in liquids, as the initial laser pulse forms a cavity bubble in which the second pulse acts on the evaporated material. LIBS is one of several analytical techniques that can be deployed in the field as opposed to pure laboratory techniques, e.g., spark OES. Recent research on LIBS is focusing on compact.

12.15 MÖSSBAUER SPECTROSCOPY

Mössbauer spectroscopy is a spectroscopic technique based on the Mössbauer effect. In its most common form, Mössbauer Absorption Spectroscopy, a solid sample is exposed to a beam of gamma radiation, and a detector measures the intensity of the beam that is transmitted through the sample, which will change depending on how many gamma-rays are absorbed by the sample. The atoms in the source emitting the gamma-rays are the same as the atoms in the sample absorbing them. Thanks to the Mössbauer effect that a significant fraction of the gamma-rays emitted by the atoms in the source do not lose any energy due to recoil and thus have almost the right energy to be absorbed by the target atoms. The gamma-ray energy is varied by accelerating the gamma-ray source through a range of velocities with a linear motor. The relative motion between the source and sample results in an energy shift due to the Doppler effect. In the resulting spectra, gamma-ray intensity is plotted as a function of the source velocity. At velocities corresponding to the resonant energy levels of the sample, some of the gamma-rays are absorbed, resulting in a drop in the measured intensity and a corresponding dip in the spectrum. The number, positions, and intensities of the dips (also called peaks) provide information about the chemical environment of the absorbing nuclei and can be used to characterize the sample.

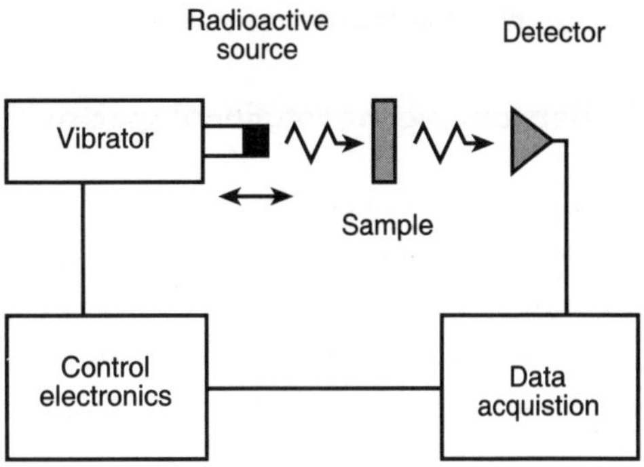

Fig. 12.41: Schematic of an experimental set-up for transmission Mossbauer spectroscopy

In order for Mössbauer absorption of gamma-rays to occur, the gamma-ray must be of the appropriate energy for the nuclear transitions of the atoms being probed, which is almost always achieved by having the same atoms of the same isotope in both the source and the target. Also, the gamma-ray energy should be relatively low, otherwise the system will have a low recoil-free fraction (see Mössbauer effect) resulting in a poor signal-to-noise ratio. Only a handful of elemental isotopes exist for which these criteria are met, so Mössbauer spectroscopy can only be applied to a relatively small group of atoms including: ^{57}Fe, ^{129}I, ^{119}Sn, and ^{121}Sb. Of these, ^{57}Fe is by far the most common element studied using the technique.

12.16 ROTATIONAL SPECTROSCOPY

Rotational spectroscopy or microwave spectroscopy studies the absorption and emission electromagnetic radiation (typically in the microwave region of the electromagnetic spectrum) by molecules associated with a corresponding change in the rotational quantum number of the molecule. The use of microwaves in spectroscopy essentially became possible due to the development of microwave technology for RADAR during World War II. Rotational spectroscopy is only really practical in the gas phase where the rotational motion is quantized. In solids or liquids the rotational motion is usually quenched due to collisions. Rotational spectrum from a molecule (to first order) requires that the molecule have a dipole moment, that is a difference between the centre of charge and the centre of mass, or equivalently a separation between two unlike charges. It is this dipole moment that enables the electric field of the light (microwave) to exert a torque on the molecule causing it to rotate more quickly (in excitation) or slowly (in de-excitation). Diatomic molecules such as Oxygen (O_2), Hydrogen (H_2), etc. do not have a dipole moment and hence no purely rotational spectrum. However, electronic excitations can lead to asymmetric charge distributions and thus provide a net dipole moment to the molecule. Under such circumstances, these molecules will exhibit a rotational spectrum. Amongst the diatomic molecules, Carbon monoxide (CO) has one of the simplest rotational spectra. As for tri-atomic molecules, Hydrogen cyanide (HC≡N) has a simple rotational spectrum for a linear molecule and Hydrogen isocyanide (HN=C:) for a nonlinear molecule. As the number of atoms increases the spectrum becomes more complex as lines due to different transitions start overlapping.

Understanding the rotational spectrum

In quantum mechanics the free rotation of a molecule is quantized, that is the rotational energy and the angular momentum can only take certain fixed values; what these values are is simply related to the moment of inertia, I, of the molecule. In general for any molecule, there are three moments of inertia: I_A, I_B and I_C about three mutually orthogonal axes A, B, and C respectively with the origin at the center of mass of the system. A linear molecule is a special case in this regard. These molecules are cylindrically symmetric and one of the moment of inertia (I_A, which is the moment of inertia for a rotation taking place along the axis of the molecule) is negligible (i.e. $I_A << I_B = I_C$).

Classification of molecules based on rotational behaviour

The general convention is to define the axes such that the axis A has the smallest moment of inertia (and hence the highest rotational frequency) and other axes such that $I_A \leq I_B \leq I_C$. Sometimes the axis A may be associated with the symmetric axis of the molecule, if any. If such is the case, then I_A need not be the smallest moment of inertia. To avoid confusion, we will stick with the former convention for the rest of the article. The particular pattern of energy levels (and hence of transitions in the rotational spectrum) for a molecule is determined by its symmetry. A convenient way to look at the molecules is to divide them into four different classes (based on the symmetry of their structure). These are,

1. **Linear molecules (or linear rotors):** As mentioned earlier, for a linear molecule $I_A << I_B = I_C$. For most of the purposes, I_A is taken to be zero. For a linear molecule, the separation of lines in the rotational spectrum can be related directly to the moment of inertia of the molecule, and for a molecule of known atomic masses, can be used to determine the bond lengths (structure) directly. For diatomic molecules this process is trivial, and can be made from a single measurement of the rotational spectrum. For linear molecules with more atoms, rather more work is required, and it is necessary to measure molecules in which more than one isotope of each atom have been substituted (effectively this gives rise to a set of simultaneous equations which can be solved for the bond lengths). Examples or linear molecules: Oxygen (O=O), Carbon monoxide (O≡C*), Hydroxy radical (OH), Carbon dioxide (O=C=O), Hydrogen cyanide (HC≡N), Carbonyl sulphide (O=C=S), Chloroethyne (HC≡CCl), Acetylene (HC≡CH).
2. **Symmetric tops (or symmetic rotors):** A symmetric top is a molecule in which two moments of inertia are the same. As a matter of historical convenience, spectroscopists divide molecules into two classes of symmetric tops, Oblate symmetric tops (saucer or disc shaped) with $I_A = I_B < I_C$ and Prolate symmetric tops (rugby football, or cigar shaped) with $I_A < I_B = I_C$. The spectra look rather different, and are instantly recognizable. As for linear molecules, the structure of symmetric tops (bond lengths and bond angles) can be deduced from their spectra. Examples of symmetric tops: Oblate: Ben-

zene (C_6H_6), Cyclobutadiene (C_4H_4), Ammonia (NH_3) Prolate: Chloroform ($CHCl_3$), Propyne ($CH_3C{\equiv}CH$).

3. **Spherical tops (or spherical rotors):** A spherical top molecule, can be considered as a special case of symmetric tops with equal moment of inertia about all three axes ($I_A = I_B = I_C$). Examples of spherical tops: Phosphorus tetramer (P_4), Carbon tetrachloride (CCl_4), Nitrogen tetrahydride (NH_4), Ammonium ion (NH_4^+), Sulphur hexafluoride (SF_6).
4. **Asymmetric tops:** As you would have guessed a molecule is termed an asymmetric top if all three moments of inertia are different. Most of the larger molecules are asymmetric tops, even when they have a high degree of symmetry. Generally for such molecules a simple interpretation of the spectrum is not normally possible. Sometimes asymmetric tops have spectra that are similar to those of a linear molecule or a symmetric top, in which case the molecular structure must also be similar to that of a linear molecule or a symmetric top. For the most general case, however, all that can be done is to fit the spectra to three different moments of inertia. If the molecular formula is known, then educated guesses can be made of the possible structure, and from this guessed structure, the moments of inertia can be calculated. If the calculated moments of inertia agree well with the measured moments of inertia, then the structure can be said to have been determined. For this approach to determining molecular structure, isotopic substitution is invaluable. Examples of asymmetric tops: Anthracene ($C_{14}H_{10}$), Water (H_2O), Nitrogen dioxide (NO_2).

A molecule is always in vibration. As the molecule vibrates, its moment of inertia changes. Further there is a fictitious force, Coriolis coupling, between the vibrational motion of the nuclei in the rotating (non-inertial) frame. However, as long as the vibrational quantum number does not change (i.e., the molecule is in only one state of vibration), the effect of vibration on rotation is not important, because the time for vibration is much greater than the time required for rotation. The Coriolis coupling is often negligible, too, if one is interested in low vibrational and rotational quantum numbers only.

Hyperfine interactions

In addition to the main structure that is observed in microwave spectra due to the rotational motion of the molecules, a whole host of further interactions are responsible for small details in the spectra, and the study of these details provides a very deep understanding of molecular quantum mechanics. The main interactions responsible for small changes in the spectra (additional splittings and shifts of lines) are due to magnetic and electrostatic interactions in the molecule. The particular strength of such interactions differs in different molecules, but in general, the order of these effects (in decreasing significance) is:

- electron spin — electron spin interaction (this occurs in molecules with two or more unpaired electrons, and is a magnetic-dipole / magnetic-dipole interaction)
- electron spin — molecular rotation (the rotation of a molecule corresponds to a magnetic dipole, which interacts with the magnetic dipole moment of the electron)

- electron spin — nuclear spin interaction (the interaction between the magnetic dipole moment of the electron and the magnetic dipole moment of the nuclei, if present).
- electric field gradient — nuclear electric quadrupole interaction (the interaction between the electric field gradient of the electron cloud of the molecule and the electric quadrupole moments of nuclei, if present).
- nuclear spin — nuclear spin interaction (nuclear magnetic moments interacting with one another).

These interactions give rise to the characteristic energy levels that are probed in "magnetic resonance" spectroscopy such as NMR and ESR, where they represent the "zero field splittings" which are always present. Fourier transform infrared (FTIR) spectroscopy can be used to experimentally study rotational spectrum.

Applications

Microwave spectroscopy is commonly used in physical chemistry to determine the structure of small molecules (such as ozone, methanol, or water) with high precision. Other common techniques for determining molecular structure, such as X-ray crystallography do not work very well for some of these molecules (especially the gases) and are not as precise. However, microwave spectroscopy is not useful for determining the structures of large molecules such as proteins.

12.17 TIPS TO REMEMBER

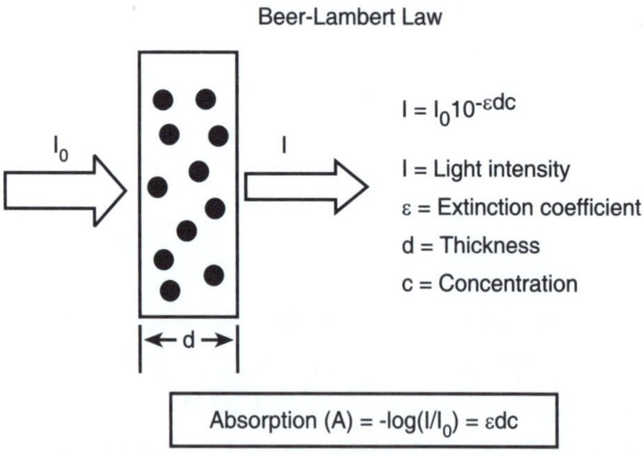

Spectrophotometry

- versatile and widely used analytical tool
- based on how substances affect radiation (i.e., light)
- advantages:
  — often non-destructive

— can be selective

— short time interval of measurement (10-14 s)

Factors Affecting Absorption

- pH (ionization of chromophore)
- redox state
- polarity or solvent effects
- orientation effects

Hypochromism of Nucleic Acids

Free nucleotides > Single stranded > Double stranded

Spectrophotometry	Comment
Infrared (IR)	Vibrational levels
Raman	
Optical Rotary Dispersion (ORD)	Polarize light
Circular Dichroism (CD)	
Nuclear Magnetic Resonance (NMR)	Magnetic moments
Electron Spin Resonance (ESR)	

12.18 VIVA VOCE

Q.1. What do you understand by spectroscopy?

Q.2. What are the basis on which spectroscopic techniques are classified?

Q.3. What is the use of FTIR?

Q.4. What is Lambert-Beer Law?

Q.5. Differentiate between single beam and dual beam spectrophotometer.

Q.6. What is hyperchromic and hypochromic effect?

Q.7. Differentiate between atomic absorption spectroscopy and atomic emmision spectroscopy.

Q.8. What are the applications of gamma spectroscopy?

Q.9. What is NMR?

Q.10. What is MALDI spectroscopy?

Q.11. Proteins gives absorption at................ whereas nucleic acid gives absorption at..................

Q.12. Explain CD spectra?

Q.13. What is spectroflurometry?

Q.14. What is the principle of Raman spectroscopy?

Q.15. In the mass spectrum, a compound with two bromine atoms will have:

A. two molecular ions, in the ratio 1:1

B. four molecular ions, in the ratio 1:1:1:1

C. two molecular ions, in the ratio 2:1

D. four molecular ions, in the ratio 1:2:2:1

E. three molecular ions, in the ratio 1:2:1.

Q.16. In the ^{1}H NMR of diethylether (CH_3CH_2–O–CH_2CH_3) the signal from the CH_2 protons will appear as:

A. a triplet

B. a multiplet consisting of 9 peaks

C. a multiplet consisting of seven peaks

D. a multiplet consisting of 6 quartets

E. a quartet

Q.17. How many peaks will be present in the proton-decoupled 13C NMR of the compound shown below:

A. 4

B. 3

C. 5

D. 7

E. 6

Q.18. The base peak in the mass spectrum of benzaldehyde is most likely to be due to:

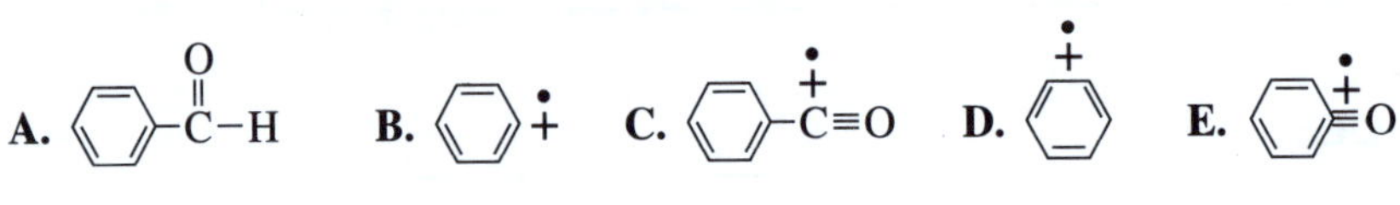

Q.19. The major bands in the infrared spectrum of ethyl cyanoacetate will be:

A. 3150 cm^{-1}; 2200 cm^{-1}; 1610 cm^{-1}

B. 3450 cm^{-1}; 2930 cm^{-1}; 1760 cm^{-1}

C. 2930 cm^{-1}; 2450 cm^{-1}; 1610 cm^{-1}

D. 2930 cm^{-1}; 2200 cm^{-1}; 1750 cm^{-1}

E. 3010 cm^{-1}; 2200 cm^{-1}; 1750 cm^{-1}

Q.20. Which of the following is not true regarding the spectra of acetophenone?

A. In the ^{1}H NMR, the CH_3 group will appear as a singlet.

B. The ^{13}C NMR will have a singlet at approximately 200 ppm.

C. The IR spectrum will display by a significant absorption at about 2400 cm^{-1}.

D. In the ^{1}H NMR, the CH_3 group will absorb at approximately at 220 nm.

E. In the mass spectrum, a significant peak will occur at m/e = 43.

CHAPTER 13

Radioisotope Techniques

13.1 BASIC PRINCIPLES OF RADIOACTIVITY

Radioactivity is a phenomenon that occurs naturally in a number of substances. Atoms of the substance spontaneously emit invisible but energetic radiations, which can penetrate materials that are opaque to visible light. The effects of these radiations can be harmful to living cells but, when used in the right way, they have a wide range of beneficial applications, particularly in medicine. Radioactivity has been present in natural materials on the earth since its formation (for example, in potassium-40 which forms part of all our bodies). However, because its radiations cannot be detected by any of the body's five senses, the phenomenon was only discovered 100 years ago when radiation detectors were developed. Nowadays we have also found ways of creating new man made sources of radioactivity; some (like iodine-131 and molybdenum-99) are incidental waste products of the nuclear power industry which nevertheless have important medical applications, whilst others (for example, fluorine-18) are specifically produced for the benefits of their medical use.

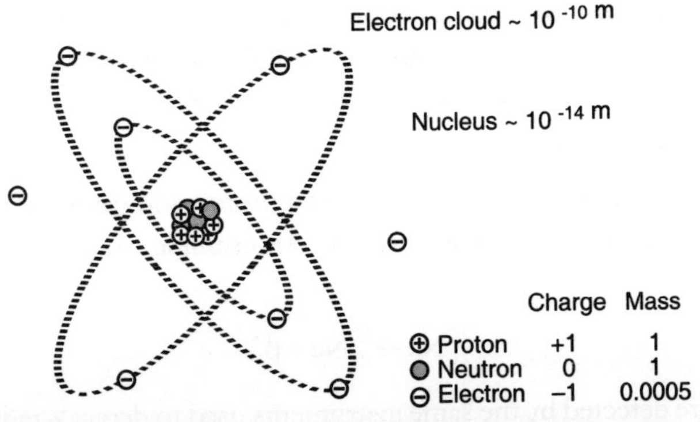

Fig. 13.1: An atom

13.2 TYPES OF RADIOACTIVE DECAY

There are several types of radioactive decay; only those more relevant to biochemists are considered below.

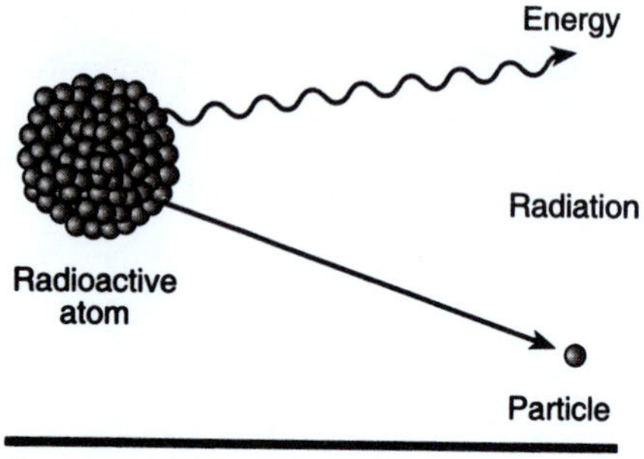

Fig. 13.2: Radioactive decay

Decay by negatron emission

In this case a neutron is converted to a proton by the ejection of a negatively charged beta (β) particle called a negatron (β^-)

$$\text{neutron} \rightarrow \text{proton} + \text{negatron}$$

To all intents and purposes a negatron is an electron, but the term negatron is preferred, although not always used, since it serves to emphasise the nuclear origin of the particle. As a result of negatron emission, the nucleus loses a neutron but gains a proton. The N/Z ratio, therefore, decreases while Z increases by 1 and A remains constant. An isotope frequently used in biological work that decays by negatron emission is ^{14}C.

$$^{14}_{6}C \rightarrow ^{14}_{7}N + \beta^-$$

Negatron emission is very important to biochemists because many of the commonly used radionuclides decay by this mechanism. Examples are: ^3H and ^{14}C, which can be used to label any organic compound, ^{35}S used to label methionine, for example, to study protein synthesis; and ^{32}P, a powerful tool in molecular biology when used as a nucleic acid label.

Decay by positron emission

Some isotopes decay by emitting positively charged β-particles referred to as positron (β^+). This occurs when a positron is converted to a neutron.

$$\text{proton} \rightarrow \text{neutron} + \text{positron}$$

Positrons are extremely unstable and have only a transient existence. Once they have dissipated their energy, they are intact with electron and are annihilated. The mass and energy of the two particles are converted to two γ-rays emitted at 180° to each other. This phenomenon is frequently described as back to back emission.

As a result of positron emission the nucleus loses a proton and gains a neutron, the N/Z ratio increases, Z decreases by 1, and A remains constant. An example of an isotope decaying by positron emission is ^{22}Na:

$$^{22}_{11}Na \rightarrow ^{22}_{10}Ne + \beta^+$$

Positron emitters are detected by the same instruments used to detect γ-radiation. They are used in biological sciences to spectacular effect in brain scanning with the technique positron emission tomography (PET scanning) used to identify active and inactive areas of the brain.

Decay by alpha-particle emission

Isotopes of elements with high atomic numbers frequently decay by emitting alpha (α)-particles. An α-particle is a helium nucleus; it consists of two protons and two neutrons ($^4He^{2+}$). Emissions of α-particle result in a considerable lightening of the nucleus, a decrease in atomic number of 2 and a decrease in the mass number of 4. Isotopes that decay by α-emission are not frequently encountered in biological work. Radium-226 (^{226}Ra) decays by α-emission to radon-222 (^{222}Rn), which is itself radioactive. Thus begins a complex decay series, which culminates in formation of ^{206}Pb:

$$^{226}_{88}Ra \rightarrow {}^{4}_{2}He^{2+} + {}^{222}_{86}Rn \rightarrow\rightarrow\rightarrow {}^{206}_{82}Pb$$

Alpha emitters are extremely toxic, if ingested, due to the large mass and the ionizing power of the atomic particle.

Electron capture

In this form of decay a proton captures an electron orbiting in the innermost K-shell:

$$\text{proton} + \text{electron} \rightarrow \text{neutron} + \text{X-ray}$$

The proton becomes a neutron and electromagnetic radiation (X-rays) is given out. The process is called electron capture decay, or sometimes K capture because the captured electron comes from the innermost shell of orbiting electrons which is known as the K-shell. After this has taken place there will be an electron missing from the K-shell and so other electrons will immediately fall down to fill the vacancy, emitting characteristic X-rays as they do so. In the same way the swapping of a proton for a neutron in the nucleus involves a change in the configuration of nucleons and the emission of the excess energy of the nucleus as gamma-rays. Iodine-125 is an example of a nuclide which decays by electron capture.

Example: $^{125}_{53}I \rightarrow {}^{125}_{52}Te + \text{X - ray}$

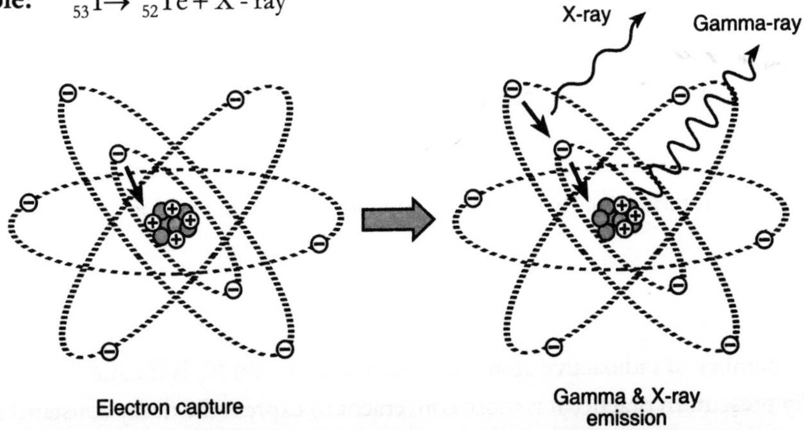

Fig. 13.3: Electron capture

Decay by emission of γ-rays

In contrast to emission of α- and β-particles, γ-emission involves electromagnetic radiation similar to, but with a shorter wavelength than, X-rays. These γ-rays result from a transformation of a nucleus of an atom (in contrast to X-rays, which are emitted as a consequence of excitation involving the orbital electrons of an atom) and frequently accompany α- and β-particles emission. Emission of γ- rays in itself leads to no change in atomic number or mass.

γ-radiation has low ionizing power but high penetration. For example, the γ-radiation from ^{60}Co penetrates 15 cm of steel. The toxicity of γ-radiation is similar to that of X-rays.

Example: $^{131}_{53}I \rightarrow ^{131}_{54}Xe + \beta^- + \gamma$

13.2.1 Radioactive decay energy

The usual unit used in expressing energy levels associated with radioactive decay is the electron volt. (One electron volt (eV) is the energy acquired by one electron in accelerating through a potential difference of 1V and is equivalent to 1.6×10^{-19} J). For the majority of isotopes, the term million or mega electron volt (MeV) is more applicable. Isotopes emitting α-particles are more normally the most energetic, falling in the range 4.0-8.0 MeV, whereas β- and γ-emitters generally have decay energies of less than 3.0 MeV.

13.2.2 Rate of radioactive decay

Radioactive decay is a spontaneous process and it occurs at a definite rate characteristic of the source. This rate always follows an exponential law. Thus the number of atoms disintegrating at any time is proportional to the number of atoms of the isotope (N) present at any time (t). Expressed mathematically, the exponential curve gives the equation.

$$-\frac{dN}{dT} \alpha N$$

or,

$$-\frac{dN}{dT} = \lambda N, \tag{13.1}$$

where λ is the decay constant, a characteristic of a given isotope defined as the fraction of an isotope decaying in unit time (t^{-1}). By integrating equation 13.1, it can be converted to a logarithmic form:

$$\text{In}\frac{N_t}{N_0} = \lambda t, \tag{13.2}$$

where N_t is the number of radioactive atoms present at time t, and N_0 is the number of radioactive atoms originally present. In practice it is more convenient to express the decay constants in terms of half-life ($t_{1/2}$). This is defined as the time taken for the activity to fall from any value to half that value. If N_t in equation 13.2 is equal to one-half of N_0 then t will equal the half-life of the isotope.

Thus

$$\ln\frac{1}{2} = -\lambda t\frac{1}{2}$$

or,

$$2.303 \log (1/2) = -\lambda t_{1/2}$$

or,

$$t_{1/2} = 0.693/\lambda.$$

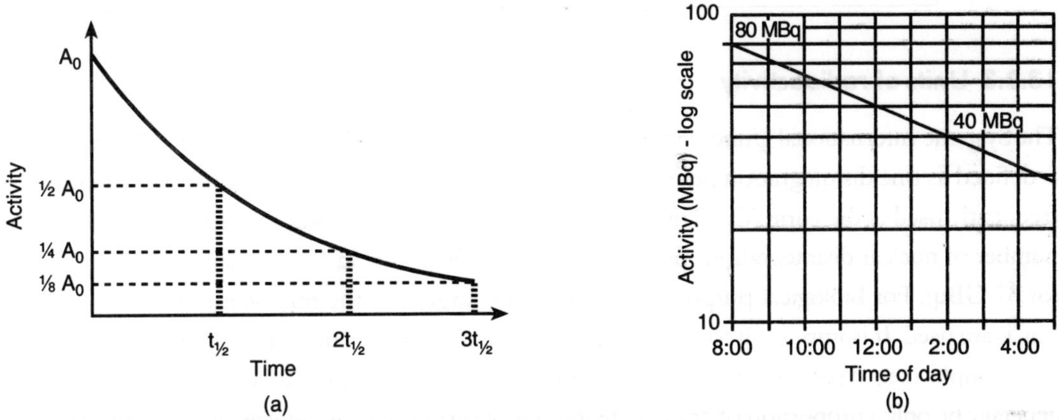

Fig. 13.4: (a) Exponential decay of activity, (b) Exponential decay on a logarithmic scale

The values of $t_{1/2}$ vary widely from 10^{19} years for lead-204 (^{204}Pb) to 3×10^{-7}s for polonium-212 (^{212}Po). The half-lives of some isotopes frequently used in biological work are given in Table 13.1. Note that two important elements, oxygen and nitrogen are missing from the table. This is because the half-lives of radioactive isotopes of these elements are too short for most biological studies (^{15}O has $t_{1/2}$ of 2.03 min, whereas ^{13}N has $t_{1/2}$ of 10.00 min). The advantages and disadvantages of working with isotopes of differing half-lives are given in Table 13.2.

Table 13.1: Half-lives of some isotopes used in biological studies

Isotopes	Half-lives
^{3}H	12.26 years
^{14}C	5760 years
^{22}Na	2.58 years
^{32}P	14.20 days
^{33}P	25.4 days
^{35}S	87.20 days
^{42}K	12.40 h
^{45}Ca	165 days
^{59}Fe	45 days
^{125}I	60 days
^{131}I	8.05 days
^{135}I	9.7 h

Table 13.2: The advantages and disadvantages of working with a short half-life isotope	
Advantages	Disadvantages
High specific activity Makes the experiment more sensitive Easier and cheaper to dispose of Lower doses likely (e.g., in diagnostic testing of human subject)	Experimental design, isotope decays during time of experiment Cost of replacement for further experiments Frequently need to calculate amount of activity remaining

13.2.3 Units of radioactivity

The Systeme International Unites (SI system) uses the becquerel (Bq) as the unit of radioactivity. This is defined as one disintegration per second (1 d.p.s.). However, an older unit, not SI system and still frequently used is the curie (Ci). This is defined as the quantity of radioactive material in which the number of nuclear disintegrations per second is the same as that in 1 g of radium, namely 3.7×10^{10} (or 37 GBq). For biological purposes this unit is too large, and the microcurie (μCi) and millicurie (mCi) are used. It is important to realize that the curie refers to the number of disintegration actually in a sample (i.e., d.p.s.) not to the disintegrations detected by the radiation counter, which will generally be only a proportion of the disintegrations occurring and are referred to as counts (i.e., c.p.s.).

Normally, in experiments with radioisotopes, a carrier of the stable isotope of the element is added. It, therefore, becomes necessary to express the amount of radioisotope present per unit mass. This is the specific activity. It may be expressed in a number of ways including disintegration rate (d.p.s. or d.p.m.), count rate (c.p.s. or c.p.m.) or curies (mCi or μCi) per unit of mass of mixture (units of mass are normally either moles or grams). An alternative method of expressing specific activity, which is not very frequently used, is atom percentage excess. This is defined as the number of radioactive atoms per total of 100 atoms of the compound. For quick reference, a list of units and definitions frequently used in radiobiology is provided in Table 13.3.

Table 13.3: Units commonly used to describe radioactivity

Unit	Abbreviation	Definition
Counts per minute or second	c.p.m. / c.p.s.	The recorded rate of decay
Disintegrations per minute or second	d.p.m. / d.p.s.	The actual rate of decay
Curie	Ci	The number of d.p.s. equivalent to 1 g of radium (3.7×10^{10} d.p.s.)
Millicurie	mCi	Ci $\times 10^{-3}$ or 2.22×10^{9} d.p.m.
Microcurie	μCi	Ci $\times 10^{-6}$ or 2.22×10^{6} d.p.m.
Becquerel (SI unit)	Bq	1 d.p.s.
Terabecquerel (SI unit)	TBq	10^{12} Bq or 27.027 Ci
Gigabecquerel (SI unit)	GBq	10^{9} Bq or 27.027 mCi

Contd.

Table 13.3: Units commonly used to describe radioactivity		
Unit	Abbreviation	Definition
megabecquerel (SIunit)	MBq	10^6 Bq or 27.027 μCi
electron volt	eV	The energy attained by an electron accelerated through a potential difference of 1 volt. Equivalent to 1.6×10^{-19} J
roentgen	R	The amount of radiation that produces 1.61×10^{15} ion-pairs kg^{-1}
rad	rad	The dose that gives an energy absorption of 0.01 J kg^{-1}
gray	Gy	The dose that gives an energy absorption of 1 J kg^{-1}. Thus 1 Gy = 100 rads.
rem	rem	The amount of radiation that gives a dose in humans equivalent to 1 roentgen of X-rays
sievert	Sv	The amount of radiation that gives a dose in humans equivalent to 1 Gy of X-rays. Thus 1 Sv = 100 rad

13.3 INTERACTION OF RADIOACTIVITY WITH MATTER

13.3.1 Alpha particles

Fig. 13.5 illustrates the passage of an alpha-particle through matter. Because alpha-particles (made up of 2 protons and 2 neutrons) are comparatively heavy and doubly charged, they cause a great deal of ionization as they collide with atomic electrons in the material, knocking them out of their atoms. Because they are so much heavier than an electron, they do this without deviating from a straight path, but each collision results in a small loss of energy to the alpha particle, so that it steadily slows down. The density of ionisation tends to increase towards the end of the particle's path. Occasionally an alpha-particle may suffer a collision with an atomic nucleus, but this is comparatively rare because the nucleus is so small. However, if an alpha-particle does hit a nucleus it will be deviated significantly from its forward path and will also send the nucleus recoiling in a different direction. The recoil nucleus can then go on to cause additional ionisation in the material. Because the ionisation produced is so dense, the alpha particle will soon lose all its energy as a result of many electron collisions and rapidly come to a stop. The distance travelled before it finally stops is called the particle's range. The range depends on the particle's energy and the material through which it travels, but for an alpha-particle it is always very short. For example an alpha-particle with an energy of 1 MeV will have a range of 5 mm in air and only 7 microns in tissue. From this it is clear that an alpha-particle source outside the body will do little harm, because all the alpha-particles will be absorbed in the superficial layers of the skin which are dead anyway. However, if an alpha source was allowed to get inside the body its radiation would be absorbed in a few cells and could produce very damaging effects. This is why alpha emitting radionuclides, such as ^{238}Pu, are so dangerous, if inhaled. Alpha emitters are not used in medicine.

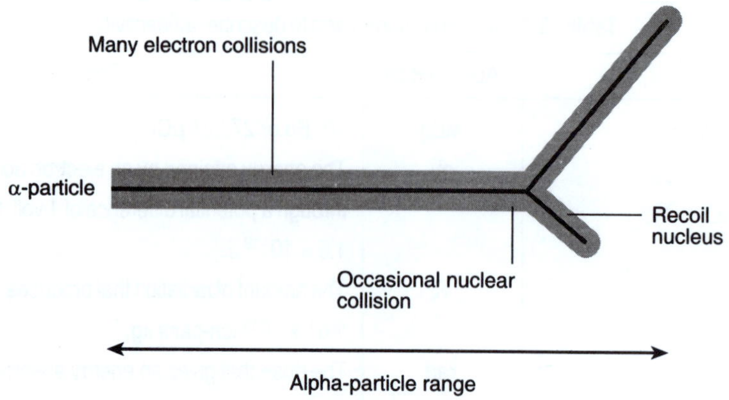

Fig. 13.5: Alpha-particle interaction with matter

Negatrons

Compared with α-particles, negatrons are very small and rapidly moving particles that carry a single negative charge. They interact with matter to cause ionization and excitation exactly as with α-particles. However, due to their speed and size, they are less likely than α-particles to interact with matter and, therefore, are less ionizing and more penetrating than α-radiation. Another difference between α-particles and negatrons is that, whereas for a given α-emitter all the particles have the same energy, negatrons are emitted over a range of energy, i.e., negatron emitters have a characteristic energy spectrum. The maximum energy level (E_{max}) varies from one isotope to another, ranging from 0.018 MeV for ^3H to 4.81 MeV for 38Cl. The difference in E_{max} affects the penetration of radiation; β-particles from ^3H can travel only a few millimetres in air, whereas those from ^{32}P can penetrate over 1 m of air. The reason for negatrons of a given isotope being emitted within an energy range was explained by W. Pauli in 1931, when he postulated that each radioactive event occurs with an energy equivalent to E_{max}, but that the energy is shared between a negatron and a neutrino. The proportion of total energy taken by the negatron and the neutrino varies for each disintegration. Neutrinos have no charge and negligible mass and do not interact with matter.

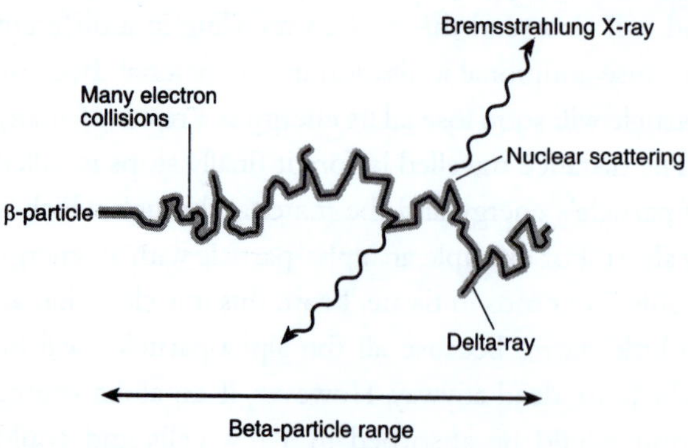

Fig. 13.6: Beta-particle interaction with matter

13.3.2 Beta-particles

Fig. 13.6 illustrates the passage of a beta-particle through matter. Because beta particles (electrons) are lighter and only singly charged, they produce less dense ionization than alpha-particles and are much more easily deviated from a straight line as they ionise atoms in the material through which they pass. Frequently the collisions with an atomic electron are sufficiently violent to cause the beta-particle to deviate through a large angle and then the atomic electron with which it collided acquires enough energy to move off on its own. This electron is called a delta-ray and it goes on to produce further ionisation. Occasionally, if a beta-particle happens to encounter an atomic nucleus in a material of high atomic number, it will be deviated very violently and in doing so gives off bremsstrahlung X-rays (from the German for breaking radiation). After a very zig-zag track, beta-particles will eventually come to rest and so, like alpha-particles, they exhibit a definite range. However, since they produce less dense ionisation, they slow down more gradually than alpha-particles and will have a longer range. For example, 600 keV is a typical beta-particle energy; this is the maximum energy of beta particles from $_{131}I$ decay and the average energy from decay of $_{90}Sr$ and its daughter $_{90}Y$. A 600 keV beta-particle has a range of 2.5 metres in air or 3 mm in tissue. Because all beta sources emit a range of beta-particle energies, rather than just a single energy, there will always be a spread in the range of particles emitted. The number of beta-particles, therefore, falls off quite rapidly with thickness of material traversed, until none remains after a thickness equal to the range of the maximum energy present. If a beta source is close to, or even inside the body, its radiation will be absorbed within a few millimetres of the source. This means that all the energy is absorbed in local tissues and, since beta-particles are moderately ionising, there is potential for damaging effects to these tissues. In diagnosis this may be looked upon as a hazard to be minimised, but it can be put to good use in therapeutic applications. For example the high energy beta particles emitted from the decay of $_{90}Sr$ (and its daughter $_{90}Y$) can be used in an external application for therapeutic doses to surface tissues. The beta-particles from $_{131}I$ are also used in therapy of the thyroid. Since iodine is concentrated by thyroid tissue, a patient administered with radioactive iodine $_{131}I$ will receive a larger radiation dose to the thyroid than to other organs. Nuclides such as $_{3}H$, which emit low energy beta particles, result in a smaller radiation dose which means that, in small amounts, they may be safely administered internally for in vivo diagnostic studies. However, since the beta-particles will not escape from the patient, measurements of the activity present must be made by collecting blood or urine samples and then counting these in the laboratory. Even then detecting the low energy beta-particles is not easy and the samples must be intimately mixed with the detecting medium in a liquid scintillation counter. Slightly higher energy beta particles, such as those from $_{35}S$, are useful for autoradiography. When tissue containing $_{35}S$ is placed on a photographic film, the beta-particles will only blacken the film locally, producing an image of the activity distribution in the sample.

13.3.3 γ-Rays and X-rays

Gamma-rays and X-rays are not particles like alpha and beta, but are examples of electromagnetic radiation (like high energy light) and consequently interact with matter in a rather different way. Fig. 17 illustrates the passage of several gamma-rays through matter. Unlike alpha and beta, where each particle undergoes many individual interactions, each gamma-ray only encounters one, or possibly two interactions and many gamma-rays pass through with no interaction at all.

Gamma-rays and X-rays do not produce ionisation directly, but instead they do so by first producing secondary electrons. These arise from two types of process; scattering and absorption. Scattering occurs through the process of Compton scattering in which a gamma-ray interacts with

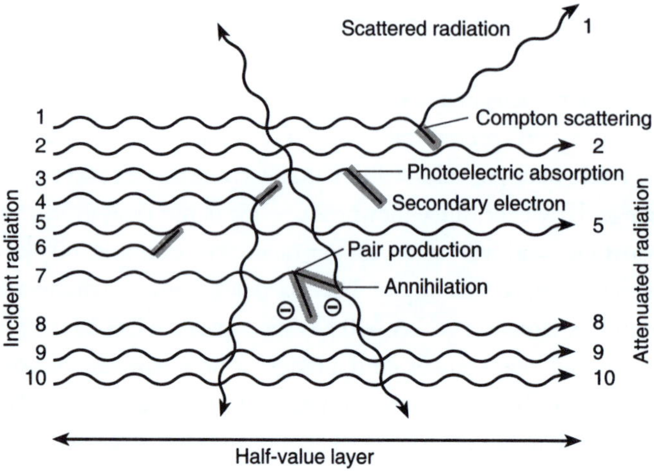

Fig. 13.7: Gamma-ray interaction with matter

a free electron in the material. The gamma-ray passes some of its energy to the electron and continues on its way as scattered radiation with a lower energy and travelling in a different direction (for example rays 1 and 4 in Fig. 13.7). There are two possible absorption processes in which the gamma-ray disappears altogether. Photoelectric absorption occurs when a gamma-ray gives up all of its energy at once to an atomic electron which is then ejected from the atom (rays 3 and 6 in Fig. 13.7). Gamma-rays with energy greater than 1 MeV can also be absorbed by pair production, in which an electron and positron pair are spontaneously produced (ray 7 in Fig. 13.7). The positron will subsequently annihilate with an atomic electron producing two gamma-rays of 511 keV. After any of these processes the secondary electrons produced go on to produce ionisation of the material, just like a beta-particle. Unlike alpha- and beta-particles, which are stopped after many interactions, gamma-rays and X-ray each undergo only a few interactions. Gamma-rays do not, therefore, have a definite range, but instead the intensity of a gamma-ray beam is attenuated by a combination of scattering and absorption processes so that it falls off steadily with distance. The distance required to halve the number of the gamma-rays is called the half-value layer (HVL). This is analogous to the half-life of radionuclide decay and the same exponential mathematics applies. In Fig. 13.7 the incident radiation consists of ten gamma-rays entering at the left. During passage through one half value layer of the material 3 of these are absorbed and 2 scattered leaving an attenuated beam containing only 5 gamma-rays. In the next half value layer, half of these would again disappear. The half value layer depends on the energy of the radiation and the nature of the material. For a gamma-ray of 140 keV the HVL in lead is 0.25 mm. Therefore, 0.25 mm of lead shielding will reduce the intensity of 140 keV radiation to half its original value, 0.5 mm will reduce it to one quarter, 1 mm (4 HVLs) to one sixteenth and 3 mm (12 HVLs) by a factor of 1/4096.

In tissue the HVL is much greater, being 47 mm for 140 keV gamma-rays. Thus, if a radionuclide such as ^{99m}Tc, which emits 140 keV gamma-rays, is present inside a person it is clear that sufficient numbers of gamma-rays will be able to penetrate body tissues to permit external detection of the whereabouts of the radionuclide for diagnostic imaging purposes. Gamma-rays that do not escape will distribute their energy throughout several organs, leading to a distributed radiation dose which is much less damaging than the local doses from beta emitters. This is why pure gamma emitting radionuclides such as ^{99m}Tc are so useful for diagnostic imaging purposes.

13.3.4 Radiation dose

When ionising radiations (alpha, beta, gamma or X-rays) pass through matter they pass on some or all of their energy to the material by ionising and exciting the atoms of the material through the processes described above. The damage done by this depends both on the energy deposited and the amount of material involved. The radiation damage increases as the amount of energy deposited increases, and decreases if it is spread throughout a greater amount of material. The radiation absorbed dose is, therefore, defined as the energy absorbed divided by the mass of material involved. One joule of energy absorbed in each kilogram of material is defined as an absorbed dose of one gray (written 1 Gy). Usually we are dealing with doses smaller than this so we use units of one thousandth of a gray, or one milligray (written 1 mGy). The concept of absorbed dose applies to all types of material but, when we need to assess the effect on biological tissues, we also need to take account of the fact that some types of radiation are more harmful than others. For example, because they are so densely ionising, alpha-particles are about twenty times as effective at killing cells as beta-particles, gamma-rays or X-rays. Therefore, when measuring the dose to biological tissues, we use a quantity called equivalent dose which is defined as the absorbed dose multiplied by a radiation weighting factor. This radiation weighting factor is 20 for alpha-particles but 1 for beta-particles, gamma-rays and X-rays. Confusingly, although equivalent dose has essentially the same units as absorbed dose, it is given a different special name of the sievert (written Sv) or millisievert (mSv). Since we never have to deal with alpha particles in medical applications it happens that the equivalent dose (in Sv or mSv) is always numerically the same as the absorbed dose (in Gy or mGy). There is one final complication to measuring the effect of radiation on a person; not all tissues in the body are equally sensitive to radiation damage. For example the bone marrow is particularly susceptible to damage whereas the skin is relatively insensitive. Therefore, in situations where different parts of the body might receive different doses, it is usual to calculate a weighted sum of the equivalent doses to each organ. The organ weighting factors take account of the susceptibility of each organ to damage. Thus bone marrow gets a larger weighting factor than skin. This weighted sum of organ doses is called the effective dose. Because the weighting factors for all organs in the body add up to one, if every organ receives the same equivalent dose, the effective dose will be the same as the equivalent dose. Therefore, the effective dose can be thought of as the uniform whole body dose which would have the same effect (in terms of the risk of doing harm) as the actual non-uniform dose. Effective dose is measured in units of sievert (Sv) or millisieverts (mSv) just the same as equivalent dose.

13.4 DETECTION AND MEASUREMENT OF RADIOACTIVITY

There are three commonly used methods of detecting and quantifying radioactivity. These are based on the ionization of gases, on the excitation of solids or solutions, and the ability of radioactivity to expose photographic emulsions, i.e., autoradiography. The following figure shows the scheme of basic radiation detector system.

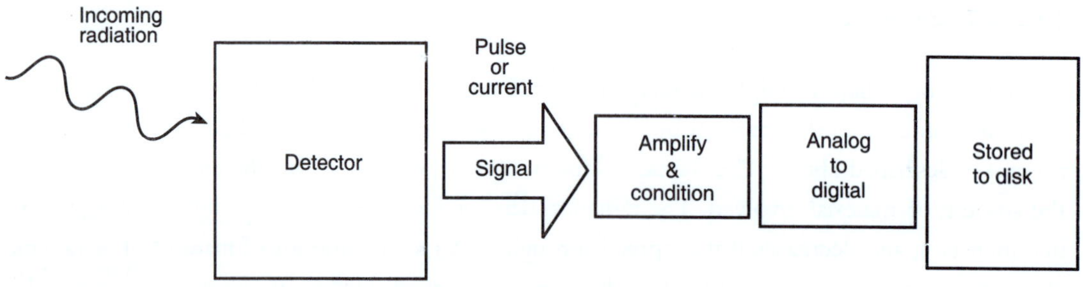

Fig. 13.8: Scheme of basic detector system

13.4.1 Methods based upon gas ionization

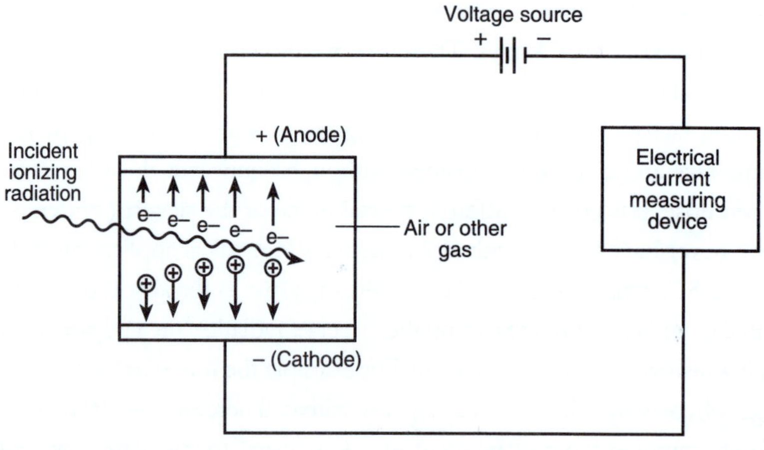

Fig. 13.9: Basic principles of gas filled detectors. The electrical charge liberated by ionising electrodes is collected by positive and negative electrodes

The effect of voltage upon ionization

As a charged particle passes through a gas, its electrostatic field dislodges orbital electrons from atoms sufficiently close to its path and causes ionization (Fig. 13.10). The ability to induce ionization decreases in the order

$$\alpha > \beta > \gamma \qquad (10\,000 : 100 : 1)$$

Accordingly, α- and β-particles may be detected by gas ionization methods, but these methods are poor for detecting γ-radiation. If ionization occurs between a pair of electrodes enclosed in a suitable chamber, a pulse (current) flows, the magnitude of which is elated to the applied potential and the number of radiation particles entering the chamber (Fig. 13.10). The various 'regions' shown in Fig. 13.10 will now be considered.

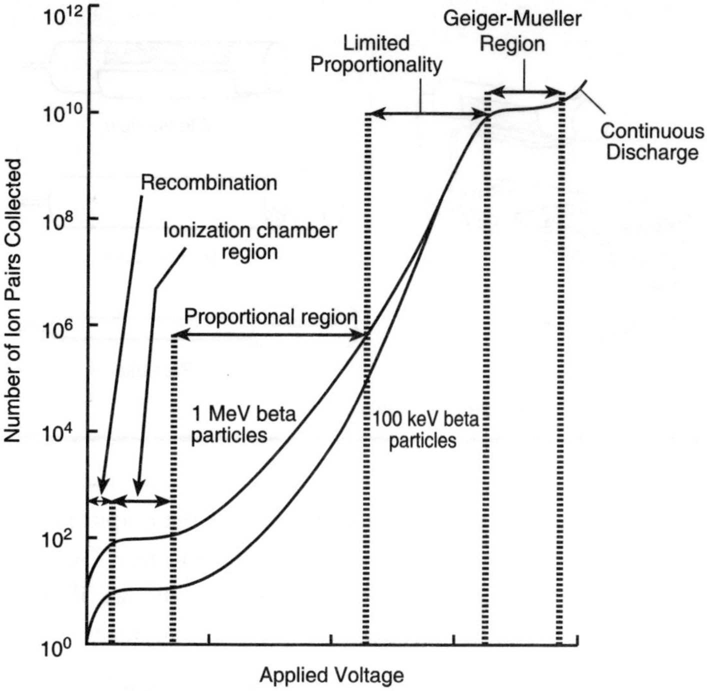

Fig. 13.10: Amount of electric charge collected after a single interaction as a function of electric potential difference (voltage) applied between the two electrodes of gas filled detectors. The lower curve shows the charge collected when 100 keV interacts, and the upper curve shows the results from the 1 MeV.

In the ionization chamber region of the curve, each radioactive particle produces only one ion-pair per collision. Hence the currents are low, and very sensitive measuring devices are necessary. This method is little used in quantitative work, but various types of electroscopes, which operate on this principle, are useful in demonstrating the properties of radioactivity. At a higher voltage level than that of the simple ionization chambers, electrons resulting from ionization move towards the anode much more rapidly; consequently they cause secondary ionization of gas in the chamber, resulting in the production of secondary electrons, which cause further ionization and so on. Hence from the original event a whole torrent of electron reaches the anode. This is the principle of gas amplification and is known as the Towensend avalanche effect, after its discoverer. As a consequence of this gas amplification, current flow is much greater. As can be seen in Fig. 13.10, in the proportional counter region the number of ion-pairs collected is directly proportional to the applied voltage until a certain voltage is reached, when a plateau occurs. Before the plateau is reached there is a region known as the limited proportion region, which is not often used in detection and quantification of radioactivity and hence will not be discussed.

The main drawback of counters that are manufactured to operate in the proportional region is that they require a very stable voltage supply because small fluctuations in voltage result in significant changes in amplification. Proportional counters are particularly useful for detection and quantification of α-emitting isotopes, but it should be noted that relatively a few such isotopes are used in biological work.

In the Geiger-Muller region all radiated β-particles, induce complete ionization of the gas in the chamber. Thus the size of the current is no longer dependent on the number of primary ions

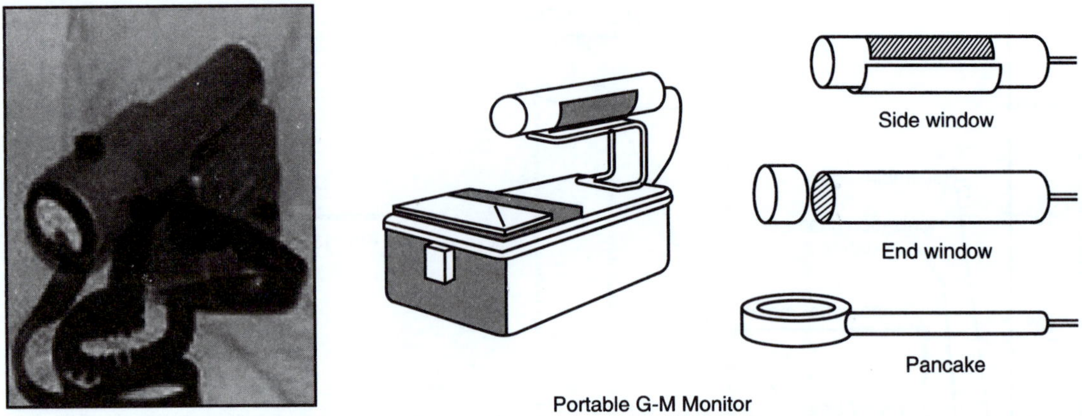

Fig. 13.11: The GM counter and its different parts

produced. Since maximum gas amplification is realized in this region, the size of the output pulse from the detector will remain the same over a considerable voltage range (the so-called Geiger-Muller plateau). The number of times this pulse is produced is measured rather than its size. Therefore, it is not possible to discriminate between different isotopes using this type of counter.

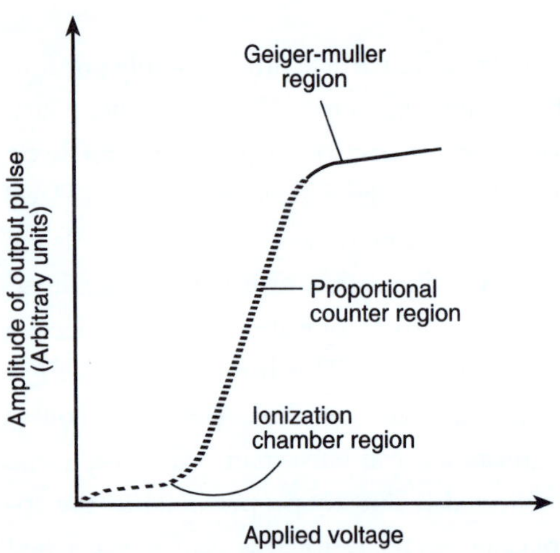

Fig. 13.12: Effect of voltage upon ionisation

Since it takes a finite time for the ion-pairs to travel to their respective electrodes, other ionizing particles entering the tube during this time fail to produce ionization and hence are not detected, thereby reducing the counting efficiency. This is referred to as the dead time of the tube and is normally 100 to 200 μs. When the ions reach the electrode they are neutralized. Inevitably some escape and produce their own ionization avalanche. Thus, if unchecked a Geiger-Muller tube would tend to give a continuous discharge. To overcome this, the tube is quenched by the addition of a suitable gas, which reduces the energy of the ions. Common quenching agents are ethanol, ethyl formate and the halogens.

Instrumentation

Counters based on gas ionization used to be the main method employed in the quantification of radioisotopes in biological samples. Currently, scintillation counting has virtually taken over. However, all laboratories use small hand-held radioactivity monitors based on gas ionization, the end-window design being the most popular type (Fig. 13.12). These counters have a thin end-window made from aluminium and can detect β-radiation from high energy (^{32}P) and weak emitters (^{14}C), but are incapable of detecting ^{3}H, because the radiation cannot penetrate the end-window. For the same reason they are not very efficient detectors of α-radiation.

End-window ionization counters are used for routine monitoring of the radioactive laboratory to check for contamination. They are useful in experimental situations where the presence or absence of radioactivity needs to be known rather than the absolute quantity, for example quick screening of radioactive gels prior to autoradiography or checking of chromatographic fractions for the labelled components.

The inability of end-window counters to detect weak β-emitters presents a problem in biosciences because ^{3}H is a very commonly used radioisotope. The problem can be overcome by using a so-called windowless counters where a gas flow is used. These instruments are rather cumbersome and need to be carried around on an object that resembles a golf trolley. They are useful for mass screening of premises for ^{3}H contamination but are rarely used as routine. Most laboratories monitor for ^{3}H by doing a wipe test regularly, i.e., using wet tissues or cotton wool to take swabs for scintillation counting.

13.4.2 Methods based upon excitation

As outlined above, radioactive isotopes interact in matter in two ways, causing ionization, which forms the basis of Geiger-Muller counting, and excitation. The latter effect leads the excited compound (known as fluor) to emit photons of light. This fluorescence can be detected and quantified. The process is known as scintillation and when the light is detected by a photomultiplier, forms the basis of scintillation counting. The electric pulse that results from the conversion of light energy to electrical energy in the photomultiplier is directly proportional to the energy of the original radioactive event. This is considerable asset of scintillation counting, since it means that two, or even more, isotopes can be separately detected and measured in the same sample, provided they have sufficiently different emission energy spectra (see next page). The mode of action of a photomultiplier is shown in Fig. 13.13.

In summary, scintillation counting provides information of two kinds:

Quantitative: The number of scintillations is proportional to the rate of decay of the sample, i.e., the amount of radioactivity.

Qualitative: The intensity of light given out and, therefore, signal from the photomultiplier is proportional to the energy of radiation.

13.4.2.1 Types of scintillation counting

There are two types of scintillation counting, which are described next page. In solid scintillation counting the sample is placed adjacent to a crystal of fluorescent material. The crystal that is normally used for γ-isotopes is sodium iodide, whereas for α-emitters zinc sulphide crystals are preferred

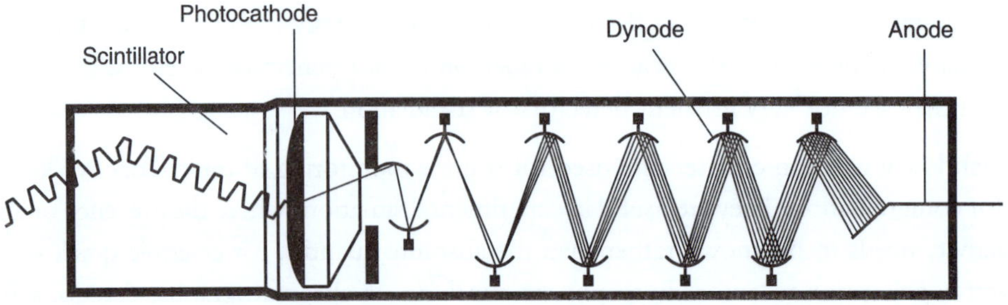

Fig. 13.13: The mode of action of a photomultiplier

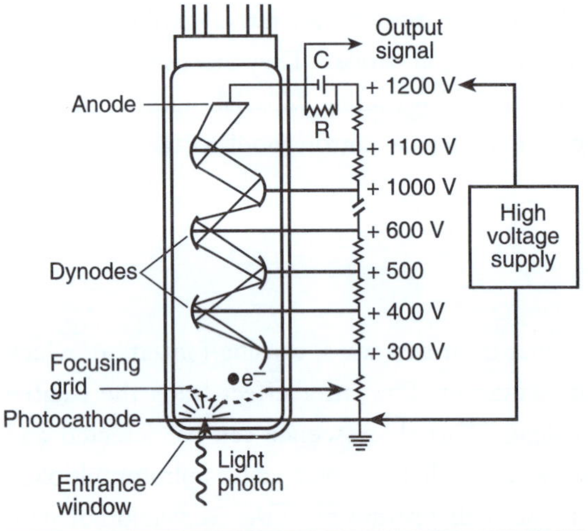

Fig. 13.14: Details and mode of action of photomultiplier tube

and for β-emitters organic scintillators such as anthracene are used. The crystals themselves are placed near to a photomultiplier, which in turn is connected to a high voltage supply and a scaler. Solid scintillation counting is particularly useful for γ-emitting isotopes. This is because these rays are electromagnetic radiation and colloid only rarely with neighbouring atoms to cause ionization or excitation. Clearly, in a crystal the atoms are densely packed, making collisions more likely. Conversely, solid scintillation counting is generally unsuitable for β-emitting isotopes such as ^{3}H and ^{14}C, because even the highest energy negatrons emitted by these isotopes would have hardly sufficient energy to penetrate the walls of the counting vials in which the samples are placed for counting. As many of the isotopes used in radioimmunoassay are γ-emitting isotopes, solid scintillation counting is frequently used in biological work.

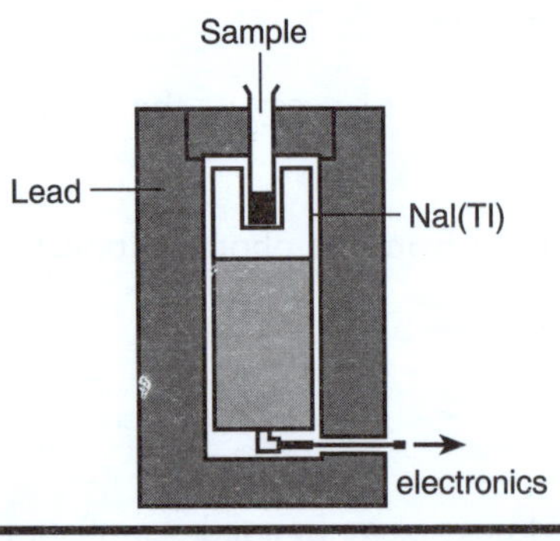

Fig. 13.15: Structure of solid scintillation counter

In details the arrangement and working of solid detectors is explained herewith by the example of NaI detector. In this the quartz window is optically coupled by a light transmitting gel to the face of evacuated electron tube (PMT) which is plugged in to the base designed to supply voltage to various electrode and collect the signal pulse from PMT anode. The sequence of events is as follows:

In liquid scintillation counting, the sample is mixed with a scintillation cocktail containing a solvent and one or more

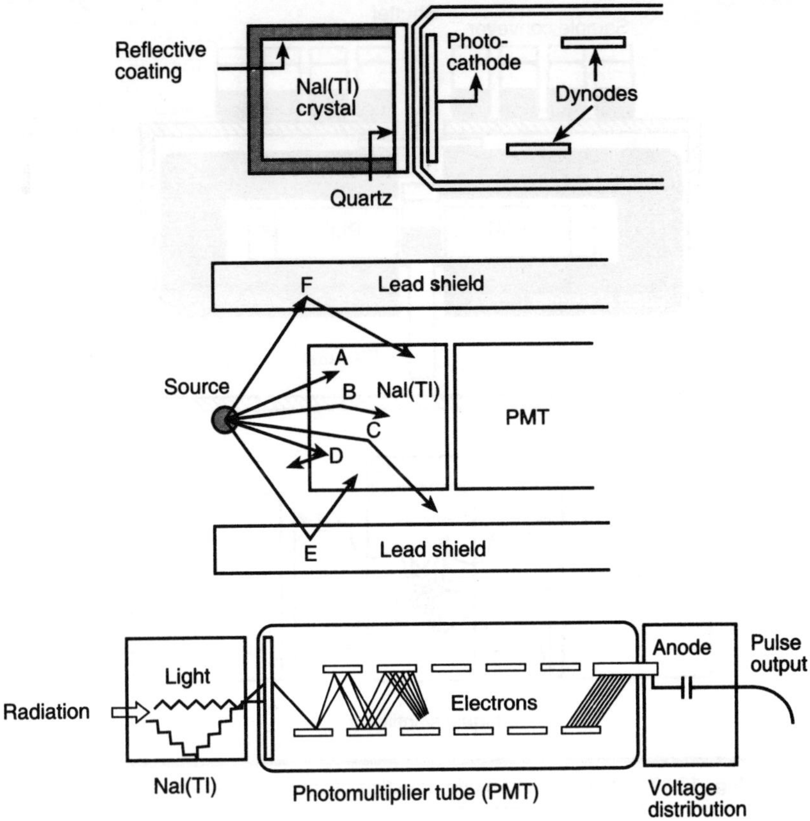

Fig. 13.16: Structure and mode of action of solid scintillation counter

fluors. This method is particularly useful in quantifying weak β-emitters such as ^{3}H, ^{14}C and ^{35}S, which are frequently used in biological work. For these isotopes, liquid scintillation counting is the usual method. Thus the remainder of this section will place particular emphasis on this technique, though it should be pointed out that most of what follow applies equally to solid scintillation counting used in the quantification of γ-emitters.

13.4.2.2 Energy transfer in liquid scintillation counting

A small number of organic solvents fluoresce when bombarded with radioactivity. The light emitted is of very short wavelength (Fig. 13.18) and is not efficiently detected by most photomultipliers; however, if a compound is dissolved that can accept the energy from the solvent and itself fluoresce at a longer wavelength, then the light can be more efficiently detected. Such a compound is known as a primary fluor and the most frequently used example is 2, 5-dephenyloxazole (PPO). Unfortunately the light emitted by PPO is not always detected with very high efficiency (depending on the photomultiplier detector) but this can be overcome by including a secondary fluor or wavelength shifter such as 1, 4-bis (5-phenyloxazol-2-yl) benzene (POPOP). Thus the energy transfer process becomes

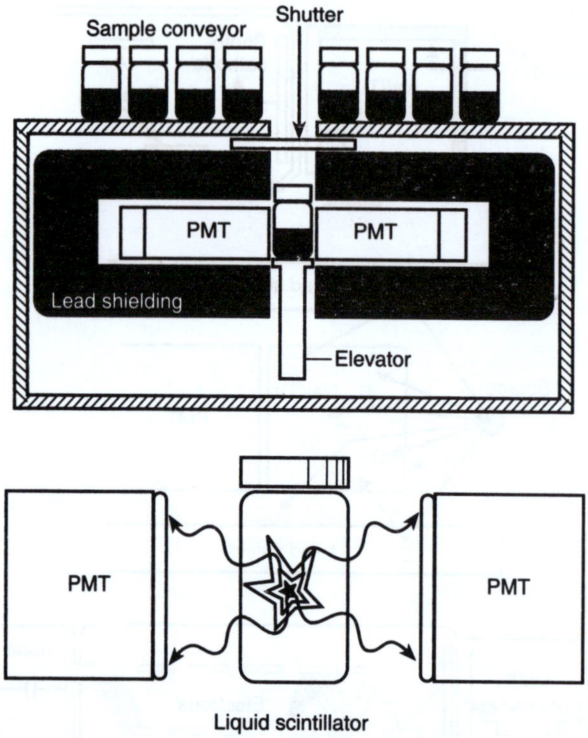

Fig. 13.17: Structure and mode of action of liquid scintillation counter

The question obviously arises as to why a primary fluor and a secondary fluor are necessary when it is the latter that emits light at the best wavelength for detection. The answer is simple that the solvent cannot transfer its energy directly to the secondary fluor.

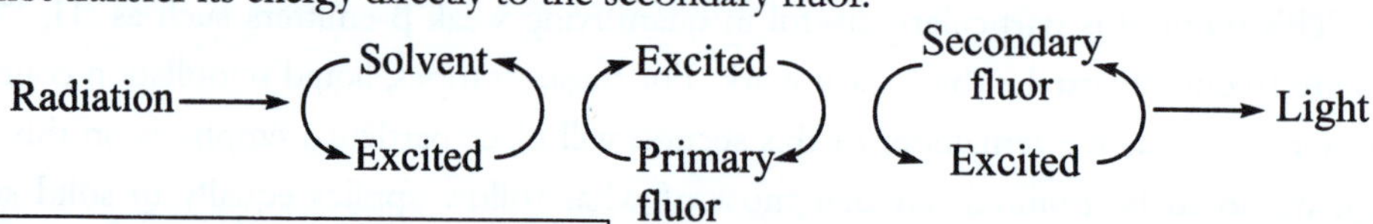

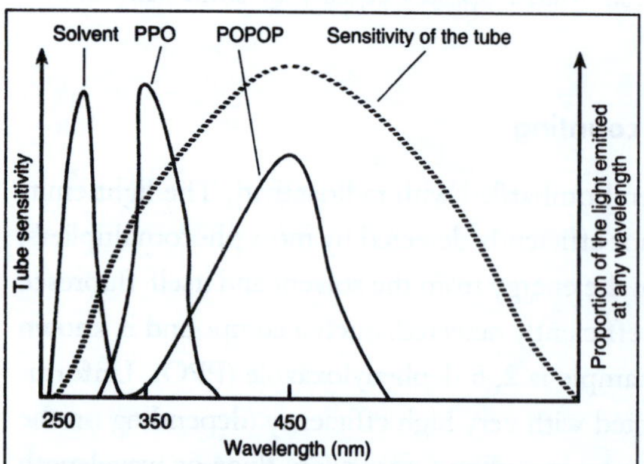

Fig. 13.18: Emission spectra of various fluors in relation to sensitivity of phototubes

PPO and POPOP were among the original fluors used in liquid scintillation counting and remain a favourable choice. However, compounds such as 2-(4′-t-butylphenyl)-5-(4″ biphenylyl)-1, 3, 4-oxadizole (BUTYL-PBD) is a better primary fluor but is quiet expensive and is affected by extremes of pH.

Most laboratories now buy their scintillation cocktails already prepared and there are many different makes and recipes on the market. Competition and an increasing awareness of health and safety mean that scintillation cocktails are gradually becoming less toxic and have a lower fire hazard. A final point: some cocktails (as just described) are designed for organic samples and others for aqueous samples (these cocktails include an emulsifier such as the detergent Triton X-100); it is important that the appropriate formulation is used.

13.4.2.3 Advantages of scintillation counting

The very fact that scintillation counting is widely used in biological work indicates that it has several advantages over gas ionization counting. These advantages are listed below.

- The rapidity of fluorescence decay (10^{-9}s), which, when compared to dead time in a Geiger-Muller tube (10^{-4}s), means much higher count rates are possible.
- Much higher counting efficiencies particularly for low energy β-emitters; over 50% efficiency is routine in scintillation counting and efficiency can rise to over 90% for high energy emitters. This is partly due to the fact that the negatrons do not have to travel through air or pass through an end-window of a Geiger-Muller tube (thereby dissipating much of the energy before causing ionization) but interact directly with the fluor; energy loss before the event that is counted is therefore minimal.
- The ability to accommodate samples of any type, including liquids, solids, suspensions and gels.
- The general ease of sample preparation (see below).
- The ability to count separately different isotopes in the same sample, which means dual labelling experiments can be carried out (see below).
- Scintillation counters are highly automated, hundreds of samples can be counted automatically and built-in computer facilities carry out many forms of data analysis, such as efficiency correction, graph plotting, radioimmunoassay calculations, etc.

13.4.2.4 Disadvantages of scintillation counting

It would not be reasonable, having outlined some of the advantages of scintillation counting, to disregard the disadvantages of the method. Fortunately, however, most of the inherent disadvantages have been overcome by improvement in instrument design. These disadvantages include the following.

- The cost per sample of scintillation counting is not insignificant; however, other factors including versatility, sensitivity and ease and accuracy outweigh this factor for most applications.
- At the high voltages applied to the photomultiplier, electronic events occur in the system that are independent of radioactivity but contribute to a high background counts. This is referred to as photomultiplier noise and can be partially reduced by cooling the photomultipliers. Since temperature affects counting efficiency, cooling also presents a controlled temperature for counting, which may be useful. Low cost photomultipliers, however, have been designed to provide greater temperature systems. Also the use of a

pulse height analyzer can be set so as to reject, electronically, most of the noise pulses that are of low energy (the threshold or gate setting). The disadvantage here is that this also rejects the low energy pulses resulting from low energy radioactivity (e.g., ^3H). Another method of reducing noise, which is incorporated into most scintillation counters, is to use coincidence counting; in this system two photomultipliers are used. These are set in coincidence such as that only when a pulse is generated in both tubes at the same time it is allowed to pass to the scaler. The chances of this happening for a pulse generated by radioactive event is very high compared to the chances of a noise event occurring in both photomultipliers during the so-called resolution time of the system, which is commonly of the order of 20 ns. In general, this system reduces photomultiplier noise to a very low level.

- The greatest disadvantage of scintillation counting is quenching. This occurs when the energy transfer process described earlier suffers interference. Correcting for this quenching contributes significantly to the cost of scintillation counting. Quenching can be any one of three kinds.
    (i) *Optical quenching*: This occurs if inappropriate or dirty scintillation vials are used. These will absorb some of the light being emitted, before it reaches the photomultiplier.
    (ii) *Chemical quenching*: This form of quenching, which occurs when anything in the sample interferes with the transfer of energy from the solvent to the primary fluor or from the primary fluor to the secondary fluor, is the most difficult form of quenching to accommodate. In a series of homogeneous sample (e.g., $^{14}CO_2$ released during metabolism of [^{14}C] glucose and trapped in alkali, which is then added to the scintillation cocktail for counting), chemical quenching may not vary greatly from sample to sample. In these cases relative counting using sample counts per minute can be compared directly. However, in the majority of biological experiments using radioisotopes, such homogeneity of samples is unlikely and it is not sufficiently accurate to use relative counting (i.e., counts per minute). Instead an appropriate method of standardization must be used. This requires the determination of counting efficiency of each sample and the conversion of counts per minute to absolute counts (i.e., disintegrations per minute), as described later. It should be noted that quenching is not such a great problem in solid (external) scintillation counting.
- *Chemiluminiscence*: This can also cause problem during liquid scintillation counting. It results from chemical reactions between components of the sample to be counted and scintillation cocktail, and produces light emission unrelated to excitation of the solvent and fluor system by radioactivity. These light emissions are generally low energy events and generally rejected by the threshold setting of the photomultiplier in the same way as photomultiplier noise. Chemiluminescence, when it is a problem, can usually be overcome by storing samples for same time before counting, to permit the chemiluminescence to decay. Many cotemporaneous instruments are able to detect chemiluminescence and subtract it or flag it on the printout.

- *Phospholuminescence*: This results from components of the sample, including the vial itself, absorbing light and re-emitting it. Unlike, chemiluminescence, which is a once-only effect, Phospholuminescence will occur on each exposure of a sample to light. Samples that are pigmented are most likely to phosphoresce. If this is a problem, samples should be adapted to dark prior to counting and the sample holder should be kept closed throughout the counting process.

Despite all the complications described above, scintillation counters are universal in biosciences departments. This is because the instrument has automated systems for calculating counting efficiency. In other words, the instruments do all the hard work.

13.4.2.5 Using scintillation counting for dual-labelled samples

A feature of scintillation counting is that the size of electric pulse produced by the conversion of light energy in the photomultiplier is related directly to the energy of the original radioactive event. Since different β-emitting isotopes have different energy spectra, it is possible to quantify two isotopes separately in a single sample, provided their energy spectra are sufficiently different. Examples of pairs of isotopes that have sufficiently different energy spectra are ^{3}H and ^{14}C, ^{3}H and ^{35}S, ^{3}H and ^{32}P, ^{14}C and ^{32}P, ^{35}S and ^{32}P. The principle of the method is illustrated with the example of a pair of ^{3}H and ^{14}C as follows:

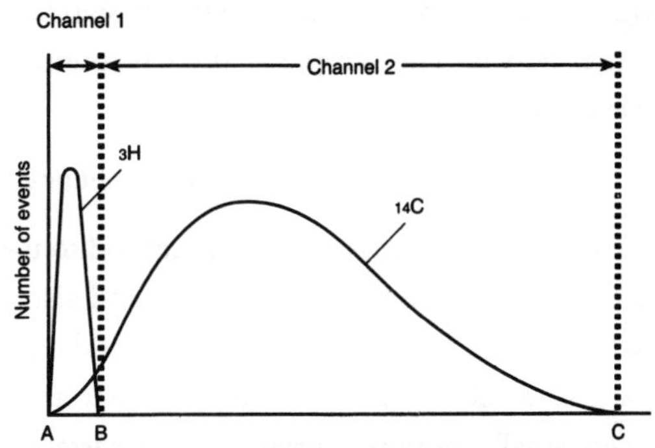

Fig. 13.19: Diagram to illustrate the principle of counting dual-labelled samples

Sample	Radioactivity (cpm) Channel A	Channel B	Channel Ratios B/A	A/B
Experimental	500	800		
^{3}H standard	16 840	25	0.0015	–
^{14}C standard	4 250	10 200	_	0.416

^{3}H in channel B = (0.0015)(500) £ 0.75 cpm

^{14}C in channel B = 800 - 0.75 » 800 cpm

^{14}C in channel A = (0.416)(800) = 333 cpm

^{3}H in channel A = 500 - 333 = 167

Total ^{3}H = 167 cpm

Total ^{14}C = 800 + 333 = 1133 cp

Most modern counters operate with a so-called multichannel analyzer. These are based on an analogue-to-digital converter, electronic signals from the photomultiplier are converted to digital signals stored in computer. Thus the entire energy spectrum is analyzed simultaneously. This greatly facilitates multi-isotope counting and in particular allows the effect of quenching on dual-label counting to be assessed adequately.

Dual-label counting has proved to be useful in many aspects of molecular biology (e.g., nucleic acid hybridization and transcription), metabolism (e.g., steroid synthesis) and drug development.

13.4.2.6 Determination of counting efficiency

As outlined above, a major problem encountered in scintillation counting is that of quenching, which makes it necessary to determine the counting efficiency of some, if not all, of the samples in a particular experiment. This can be done by one of several methods of standardization, all of which apply to both solid and liquid scintillation counting, though again in this section emphasis is placed on the latter method.

Internal standardization: The sample is counted (and gives a reading, say, A c.p.m.), removed from the counter and a small amount of standard material of known disintegrations per minute (B d.p.m.) is added. The sample is then recountered (C c.p.m.) and the counting efficiency of the sample calculated:

$$\text{counting efficiency} = [100\ (C - A)/B]\%$$

It is obviously necessary in this method to use an internal standard (the spike) that contains the same isotope as the one being counted and also to ensure that the standard itself does not act as a quenching agent. Suitable 14C-labelled standards include [^{14}C] toluene, [^{14}C] hexadecane, [^{3}H] benzoic acid and $^{3}H_2O$ (benzoic acid and water are themselves quenching agents and must be used in only very small amounts). Internal standardization is simple and reliable and corrects adequately for all types of quenching. Carefully carried out, it is the most accurate way of correcting for quenching. On the other hand, it demands very accurate pipetting when the standard is added, and it is time

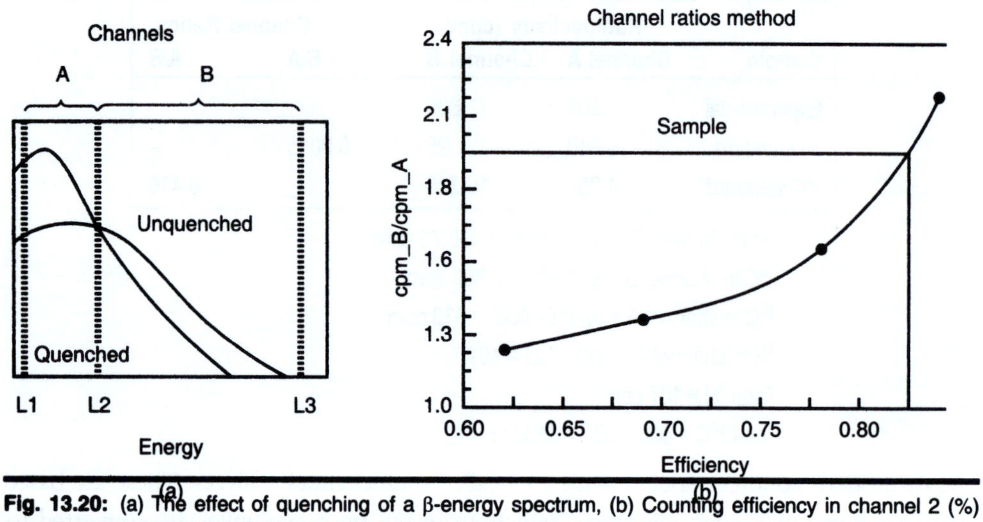

Fig. 13.20: (a) The effect of quenching of a β-energy spectrum, (b) Counting efficiency in channel 2 (%)

consuming because each sample must be counted twice. It also means that the sample cannot be recounted in the event of error because it will be contaminated with the standard. Moreover, time elapses between the first and second count and changes in sample quenching characteristics can also occur, which can lead to considerable inaccuracies. However, it is the means by which the following two methods are calibrated.

1. Channels ratio

When a sample in a scintillation counter is quenched, the scintillation process is less efficient; less light is produced for a given quantum energy of radiation. Thus the energy spectrum for a quenched sample appears to be lower than that for an unquenched sample (Fig. 13.20a). The higher the degree of quenching, the more pronounced is the resulting decrease in the spectrum. This fact is applied in the channels ratio method for determining counter efficiency. The method involves the preparation of a calibration curve based on counting in two channels that cover different, but overlapping, parts of the spectrum. As a sample is quenched, and the spectrum shifts to gradually lower apparent energies, the ratio of counts in each channel can easily be determined.

The efficiency is then plotted against channels ratio to form the standard curve (Fig. 13.20b). It is important to realize that a standard curve applies to only one set of circumstances–one radioisotope, counter and scintillation fluid.

Once the standard curve has been prepared, the efficiency of counting experimental samples can be determined. Samples are counted in the same two channels, the ratio is calculated, put into graph and the efficiency read. In practice all the data can be stored in the counter's computer and corrected values are printed automatically.

Multichannel scintillation counters operate on the same principle but the whole shape and position of the spectrum is analyzed. This is given a digital parameter that relates to counting efficiency. Manufacturers have developed their own title for such parameters, for example LKB Instruments' Automatic Quench Compensation or Packard's Automatic Efficiency Control. These systems have greater precision than the two-channels approach, as the whole of the spectrum is used for analysis. The channels ratio method is suitable for all types and even high degrees of quenching. Furthermore, counting in more than one channel is simultaneous and this method is, therefore, less time consuming than either internal or external standardization. It is also, in practice, an acceptably accurate method for determining counting efficiency, provided care is taken in the preparation of the calibration curve. However, it is notoriously inaccurate at low count rates, because the error on the counts per minute is high and there will be a larger error on the channels ratio because it is calculated from two values of the counts per minute. It is also inaccurate for very highly quenched samples. For these reasons the method that follows is most frequently the procedure of choice.

2. External standardization

Instruments have a γ-emitting external standard build into the counter. Under the control of the counter each sample to be counted is exposed to this external source, which is automatically shifted

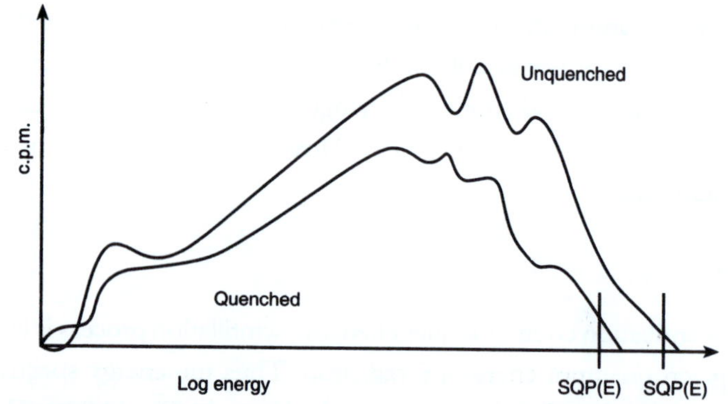

Fig. 13.21: The external standard for estimating counting efficiency. The external source irradiates the sample. The counter analyses the spectrum, which shifts to lower energies if the sample is quenched. The SQP(E), i.e., standard quench parameter (external), expressed without units, is derived from the energy axis and relates to the extend of quench. The greater the quench in the sample, the lower the SQP (E) and the lower the counting efficiency.

from a lead shield to the counting chamber. The γ-radiation penetrates the vial and excites the scintillation fluid. The resulting spectrum is unique to the source and is significantly different from that produced by the sample in the vial. The γ-source used (e.g., ^{137}Cs, ^{133}Ba or ^{226}Ra) varies according to the make of the instrument. The spectrum obtained by 226Ra is shown in Fig. 13.21.

Table 13.4: Recorded radioactivity from a ^{14}C standard sample with increasing quench detected by an external standard

Sample	c.p.m. parameter	External quench (%)	Counting efficiency
^{14}C standard (203600 d.p.m.) unquenched	194930	810	95.7
^{14}C standard (203600 d.p.m. increasing quench)	146141	422	93.5
	181171	207	89.0
	167731	126	82.4
	145879	76	71.6
	126913	55	62.3
	108641	42	53.3
	96103	37	47.2

Quenching agents present in scintillation fluid will significantly affect the spectrum obtained. The instrument analyses this spectrum and assigns a quench parameter to it. The precise method used depends on the make of counter; LKB. Instruments refer to a standard quench parameter for the external source (SQP(E)) based on a point on the energy axis (Fig. 13.21). Other manufacturers use slightly different approaches but the principle is same. The spectrum for the external standard varies according to the degree of quench to the vial and, therefore, the efficiency of the counting of the internal experimental sample.

As for the channels ratio method, a standard curve is required, i.e., a range of quench standards is counted and the external spectrum analyzed in each case. The resulting data is used to prepare a standard curve that is held in instrument's computer. Unknown samples are then counted in the same way, the efficiency read from the standard curve and the sample counts correct.

The external standard approach is now routine in most laboratories, the main advantage over the channels ratio method being that it is suited to samples with low count rates. However, it is not without disadvantages. A standard curve is required for each set of circumstances (as with the channels ratio) and the user can be lulled into a false sense of security. The system is so highly automated that it is easy to lose the sight of the basic principle and the method is not always appropriate. A case in point is the counting of ^{3}H precipitated on to filters counted in scintillation fluid. The external standard method will calculate the degree of quench in the fluid (which will probably be very low) but will not take into account the poor preparation of ^{3}H β-particles from the filter into the scintillation fluid; artificially high efficiencies will be recorded.

In all cases where an automated procedure for calculating counting efficiencies is employed it is prudent to count a few prepared samples in which the the true amount of radioactivity is known.

13.4.2.7 Sample preparation

It is impossible here to give details of all aspects of sample preparation for scintillation counting. Major considerations are outlined.

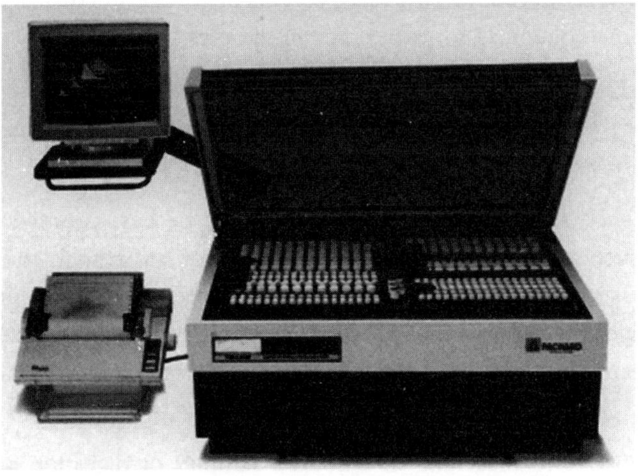

Fig. 13.22: Liquid Scintillator counter

In solid scintillation counting, sample preparation is easy and only involves transferring the sample to a glass or plastic vial (or tube) compatible with the counter. In liquid scintillation counting, sample preparation is more complex and starts with the decision on that type of the sample vial to be used. These may be glass, low potassium glass (with low levels of ^{40}K that reduce background count) or polyethylene. The last of these types are cheaper but are not suitable for cleaning and reuse, whereas glass vials can be reused many times provided they are thoroughly cleaned. Polyethylene bottles give better light transfer and result in slightly higher counting efficiencies, but are inclined to exhibit more phosphorescence than do glass vials. The recent trend is towards mini-vials, which use far smaller volumes of expensive scintillation cocktails. Modern counters are able to accept many types of vial; the smallest vial possible should be used (within the obvious constraints of sample volume) to save costs and in consideration of environmental issues, as scintillation fluids are toxic. Some counters are designed to accept very small sample in special polythene bags split into an array of many compartments; these are particularly useful to, for example, the pharmaceutical industry where there are laboratories that do large numbers of receptor building assay.

Now we will consider scintillation cocktails employed for counting. Toluene-based cocktails are the most efficient, but will not accept aqueous samples, because toluene and water are immiscible and massive quenching results. Cocktails based on 1,4-dioxane and naphthalene that can accommodate up to 20% (v/v) water can be used, but they have largely been phased out due to toxicity. Emulsifier-based cocktails are the most frequently used for counting aqueous samples. They contain an emulsifier such as Triton X-100 and can accept up to 50 % water (v/v); however, phase transitions occur from single phase to two phase gel, as the water content increases. Accurate counting cannot be done if the samples are in two-phase state. Many ready-made cocktails are on the market and are sold with precise instructions regarding sample conditions.

It should be noted that the efficiency of scintillation counting varies with sample volume, though this is less of a problem in modern counters. Nevertheless, care should be taken that sample vials in a given series of counts contain the same volume of sample and that all instruments calibration is done using the same volume as for experimental samples.

If colour quenching is a problem it is possible to bleach samples before counting. Care should be taken, however, since bleaching agents such as hydrogen peroxide can give rise to chemiluminescence in some scintillation cocktails.

Solid samples, such as plant and animal tissues, may be best counted after solubilisation by quaternary amines such as NCS solubilisers or Soluene. Not surprisingly these solutions are highly toxic and great care is required. The sample is added to the counting vial containing a small amount of solubiliser and digestion is allowed to proceed. When digestion is complete, scintillation cocktail is added and the sample is counted. Again, chemiluminescence can be a problem with tissue solubilisers.

A suitable alternative to bleaching of coloured samples or digestion of tissues is the use of combustion techniques. Here sample are combusted in an atmosphere of oxygen, usually in a commercially available combustion apparatus. Thus samples containing ^{14}C would be combusted to $^{14}CO_2$, which is collected in a trapping agent such as sodium hydroxide and then counted; ^{3}H-containing samples are converted to $^{3}H_2O$ for counting.

As indicated earlier, only important considerations in sample are discussed above and details are not given. However, it is worthy of comment that almost any type of radioactive sample containing β-emitting isotopes can be prepared for counting in a liquid scintillation counter by one method or another, including cuttings from paper chromatographs or membrane filters, again illustrating the versatility and importance of this technique for quantifying radioactivity.

Cerenkov counting

The Cerenkov effect occurs when a particle passes through a substance with a speed higher than that of light passing through the same substance. If a β-emitter has a decay energy in excess of 0.5 MeV, then this causes water to emit a bluish white light usually referred to as Cerenkov light. It is possible to detect this light using a typical liquid scintillation counter.

Since there is no requirement for organic solvents and fluors, this technique is relatively cheap, sample preparation is very easy, and there is no problem in chemical quenching. Table 13.5 lists some isotopes that are suitable for this detection method. Most work has been done on ^{32}P, which has 80% of its energy spectrum above the Cerenkov threshold and which can be detected at around 40% efficiency. It may be noted from Table 14.6 that, as the proportion of the energy spectrum above 0.5 MeV increases, so too does the detection efficiency.

Table 13.5: Some isotopes suitable for Cerenkov counting

Radioisotope	E_{max} (MeV)	% of spectrum above 0.5 MeV	Counting efficiency (%)
^{22}Na	1.39	60	30
^{32}P	1.71	80	40
^{36}Cl	0.71	30	10
^{42}K	3.5	90	80

Scintillation proximity assay

Scintillation proximity assay (SPA) is an application of scintillation counting that facilitates automation and rapid throughput of experiments. It is, therefore, highly suited to work such as screening for biological activity in new drugs. SPA beads are constructed from polystyrene (or sometimes other materials) that combine a binding site for a molecule of interest with a scintillation. You need to remember that some types of radiation do not travel far, in particular β-particles from weak energy emitters such as ^{3}H and ^{14}C. If molecules containing such radioisotopes are in solution with a suspension of SPA beads, the radiation does not stimulate the scintillation in the beads and cannot be detected efficiently by a scintillation counter. This is because the radiation is absorbed by the solution; it does not reach the scintillant. If, on the other hand, the radioisotope becomes bound to the bead, it is close enough to stimulate the scintillant in the bead, so light is given out and the isotope is detected.

There are many applications of this technology such as enzyme assays and receptor binding, indeed any situation where we want to investigate the interaction between two molecules. Take receptor binding as an example. In this case a receptor for a particular ligand (such as a drug or hormone) is attached to the SPA beads. The ligand is radiolabelled and mixed with the beads. Any ligand that binds will stimulate the scintillation and be counted, if the researcher wishes to investigate chemicals that might interface with this binding (which is the mode of action of many medicines), that can be added at increasing concentration to study the effect and, for example, determine optimum dosage.

Advantages of scintillation proximity assay are:

1. Versatile: use with enzyme assays, receptors, any molecular interactions
2. Works with a range of appropriate isotopes as ^{3}H, ^{14}C, ^{35}S and ^{33}P
3. No need for separation step (e.g., free from bound ligand)
4. Less manipulation, therefore, reduced toxicity
5. Amenable to automation.

13.4.3 Methods based upon exposure of photographic emulsions

Ionizing radiation acts upon a photographic emulsion to produce a latent image much as does visible light. For a photograph, a radiation source, an object to be imaged and a photographic emulsion are required. For an autoradiograph, a radiation source (i.e. radioactivity) emanating from within the material to be imaged (the object) is required, along with a sensitive emulsion. The emulsion consists of a large number of silver halide crystal embedded in a solid phase such as gelatin. As energy from the radioactive material is dissipated in the emulsion, the silver halide becomes negatively charged and is reduced to metallic silver, thus forming a particular latent image. Photographic developers are designed to show these silver grains as a blackening of the film, and the fixers remove any remaining silver halide. Thus a permanent image of the location of the original radioactive event remains.

This process, which is known as autoradiography, is very sensitive and has been used in a wide variety of biological experiments. These unusually involve a requirement to locate the distribution of radioactivity in biological specimens of different types. For instance, the sites of localization of radiolabelled drug throughout the body of an experimental animal can be determined by placing whole-body sections of the animal in close contact with a sensitive emulsion such as an X-ray plate. After a period of exposure, the plate, upon development, will show an image of the section in tissues and organs in which radioactivity was present. Similarly, radioactive metabolites isolated and separated by chromatographic or electrophoretic techniques during metabolites studies can be located on the chromatograph and electrophoretograph and the radioactive spots can subsequently be recovered for counting and identification.

The technique of autoradiography has become more important with recent developments in molecular biology. Consequently, more detail is given below on some important aspects of the technique.

Suitable isotopes

In general, weak β-emitting isotopes (e.g., ^3H, ^{14}C and ^{35}S) are most suitable for autoradiography, particularly for cell and tissue localization experiments. This is because, as a result of the low energy of the negatrons, the ionizing track of the isotope will be short and a discrete image will result. This is particularly important when radioactivity associated with subcellular organelles is being located. For this ^3H is the best radioisotope, since its energy will all be quickly dissipated within the emulsion. Electron microscopy can then be used to locate the image in the developed film. For location within whole organisms or tissues, either ^{14}C or ^3H is suitable; more energetic isotopes (e.g., ^{32}P) are less suitable because their higher energy negatrons produce much longer track lengths and result in less discrete images that are not sufficiently discriminatory for microscopic location. Conversely, for location of, for example, DNA bands in an electrophoretic gel, ^{32}P is useful. In this case low energy ^3H negatrons would largely dissipate their energy within the gel (and in the wrapping around the gel, which is usually necessary to prevent the gel sticking to the emulsion), thereby reducing sensitivity to a low level. However, the more energetic ^{32}P negatrons will leave the gel and produce a strong image. If very thin gels are prepared, then ^{35}S or ^{14}C can be detected with high resolution, for example, DNA sequencing gels where ^{35}S is used as the label.

Choice of emulsion and film

A variety of emulsions is available with different pacing densities of the silver halide crystals. Care must be taken to choose an emulsion suitable for the purposes of experiment, since the sensitivity of the emulsion will affect the resolution obtained. Manufacturers' literature should be consulted and their advice sought if one is in any doubt. X-ray film is generally for macroscopic samples such as whole-body sections of small mammals, chromatographs or electrphoretographs. When light or electron microscopic detection of the location of the image in the emulsion is required (cellular and subcellular localization of radioactivity), very sensitive films are necessary, as is a very close apposition of sample and film. In these cases a stripping film technique can be used in which the film is supplied attached to a support. It is stripped from this and applied directly to the sample. Alternatively, liquid emulsions are prepared by melting strips of emulsion by heating them to around 60°C. Then either the emulsion is poured onto the sample or the sample attached to a support is dipped into the emulsion. The emulsion is then allowed to set before being dried. Such a method is often referred to as a dipping-film method and is preferred when very thin films are required.

Background

Accidental exposure to light, chemicals in the sample, natural background radioactivity (particularly ^{40}K in glass) and even pressure applied during handling and storage of film will cause a background fog (i.e., latent image) on the developed film. This can be problematic, in high resolution work (e.g., involving microscopy) and care must be taken at all times to minimize its effect. Background will always increase during exposure time, which should, therefore, always be kept to a minimum.

Time of exposure and film processing

The time of exposure depends upon the isotope, sample type, level of activity, film type and purpose of the experiment. The same applies to the processing of the film in order to display the image. Generally the process must be adapted to a given purpose, and a great deal of trial and error is often involved in arriving at the most suitable procedures.

Direct autoradiograph

In direct autoradiography, the X-ray film or emulsion is placed as close as possible to the sample and exposed at any convenient temperature. Quantitative images are produced until saturation is reached. The approach provides high resolution but limited sensitivity; isotopes of energy equal to or higher than ^{14}C (E_{max} = 0.156 MeV) are required.

Fluorography

Many of the currently popular methods in molecular biology involve separation of macromolecules or fractions of macromolecules by gel electrophoresis. The separated macromolecules or fractions from bands in the electrophoretograph that must be located. This is often achieved by radiolabelling the macromolecules with ^3H or ^{14}C and subjecting the gel to autoradiography. Because these are weak β-emitters, much of their energy is lost in the gel and long exposure times are necessary even when very high specific activity source are used. However, if a fluor (e.g., PPO or sodium silicate) is infiltrated into the gel, and the gel dried and then placed in contact with a preflashed film, sensitivity can be increased by several orders of magnitude. This is because the negatrons emitted from the isotope will cause the fluor to become excited and emit light, which will react with the film. Thus use is made of both the ionising and the excited effects of radioactivity in fluorography.

Intensifying screens

When ^{32}P-labelled or γ-isotope-labelled samples (e.g., [^{32}P] DNA or ^{128}I-labelled protein fractions in gels) are to be located, the opposite problem to that presented by low energy isotopes prevails. These much more penetrating particles and rays cause little reaction with the film as they penetrate right through it, producing a poor image. The image can be greatly improved by placing, on the other side of the film from the sample, a thick intensifying screen consisting of a solid phosphor. Negatrons penetrating the film cause the phosphor to fluoresce and emit light, which superimposes its image on the film. There is, therefore, an increase in sensitivity but a parallel reduction in resolution due to the spread of light emanating from the screen.

Low temperature exposure

If the energy of ionizing radiation is converted to light (i.e., with fluorography or intensifying screens) the kinetics of the film's response are affected. The light is of low intensity and a back reaction occurs that cancels the forming of latent image. Exposure at low temperature (–70°C) slows this back reaction and will, therefore, provide higher sensitivity. There is no point in doing direct autoradiography at low temperature as the kinetics of the film response are different. There is nothing to be gained by exposing preflashed film at low temperature.

Preflashing

As described above, the response of a photographic emulsion to radiation is not linear and usually involves a slow initial phase (lag) followed by a linear phase. Sensitivity of the phase may be increased by preflashing. This involves a millisecond light flash prior to the sample being brought into juxtaposition with the film and is often used where high sensitivity is required or if results are to be quantified.

Quantification

As indicated earlier, autoradiography is usually used to locate rather than to quantify radioactivity. However, it is possible to obtain quantitative data directly from autoradiographs by using a densitometer, which records the intensity of the image. This in turn is related to the amount of radioactivity in the original sample. There are many varieties of densitometers available and the choice made will depend on the purpose of the experiment. Quantification is not reliable at low or high levels of exposure because of the lag phase (i.e., the back reaction, as described above) or saturation, respectively; however, preflashing combined with fluorography or intensifying screens obviates the problem for small amount of radioactivity. In this case all photons contribute equally to the image of the pre-exposed film.

13.5 INHERENT ADVANTAGES AND RESTRICTIONS OF RADIOTRACER EXPERIMENT

Perhaps the greatest advantage of radiotracer methods over other chemical and physical methods is their sensitivity. For example, a dilution factor of 10^{12} can be tolerated without the detection of ^{3}H-labelled compounds being jeopardised. It is thus possible to detect the occurrence of metabolic substances that are normally present in tissues at such low concentrations as to defy the most sensitive chemical method of identification. A second major advantage of using radiotracers is that they enable studies in vivo to be carried out to a far greater degree than can any other technique.

In spite of these significant advantages, certain restrictions have to be appreciated. First, although they undergo the same reactions, different isotopes may do so at different rates. This effect is known as the isotope effect. The different rates are approximately proportional to the differences in mass between the isotopes. The extreme case is the isotopes ^{1}H and ^{3}H, the effect being small for ^{12}C and ^{14}C and almost insignificant for ^{33}P and ^{32}P. Secondly, the amount of activity employed must be kept to the minimum necessary to permit reasonable counting rates in the samples to be analyzed, otherwise the radiation from the tracer may elicit response from the experimental organism and hence distort the results. A third consideration is that, in order to administer the tracer, the normal chemical level of the compound in the organism is automatically exceeded. The results are, therefore, always open to question.

Table 13.6: The relative merits of commonly used b-emitters

Isotope	Advantages	Disadvantages
^{3}H	Safety High specific activity possible Wide choice of positions in organic compounds Very high resolution in autoradiography	Low efficiency of detection Isotope exchange with environment Isotope effect
^{14}C	Safety Wide choice of labelling position in organic compounds Good resolution in autoradiography	Low specific activity
^{32}S	High specific activity Good resolution in autoradiography	Short half-life Relatively long biological half-life
^{33}P	High specific activity Good resolution in autoradiography Less hazardous than ^{32}P	Lower specific activity than ^{32}P Less sensitive than ^{32}P Cost
^{32}P	Ease of detection Short half-life simplifies disposal Cerenkov counting	Short half-life affects costs and experimental design External radiation hazard Poor resolution in autoradiography

13.6 OTHER PRINCIPLE ASPECTS OF COUNTING RADIOACTIVITY AND ANALYSIS OF DATA

Assessing the presence and amount of radioactivity is invariably accomplished by radiations emitted. The difference between amount of radiations emitted and detected actually depends on number of factors which can be classified into two categories as follows.

1. Counter characteristic

(a) **Background count:** Radiation counter of all types always register an "unwanted" background count, even in the absence of radioactive material in the apparatus. This may be due to such sources as cosmic radiation, natural radioactivity in the vicinity, nearby X-ray generators, and/or circuit noises and may interfere in a measurement. By means of the various methods already outlined and the use the lead shielding, this background radiation may be considerably reduced, but its values must always be recorded and accounted for in all experiments. Some commercial experiments have automatic background subtraction facilities.

(b) **Dead time:** Detector systems (like GM Counter) designed for quantitative measurement has electronic equipment to count and record the number of pulses received from detector. All such equipments require a finite time to record a pulse, and during this time it is "dead" to any other pulse which may arrive. Thus at very high count rates in Geiger-Muller counting, counts are lost due to the dead time of the Geiger-Muller tube. Correction tables are available and those should be used when necessary to correct for lost counts. Dead time is not a problem in scintillation counting.

(c) **Geometry:** Those factors which determine whether the radiation misses or hits the detector are collectively referred to as counting geometry. These factors are:

1. Geometric relationship between the source of the radiation and detector.
2. The presence of material which absorbs or scatters the emitted radiation.

Therefore, when samples with an end-window ionization counter, such as a Geiger-Muller tube, are compared, it is important to standardize the position of the sample in relation to the tube, otherwise the fraction of the emitted radiation entering the tube may vary and hence will vary the observed count.

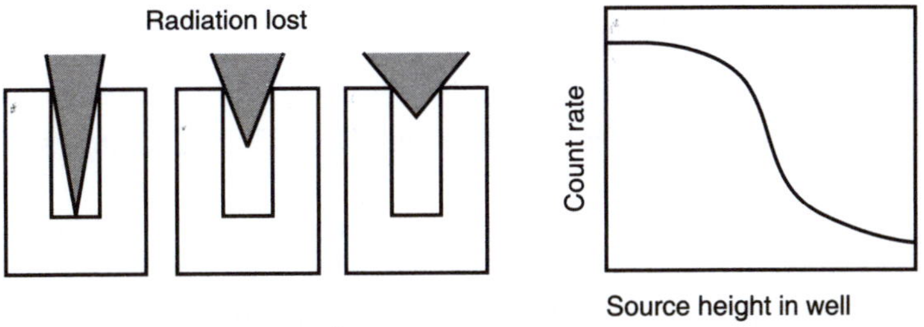

Fig. 13.23: Variation in count rate with source position

2. Sample and isotope characteristic

(a) **Self-absorption:** Self-absorption is primarily a problem with low energy β-emitters; radiation is absorbed by the sample itself. Self-absorption can be a serious problem in the counting of low energy radioactivity by scintillation counting if the sample is particulate or is, for instance, stuck to membrane filter. Care should be taken to ensure comparability of samples because the methods of standardization outlined earlier will not correct self-absorption effects. When homogeneity is not possible, particulate samples should be digested or otherwise solubilised prior to counting. Self-absorption is a major problem with Geiger-Muller counting and significantly reduces sensitivity and reliability. It is very difficult to count low energy emitters reliably with these counters and this was a major factor in the switch to scintillation counting.

(b) **Half-life:** The half-life of an isotope may be short and, if so, this must be allowed for analysis of the data.

(c) **Statistics:** The emission of radioactivity is a random process. This can be demonstrated readily by making repeated measurements of the activity of long-lived isotope, each for an identical period of time. The resulting counts will not be the same but will vary over a range of values, with clustering near the centre of the range. If a sufficiently large number of such measurements is made and the data are plotted, a normal distribution curve will be obtained. For a single, therefore, we cannot obtain a true count. Instead, we take the mean of a large number of counts as being very close to the true count. However, the accuracy of this mean will depend on the spread or standard deviation (σ) for the data. Statistical theory states that, for a normal distribution 68.2% of values obtained lie $\pm 1\sigma$, and 95.5% lie $\pm 2\sigma$ from the mean ($\bar{x}$).

Clearly, if we wish to compare samples, and in particular to state that two samples contain different amounts of radioactivity, then we need to take account of the counting statistics. Fortunately Poisson mathematics makes the task relatively easy as

$$\sigma = \sqrt{\text{total count taken}}$$

or,

$$\sigma = \sqrt{\frac{\text{count rate}}{\text{time}}}.$$

Therefore, to quote a figure for normal distribution with 95.5% certainty, state:

$$\text{total counts} \pm 2\sqrt{\text{total counts}}.$$

For example, if 1600 counts are recorded, this can be expressed as 1600 ± 80. There is, therefore, a 95.5% chance that the true figure lies between 1520 and 1680.

When data are expressed as d.p.m. or c.p.m, again using 95.5% certainty, then error on

$$\text{count rate} = 2\sqrt{\left(\frac{\text{count rate}}{\text{time}}\right)}$$

Using the sample example, if the 1600 counts were obtained in 1 min:

$$\text{error on rate} = 2\sqrt{\left(\frac{1600}{1}\right)} = 80$$

and, therefore, 1600 c.p.m. ± 80 c.p.m.

If the 1600 counts were obtained in 10 min:

$$\text{error on count rate} = 2\sqrt{\left(\frac{160}{1}\right)} = 8$$

and therefore 160 c.p.m. ± 8 c.p.m. Note that the error is the same. This is because effectively the same number of measurements, i.e., counts, has been taken in each case.

If we had recorded 160 counts in only 1 min, then:

$$\text{error on count rate} = 2\sqrt{\left(\frac{160}{1}\right)} = 25$$

and, therefore, 160 c.p.m. ± 25 c.p.m.

Consider other simple examples for a series of 1 min counts:

counts = 100 $\sigma = \sqrt{100}$, therefore, ± 10% error at 68.2% certainty

counts = 1000 $\sigma = \sqrt{1000}$, therefore, ± 3% error at 68.2% certainty

counts = 10000 $\sigma = \sqrt{10000}$, therefore, ± 1% error at 68.2% certainty.

In summary, the counts per minute data become more accurate for higher count rates and/or longer counting times. It is a common practice to count 10000 counts or for 10 min, whichever is the quicker, although for very low count rates longer counting times are required.

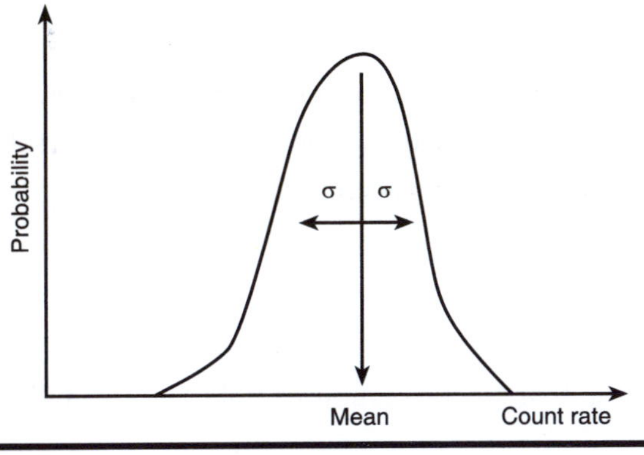

Fig. 13.24: Distribution of count rate around a mean, showing standard deviation

3. Supply, storage and purity of radiolabelled compound

There are several suppliers of radiolabelled compounds, the main ones being Amersham Biosciences plc. Du Pont, NEN and ICN. The suppliers usually include details of the best storage conditions and quality control data with their production. This is because several types of decompositions can occur; for example internal decomposition resulting from radioactive decay such as $^{14}C \rightarrow {}^{14}N$, and external decomposition where emitted radiation is absorbed by other radioactive molecules, causing impurities. The extent to which decomposition occurs is dependent on many factors such as temperature, energy of radiation, concentration and the formulation of the compound. It is, therefore, imperative to store radioisotopes by the method recommended by the supplier and to maintain sterility of the stock. If necessary, chromatographic procedures will be required to check on the purity of the labelled compound.

4. Specific activity

The specific activity of a radioisotope defines its radioactivity related to the amount of material (e.g., Bq mol^{-1}, Ci $mmol^{-1}$ or d.p.m. μmol^{-1}). Suppliers offer a range of specific activities for their compound, the highest often being the most expensive. The advantages of using a very high specific activity are as follows:

- Products of a reaction using the labelled precursor can be produced at high specific activity (e.g., for DNA probes).
- Small quantities of radiolabelled compound can be added such that the equilibrium of metabolic concentrations is not unduly pertubated.
- Calculating the amount of substance required to make up radioactive solutions of known specific activity is simplified, as the contribution to concentration made by the stock radiolabelled solution is often negligible.

Sometimes, however, it is not necessary to purchase the highest specific activity available. For example, enzyme assays in vitro often require a relatively high substrate concentration and so specific activity may need to be lowered. Consider the example below:

[^{3}H]Leucine is purchased with a specific activity of 5.55 TBq $mmol^{-1}$ (150 Ci $mmol^{-1}$) and a concentration of 9.25 MBq 250 mm^{-3} (250 μCi 250 mm^{-3}). A 10 cm^3 solution of 250 mM and 3.7 kBq cm^{-3} (0.1μCi cm^{-3}) is required. It is made up as follows:

- 10 cm^3 at 3.7 kBq cm^{-3} is 37 kBq (1μCi), therefore, pipette 1 mm^3 of stock radioisotope into a vessel (or, to be more accurate, pipette 100 mm^3 of a × 100 dilution of stock in water).
- Add 2.5 cm^3 of a 1 M stock solution of cold leucine, and make up to 10 cm^3 with distilled water.

There is no need to take into account the amount of unlabelled leucine in the [^{3}H] leucine preparation; it is a negligible quantity due to the high specific activity. If necessary (e.g., to manipulate solutions of relatively low specific activity), however, the following formula can be applied:

$$W = Ma\,[(1/A') - (1/A)],$$

where, W is the mass of cold carrier required (mg), M is the amount of radioactivity present (M Bq), a is the molecular weight of the compound, A is the original specific activity (M Bq $mmol^{-1}$), and A′ is the required specific activity (M Bq $mmol^{-1}$).

5. The choice of radionuclide

This is the complex question depending on the precise requirements of the experiment. A summary of some of the key features of radioisotopes commonly used in biological work is shown in the table in above sections.

13.7 PRECAUTIONARY MEASURES WITH RADIOISOTOPES

The greatest practical disadvantages of using radioisotopes is their toxicity; they produce ionizing radiations. When absorbed, radiation causes ionisation and free radicals from that interact with the cell's macromolecules, causing mutation of DNA and hydrolysis of proteins. The toxicity of radiation is dependent not simply on the amount present but on the amount of absorbed radiation and its biological effect. There are, therefore, a series of additional units used to describe these parameters. Originally, radiation hazard was measured in terms of exposure, i.e., a quantity expressing the amount of ionization in air. The unit of exposure is the roentgen (R), which is the amount of radiation that produces 1.61×10^{15} ion-pairs (kg air)$^{-1}$ (or 2.58×10^{-4} coulombs (kg air)$^{-1}$).

The amount of energy required to produce an ion-pair in air is 5.4×10^{-18} joules (J) and so the amount of energy absorbed by air with an exposure of 1 R is:

$$1.61 \times 10^{15} \times 5.4 \times 10^{-18} = 0.00869 \text{ J (kg air)}^{-1}.$$

Although the roentgen has been used as a unit of radiation hazard, it is now considered inadequate for two reasons: first, it is defined with reference to X-rays (or γ-rays) only; and, secondly, the amount of ionization or energy absorption in different types of material, including living tissue, is likely to be different from that in air.

The concept of radiation absorbed dose (rad) was introduced to overcome these restrictions. The rad is defined as the dose of radiation that gives an energy absorption of 0.01 J (kg absorber)$^{-1}$; this has now been changed to the gray, an SI unit, representing absorption of 1 J kg^{-1} (i.e., 100 rads).

The gray (Gy) is a useful unit, but it still does not adequately describe the hazard to living organisms. This is because different types of radiation are associated with differing degrees of biological hazard. It is, therefore, necessary to introduce a correction factor, known as the weighting factor (W), which is calculated by comparing the biological effects of any type of radiation with that of X-rays. The unit of absorbed dose, which takes into account the weighting factor is the sievert (Sv) and is known as the equivalent dose. Thus:

$$\text{equivalent dose (Sv)} = \text{Gy} \times \text{W}.$$

The majority of isotopes used in biological research emit β-radiation. This is considered to have a biological effect that is very similar to X-rays and has a weighting factor of 1. Therefore, for β-radiation, Gy = Sv. Alpha particles, with their stronger ionizing power, are much more toxic and have a weighting factor of 20. Therefore, for α-radiation, 1 Gy = 20 Sv. It is likely that, as our knowledge of the biological effectiveness of different forms of radiation progresses, so the quality factor for different types of radiation may change in the future. Absorbed dose from known sources can be calculated from knowledge of the rate of decay of the source, the energy of radiation, the penetrating power of the radiation and the distance between the source and the laboratory-worker. As the radiation is emitted from a source in all directions, the level of irradiation is related to the area of sphere, $4\pi r^2$. Thus the absorbed dose is inversely related to the square of the distance from the source (r); or, if put another way, when the distance is doubled, the dose is quartered.

A useful formula is

$$\text{dose}_1 \times \text{dis tan ce}_1^2 = \text{dose}_2 \times \text{dis tan ce}_2^2$$

The rate at which dose is delivered is referred to as the dose rate, expressed in Sv h^{-1}. It can be used to calculate your total dose. For example, a source may be delivering 10 μSv h^{-1}. If you work with the source for 6 h, your total dose would be 60 μSv.

Currently the dose limit for workers exposed to radiation is 15 mSv in a year to the whole body, but thus is rarely ever approached by biologists because the levels of radiation used are so low. Limits are set for individual organs. The most important of these to know are for hands (500 mSv $year^{-1}$) and for lens of the eye (150 mSv $year^{-1}$). Dose limits are constantly under review and, although dose limits are set, it is against internationally agreed guidelines to work such a limit, i.e., to assume that all is satisfactory if the limit is not exceeded. Instead, the ALARA principle is applied, to work always to a dose limit that is as low as reasonably achievable. Work that may cause a worker to exceed three-tenths or one-tenth of the dose limit must be carried out in a controlled area or a supervised area, respectively. In practice, work in the biosciences rarely involves a worker receiving a measurable dose. Supervised areas are common but not always required (e.g., for ^3H or ^{14}C experiment). Controlled areas are required in only certain circumstances, for example, for isotopes or radioiodination work. A major problem, however, in biosciences is the internal radiation hazard. This is caused by radiation entering the body, for example by inhalation, ingestion, absorption or puncture. This is a likely source of hazard where work involves manipulations of radioactive liquids and gases; most works in biology involves manipulations of radioactive liquids. Control of contamination is assisted by

- complying with local rules, written by an employer,
- conscientious personal conduct in the laboratory,
- regular monitoring,
- carrying out work in some kind of contaminant.

Calculating the dose received following the ingestion of a radioisotope is complex. Detailed information is published by the International Commission on Radiological Protection and assessments, for example, for experiments on human volunteers, can be obtained from the National Radiological Protection Board. However, one relatively simple concept is the annual limit on intake (ALI). The ingestion of one ALI results in a person receiving a dose limit to the whole body or to a particular organ. Some ALIs are shown in Table 13.7. Management of radiation protection is similar in most countries. In the USA, there is a Code of Federal Regulations. In the UK there is the Radioactive Substances Act (1993) and the Ionizing Radiations Regulations (1999). Every institution requires certification (monitored by the Environmental Protection Agency in the USA or the Environmental Agency in the UK) and employs a Radiation Protection Advisor.

When handling radioisotopes the rule is to:

- maximize the distance between yourself and the source,
- minimize the time of exposure and
- maintain shielding at all times.

Table 13.7: Annual limits on intake (ALI) for some commonly used isotopes

Nuclide	ALI (MBq)
^{3}H	480
^{14}C	34
^{32}P	6.3
^{125}I	1.3

13.8 APPLICATIONS OF RADIOISOTOPES IN THE BIOLOGICAL SCIENCE

13.8.1 Investigating aspects of metabolism

1. Metabolic pathways

Radioisotopes are frequently used for tracing metabolic pathways. This usually involves adding a radioactive substrate, taking samples of the experimental material at various times, extracting and chromatographically or otherwise separating the products. Radioactivity detectors can be attached to gas-liquid chromatography or high performance liquid chromatography columns to monitor radioactivity coming off the column during separation. Alternatively, radioactivity can be located on paper or thin-layer chromatography with either a Geiger-Muller chromatograph scanner or with autoradiography. If it is suspected that a particular compound is metabolized by a pathway, then radioisotope can also be used to confirm this. For instance, it is possible to predict the fate of individual carbon atoms of [^{14}C] acetate through the tricarboxylic cycle, or Krebs cycle. Methods have been developed whereby intermediates can be ascertained. This is the so-called specific labelling pattern. Should the actual pattern coincide with the theoretical pattern, then this is a very good evidence for the mode of operation of the Krebs cycle.

Another example of the use of radioisotopes to confirm the mode of operation, or otherwise, of a metabolic pathway is in studies carried out on glucose catabolism. There are numerous ways whereby glucose can be oxidized, the two most important ones in aerobic organisms being glycolysis followed by Krebs cycle together with the pentose phosphate pathway. Frequently, organisms or tissues posses the necessary enzymes for both pathways to occur and it is of interest to establish the relative contribution of each to glucose oxidation. Both pathways involve the complete oxidation of glucose to carbon dioxide, but the origin of the carbon dioxide in terms of the six carbon atoms of glucose is different (at least in the initial stages of respiration of exogenously added substrate), thus it is possible to trap the carbon dioxide evolved during the respiration of specifically labelled glucose (e.g., [$^{6\text{-}14}C$] glucose or [$^{1\text{-}14}C$] glucose in which only the C-6 atom is radioactive) and obtain an evaluation of the contribution of each pathway to glucose oxidation.The use of radioisotopes in studying the operation of Krebs cycle or in evaluating the pathway of glucose catabolism are just two examples of how such isotopes can be used to confirm metabolic pathways.

2. Metabolic turnover times

Radioisotopes provide a convenient method of ascertaining turnover times for particular compounds. As an example, the turnover of proteins in rats will be considered. A group of rats is injected with a radioactive amino acid and left for 24 h, during which time most of the amino acid is assimilated into proteins. The rats are then killed at suitable time intervals and radioactivity in organs or tissues of interest is determined. In this way it is possible to ascertain the rate of metabolic turnover of protein. Using this sort of method, it has been shown that liver protein is turned over in 7-14 days, while skin and muscle proteins are turned over every 8-12 weeks, and collagen is turned over at a rate of less than 10% per annum.

3. Studies of absorption, accumulation and translocation

Radioisotopes have been very widely used in this study of the mechanisms and rates of absorption, accumulation and translocation of inorganic and organic compounds by both plants and animals. Such experiments are generally simple to perform and can also yield evidence on the route of translocation and sites of accumulation of molecules of biological interest.

4. Pharmacological studies

Another field where radioisotopes are widely used is in the development of new drugs. This is a particularly complicated process, because, besides showing whether a drug has a desirable effect, much more must be ascertained before it can be used in the treatment of clinical conditions. For instance, the site of drug accumulation, the rate of accumulation, the rate of metabolism and the metabolic products must all be determined. In each of these areas of study, radiotracers are extremely useful, if not indispensable. For instance, autoradiography on whole sections of experimental animals yields information on the site and rate of accumulation, while typical techniques used in metabolic studies can be used to follow the rate and products of metabolism.

13.8.2 Analytical applications

1. Enzyme and ligand binding studies

Virtually any enzyme reaction can be assayed using radiotracer methods provided that a radioactive form of the substrate is available. Radiotracer-based enzyme assays are more expensive than other methods, but frequently have the advantage of a higher degree of sensitivity. Radioisotopes have also been used in the study of the mechanism of enzyme action and in the studies of ligand binding to membrane receptors.

2. Isotope dilution analysis

There are many compounds present in living organisms that cannot be accurately assayed by conventional means because they are present in such low amounts and in mixtures of similar compounds. Isotope dilution analysis offers a convenient and accurate way of overcoming this problem and avoids the necessity of quantitative isolation. For instance, if the amount of iron in a protein preparation is to be determined, this may be difficult using normal methods, but it can also be done if a source of ^{59}Fe is available. This is mixed with the protein and a sample of iron is subsequently isolated, assayed for total iron and the radioactivity is determined.

If the original specific activity was 10000 d.p.m. (10 mg)$^{-1}$ and the specific activity of the isolated iron was 9000 d.p.m (10 mg)$^{-1}$, then the difference is due to the iron in the protein (x), i.e.,

$$\frac{9000}{1} = \frac{10000}{10 + x}$$

$$\therefore \qquad x = 1.1 \text{ mg.}$$

This technique is widely used, for instance, in studies on trace elements.

3. Radioimmunoassay

One of the most significant advances in biochemical technique in recent years has been the development of the radioimmunoassay. This technique is discussed in Section 7.7 and is not elaborated upon here.

4. Radiodating

A quite different analytical use for radioisotopes is in the dating (i.e., determining the age) of rocks, fossils and sediments. In this technique it is assumed that the proportion of an element that is naturally radioactive has been the same throughout time. From the time of fossilization or deposition the radioactive isotope will decay. By determining the amount of radioisotope remaining (or by examining the amount of the decay product) and from the knowledge of the half-life, it is possible to date the sample. For instance, if the radioisotope normally comprises 1% of the element and it is found that the sample actually contains 0.25%, then two half-lives can be assumed to have elapsed since deposition. If the half-life is one million years then the sample can be dated as being two million years old.

13.8.3 Other applications

1. Molecular biology techniques

Recent advances in molecular biology that have led to advances in genetic manipulation have dependent heavily upon use of radioisotope in DNA and RNA sequencing, DNA replication, transcription, synthesis of complementary DNA, recombinant DNA technology and many similar studies. Many of these techniques are more fully discussed in other chapters of the book.

2. Clinical diagnosis

Radioisotopes are very widely used in medicine, in particular for diagnostic tests. Lung function tests routinely made using xenon-133 (^{133}Xe) are particularly useful in diagnosis of malfunctions of lung ventilation. Kidney function tests using [133] iodohippuric acid are used in diagnoses of kidney infection, kidney blockages or imbalance of function between the two kidneys. Various aspects of haematology are also studied by using radioisotopes. These include such aspects as blood cell lifetimes, blood volumes and blood circulation times, all of which may vary in particular clinical conditions.

3. Ecological studies

The bulk of radiotracer work is carried out in biochemical, clinical or pharmacological laboratories; nevertheless, radiotracers are also of use to ecologists. In particular, migratory patterns and behaviour patterns of many animals can be monitored using radiotracers. Another ecological application is in the examination of food chains where the primary producers can be made radioactive and the path of radioactivity followed throughout the resulting food chain.

4. Sterlisation of food and equipment

Very strong γ-emitters are now widely used in the food industry for sterlisation of prepacked foods such as milk and meats. Normally either ^{60}Co or ^{137}Ce is used, but care has to be taken in some cases to ensure that the food product itself is not affected in any way. Thus doses often have to be reduced to an extent where sterlisation is not complete but nevertheless food spoilage can be greatly reduced. ^{60}Co and ^{137}Ce are also used in sterlisation of plastic disposable equipment such as Petridishes and syringes, and in sterlisation of drugs that are administered by injection.

5. Mutagens

Radioisotopes may cause mutations, particularly in micro-organisms. In various microbiological studies mutants are desirable, especially in industrial microbiology. For instance, development of new strains of a micro-organism that produce higher yields of a desired microbial product frequently involve mutagenesis by radioisotopes.

13.9 TIPS TO REMEMBER

Isotopes

- species of the same chemical element with different atomic masses (i.e., different numbers of neutrons)
- some isotopes are unstable and tend to rearrange or disintegrate
- disintegration involves the release of subatomic particles or electromagnetic radiation
- radioactive isotopes are called radioisotopes or radionuclides.

Specific Activity

- radioactivity per unit mass (e.g., Bq/mmole)
- reflects mixture of radioactive and nonradioactive isotopes.

Radiochemical Purity

- % of radioactivity in compound of interest.

Types of Radiation Detectors

Physical composition

- Gas-filled detectors
- Solid-state (semiconductor) detectors
- Organic scintillators (liquid & plastic)
- Inorganic scintillators.

Scintillators operate with a photo-sensor (i.e., another detector)

In gas-filled detectors, the detector measures the electrical charge or current resulting from ionization of the gas by the radiation. Gas-filled detectors maintain an electric field within the gas. The field causes electrons and ions created by the incident radiation to migrate to their respective collecting electrodes.

In scintillation detectors, the detector measures light flashes produced in the scintillator by the radiation. The light sensitive portion of the photomultiplier is its photocathode. The photocathode converts incident light photons into low-energy electrons. The electron mutiplier section of the photomultiplier tube acts as an amplifier boosting the electrical signal produced by these electrons sufficiently to be measured.

The amount of radioactivity in a source is measured by the rate at which atoms undergo radioactive disintegration. The natural unit for this is disintegrations per second (dps) and, in honour of the discoverer of radioactivity A.H. Becquerel, this has been given the special unit name of the becquerel. One becquerel (written 1 Bq) is equal to 1 disintegration per second (1 dps), but we commonly encounter much larger quantities, so we use the following:

1 dps	= one becquerel, written as 1 Bq
1 thousand dps	= 103 becquerels = one kilobecquerel, written as 1 kBq
1 million dps	= 106 becquerels = one megabecquerel, written as 1 MBq
1 billion dps	= 109 becquerels = one gigabecquerel, written as 1 GBq

Scintillation proximity assay (SPA) is an application of scintillation counting that facilitates automation and rapid throughput of experiments. It is, therefore, highly suited to work such as screening for biological activity in new drugs. SPA beads are constructed from polystyrene (or sometimes other materials) that combine a binding site for a molecule of interest with a scintillation.

13.10 VIVA VOCE

Q.1. What are radioisotopes?

Q.2. What is radioactivity?

Q.3. What is scintillation proximity assay?

Q.4. How radioactivity is detected and measured?

Q.5. What are the factors which effect the radioactivity detection and measurement?

Q.6. What is half-life and how does it affect radioactivity measurement?

Q.7. What is the principle of gas-filled ionization chambers?

Q.8. What is PMT and what is its light sensitive region?

Q.9. What are scintillation counters?

Q.10. Differentiate between solid scintillation and liquid scintillation counters.

Q.11. What are semiconductor scintillation counters?

Q.12. Explain the phenomenon of electron capture.

Q.13. What is GM counter?

Q.14. Explain dead time.

Q.15. How can you reduce background radiations?

Q.16. What is autoradiography?

Q.17. What are the precautionary measures which must be taken while working with radioisotopes?

Q.18. What is one becquerel?

Q.19. List advantages and disadvantages of scintillation counting.

Q.20. List applications of radioactive isotopes.

CHAPTER 14

Protein Purification

14.1 INTRODUCTION

Protein purification is a series of processes intended to isolate a single type of protein from a complex mixture. Protein purification is vital for the characterization of the function, structure and interactions of the protein of interest. The starting material is usually a biological tissue or a microbial culture. The various steps in the purification process may free the protein from a matrix that confines it, separate the protein and non-protein parts of the mixture, and finally separate the desired protein from all other proteins. Separation of one protein from all others is typically the most laborious aspect of protein purification. Separation steps exploit differences in protein size, physico-chemical properties and binding affinity.

Purification may be *preparative* or *analytical.* Preparative purifications aim to produce a relatively large quantity of purified proteins for subsequent use. Examples include the preparation of commercial products such as enzymes (e.g., lactase), nutritional proteins (e.g., soy protein isolate), and certain biopharmaceuticals (e.g., insulin). Analytical purification produces a relatively small amount of protein for a variety of research or analytical purposes, including identification, quantification, and studies of the protein's structure, post-translational modifications and function. Among the first purified proteins were urease and Concanavalin-A.

14.2 PRINCIPLE STRATEGIES

Most purification protocols require more than one step to achieve desired level of purity. Each step in the process will cause some loss of product, a yeild of 80% in each step is assumed, and therefore, it is advisable to have as few purification steps as possible. Choice of a starting material is key to the design of a purification process. In a plant or animal, a particular protein usually is not distributed homogeneously throughout the body; different organs or tissues have higher or lower concentrations of the protein. Use of only the tissues or organs with the highest concentration decreases the volumes needed to produce a given amount of purified protein. If the protein is present in low abundance, or if it has a high value, scientists may use recombinant DNA technology to develop cells that will produce large quantities of the desired protein (this is known as an expression system). Recombinant expression allows the protein to be tagged, e.g., by a His-tag, to facilitate purification, which means that the purification can be done in fewer steps. In addition to this recombinant expression usually starts with a higher fraction of the desired protein than is present in a natural source.

With background information, assays and sample procedures in place the three phase purification strategies can be considered. The purification has three phases of capture, intermediate purification and polishing, each with specific objective. In capture phase the objectives are to isolate, concentrate and stabilize the target procedure. During intermediate purification phase the objective is to remove bulk impurities such as other proteins, nucleic acids, endotoxins and viruses. In polishing phase the objective is to remove any trace impurities and closely related substances. Therefore selection and optimum combination of purification techniques for capture, intermediate purification and polishing is necessary to ensure good yeild and pure product.

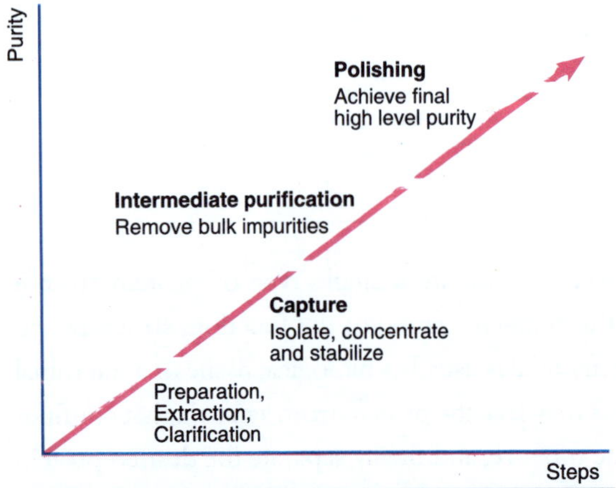

Fig. 14.1: Preparation and three-phase purification strategy

The final purification process ideally consists of sample preparation, including extraction and clarification when required followed by above described three phases of purification. The number of steps will always depend on purity required and intended use of protein.

An analytical purification generally utilizes three properties to separate proteins. First, proteins may be purified according to their isolectric points by running them through a pH graded gel or an ion exchange column. Second, proteins can be separated according to their size or molecular weight via size exclusion chromatography or by SDS-PAGE (sodium dodecyl sulphate-polyacrylamide gel electrophoresis) analysis. Proteins are often purified by using 2D-PAGE and are then analysed by peptide mass fingerprinting to establish the protein identity. This is very useful for scientific purposes and the detection limits for protein are nowadays very low and nanogram amounts of protein are sufficient for their analysis.

14.3 SELECTION AND COMBINATION OF PURIFICATION TECHNIQUES

The aim of this combination is to evolve a fastest route to a product of required purity. For any chromatographic separation each different technique will offer a different performance with respect to recovery, resolution, speed and capacity. A technique can be optimized to focus on one of these parameters; for example, resolution to achieve the best between two parameters such as speed and capacity. A separation optimized for one of these parameters will produce results quite different in appearance from those produced using the same technique but focused on alternative parameters. Therefore, it is always preferable to select a technique to meet the objectives in purification step.

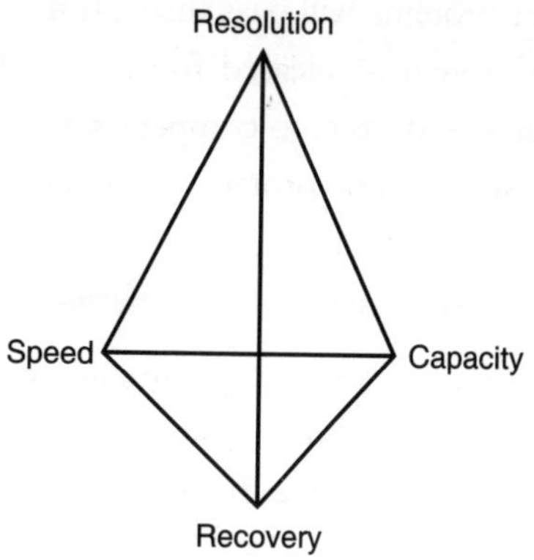

Fig. 14.2: Selection and combination of purification techniques so as to obtain a balance between resolution, speed, capacity and recovery

Capacity in the simple model shown represents the amount of target protein loaded during the purification. In some cases the amount which can be loaded may be limited by volume (as in gel filtration) or by large volume of contaminants rather than the amount of target proteins.

Speed is of highest importance at the beginning where contaminants like proteases must be removed as quickly as possible.

Recovery becomes increasingly important as the purification progresses because of the increased value of the purified product. The recovery is influenced by destructive processes in the samples and unfavourable conditions on the column.

Resolution is achieved by the selectivity of the technique and efficiency of the chromatographic matrix to produce narrow peaks. In general, resolution is most difficult to achieve in final phases of purification where target protein and impurities have very similar properties.

Every technique offers a balance between capacity, speed, resolution and recovery and should be selected to meet the objectives of each purification step. In general, the optimization of any of these parameters can be achieved only at the expense of other and purification step will be a compromise. The importance of each parameter will vary depending upon whether a purification step is used for capture, intermediate purification or polishing. This will steer the optimization of critical parameters as well as for the selection of suitable media for the step.

14.4 PURIFICATION OF A TAGGED PROTEIN

Adding a tag to the protein gives the protein a binding affinity it would not otherwise have. Usually the recombinant protein is the only protein in the mixture with this affinity, which aids in separation. The most common tag is the Histidine-tag (His-tag) that has affinity towards nickel or cobalt ions. Thus by immobilizing nickel or cobalt ions on a resin, an affinity support that specifically binds to histidine-tagged proteins can be created. Since the protein is the only component with a His-tag, all other proteins will pass through the column, and leave the His-tagged protein bound to the resin. The protein is released from the column in a process called elution, which in this case involves adding imidazole, to compete with the His-tags for nickel binding, as it has a ring structure similar to histidine. The protein of interest is now the only protein component in the eluted mixture, and can easily be separated from any minor unwanted contaminants by a second step of purification, such as size exclusion chromatography or RP-HPLC.

Another way to tag proteins is to add an antigen peptide to the protein, and then purify the protein on a column containing immobilized antibody. This generates a very specific interaction usually only binding the desired protein.

When the tags are not needed anymore, they can be cleaved off by a protease. This often involves engineering a protease cleavage site between the tag and the protein.

14.5 EVALUATING PURIFICATION YIELD

The most general method to monitor the purification process is by running a SDS-PAGE of the different steps. This method only gives a rough measure of the amounts of different proteins in the mixture, and it is not able to distinguish between proteins with similar molecular weight.

If the protein has a distinguishing spectroscopic feature or an enzymatic activity, this property can be used to detect and quantify the specific protein, and thus to select the fractions of the separation, that contains the protein. If antibodies against the protein are available, then western blotting and ELISA can specifically detect and quantify the amount of desired protein. Some proteins function as receptors and can be detected during purification steps by a ligand binding assay, often using a radioactive ligand.

In order to evaluate the process of multistep purification, the amounts of the specific protein have to be compared to the amount of total protein. The latter can be determined by the Bradford total protein assay or by absorbance of light at 280 nm, however some reagents used during the purification process may interfere with the quantification. For example, imidazole (commonly used for purification of polyhistidine-tagged recombinant proteins) is an amino acid analogue and at low concentrations will interfere with the bicinchoninic acid (BCA) assay for total protein quantification. Impurities in low-grade imidazole will also absorb at 280 nm, resulting in an inaccurate reading of protein concentration from UV absorbance.

14.6 METHODS OF PROTEIN PURIFICATION

The methods used in protein purification, can roughly be divided into analytical and preparative methods. The distinction is not exact, but the deciding factor is the amount of protein, that can practically be purified with that method. Analytical methods aim to detect and identify a protein in a mixture, whereas preparative methods aim to produce large quantities of the protein for other purposes, such as structural biology or industrial use. In general, the preparative methods can be used in analytical applications, but not the other way around.

14.6.1 Extraction

Depending on the source, the protein has to be brought into solution by breaking the tissue or cells containing it. There are several methods to achieve this: Repeated freezing and thawing, sonication, homogenization by high pressure or permeabilization by organic solvents. The method of choice depends on how fragile the protein is and how sturdy the cells are. After this extraction process soluble proteins will be in the solvent, and can be separated from cell membranes, DNA, etc. by centrifugation. The extraction process also extracts proteases, which will start digesting the proteins in the solution. If the protein is sensitive to proteolysis, it is usually desirable to proceed quickly, and keep the extract cooled, to slow down proteolysis.

14.6.2 Precipitation and differential solubilisation

In bulk protein purification, a common first step to isolate proteins is precipitation with ammonium sulphate $(NH_4)_2SO_4$. This is performed by adding increasing amounts of ammonium sulphate and collecting the different fractions of precipatate protein. One advantage of this method is that it can be performed inexpensively with very large volumes.

The first proteins to be purified are water-soluble proteins. Purification of integral membrane proteins requires disruption of the cell membrane in order to isolate any one particular protein from others that are in the same membrane compartment. Sometimes a particular membrane fraction can be isolated first, such as isolating mitochondria from cells before purifying a protein located in a mitochondrial membrane. A detergent such as sodium dodecyl sulphate (SDS) can be used to dissolve cell membranes and keep membrane proteins in solution during purification; however, because SDS causes denaturation, milder detergents such as Triton X-100 or CHAPS can be used to retain the protein's native conformation during purification.

14.6.3 Ultracentrifugation

Centrifugation is a process that uses centrifugal force to separate mixtures of particles of varying masses or densities suspended in a liquid. When a vessel (typically a tube or bottle) containing a mixture of proteins or other particulate matter, such as bacterial cells, is rotated at high speeds, the angular momentum yields an outward force to each particle that is proportional to its mass. The tendency of a given particle to move through the liquid because of this force is offset by the resistance the liquid exerts on the particle. The net effect of "spinning" the sample in a centrifuge is that massive, small, and dense particles move outward faster than less massive particles or particles with more "drag" in the liquid. When suspensions of particles are "spun" in a centrifuge, a "pellet" may form at the bottom of the vessel that is enriched for the most massive particles with low drag in the liquid. The remaining, non-compacted particles still remaining mostly in the liquid are called the "supernatant" and can be removed from the vessel to separate the supernatant from the pellet. The rate of centrifugation is specified by the angular acceleration applied to the sample, typically measured in comparison to the g. If samples are centrifuged long enough, the particles in the vessel will reach equilibrium wherein the particles accumulate specifically at a point in the vessel where their buoyant density is balanced with centrifugal force. Such an "equilibrium" centrifugation can allow extensive purification of a given particle.

Sucrose gradient centrifugation — A linear concentration gradient of sugar (typically sucrose, glycerol, or Percoll) is generated in a tube such that the highest concentration is on the bottom and lowest on top. A protein sample is then layered on top of the gradient and spun at high speeds in an ultracentrifuge. This causes heavy macromolecules to migrate towards the bottom of the tube faster than lighter material. During centrifugation in the absence of sucrose, as particles move farther and farther from the centre of rotation, they experience more and more centrifugal force (the further they move, the faster they move). The problem with this is that the useful separation range within the vessel is restricted to a small observable window. Spinning a sample twice as long does not mean the particle of interest will go twice as far; in fact, it will go significantly farther. However, when the proteins are moving through a sucrose gradient, they encounter liquid of increasing density and viscosity. A properly designed sucrose gradient will counteract the increasing centrifugal force, so the particles move in close proportion to the time they have been in the centrifugal field. Samples separated by these gradients are referred to as "rate zonal" centrifugations. After separating the protein/particles, the gradient is then fractionated and collected.

14.6.4 Chromatographic methods

Usually a protein purification protocol contains one or more chromatographic steps. Various chromatographic methods in details we have already come across in Chapter 9. The basic procedure in chromatography is to flow the solution containing the protein through a column packed with various materials. Different proteins interact differently with the column material, and can thus be separated by the time required to pass the column, or the conditions required to elute the protein from the column. Usually proteins are detected as they are coming off the column by their absorbance at 280 nm. Many different chromatographic methods exist:

1. Size exclusion chromatography

Chromatography can be used to separate protein in solution or denaturing conditions by using porous gels. This technique is known as size exclusion chromatography. The principle is that smaller molecules have to traverse a larger volume in a porous matrix. Consequentially, proteins of a certain range in size will require a variable volume of eluant (solvent) before being collected at the other end of the column of gel.

In the context of protein purification, the eluant is usually pooled in different test tubes. All test tubes containing no measurable trace of the protein to purify are discarded. The remaining solution is thus made of the protein to purify and any other similarly-sized proteins.

2. Ion exchange chromatography

Ion exchange chromatography separates compounds according to the nature and degree of their ionic charge. The column to be used is selected according to its type and strength of charge. Anion exchange resins have a positive charge and are used to retain and separate negatively charged compounds, while cation exchange resins have a negative charge and are used to separate positively charged molecules.

Before the separation begins a buffer is pumped through the column to equilibrate the opposing charged ions. Upon injection of the sample, solute molecules will exchange with the buffer ions as each competes for the binding sites on the resin. The length of retention for each solute depends upon the strength of its charge. The most weakly charged compounds will elute first, followed by those with successively stronger charges. Becauses of the nature of the separating mechanism, pH, buffer type, buffer concentration, and temperature all play important roles in controlling the separation.

Ion exchange chromatography is a very powerful tool for use in protein purification and is frequently used in both analytical and preparative separations.

3. Affinity chromatography

Affinity Chromatography is a separation technique based upon molecular conformation, which frequently utilizes application specific resins. These resins have ligands attached to their surfaces which are specific for the compounds to be separated. Most frequently, these ligands function in a fashion similar to that of antibody-antigen interactions. This "lock and key" fit between the ligand and its target compound makes it highly specific, frequently generating a single peak, while all else in the sample is unretained.

Many membrane proteins are glycoproteins and can be purified by lectin affinity chromatography. Detergent-solubilized proteins can be allowed to bind to a chromatography resin that has been modified to have a covalently attached lectin. Proteins that do not bind to the lectin are washed away and then specifically bound glycoproteins can be eluted by adding a high concentration of a sugar that competes with the bound glycoproteins at the lectin binding site. Some lectins have high affinity binding to oligosaccharides of glycoproteins that is hard to compete with sugars, and bound glycoproteins need to be released by denaturing the lectin.

4. Metal binding

A common technique involves engineering a sequence of 6 to 8 histidines into the C-terminal of the protein. The polyhistidine binds strongly to divalent metal ions such as nickel and cobalt. The protein can be passed through a column containing immobilized nickel ions, which binds the polyhistidine tag. All untagged proteins pass through the column. The protein can be eluted with imidazole, which competes with the polyhistidine tag for binding to the column, or by a decrease in pH (typically to 4.5), which decreases the affinity of the tag for the resin. While this procedure is generally used for the purification of recombinant proteins with an engineered affinity tag (such as a 6xHis-tag or Clontech's HAT tag), it can also be used for natural proteins with an inherent affinity for divalent cations.

5. Immunoaffinity chromatography

Immunoaffinity chromatography uses the specific binding of an antibody to the target protein to selectively purify the protein. The procedure involves immobilizing an antibody to a column material, which then selectively binds the protein, while everything else flows through. The protein can be eluted by changing the pH or the salinity. Because this method does not involve engineering in a tag, it can be used for proteins from natural sources.

6. HPLC

High performance liquid chromatography or high pressure liquid chromatography is a form of chromatography applying high pressure to drive the solutes through the column faster. This means that the diffusion is limited and the resolution is improved. The most common form is "reversed phase" HPLC, where the column material is hydrophobic. The proteins are eluted by a gradient of increasing amounts of an organic solvent, such as acetonitrile. The proteins elute according to their hydrophobicity. After purification by HPLC the protein is in a solution that only contains volatile compounds, and can easily be lyophilized. HPLC purification frequently results in denaturation of the purified proteins and is thus not applicable to proteins that do not spontaneously refold.

14.7 CONCENTRATION OF THE PURIFIED PROTEIN

At the end of protein purification, the protein often has to be concentrated. Different methods exist for this purpose is:

1. Lyophilization

If the solution does not contain any other soluble component than the protein in question the protein can be lyophilized (dried). This is commonly done after a HPLC run. This simply removes all volatile components leaving the proteins behind.

2. Ultrafiltration

Ultrafiltration concentrates a protein solution using selective permeable membranes. The function of the membrane is to let the water and small molecules pass through while retaining the protein. The solution is forced against the membrane by mechanical pump or gas pressure or centrifugation.

14.8 ANALYSIS OF ISOLATED PROTEINS

This analysis is done by following techniques:

1. Denaturing-Condition Electrophoresis

Gel electrophoresis is a common laboratory technique that can be used both as preparative and analytical methods. In details it is explained in Chapter 8. The principle of electrophoresis relies on the movement of a charged ion in an electric field. In practice, the proteins are denatured in a solution containing a detergent (SDS). In these conditions, the proteins are unfolded and coated with negatively charged detergent molecules. The proteins in SDS-PAGE are separated on the sole basis of their size.

In analytical methods, the protein migrates as bands based on size. Each band can be detected using stains such as Coomassie blue dye or silver stain. Preparative methods to purify large amounts of protein, require the extraction of the protein from the electrophoretic gel. This extraction may involve excision of the gel containing a band, or eluting the band directly off the gel as it runs off the end of the gel.

In the context of a purification strategy, denaturing condition electrophoresis provides an improved resolution over size exclusion chromatography, but does not scale to large quantity of proteins in a sample as well as the late chromatography columns.

2. Non-Denaturing-Condition Electrophoresis

An important non-denaturing electrophoretic procedure for isolating bioactive metalloproteins in complex protein mixtures is termed 'quantitative native continuous polyacrylamide gel electrophoresis' (QNC-PAGE). In details non-denaturing electrophoresis have been explained in Chapter 8.

14.9 TIPS TO REMEMBER

- Purification may be preparative or analytical.
- Each step in the purification process will cause some loss of product, a yeild of 80% in each step is assumed, and therefore, it is advisable to have as few purification steps as possible.
- The purification has three phases of capture, intermediate purification and polishing each with specific objective.
- Every technique offers a balance between capacity, speed, resolution and recovery and should be selected to meet the objectives of each purification step.

14.10 VIVA VOCE

Q.1. Why protein purification is necessary?

Q.2. What are the factors which may affect the purification process?

Q.3. What all precautions one should take while selecting technique for purification?

Q.4. What is three-phase strategy?

Q.5. What are the steps in purification of proteins?

Q.6. How one can achieve balance between different parameters during purification?

Q.7. What do you understand by salting out of proteins?

Q.8. How one can concentrate the dilute yield after purification?

Q.9. Differentiate between analytical and preparative purification.

Q.10. What are different technniques employed for protein purification?

CHAPTER 15 Flow Cytometry

15.1 INTRODUCTION

The fluorescent antibody techniques described are extremely valuable qualitative tools, but they do not give quantitative data. This shortcoming was remedied by development of the flow cytometer, which was designed to automate the analysis and separation of cells stained with fluorescent antibody. The flow cytometer uses a laser beam and light detector to count single intact cells in suspension. Every time a cell passes the laser beam, light is deflected from the detector, and this interruption of the laser signal is recorded. Those cells having a fluorescently tagged antibody bound to their cell surface antigens are excited by the laser and emit light that is recorded by a second detector system located at a right angle to the laser beam. The simplest form of the instrument counts each cell as it passes the laser beam and records the level of fluorescence the cell emits; an attached computer generates plots of the number of cells as the ordinate and their fluorescence intensity as the abscissa. More sophisticated versions of the instrument are capable of sorting populations of cells into different containers according to their fluorescence profile. Use of the instrument to determine which and how many members of a cell population bind fluorescently labelled antibodies is called analysis; use of the instrument to place cells having different patterns of reactivity into different containers is called cell sorting.

The flow cytometer has multiple applications to clinical and research problems. A common clinical use is to determine the kind and number of white blood cells in blood samples. By treating appropriately processed blood samples with a fluorescently labelled antibody and performing flow cytometric analysis, one can obtain the following information:

- How many cells express the target antigen as an absolute number and also as a percentage of cells passing the beam. For example, if one uses a fluorescent antibody specific for an antigen present on all T cells, it would be possible to determine the percentage of T cells in the total white blood cell population. Then, using the cell-sorting capabilities of the flow cytometer, it would be possible to isolate the T-cell fraction of the leukocyte.

- The distribution of cells in a sample population according to antigen densities as determined by fluorescence intensity. It is thus possible to obtain a measure of the distribution of antigen density within the population of cells that possess the antigen. This is a powerful feature of the instrument, since the same type of cell may express different levels of antigen depending upon its developmental or physiological state.
- The size of cells. This information is derived from analysis of the light-scattering properties of members of the cell population under examination.

Flow cytometry also makes it possible to analyze cell populations that have been labelled with two or even three different fluorescent antibodies. For example, if a blood sample is reacted with a fluorescein-tagged antibody specific for T cells, and also with a phycoerythrin-tagged antibody

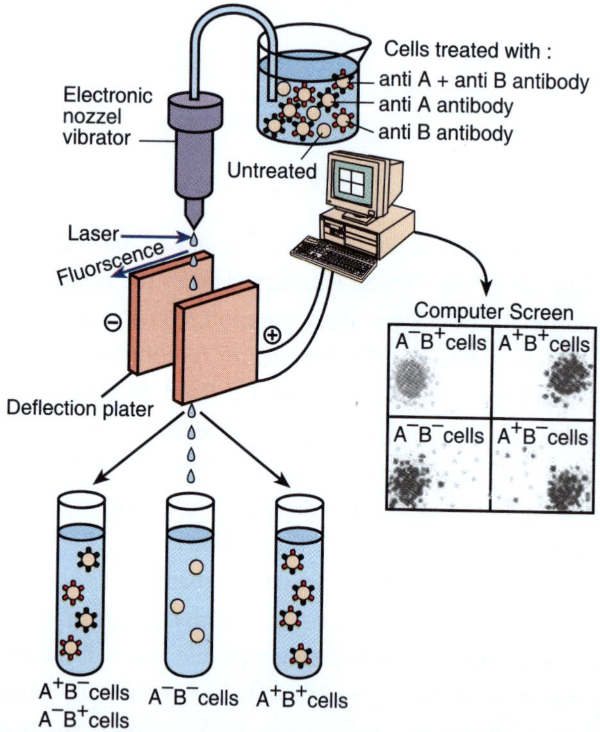

Fig. 15.1: Separation of fluorochrome-labelled cells with the flow cytometer. In the example shown, a mixed cell population is stained with two antibodies, one specific for surface antigen A and the other specific for surface antigen B. The anti-A antibodies are labelled with fluorescein (green) and the anti-B antibodies with rhodamine (red). The stained cells are loaded into the sample chamber of the cytometer. The cells are expelled, one at a time, from a small vibrating nozzle that generates microdroplets, each containing no more than a single cell. As it leaves the nozzle, each droplet receives a small electrical charge, and the computer that controls the flow cytometer can detect exactly when a drop generated by the nozzle passes through the beam of laser light that excites the fluorochrome. The intensity of the fluorescence emitted by each droplet that contains a cell is monitored by a detector and displayed on a computer screen. Because the computer tracks the position of each droplet, it is possible to determine when a particular droplet will arrive between the deflection plates. By applying a momentary charge to the deflection plates when a droplet is passing between them, it is possible to deflect the path of a particular droplet into one or another collecting vessel. This allows the sorting of a population of cells into subpopulations having different profiles of surface markers.

specific for B cells, the percentages of B and T cells may be determined simultaneously with a single analysis. Numerous variations of such "two-colour" analyses are routinely carried out, and "three-colour" experiments are common. Aided by appropriate software, highly sophisticated versions of the flow cytometer can even perform "five-colour" analyses.

15.2 PRINCIPLES OF FLOWCYTOMETRY

15.2.1 Fluidics system

One of the fundamentals of flow cytometry is the ability to measure the properties of individual particles. When a sample in solution is injected into a flow cytometer, the particles are randomly distributed in three-dimensional space. The sample must, therefore, be ordered into a stream of single particles that can be interrogated by the machine's detection system. This process is managed by the fluidics system. Essentially, the fluidics system consists of a central channel/core through which the sample is injected, enclosed by an outer sheath that contains faster flowing fluid. As the sheath fluid moves, it creates a massive drag effect on the narrowing central chamber. This alters the velocity of the central fluid whose flow front becomes parabolic with greatest velocity at its centre and zero velocity at the wall (see Fig. 1). The effect creates a single file of particles and is called hydrodynamic focusing. Under optimal conditions (laminar flow) the fluid in the central chamber will not mix with the sheath fluid.

The flow characteristics of the central fluid can be estimated using Reynolds Number (R_e):

$$R_e = \frac{pVD}{\mu},$$

where D = tube diameter; V = mean velocity of fluid; p = density of fluid; and μ = viscosity of fluid.

When $R_e < 2300$, flow is always laminar. When $R_e > 2300$, flow can be turbulent, which accelerates diffusion. Without hydrodynamic focusing the nozzle of the instrument (typically 70 μM) would become blocked, and it would not be possible to analyze one cell at a time.

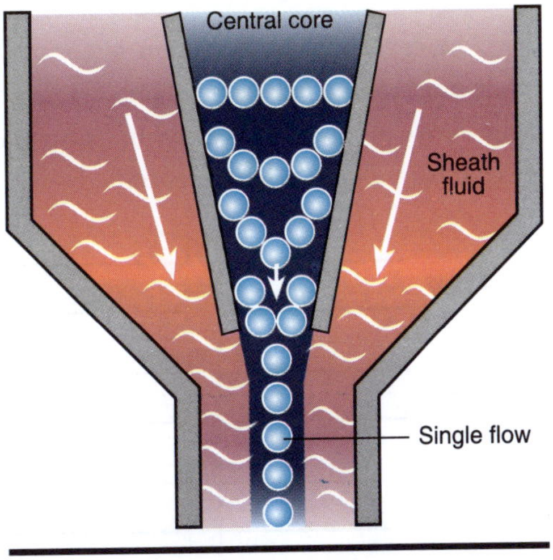

Fig. 15.2: Hydrodynamic focusing produces a single stream of particles

15.2.2 Optics and detection

After hydrodynamic focusing, each particle passes through one or more beams of light. Light scattering or fluorescence emission (if the particle is labelled with a fluorochrome) provides information about the particle's properties. The laser and the arc lamp are the most commonly used light sources in modern flow cytometry. Lasers produce a single wavelength of light (a laser line) at one or more discrete frequencies (coherent light). Arc lamps tend to be less expensive than lasers and exploit the colour emissions of an ignited gas within a sealed tube.

However, this produces unstable incoherent light of a mixture of wavelengths, which needs subsequent optical filtering.

Light that is scattered in the forward direction, typically up to 20° offset from the laser beam's axis, is collected by a lens known as the *forward scatter channel* (FSC). The FSC intensity roughly equates to the particle's size and can also be used to distinguish between cellular debris and living cells.

Light measured approximately at a 90° angle to the excitation line is called *side scatter.* The *side scatter channel* (SSC) provides information about the granular content within a particle. Both FSC and SSC are unique for every particle, and a combination of the two may be used to differentiate different cell types in a heterogeneous sample.

Fluorescence measurements taken at different wavelengths can provide quantitative and qualitative data about fluorochrome-labelled cell surface receptors or intracellular molecules such as DNA and cytokines. Flow cytometers use separate fluorescence (FL) channels to detect light emitted. The number of detectors will vary according to the machine and its manufacturer. Detectors are either silicon photodiodes or photomultiplier tubes (PMTs). Silicon photodiodes are usually used to measure forward scatter when the signal is strong. PMTs are more sensitive instruments and are ideal for scatter and fluorescence readings.

The specificity of detection is controlled by optical filters, which block certain wavelengths while transmitting (passing) others. There are three major filter types. 'Long pass' filters allow through light above a cut-off wavelength, 'short pass' permit light below a cut-off wavelength and 'band pass' transmit light within a specified narrow range of wavelengths (termed a bandwidth).

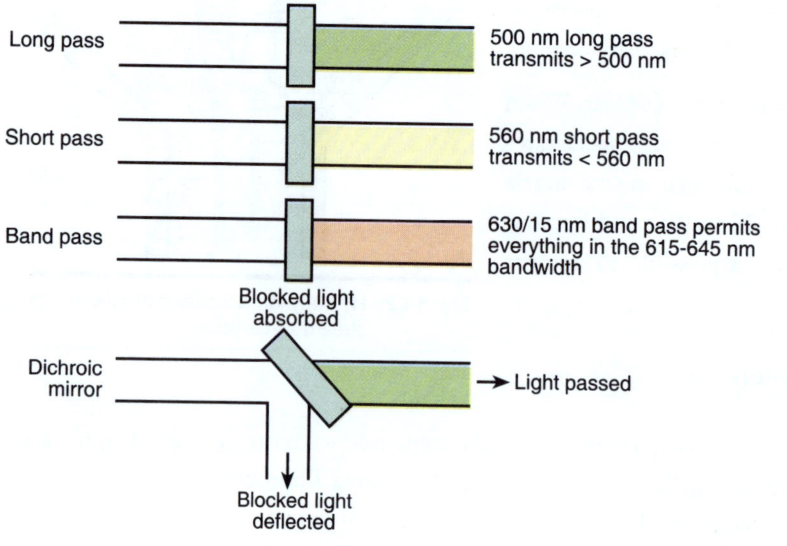

Fig. 15.3: Different types of optical filters

All these filters block light by absorption. When a filter is placed at a 45° angle to the oncoming light it becomes a dichroic filter/mirror. As the name suggests, this type of filter performs two functions; first, to pass specified wavelengths in the forward direction and, second, to deflect blocked light at a 90° angle. To detect multiple signals simultaneously, the precise choice and order of optical filters will be an important consideration.

15.2.3 Signal processing

When light hits a photodetector a small current (a few microamperes) is generated. Its associated voltage has an amplitude proportional to the total number of light photons received by the detector. This voltage is then amplified by a series of linear or logarithmic amplifiers, and by analog to digital convertors (ADCs), into electrical signals large enough (5-10 volts) to be plotted graphically.

Log amplification is normally used for fluorescence studies because it expands weak signals and compresses strong signals, resulting in a distribution that is easy to display on a histogram. Linear scaling is preferable where there is not such a broad range of signals, e.g., in DNA analysis. The measurement from each detector is referred to as a 'parameter', e.g., forward scatter, side scatter or fluorescence. The data acquired in each parameter are known as the 'events' and refer to the number of cells displaying the physical feature or marker of interest.

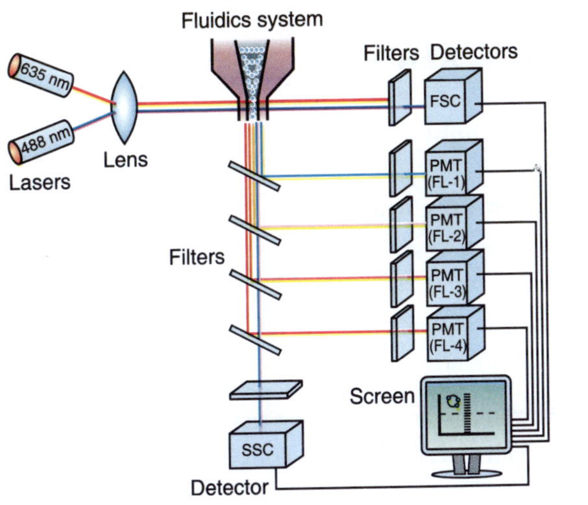

Fig. 15.4: Schematic overview of a typical flow cytometer set-up

15.2.4 Electrostatic cell sorting

A major application of flow cytometry is to separate cells according to subtype or epitope expression for further biological studies. This process is called cell sorting or FACSTM analysis. After the sample is hydrodynamically focused, each particle is probed with a beam of light. The scatter and fluorescence signal is compared to the sort criteria set on the instrument. If the particle matches the

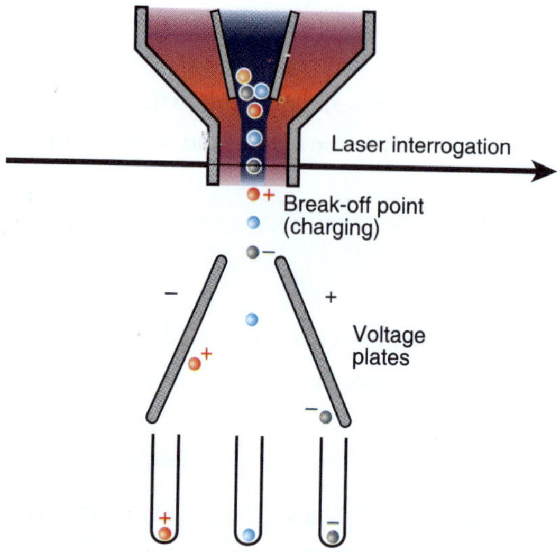

Fig. 15.5: Electrostatic flow sorting

selection criteria, the fluid stream is charged as it exits the nozzle of the fluidics system. Electrostatic charging actually occurs at a precise moment called the 'break-off point', which describes the instant droplet containing the particle of interest separated from the stream.

To prevent the break-off point happening at random distances from the nozzle and to maintain consistent droplet sizes, the nozzle is vibrated at high frequency. The droplets eventually pass through a strong electrostatic field, and are deflected left or right based on their charge (Fig. 15.4). The speed of flow sorting depends on several factors including particle size and the rate of droplet formation. A typical nozzle is between 50-70 μm in diameter and, depending on the jet velocity from it, can produce 30,000-100,000 droplets per second, which is ideal for accurate sorting. Higher jet velocities risk the nozzle becoming blocked and will also decrease the purity of the preparation.

15.3 PRINCIPLES OF FLUORESCENCE

15.3.1 Fluorochromes and light

Fluorochromes are essentially dyes, which accept light energy (e.g., from a laser) at a given wavelength and re-emit it at a longer wavelength. These two processes are called *excitation* and *emission.* The process of emission follows extremely rapidly, commonly in the order of nanoseconds, and is known as fluorescence. Before considering the different types of fluorochrome available for flow cytometry, it is necessary to understand the principles of light absorbance and emission.

Light is a form of electromagnetic energy that travels in waves. These waves have both frequency and length, the latter of which determines the colour of light. The light that can be visualized by the

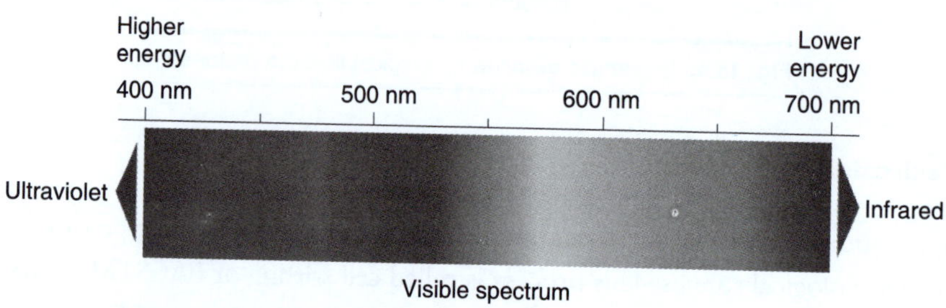

Fig. 15.6: The electromagnetic spectrum

human eye represents a narrow wavelength band (380-700 nm) between ultraviolet (UV) and infrared (IR) radiation (Fig. 15.5). Sunlight, for example, contains UV and IR light that, although invisible to the eye, can still be felt as warmth on the skin and measured scientifically using photodetectors. The visible spectrum can further be subdivided according to colour, often remembered by the mnemonic 'ROY G BV' standing for red, orange, yellow, green, blue and violet. Red light is at the longer wavelength end (lower energy) and violet light at the shorter wavelength end (higher energy).

15.3.2 Stokes Shift

When light is absorbed by a fluorochrome, its electrons become excited and move from a resting state (1) to a maximal energy level called the 'excited electronic singlet state' (2). The amount of energy required will differ for each fluorochrome and is depicted in Fig. 15.6 as $E_{excitation}$. This state only lasts for 1-10 nanoseconds because the fluorochrome undergoes internal conformational change and, in doing so, releases some of the absorbed energy as heat. The electrons subsequently fall to a lower, more stable, energy level called the 'relaxed electronic singlet state' (3). As electrons steadily move back from here to their ground state they release the remaining energy ($E_{emission}$) as fluorescence (4).

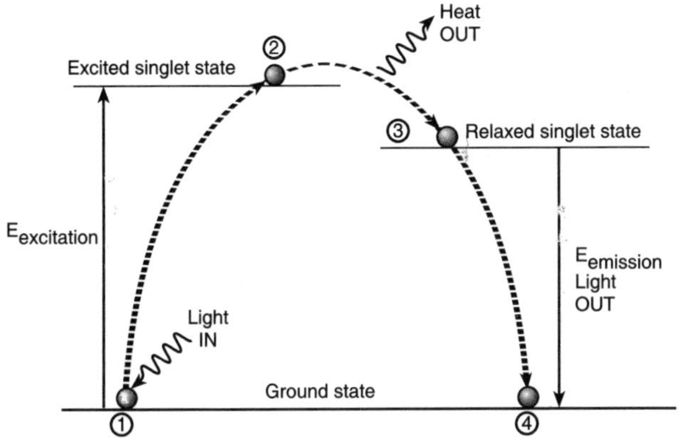

Fig. 15.7: Stokes Shift

As $E_{emission}$ contains less energy than was originally put into the fluorochrome it appears as a different colour of light to $E_{excitation}$. Therefore, the emission wavelength of any fluorochrome will always be longer than its excitation wavelength. The difference between $E_{excitation}$ and $E_{emission}$ is called Stokes Shift and this wavelength value essentially determines how good a fluorochrome is for fluorescence studies. After all, it is imperative that the light produced by emission can be distinguished from the light used for excitation. This difference is easier to detect when fluorescent molecules have a large Stokes Shift.

15.3.3 Maximal absorbance and maximal emission

The wavelength of excitation is critical to the total photons of light the fluorochrome will absorb. FITC (fluorescein isothiocyanate), for example, will absorb light within the range 400-550 nm but the closer the wavelength is to 490 nm (its peak or maximum), the greater the absorbance is. In turn, the more photons absorbed, the more intense the fluorescence emission will be. These optimal conditions are termed maximal absorbance and maximal emission wavelengths. Maximal absorbance usually defines the laser spectral line that is used for excitation. In the case of FITC, its maximum falls within the blue spectrum. Therefore, the blue Argon-ion laser is commonly used for this fluorochrome, as it excites at 488 nm, close to FITC's absorbance peak of 490 nm.

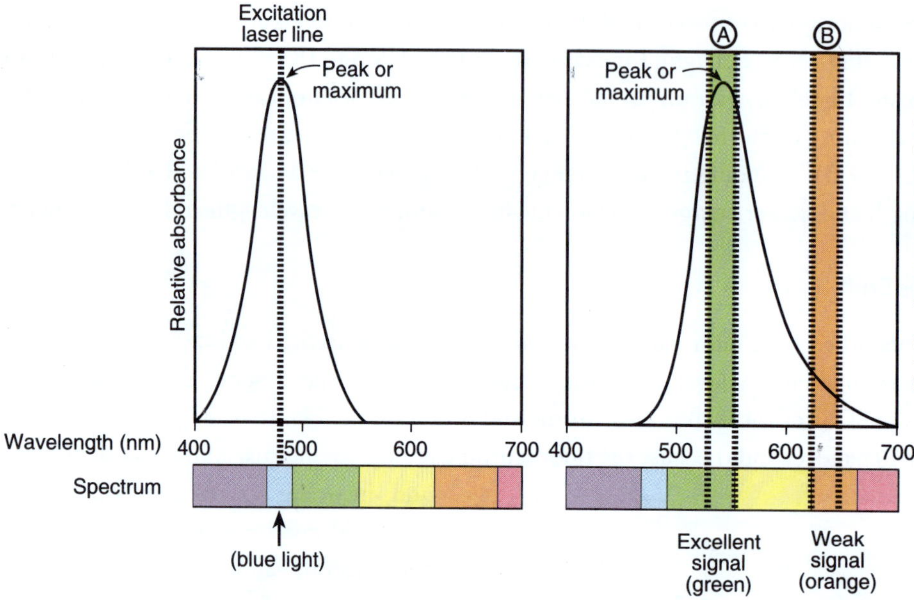

Fig. 15.8: Light absorbance (LEFT) and light emission (RIGHT) of FITC

FITC emits fluorescence over the range 475-700 nm peaking at 525 nm, which falls in the green spectrum. If filters are used to screen out all light other than that measured at the maximum via channel A (see Figure 8), FITC will appear green. Hence, 'fluorescence colour' usually refers to the colour of light a fluorochrome emits at its highest stable excited state. However, if FITC fluorescence is detected only via channel B (see Fig. 15.8), it will appear orange and be much weaker in intensity. How the flow cytometer is set up to measure fluorescence will ultimately determine the colour of a fluorochrome.

15.3.4 Why use a fluorescent probe?

The purpose of a fluorescent probe, such as a fluorochrome-conjugated antibody, is to directly target an epitope of interest and to allow its biological and biochemical properties to be measured more easily by the flow cytometer. Fluorescent probes are useful in a wide range of applications including identifying and quantifying distinct populations of cells, cell surface receptors or intracellular organelles; cell sorting; immunophenotyping; calcium influx experiments; determining nucleic acid content; measuring enzyme activity, and for apoptosis studies. By changing the excitation light and using more than one fluorochrome, it is possible to analyze several parameters of the sample at any one time. This forms the basis of multicolour fluorescence studies.

15.3.5 Which fluorochromes are useful for flow cytometry?

There are dozens of fluorescent molecules (fluorochromes) with a potential application in flow cytometry. The list is ever growing but it is not the scope of this chapter to cover them all. Instead, some of the most useful fluorochromes for surface or intracellular epitope detection are described in table 15.1, including the very latest in fluorescent probe technology–tandem dyes. There is enough variation in the two tables to cover most researchers' needs.

Single dyes

Some of these single dyes, e.g., FITC have been in use for the past 30 years but are now facing competition from alternatives like Alexa Fluor® dyes, which offer the user greater photostability and increased fluorescence.

Tandem dyes

In a tandem dye, a small fluorochrome takes a 'piggy-back' ride on another larger fluorochrome. When the first dye is excited and reaches its maximal singlet state, all its energy transfers to the second dye (an acceptor molecule), located in close proximity. This activates the second fluorochrome, which then produces the fluorescence emission. The process is called FRET (fluorescence resonance energy transfer). It is a clever way to achieve higher Stokes Shifts and, therefore, increases the number of colours that can be analyzed from a single laser wavelength. The majority of tandem dyes have been manufactured for the standard 488 nm laser, which is found in most flow cytometers. Tandem dyes are very useful for multicolour fluorescence studies especially in combination with single dyes. For example, Alexa Fluor® 488, Phycoerythrin, PerCP-Cy5.5 and PE-Cy7 can all be excited at 488 nm, but will produce green, yellow, purple and infrared emissions respectively, which can be measured using separate detectors.

Table 15.1: List of fluorescent dyes used for flowcytometry

Single dyes				
Dye	Laser excitation line (nm)	Maximal absorbance (nm)	Maximal emission (nm)	Fluorescence colour
Alexa Fluor® 405	405, 407	401	421	
Alexa Fluor® 430	405, 407	433	541	
Alexa Fluor® 488	488	495	519	
Alexa Fluor® 633	633, 635, 647	632	647	
Alexa Fluor® 647	633, 635, 647	650	665	
Alexa Fluor® 660	633, 635, 647	663	690	
Alexa Fluor® 680	633, 635, 647	679	702	
Alexa Fluor® 700	633, 635, 647	702	723	Infrared
APC	633, 635, 647	650	661	
FITC	488	490	525	
Pacific Blue™	405, 407	410	455	
PerCP	488	490	675	
Phycoerythrin	488	490, 565	578	

Contd.

Table 15.1: List of fluorescent dyes used for flowcytometry

Single dyes

Dye	Laser excitation line (nm)	Maximal absorbance (nm)	Maximal emission (nm)	Fluorescence colour
APC-Alexa Fluor® 750	633, 635, 647	650	779	Infrared
APC-Cy 5.5	633, 635, 647	650	695	
APC-Cy 7	633, 635, 647	650	785	Infrared
PerCP-Cy 5.5	488	496, 546	695	
PE-Alexa Fluor® 610	488	496, 546	627	
PE-Alexa Fluor® 647	488	496, 546	667	
PE-Alexa Fluor® 680	488	496, 546	702	
PE-Alexa Fluor® 700	488	496, 546	723	Infrared
PE-Alexa Fluor® 750	488	496, 546	779	Infrared
PE-Cy 5.5	488	496, 546	695	
PE-Cy 5	488	496, 546	667	
PE-Cy 7	488	496, 546	785	Infrared
PE-Texas Red®	488	496, 546	615	

Abbreviations

APC	Allophycocyanin
FITC	Fluorescein isothiocyanate
PE	Phycoerythrin
PerCP	Peridinin-chlorophyll-protein complex

Note. Phycoerythrin (PE) is same as R-Phycoerythrin (RPE)

15.3.6 Fluorescence compensation

One consideration to be aware of when performing multicolour fluorescence studies is the possibility of spectral overlap. When two or more fluorochromes are used during a single experiment there is a chance that their emission profiles will coincide, making measurement of the true fluorescence emitted by each difficult. This can be avoided by using fluorochromes at very different ends of the spectrum, e.g., Alexa Fluor® 405 and Phycoerythrin; however, this is not always practical.

Instead, a process called fluorescence compensation is applied during data analysis, which calculates how much interference (as a %) a fluorochrome will have in a channel that was not assigned specifically to measure it. Figure 8 helps to explain the concept.

The graphs show the emission profiles of two imaginary fluorochromes 'A' and 'B' which are being detected in FL-1 and FL-2 channels respectively. Because the emission profiles are so close together, a portion of fluorochrome A spill over into FL-2 (red shade) and conversely, some of fluorochrome B reaches FL-1 (dark blue shade).

To calculate how much compensation needs to be applied to the data set if both dyes are used simultaneously, some control readings must first be taken. Fluorochrome A should be run through the flow cytometer on its own and the % of its total emission that is detectable in FL-2 (spillover) determined. The procedure should be repeated with fluorochrome B, except that this time FL-1 is spillover.

Suppose the results are:

	Spillover Fluorescence	
	FL-1	FL-2
Fluorochrome A	N/A	17%
Fluorochrome B	5%	N/A

This means that when the two fluorochromes are used for a dual-colour experiment, the true reading for fluorochrome A in FL-1 = (total fluorescence measured in FL-1) minus (5% of fluorochrome B's total fluorescence)

Similarly, the true reading for fluorochrome B in FL-2 = (total fluorescence measured in FL-2) minus (17% of fluorochrome A's total fluorescence)

Fortunately, modern flow cytometry analytical software applies fluorescence compensation mathematics automatically, which simplifies matters considerably.

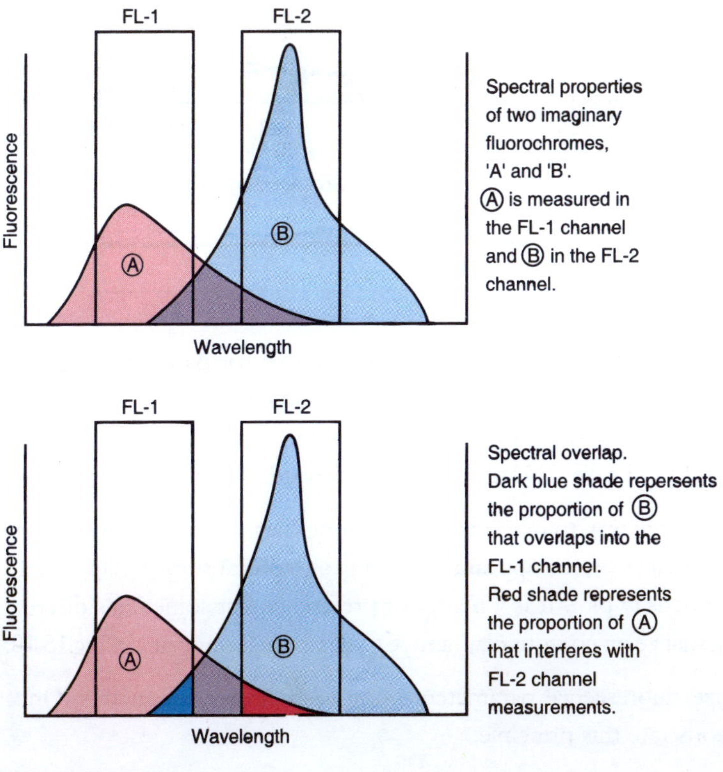

Fig. 15.9: Fluorescence compensation

15.4 DATA ANALYSIS

15.4.1 Gates and regions

An important principle of flow cytometry data analysis is to selectively visualize the cells of interest while eliminating results from unwanted particles, e.g., dead cells and debris. This procedure is called gating. Cells have traditionally been gated according to physical characteristics. For instance, subcellular debris and clumps can be distinguished from single cells by size, estimated by forward scatter. Also, dead cells have lower forward scatter and higher side scatter than living cells. Lysed whole blood cell analysis is the most common application of gating, and Fig. 15.10 depicts typical graphs for SSC versus FSC when using large cell numbers. The different physical properties of granulocytes, monocytes and lymphocytes allow them to be distinguished from each other and from cellular contaminants.

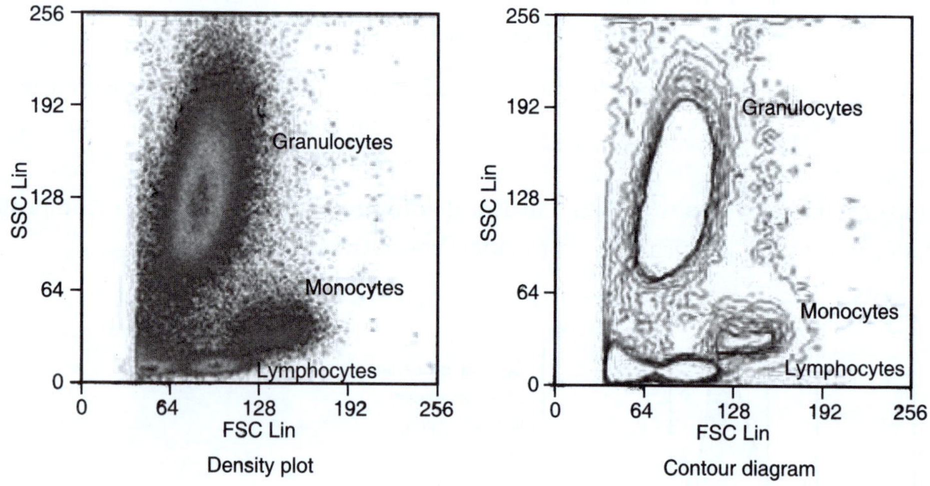

Fig. 15.10: Analysis of lysed whole blood using FSC/SSC

On the density plot, each dot or point represents an individual cell that has passed through the instrument. Yellow/green hotspots indicate large numbers of events resulting from discrete populations of cells. The colours give the graph a three-dimensional feel. After a little experience, discerning the various subtypes of blood cells is relatively straightforward.

Contour diagrams are an alternative way to demonstrate the same data. Joined lines represent similar numbers of cells. The graph takes on the appearance of a geographical survey map, which, in principle, closely resembles the density plot. It is a matter of preference but sometimes discrete populations of cells are easier to visualize on contour diagrams, e.g., compare monocytes in Fig.15.10.

Newer gating strategies utilize fluorescence parameters along with scatter parameters. Once again, blood can be used to demonstrate this principle.

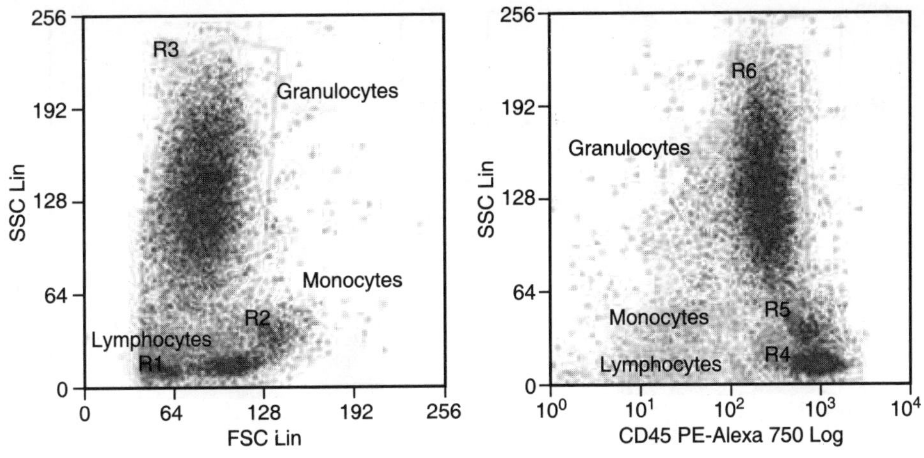

Fig. 15.11: Lysed whole blood analysis using scatter and fluorescence

Above on the left is a FSC/SSC plot for human lysed whole blood using smaller numbers of cells than in Figure 15.10. The lymphocytes, monocytes and granulocytes have been gated as region 1 (R1), region 2 (R2) and region 3 (R3), respectively. 'Region' simply refers to an area drawn on a plot displaying flow cytometry data.

On the right the same cells are now plotted as SSC on the y-axis versus CD45 fluorescence on the x-axis. CD45 is a marker expressed on all white blood cells at varying intensities but is absent on red blood cells. In relative terms, lymphocytes have a low SSC and high CD45 count (R4), granulocytes have a high SSC and low CD45 count (R6), while monocytes are somewhere in between the other two (R5). The major difference between the lymphocytes gated in R1 and those gated in R4 is the absence of red blood cells in the latter, making it a much purer preparation. This highlights the usefulness of gating strategies that combine a scatter parameter with a fluorescence parameter.

15.4.2 Single-parameter histograms

These are graphs that display a single measurement parameter (relative fluorescence or light scatter intensity) on the x-axis and the number of events (cell count) on the y-axis.

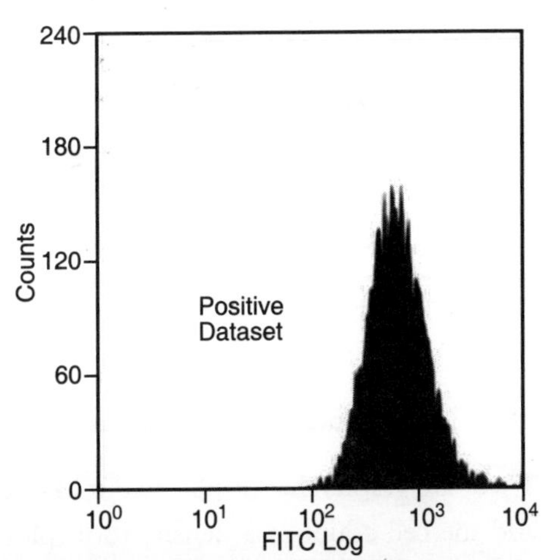

Fig. 15.12: A single-parameter histogram

The histogram in Fig.15.12 looks very basic but is useful for evaluating the total number of cells in a sample that possess the physical properties selected for or which express the marker of interest. Cells with the desired characteristics are known as the positive data set.

Ideally, flow cytometry will produce a single distinct peak that can be interpreted as the positive data set.

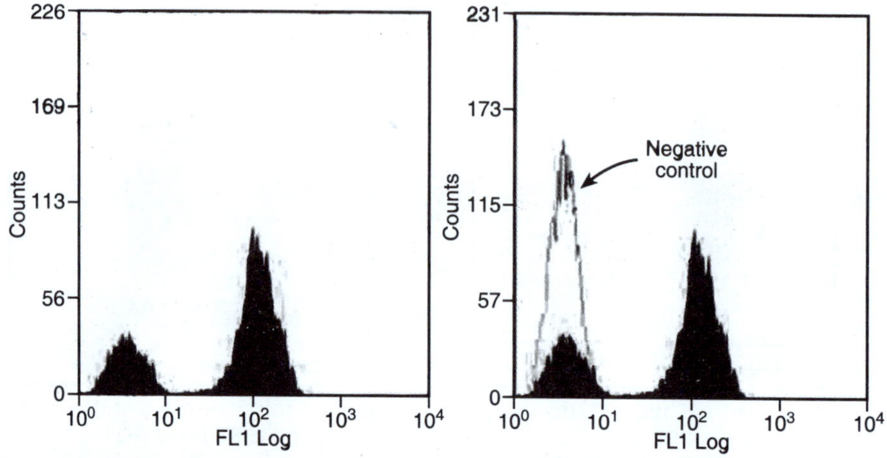

Fig. 15.13: Determining positive dataset. LEFT, using rat anti-mouse F4/80 conjugated to FITC to stain mouse peritoneal macrophages produces two peaks. RIGHT, by running an appropriate isotype control (rat IgG2b negative control conjugated to FITC) and overlaying its image on the histogram (black outline) the positive data set is identified as the taller red peak on the right.

However, in many situations, flow analysis is performed on a mixed population of cells resulting in several peaks on the histogram. In order to identify the positive data set, flow cytometry should be repeated in the presence of an appropriate negative isotype control (see Fig. 15.13).

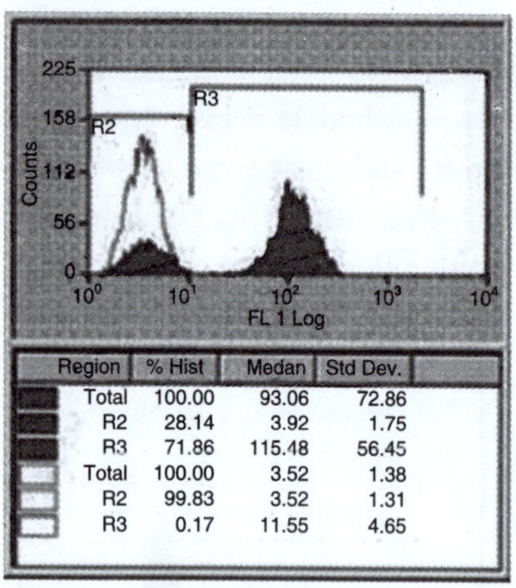

Region	% Hist	Medan	Std Dev.
Total	100.00	93.06	72.86
R2	28.14	3.92	1.75
R3	71.86	115.48	56.45
Total	100.00	3.52	1.38
R2	99.83	3.52	1.31
R3	0.17	11.55	4.65

Fig. 15.14: Statistical analysis

Analytical software packages that accompany flow cytometry instruments make measuring the % of positive-staining cells in histograms easy. For example, the F4/80 histogram is shown again below with statistics for R2 and R3 (known on this type of graph as 'bar regions').

In Fig. 15.14, 99.83% of the negative control (blue outline) is in R2. 28.14% of cells (red shade) 'stain negative' for F4/80 (R2) compared to 71.86% in the positive data set (R3). Additional statistics about the peaks (median and standard deviation) is also provided automatically here but this will vary with the software. A similar type of analysis will be generated for two-parameter histograms.

15.4.3 Two-parameter histograms

These are graphs that display two measurement parameters, one on the x-axis and one on the y-axis, and the cell count as a density (dot) plot or contour map. The parameters could be SSC, FSC or fluorescence. Some examples of two-parameter histograms were illustrated in Figs. 9 and 10. Another example is the dual-colour fluorescence histogram presented below. Lymphocytes were stained with anti-CD3 in the FITC channel (x-axis) and anti-HLA-DR in the PE channel (y-axis). CD3 and HLA-DR are markers for T cells and B cells, respectively.

In Fig. 15.15, R2 encompasses the PE-labelled B cells - note their positive shift along the PE axis. R5 contains the FITC-labelled T cells (positively shifted along the FITC axis). The top right quadrant contains a few 'activated T cells' (about 4% in this sample) that possess some HLA-DR expression also. As these stain with both antibody markers they are grouped in their own region (R3). R4 contains cells negative for both FITC and PE (no shift). Currently, flow cytometry can be performed on samples labelled with up to 17 fluorescence markers simultaneously. Therefore, a single experiment can yield a large set of data for analysis using various two-parameter histograms.

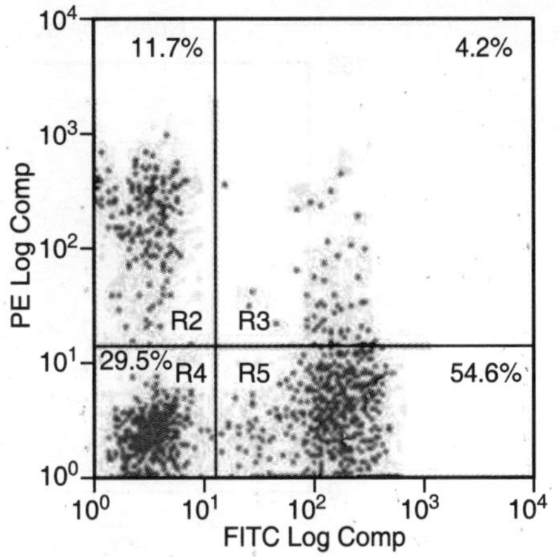

Fig. 15.15: Two-parameter (dual-colour fluorescence) histogram

15.5 INTRACELLULAR ANTIGENS

Staining intracellular antigens like cytokines can be difficult because antibody-based probes cannot pass sufficiently through the plasma membrane into the interior of the cell. To improve the situation, cells should first be fixed in suspension and then permeabilized before adding the fluorochrome. This allows probes to access intracellular structures while leaving the morphological scatter characteristics of the cells intact. Many commercial kits are available today that provide the reagents to carry out these crucial steps, e.g., LeucopermTM (see Fig. 15.16).

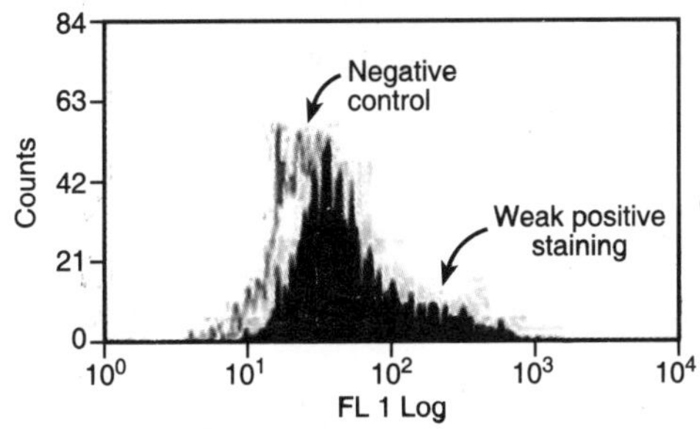

(a) Before

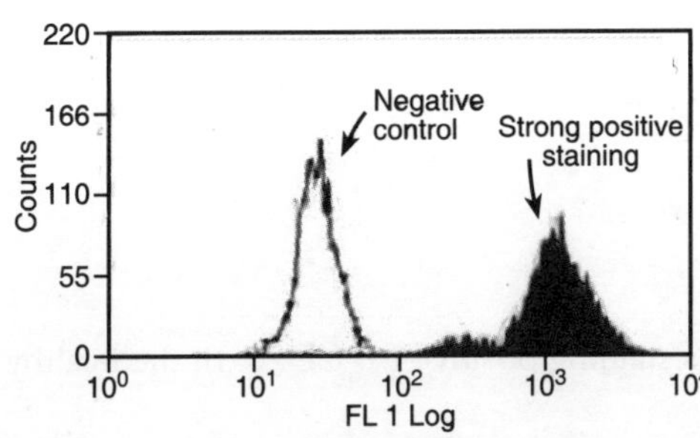

(b) After

Fig. 15.16: LeucopermTM used in conjunction with an antibody that recognizes MOMA-2, an intracellular antigen in mouse macrophages and monocytes. After fixation and permeabilization (b), notice how distinctive the positive data set becomes.

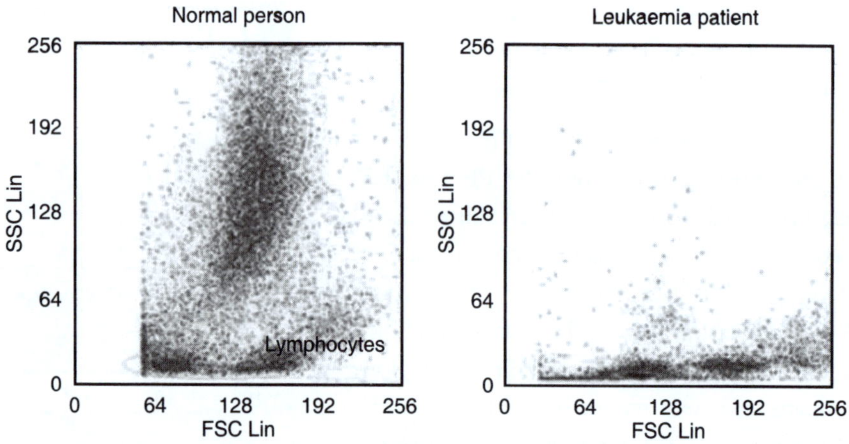

Fig. 15.17: Immunophenotyping

15.6 IMMUNOPHENOTYPING

All normal cells express a variety of cell surface markers, dependent on the specific cell type and degree of maturation. However, abnormal growth may interfere with the natural expression of markers resulting in overexpression of some and under-representation of others. Flow cytometry can be used to immunophenotype cells and thereby distinguish between healthy and diseased cells. It is unsurprising that today immunophenotyping is one of the major clinical applications of flow cytometry, and is used to aid the diagnosis of myelomas, lymphomas and leukaemias. It can also be used to monitor the effectiveness of clinical treatments. The differences between the blood profiles of a healthy individual and one suffering from leukaemia, for instance, are very dramatic. This can be seen from the FSC vs. SSC plots in Fig. 15.17. In the healthy person the cell types are clearly defined, whereas blood from a leukemia patient is abnormal and does not follow the classic profile.

Testing the patient's lymphocytes for specific cell surface markers also reveals more about the condition.

CD3

A normal person has a significant proportion of CD3-positive lymphocytes. In the patient with leukaemia, staining for CD3 is absent.

CD20

In the leukaemia patient there are a large number of cells staining positive for CD20. In the healthy person only a few stain positive.

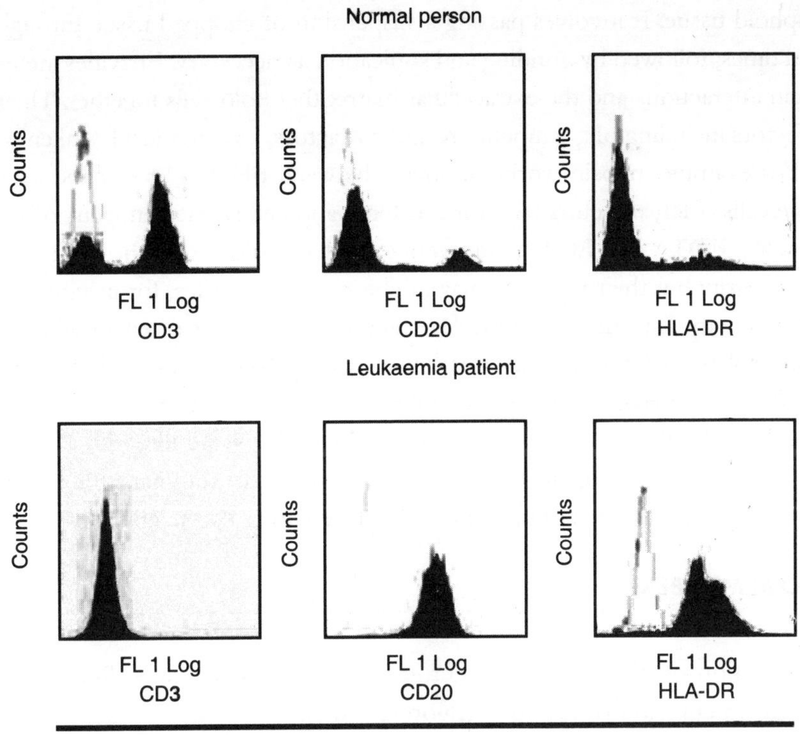

Fig. 15.18: Diagnosing leukaemia

HLA-DR

The leukaemia patient is HLA-DR-positive. In the normal person only a small number of cells stain positive. Being CD3-negative, CD20-positive and HLA-DR-positive, a clinician could diagnose with certainty that this patient is suffering from a B cell lineage leukaemia or lymphoma. The precise classification of disease may be determined using further antibodies.

15.7 SAMPLE PREPARATION

Single cells must be suspended at a density of 105-107 cells/ml to prevent the narrow bores of the flow cytometer and its tubing from clogging up. The concentration also influences the rate of flow sorting, which typically progresses at 2000-20,000 cells/second. However, higher sort speeds may decrease the purity of the preparation. Phosphate buffered saline (PBS) is a common suspension buffer and the most straightforward samples for flow cytometry include non-adherent cells from culture, water-borne micro-organisms, bacteria and yeast. Even whole blood is easy to use – red cells are usually removed by a simple lysis step; it is then possible to quickly identify lymphocytes, granulocytes and monocytes by their FSC/SSC characteristics. However, researchers may also wish to analyze cells from solid tissues, e.g., liver or tumours. In order to produce single cells, the solid material must be disaggregated. This can be done either mechanically or enzymatically. Mechanical disaggregation is suitable for loosely bound structures, e.g., adherent cells from culture, bone marrow and lymphoid tissue. It involves passing a suspension of chopped tissue through a fine-gauge needle several times, followed by grinding and sonication as necessary. Enzymes are used to disrupt protein-protein interactions and the extracellular matrix that hold cells together. Their action is dependent on factors including pH, temperature and co-factors, so care must be taken when choosing an enzyme. For example, pepsin works optimally between pH 1.5-2.5 but the acidic conditions would damage cells if left unneutralized for too long, and cell surface antigens of interest may be lost. Chelators like EDTA and EGTA can remove divalent cations responsible for maintaining cell function and integrity but their presence may inhibit certain enzymes, for instance, collagenase requires Ca^{2+} for activity. Enzymatic and mechanical disaggregation is often a trial and error process to optimize the isolation of the epitope under investigation. To study intracellular components, e.g., cytokines by flow cytometry, the plasma membrane of the cell must be permeabilized to allow dyes or antibody molecules through while retaining the cell's overall integrity. Low concentrations (up to 0.1%) of non-ionic detergents like saponin are suitable. In summary, the method for sample preparation will depend on the starting material and the nature of the epitope.

15.8 TIPS TO REMEMBER

- Absorption of photon elevates chromophore to excited state. Return to ground state results in emission of radiation (fluorochrome).
- A light or excitation source, typically a laser that emits light at a particular wavelength.
- A detector, in this case a photomultiplier tube, which is able to measure the brief flashes of light emitted as cells flow past the laser beam.

Flow-cytometer Instrument components

- **Electronics:** Control, pulse collection, pulse analysis, triggering, time delay, data display, gating, sort control, light and detector control
- **Optics:** Light source(s), detectors, spectral separation
- **Fluidics:** Specimen, sorting, rate of data collection
- **Data Analysis:** Data display & analysis, multivariate/simultaneous solutions, identification of sort populations, quantitation.

Data presentation format

- Histogram
- Dot plot
- Contour plot
- 3D plots
- Dot plot with projection
- Overviews (multiple histograms).

15.9 VIVA VOCE

Q.1. What properties of a cell or particle can be measured by a flow cytometer?

Q.2. What light source is used in most flow cytometers?

Q.3. What are the three main systems in a flow cytometer?

Q.4. What type of biological sample is best suited for flow cytometric analysis?

Q.5. What is the name given to the portion of the fluid stream where the cells are located?

Q.6. When cells labelled with fluorescent molecules pass through the focused laser beam, what two types of light signals are generated?

Q.7. Light emitted from a particle is collected by ___________.

Q.8. The electronic signal produced by the detectors is proportional to ___________.

Q.9. All flow cytometric measurements can be made simultaneously on a single cell. (True/False)

Q.10. Particles must be in single-cell suspension before flow-cytometric analysis. (True/False)

Q.11. The purpose of the fluidics system in a flow cytometer is ___________.

Q.12. What two factors can affect illumination of the particles within the laser beam?

Q.13. How many cells or particles should pass through the laser beam at a given time?

Q.14. The particle suspension is injected into ______________ within the ______________.

Q.15. The process of centring the sample core within the sheath fluid is known as ___________.

Q.16. Which regulator controls the diameter of the sample core?

Q.17. What are the three possible pressure settings for a benchtop flow-cytometer?

Q.18. Increasing sample pressure ______________ the sample flow rate and the ______________ of the sample core.

Q.19. Good data resolution is required for DNA studies. What flow rate is recommended?

Q.20. A wider sample core decreases resolution. (True/False)

Q.21. A high flow rate can be used when performing qualitative measurements. (True/False)

Q.22. What does the horizontal axis in a histogram represent?

Q.23. A dot plot can be used to display _________ parameters.

CHAPTER 16

DNA Microarray

16.1 INTRODUCTION

A DNA microarray (also commonly known as *gene* or *genome* chip, DNA chip, or gene array) is a collection of microscopic DNA spots, commonly representing single genes, arrayed on a solid surface by covalent attachment to chemically suitable matrices. DNA arrays are different from other types of microarray, only in that they either measure DNA or use DNA as part of its detection system. Qualitative or quantitative measurements with DNA microarrays utilize the selective nature of DNA-DNA or DNA-RNA hybridization under high-stringency conditions and fluorophore-based detection. DNA arrays are commonly used for expression profiling, i.e., monitoring expression levels of thousands of genes simultaneously, or for comparative genomic hybridization. Arrays of DNA can either be spatially arranged, as in commonly known *gene* or *genome* chip, DNA chip, or gene array, or can be specific DNA sequences tagged or labelled such that they can be independently identified in solution. The traditional solid-phase array is a collection of microscopic DNA spots attached to a solid surface, such as glass, plastic or silicon chip. The affixed DNA segments are known as probes (although some sources will use different nomenclature such as reporters), thousands of which can be placed in known locations on a single DNA microarray. Microarray technology evolved from Southern blotting, whereby fragmented DNA is attached to a substrate and then probed with a known gene or fragment. Applications of these arrays include:

1. mRNA or gene expression profiling—Monitoring expression levels for thousands of genes simultaneously is relevant to many areas of biology and medicine, such as studying treatments, disease, and developmental stages. For example, microarrays can be used to identify disease genes by comparing gene expression in diseased and normal cells.
2. Comparative genomic hybridization (Array CGH)—Assessing large genomic rearrangements.
3. SNP detection arrays—Looking for single nucleotide polymorphism in the genome of populations.

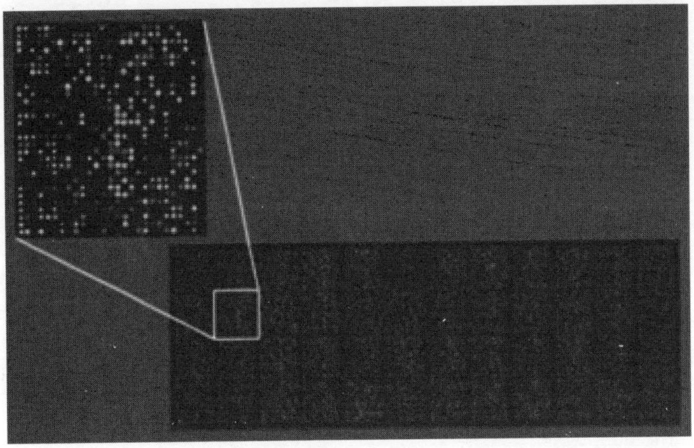

Fig. 16.1: Example of an approximately 40,000 probe spotted oligo microarray with enlarged inset to show in details

4. Chromatin immunoprecipitation (ChIP) studies—Determining protein binding site occupancy throughout the genome, employing ChIP-on-chip technology.

16.2 FABRICATION

Microarrays can be fabricated using a variety of technologies, including printing with fine-pointed pins onto glass slides, photolithography using pre-made masks, photolithography using dynamic micromirror devices, ink-jet printing, or electrochemistry on microelectrode arrays.

DNA microarrays can be used to detect RNAs that may or may not be translated into active proteins. Scientists refer to this kind of analysis as "expression analysis" or expression profiling. Since there can be tens of thousands of distinct probes on an array, each microarray experiment can accomplish the equivalent number of genetic tests in parallel. Arrays have, therefore, dramatically accelerated many types of investigations. The use of microarrays for gene expression profiling was first published in 1995 (Science) and the first complete eukaryotic genome (*Saccharomyces cerevisiae*) on a microarray was published in 1997 (*Science*).

1. Spotted microarrays

In spotted microarrays (or two-channel or two-colour microarrays), the probes are oligonucleotides, cDNA or small fragments of PCR products that correspond to mRNAs and are spotted onto the microarray surface. This type of array is typically hybridized with cDNA from two samples to be compared (e.g., diseased tissue versus healthy tissue) that are labelled with two different fluorophores (e.g., Rhodamine (Cyanine 5, red) and Fluorescein (Cyanine 3, green)). The two samples are mixed and hybridized to a single microarray that is then scanned in a microarray scanner to visualize fluorescence of the two fluorophores. Relative intensities of each fluorophore are then used to identify up-regulated and down-regulated genes in ratio-based analysis. Absolute levels of gene

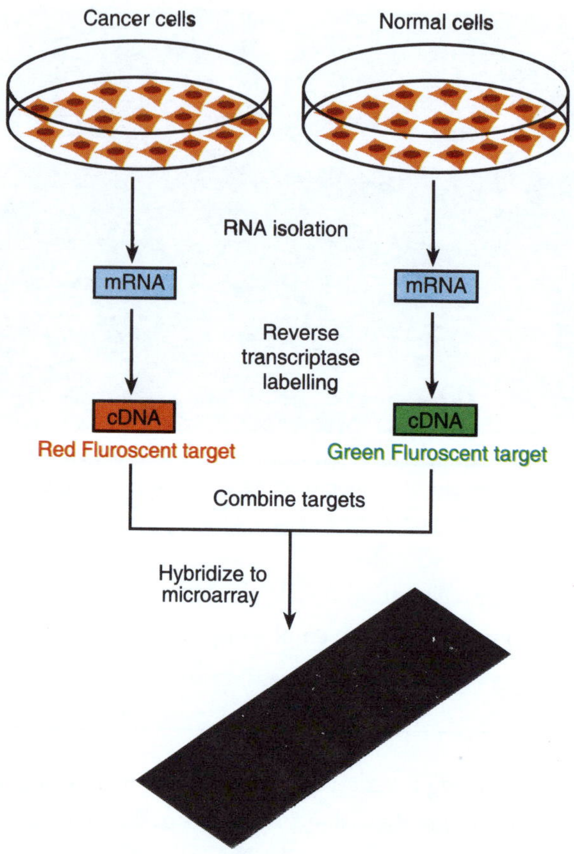

Fig. 16.2: Diagram of typical dual-colour microarray experiment

expression cannot be determined in the two-colour array, but relative differences in expression among different spots (= genes) can be estimated with some oligonucleotide arrays.

Fig. 16.3: Two Affymetrix chips

2. Oligonucleotide microarrays

In oligonucleotide microarrays (or single-channel microarrays), the probes are designed to match parts of the sequence of known or predicted mRNAs. There are commercially available designs that cover complete genomes from companies such as GE Healthcare, Affymetrix, Ocimum Biosolutions, or Agilent. These microarrays give estimations of the absolute value of gene expression and, therefore, the comparison of two conditions requires the use of two separate microarrays.

Oligonucleotide Arrays can be either produced by piezoelectric deposition with full length oligonucleotides or in situ synthesis.

Long Oligonucleotide Arrays are composed of 60-mers, or 50-mers and are produced by ink-jet printing on a silica substrate. Short Oligonucleotide Arrays are composed of 25-mer or 30-mer and are produced by photolithographic synthesis (Affymetrix) on a silica substrate or piezoelectric deposition (GE Healthcare) on an acrylamide matrix. More recently, Maskless Array Synthesis from NimbleGen Systems has combined flexibility with large numbers of probes. Arrays can contain up to 390,000 spots, from a custom array design. New array formats are being developed to study specific pathways or disease states for a systems biology approach.

Oligonucleotide microarrays often contain control probes designed to hybridize with RNA spike-ins. The degree of hybridization between the spike-ins and the control probes is used to normalize the hybridization measurements for the target probes.

16.3 GENOTYPING MICROARRAYS

DNA microarrays can also be used to read the sequence of a genome in particular positions. SNP microarrays are a particular type of DNA microarrays that are used to identify genetic variation in individuals and across populations. Short oligonucleotide arrays can be used to identify the single nucleotide polymorphisms (SNPs) that are thought to be responsible for genetic variation and the source of susceptibility to genetically caused diseases. Generally termed genotyping applications, DNA microarrays may be used in this fashion for forensic applications, rapidly discovering or measuring genetic predisposition to disease, or identifying DNA-based drug candidates.

These SNP microarrays are also being used to profile somatic mutations in cancer, specifically loss of heterozygosity events and amplifications and deletions of regions of DNA. Amplifications and deletions can also be detected using comparative genomic hybridization, or aCGH, in conjunction with microarrays, but may be limited in detecting novel Copy Number Polymorphisms, or CNPs, by probe coverage.

Resequencing arrays have also been developed to sequence portions of the genome in individuals. These arrays may be used to evaluate germline mutations in individuals, or somatic mutations in cancers. Genome tiling arrays include overlapping oligonucleotides designed to blanket an entire genomic region of interest. Many companies have successfully designed tiling arrays that cover whole human chromosomes.

16.4 MICROARRAYS AND BIOINFORMATICS

1. Experimental Design

Due to the biological complexity of gene expression, the considerations of experimental design that are discussed in the expression profiling article are of critical importance if statistically and biologically valid conclusions are to be drawn from the data.

Fig. 16.4: Gene expression values from microarray experiments can be represented as heat maps to visualize the result of data analysis

There are three main elements to consider when designing a microarray experiment. First, replication of the biological samples is essential for drawing conclusions from the experiment. Second, technical replicates (two RNA samples obtained from each experimental unit) help to ensure precision and allow for testing differences within treatment groups. The technical replicates may be two independent RNA extractions or two aliquots of the same extraction. Third, spots of each cDNA clone or oligonucleotide are present at least as duplicates on the microarray slide, to provide a measure of technical precision in each hybridization. It is critical that information about the sample preparation and handling is discussed in order to help identify the independent units in the experiment as well as to avoid inflated estimates of significance.

2. Standardization

The lack of standardization in arrays presents an interoperability problem in bioinformatics, which hinders the exchange of array data. Various grass-roots open-source projects are attempting to facilitate the exchange and analysis of data produced with non-proprietary chips.

1. The "Minimum Information About a Microarray Experiment" (MIAME) checklist helps define the level of detail that should exist and is being adopted by many journals as a requirement for the submission of papers incorporating microarray results. MIAME describes the minimum required information for complying experiments, but not its format. Thus, as of 2007, whilst many formats can support the MIAME requirements there is no format which permits verification of complete semantic compliance.
2. The "MicroArray Quality Control (MAQC) Project" is being conducted by the FDA to develop standards and quality control metrics which will eventually allow the use of MicroArray data in drug discovery, clinical practice and regulatory decision-making.
3. The MicroArray and Gene Expression (MAGE) group is working on the standardization of the representation of gene expression data and relevant annotations.

3. Statistical analysis

The analysis of DNA microarrays poses a large number of statistical problems, including the normalization of the data. There are dozens of proposed normalization methods in the published literature; as in many other cases where authorities disagree, a sound conservative approach is to try a number of popular normalization methods and compare the conclusions reached: how sensitive are the main conclusions to the method chosen?

From a hypothesis-testing perspective, the large number of genes present on a single array means that the experimenter must take into account a multiple testing problem; even if the statistical P-value assigned to a given gene indicates that it is extremely unlikely that differential expression of this gene was due to random rather than treatment effects, the very high number of genes on an array makes it likely that differential expression of some genes represents false positives or false negatives. Statistical methods tailored to microarray analyses have recently become available that assess statistical power based on the variation present in the data and the number of experimental replicates, and can help minimize type I and type II errors in the analyses.

A basic difference between microarray data analysis and much traditional biomedical research is the dimensionality of the data. A large clinical study might collect 100 data items per patient for thousands of patients. A medium-size microarray study will obtain many thousands of numbers per sample for perhaps a hundred samples. Many analysis techniques treat each sample as a single point in a space with thousands of dimensions, then attempt by various techniques to reduce the dimensionality of the data to something humans can visualize.

4. Relation between probe and gene

The relation between a probe and the mRNA that it is expected to detect is problematic. On the one hand, some mRNAs may cross-hybridize probes in the array that are supposed to detect another mRNA. On the other hand, probes that are designed to detect the mRNA of a particular gene may be relying on genomic EST information that is incorrectly associated with that gene.

16.5 PUBLIC DATABASES OF MICROARRAY DATA

Database	Microarray Experiment Sets	Sample Profiles	As of Date
Gene Expression Omnibus - NCBI	5366	134669	April 1, 2007
Stanford Microarray database	12742	?	April 1, 2007
UNC Microarray database	~31	2093	April 1, 2007
MUSC database	~45	555	April 1, 2007
ArrayExpress at EBI	1643	136	April 1, 2007
caArray at NCI	41	1741	November 15, 2006

16.6 ONLINE MICROARRAY DATA ANALYSIS PROGRAMS AND TOOLS

Several Open Directory Project categories list online microarray data analysis programs and tools:

- **Bioinformatics: Online Services:** Gene Expression and Regulation at the Open Directory Project

- **Gene Expression:** Databases at the Open Directory Project
- **Gene Expression:** Software at the Open Directory Project
- **Data Mining:** Tool Vendors at the Open Directory Project
- **Bioconductor:** Open source and open development software project for the analysis and comprehension of genomic data
- **Genevestigator:** Web-based database and analysis tool to study gene expression across large sets of tissues, developmental stages, drugs, stimuli, and genetic modifications.

16.7 TIPS TO REMEMBER

- A DNA microarray (also commonly known as *gene* or *genome* chip, DNA chip, or gene array) is a collection of microscopic DNA spots, commonly representing single genes, arrayed on a solid surface by covalent attachment to chemically suitable matrices.
- The use of microarrays for gene expression profiling was first published in 1995 (*Science*) and the first complete eukaryotic genome (*Saccharomyces cerevisiae*) on a microarray was published in 1997 (*Science*).

16.8 VIVA VOCE

Q.1. Define microarray.

Q.2. What do you understand by fabrication?

Q.3. Differentiate between spotted microarray and oligonucleotide microarray.

Q.4. What is SNP microarray?

Q.5. What is standardization in microarray and why is it necessary?

Q.6. Name four public microarray databases.

CHAPTER 17

RNA Interference (RNAi)

17.1 INTRODUCTION

Cells use dicer to trim double stranded RNA to form small interfering RNA or microRNA. An exogenous dsRNA or endogenous pre-miRNA can be processed by dicer and incorporated into the RNA-induced silencing complex (RISC), which targets single-stranded messenger RNA molecules and triggers translational repression; incorporation into the RNA-induced transcriptional silencing complex (RITS) induces genome maintenance activities such as histone methylation and chromatin reorganization.

RNA interference (also called "RNA-mediated interference", abbreviated **RNAi**) is a mechanism for RNA-guided regulation of gene expression in which double-stranded ribonucleic acid inhibits the expression of genes with complementary nucleotide sequences. Conserved in most eukaryotic organisms, the RNAi pathway is thought to have evolved as a form of innate immunity against viruses and also plays a major role in regulating development and genome maintenance.

The RNAi pathway is initiated by the enzyme dicer, which cleaves double-stranded RNA (dsRNA) to short double-stranded fragments of 20-25 base pairs. One of the two strands of each fragment, known as the *guide strand*, is then incorporated into the RNA-induced silencing complex (RISC) and base-pairs with complementary sequences. The well-studied outcome of this recognition event is a form of post-transcriptional gene silencing. This occurs when the guide strand base pairs with a messenger RNA (mRNA) molecule and induces degradation of the mRNA by argonaute, the catalytic component of the RISC complex. The short RNA fragments are known as small interfering RNA (siRNA), when they derive from exogenous sources and microRNA (miRNA), when they are produced from RNA-coding genes in the cell's own genome. The RNAi pathway has been particularly well-studied in certain model organisms such as the nematode worm *Caenorhabditis elegans*, the fruit fly *Drosophila melanogaster*, and the flowering plant Arabidopsis thaliana.

The selective and robust effect of RNAi on gene expression makes it a valuable research tool, both in cell culture and in living organisms; synthetic dsRNA introduced into cells can induce suppression of specific genes of interest. RNAi may also be used for large-scale screens that systematically shut down each gene in the cell, which can help identify the components necessary for a particular cellular process or an event such as cell division. Exploitation of the pathway is also a promising tool in biotechnology and medicine.

Historically, RNA interference was known by other names, including post transcriptional gene silencing, transgene silencing, and quelling. Only after these apparently-unrelated processes were fully understood it became clear that they all described the RNAi phenomenon. RNAi has also been confused with antisense suppression of gene expression, which does not act catalytically to degrade mRNA but instead involves single-stranded RNA fragments physically binding to mRNA and blocking translation. In 2006, Andrew Fire and Craig C. Mello shared the Nobel Prize in Physiology or Medicine for their work on RNA interference in the nematode worm *C. elegans*, which they published in 1998.

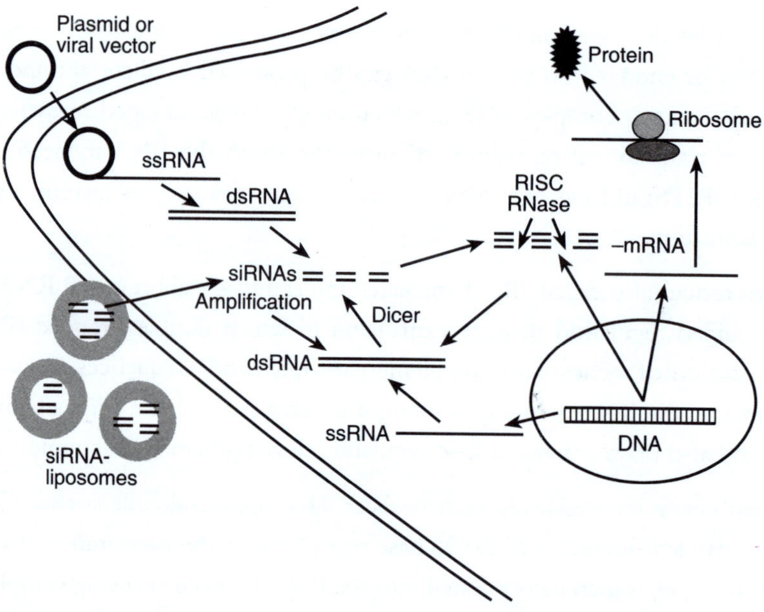

Fig. 17.1: Pathways of RNAi. Cells produce single-stranded RNA (ssRNA), which provides a template for the formation of dsRNA, which involves the activity of RNA-dependent RNA polymerases. The dsRNA is then cleaved by a protein called Dicer to form small 21-23 nucleotide siRNAs. The siRNAs (blue) then associate with the specific mRNA targeted by their nucleotide sequence (red) in a nucleic acid-protein complex called RISC, which includes RNase activity that degrades the mRNA at sites not bound by the siRNAs. The synthesis of the protein encoded by the mRNA targeted by the siRNAs is prevented, and that protein is selectively depleted from the cell. RNAi-mediated silencing can be induced experimentally by introducing synthetic siRNAs into cells using various transfection methods including liposomes (bottom left). Viral vectors can also be used to express dsRNAs against a specific gene, which are then acted upon by Dicer.

17.2 CELLULAR MECHANISMS

RNAi is an RNA-dependent gene silencing process that is mediated by the RNA-induced silencing complex (RISC) and is initiated by short double-stranded RNA molecules in the cytoplasm, where they interact with the catalytic RISC component argonaute. When the dsRNA is exogenous, coming from infection by a virus with an RNA genome or laboratory manipulations, the RNA is imported directly into the cytoplasm and cleaved to short fragments by the enzyme dicer. The initiating dsRNA can also be endogenous, as in pre-microRNAs expressed from RNA-coding genes in the genome. The primary transcripts from such genes are first processed to the characteristic stem-loop structure of pre-miRNA in the nucleus, then exported to the cytoplasm to be cleaved by dicer. Thus the two pathways for exogenous and endogenous dsRNA converge at the RISC complex, which mediates gene silencing effects.

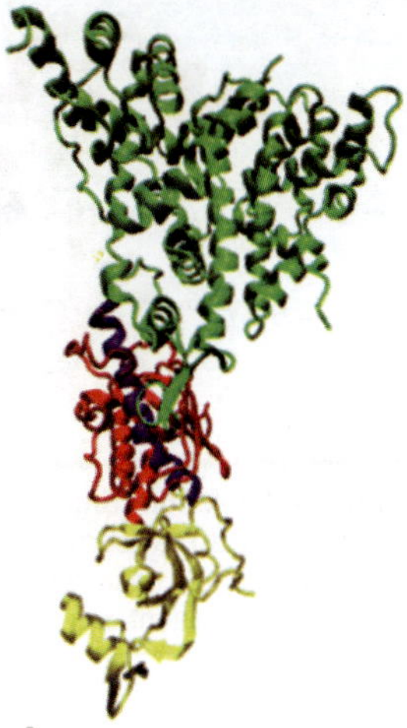

Fig. 17.2: The dicer protein from Giardia intestinalis, which catalyzes the cleavage of dsRNA to siRNAs. The RNase domains are coloured green, the PAZ domain yellow, the platform domain red, and the connector helix blue. The distance between the RNase and PAZ domains, determined by the length and angle of the connector helix, may determine the length of siRNA molecules produced by dicer variants.

17.2.1 dsRNA cleavage

Exogenous dsRNA initiates RNAi by activating the ribonuclease protein dicer, which binds and cleaves double-stranded RNAs (dsRNAs) to produce double-stranded fragments of 20-25 base pairs with a few unpaired overhang bases on each end. Bioinformatics studies on the genomes of multiple organisms suggest this length maximizes target-gene specificity and minimizes non-specific effects. These short double-stranded fragments are called small interfering RNAs (siRNAs). These siRNAs are then separated into single strands and integrated into an active RISC complex. After integration into the RISC, siRNAs base-pair to their target mRNA and induce cleavage of the mRNA, thereby preventing it from being used as a translation template.

Exogenous dsRNA is detected and bound by an effector protein known as RDE-4 in *C. elegans* and R2D2 in *Drosophila* that stimulates dicer activity. This protein only binds long dsRNAs, but the mechanism producing this length specificity is unknown. These RNA-binding proteins then facilitate transfer of cleaved siRNAs to the RISC complex. This initiation pathway may be amplified by the cell through the synthesis of a population of 'secondary' siRNAs using the dicer-produced initiating or 'primary' siRNAs as templates. These siRNAs are structurally distinct from dicer-produced siRNAs and appear to be produced by an RNA-dependent RNA polymerase (RdRP).

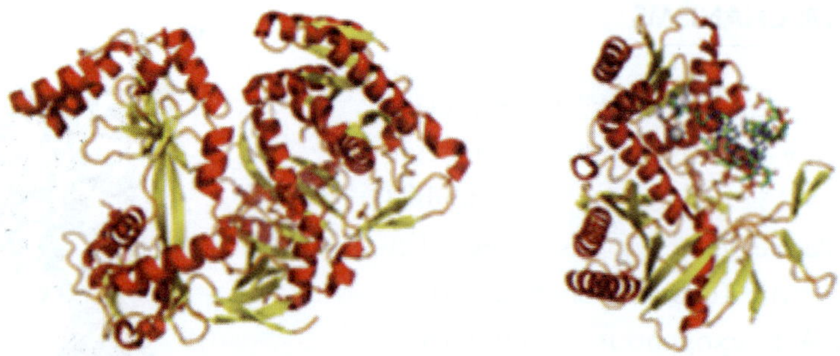

Fig. 17.3: *Left:* A full-length argonaute protein from the archaea species *Pyrococcus furiosus*. *Right:* The PIWI domain of an argonaute protein in complex with double-stranded RNA. The base-stacking interaction between the 5′ base on the guide strand and a conserved tyrosine residue (light blue) is highlighted; the stabilizing divalent cation (magnesium) is shown as a gray sphere.

17.2.2 microRNA

MicroRNAs (miRNAs) are genomically encoded non-coding RNAs that help regulate gene expression, particularly during development. The phenomenon of RNA interference, broadly defined, includes the endogenously induced gene silencing effects of miRNAs as well as silencing triggered by foreign dsRNA. Mature miRNAs are structurally similar to siRNAs produced from exogenous dsRNA, but miRNAs must first undergo extensive post-transcriptional modification. An miRNA is expressed from a much longer RNA-coding gene as a primary transcript known as a *pri-miRNA*, which is processed in the cell nucleus to a 70-nucleotide stem-loop structure called a *pre-miRNA* by the microprocessor complex. This complex consists of an RNase III enzyme called *Drosha* and a dsRNA-binding protein *Pasha*. The dsRNA portion of this pre-miRNA is bound and cleaved by dicer to produce the mature miRNA molecule that can be integrated into the RISC complex; thus, miRNA and siRNA share the same cellular machinery downstream of their initial processing.

The siRNAs derived from long dsRNA precursors differ from miRNAs; in that miRNAs, especially those in animals, typically have incomplete base pairing to a target and inhibit the translation of many different mRNAs with similar sequences. In contrast, siRNAs typically base-pair perfectly and induce mRNA cleavage only in a single, specific target. In *Drosophila* and *C. elegans*, miRNA and siRNA are processed by distinct argonaute proteins and dicer enzymes.

17.2.3 RISC activation and catalysis

The catalytically-active components of the RISC complex are endonucleases called argonaute proteins, which cleave the target mRNA strand complementary to their bound siRNA. As the fragments produced by dicer are double-stranded, they could each in theory produce a functional siRNA. However, only one of the two strands, which is known as the *guide strand*, binds the argonaute protein and directs gene silencing. The other *anti-guide strand* or *passenger strand* is degraded during RISC activation. Although it was first believed that an ATP-dependent helicase separated these two

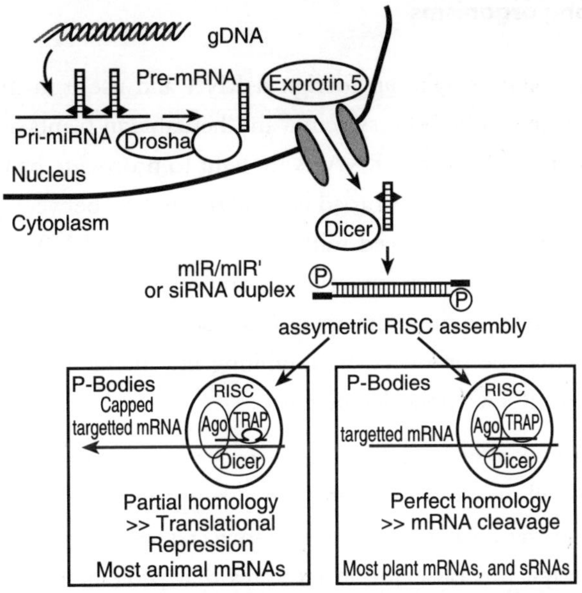

Fig. 17.4: Illustration of the major differences between plant and animal gene silencing. Natively expressed microRNA or exogenous small interfering RNA is processed by dicer and integrated into the RISC complex, which mediates gene silencing. In general, miRNAs in plants match perfectly to their gene targets and induce direct messenger RNA cleavage, while miRNAs in animals often have less than perfect base pairing to a number of targets, and induce translational repression.

strands, the process is actually ATP-independent and performed directly by the protein components of RISC. The strand selected as the guide tends to be that with a more stable 5′-end, but strand selection is unaffected by the direction in which dicer cleaves the dsRNA before RISC incorporation. Instead, the R2D2 protein may serve as the differentiating factor by binding the less-stable 5′-end of the passenger strand.

The structural basis for binding of RNA to the argonaute protein was examined by X-ray crystallography of the binding domain of an RNA-bound argonaute protein. Here, the phosphorylated 5′-end of the RNA strand enters a conserved basic surface pocket and makes contacts through a divalent cation such as magnesium and by aromatic stacking between the 5′ nucleotide in the siRNA and a conserved tyrosine residue. This site is thought to form a nucleation site for the binding of the siRNA to its mRNA target.

It is not understood how the activated RISC complex locates complementary mRNAs within the cell. Although the cleavage process has been proposed to be linked to translation, translation of the mRNA target is not essential for RNAi-mediated degradation. Indeed, RNAi may be more effective against mRNA targets that are not translated. Argonaute proteins, the catalytic components of RISC, are localized to specific regions in the cytoplasm called P-bodies (also cytoplasmic bodies or GW bodies), which are regions with high rates of mRNA decay; miRNA activity is also clustered in P-bodies. Disruption of P bodies in cells decreases the efficiency of RNA interference, suggesting that they are the site of a critical step in the RNAi process.

17.2.4 Variation among organisms

Organisms vary in their ability to take up foreign dsRNA and use it in the RNAi pathway. The effects of RNA interference can be both systemic and heritable in plants and *C. elegans*, although not in *Drosophila* or mammals. In plants, RNAi is thought to propagate by the transfer of siRNAs between cells through plasmodesmata. A broad general distinction between plants and animals lies in the targeting of endogenously produced miRNAs; in plants, miRNAs are usually perfectly or nearly perfectly complementary to their target genes and induce direct mRNA cleavage by RISC, while animals' miRNAs tend to be more divergent in sequence and induce translational repression. This translational effect may be produced by inhibiting the interactions of translation initiation factors with the messenger RNA's polyadenine tail.

Some eukaryotic protozoa such as *Leishmania major* and *Trypanosoma cruzi* lack the RNAi pathway entirely. Most or all of the components are also missing in some fungi, most notably the model organism *Saccharomyces cerevisiae*. Certain ascomycetes and basidiomycetes are also missing RNA interference pathways; this observation indicates that proteins required for RNA silencing have been lost independently from many fungal lineages, possibly due to the evolution of a novel pathway with similar function, or to the lack of selective advantage in certain niches.

17.3 BIOLOGICAL FUNCTIONS

17.3.1 Immunity

RNA interference is a vital part of the immune response to viruses and other foreign genetic material, especially in plants where it may also prevent self-propagation by transposons. Plants such as *Arabidopsis thaliana* express multiple dicer homologues that are specialized to react differently when the plant is exposed to different types of viruses. Even before the RNAi pathway was fully understood, it was known that induced gene silencing in plants could spread throughout the plant in a systemic effect, and could be transferred from stock to scion plants via grafting. This phenomenon has since been recognized as a feature of the plant innate immune system, and allows the entire plant to respond to a virus after an initial localized encounter. In response, many plant viruses have evolved elaborate mechanisms that suppress the RNAi response in plant cells. These include viral proteins that bind short double-stranded RNA fragments with single-stranded overhang ends, such as those produced by the action of dicer. Some plant genomes also express endogenous siRNAs in response to infection by specific types of bacteria. These effects may be part of a generalized response to pathogens that downregulates any metabolic processes in the host that aid the infection process.

Although animals generally express fewer variants of the dicer enzyme than plants, RNAi in some animals has also been shown to produce an antiviral response. In both juvenile and adult *Drosophila*, RNA interference is important in antiviral innate immunity and is active against pathogens such as Drosophila X virus. A similar role in immunity may operate in *C. elegans*, as argonaute proteins are upregulated in response to viruses, and worms that overexpress components of the RNAi pathway are resistant to viral infection. The role of RNA interference in mammalian innate immunity is poorly understood, and relatively little data is available. However, the existence of viruses that encode genes able to suppress the RNAi response in mammalian cells may be evidence in favour of an RNAi-dependent mammalian immune response. However, this hypothesis of RNAi-mediated immunity in mammals has been challenged as poorly substantiated. Alternative functions for RNAi in mammalian viruses also exist, such as miRNAs expressed by the herpes virus that may act as heterochromatin organization triggers to mediate viral latency

17.3.2 Genome maintenance

Components of the RNA interference pathway are used in many eukaryotes in the maintenance of the organisation and structure of their genomes. Modification of histones and associated induction of heterochromatin formation serve to downregulate genes pre-transcriptionally; this process is referred to as RNA-induced transcriptional silencing (RITS), and is carried out by a complex of proteins called the RITS complex. In fission yeast this complex contains argonaute, a chromodomain protein Chp1, and a protein called Tas3 of unknown function. As a consequence, the induction and spread of heterochromatic regions requires the argonaute and RdRP proteins. Indeed, deletion of these genes in the fission yeast *S. pombe* disrupts histone methylation and centromere formation, causing slow or stalled anaphase during cell division. In some cases, similar processes associated with histone modification have been observed to transcriptionally upregulate genes.

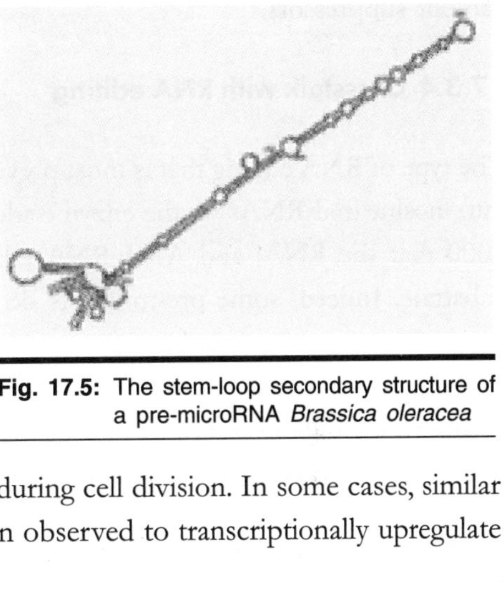

Fig. 17.5: The stem-loop secondary structure of a pre-microRNA *Brassica oleracea*

The mechanism by which the RITS complex induces heterochromatin formation and organization is not well understood, and most studies have focused on the mating-type region in fission yeast, which may not be representative of activities in other genomic regions or organisms. In maintenance of existing heterochromatin regions, RITS forms a complex with siRNAs complementary to the local genes and stably binds local methylated histones, acting co-transcriptionally to degrade any nascent pre-mRNA transcripts that are initiated by RNA polymerase. The formation of such a heterochromatin region, though not its maintenance, is dicer-dependent, presumably because dicer is required to generate the initial complement of siRNAs that target subsequent transcripts. Heterochromatin maintenance has been suggested to function as a self-reinforcing feedback loop, as new siRNAs are formed from the occasional nascent transcripts by RdRP for incorporation into local RITS complexes. The relevance of observations from fission yeast mating-type regions and centromeres to mammals is not clear, as heterochromatin maintenance in mammalian cells may be independent of the components of the RNAi pathway.

17.3.3 miRNAs and gene regulation

Endogenously expressed miRNAs, including both intronic and intergenic miRNAs, are most important in translational repression and in the regulation of development, especially the timing of morphogenesis and the maintenance of undifferentiated or incompletely differentiated cell types such as stem cells. The role of endogenously expressed miRNA in downregulating gene expression was first described in C. elegans in 1993. In plants this function was discovered when the "JAW microRNA" of *Arabidopsis* was shown to be involved in the regulation of several genes that control plant shape. In plants, the majority of genes regulated by miRNAs are transcription factors; thus miRNA activity is particularly wide-ranging and regulated entire gene networks during development by modulating the expression of key regulatory genes, including transcription factors as well as F-box proteins. In many organisms, including humans, miRNAs have also been linked to the formation of tumours and dysregulation of the cell cycle. Here, miRNAs can function as both oncogenes and tumour suppressors.

17.3.4 Crosstalk with RNA editing

The type of RNA editing that is most prevalent in higher eukaryotes converts adenosine nucleotides into inosine in dsRNAs via the enzyme adenosine deaminase (ADAR). It was originally proposed in 2000 that the RNAi and A→I RNA editing pathways might compete for a common dsRNA substrate. Indeed, some pre-miRNAs do undergo A→I RNA editing, and this mechanism may regulate the processing and expression of mature miRNAs. Furthermore, at least one mammalian ADAR can sequester siRNAs from RNAi pathway components. Further support for this model comes from studies on ADAR-null *C. elegans* strains indicating that A→I RNA editing may counteract RNAi silencing of endogenous genes and transgenes.

17.3.5 Related prokaryotic systems

Gene expression in prokaryotes is influenced by an RNA-based system similar in some respects to RNAi. Here, RNA-encoding genes control mRNA abundance or translation by producing a complementary RNA that binds to an mRNA by base pairing. However, these regulatory RNAs are not generally considered to be analogous to miRNAs because the dicer enzyme is not involved. It has been suggested that CRISPR systems in prokaryotes are analogous to eukaryotic RNA interference systems, although none of the protein components are orthologous.

17.4 EVOLUTION

Based on parsimony-based phylogenetic analysis, the most recent common ancestor of all eukaryotes most likely already possessed an early RNA interference pathway; the absence of the pathway in certain eukaryotes is thought to be a derived characteristic. The ancestral RNAi system probably contained at least one dicer-like protein, one argonaute, one PIWI protein, and an RNA dependent

RNA polymerase that may have also played other cellular roles. A large-scale comparative genomics study likewise indicates that the eukaryotic crown group already possessed these components, which may then have had closer functional associations with generalized RNA degradation systems such as the exosome. This study also suggests that the RNA-binding argonaute protein family, which is shared among eukaryotes, most archaea, and at least some bacteria (such as *Aquifex aeolicus*), is homologous to and originally evolved from components of the translation initiation system.

The ancestral function of the RNAi system is generally agreed to have been immune defense against exogenous genetic elements such as transposons and viral genomes. Related functions such as histone modification may have already been present in the ancestor of modern eukaryotes, although other functions such as regulation of development by miRNA are thought to have evolved later.

RNA interference genes, as components of the antiviral innate immune system in many eukaryotes, are involved in an evolutionary arms race with viral genes. Some viruses have evolved mechanisms for suppressing the RNAi response in their host cells, an effect that has been noted particularly for plant viruses. Studies of evolutionary rates in *Drosophila* have shown that genes in the RNAi pathway are subject to strong directional selection and are among the fastest-evolving genes in the *Drosophila genome.*

17.5 GENE KNOCKDOWN

A wild-type adult *Caenorhabditis elegans* nematode worm, grown under RNAi suppression of a nuclear hormone receptor involved in desaturase regulation. These worms have abnormal fatty acid metabolism but are viable and fertile. The RNA interference pathway is often exploited in experimental biology to study the function of genes in cell culture and in vivo in model organisms. Double-stranded RNA is synthesized with a sequence complementary to a gene of interest and introduced into a cell or organism, where it is recognized as exogenous genetic material and activates the RNAi pathway. Using this mechanism, researchers can cause a drastic decrease in the expression of a targeted gene. Studying the effects of this decrease can show the physiological role of the gene product. Since RNAi may not totally abolish expression of the gene, this technique is sometimes referred to as a "knockdown", to distinguish it from "knockout" procedures in which expression of a gene is entirely eliminated.

Extensive efforts in computational biology have been directed toward the design of successful dsRNA reagents that maximize gene knockdown but minimize "off-target" effects. Off-target effects arise when an introduced RNA has a base sequence that can pair with and thus reduce the expression of multiple genes at a time. Such problems occur more frequently when the dsRNA contains repetitive sequences. It has been estimated from studying the genomes of *H. sapiens, C. elegans*, and *S. pombe* that about 10% of possible siRNAs will have substantial off-target effects. A multitude of software tools have been developed implementing algorithms for the design of general, mammal-specific, and virus-specific siRNAs that are automatically checked for possible cross-reactivity.

Depending on the organism and experimental system, the exogenous RNA may be a long strand designed to be cleaved by dicer, or short RNAs designed to serve as siRNA substrates. In most mammalian cells, shorter RNAs are used because long double-stranded RNA molecules induce the mammalian interferon response, a form of innate immunity that reacts nonspecifically to foreign genetic material. Mouse oocytes and cells from early mouse embryos lack this reaction to exogenous dsRNA and are, therefore, a common model system for studying gene-knockdown effects in mammals. Specialized laboratory techniques have also been developed to improve the utility of RNAi in mammalian systems by avoiding the direct introduction of siRNA, for example, by stable transfection with a plasmid encoding the appropriate sequence from which siRNAs can be transcribed, or by more elaborate lentiviral vector systems allowing the inducible activation or deactivation of transcription, known as *conditional RNAi*.

17.5.1 Functional genomics

Most functional genomics applications of RNAi in animals have used *C. elegans* and *D. melanogaster*, as these are the common model organisms in which RNAi is most effective. *C. elegans* is particularly useful for RNAi research for two reasons: firstly, the effects of the gene silencing are generally heritable, and secondly, because delivery of the dsRNA is extremely simple. Through a mechanism whose details are poorly understood, bacteria such as *E. coli* that carry the desired dsRNA can be fed to the worms and will transfer their RNA payload to the worm via the intestinal tract. This "delivery by feeding" is just as effective at inducing gene silencing as more costly and time-consuming delivery methods, such as soaking the worms in dsRNA solution and injecting dsRNA into the gonads. Although delivery is more difficult in most other organisms, efforts are also underway to undertake large-scale genomic screening applications in cell culture with mammalian cells.

Approaches to the design of genome-wide RNAi libraries can require more sophistication than the design of a single siRNA for a defined set of experimental conditions. Artificial neural networks are frequently used to design siRNA libraries and to predict their likely efficiency at gene knockdown. Mass genomic screening is widely seen as a promising method for genome annotation and has triggered the development of high-throughput screening methods based on microarrays. However, the utility of these screens and the ability of techniques developed on model organisms to generalize to even closely-related species has been questioned; for example, from *C. elegans* to related parasitic nematodes. Functional genomics using RNAi is a particularly attractive technique for genomic mapping and annotation in plants because many plants are polyploid, which presents substantial challenges for more traditional genetic engineering methods. For example, RNAi has been successfully used for functional genomics studies in the hexaploid wheat *Triticum aestivum*, as well as more common plant model systems *Arabidopsis thaliana* and *Zea mays*.

17.6 TECHNICAL CONSIDERATIONS IN THE USE OF RNA INTERFERENCE

In several respects the approaches for silencing gene expression using RNAi methods are similar to those used for antisense DNA-mediated suppression of gene expression. In principle, any cloned gene can be targeted by designing RNA oligonucleotides or RNA-expressing viral vectors with sequences complementary to the mRNA transcribed from the target gene. The present section of this chapter is intended to provide readers who are planning to use, or have just begun to use, RNAi technology into their experimental tool kit with practical information on designing and performing experiments using RNAi methods. In addition, we provide descriptions of emerging technical approaches for selective gene silencing in vitro and in vivo using RNAi. More detailed information on technical aspects of RNAi technology can be found on several different websites including: http://www.ambion.com; http://www.imgenex.com; http://www.genetherapysystems.com.

17.6.1 Design and Synthesis of Small Interfering RNAs

Double-stranded RNAs of 20-23 nucleotides (siRNAs), which can be synthesized in large quantities and transfected into cells, are the most commonly used reagents for RNAi in cultured cells. All that is needed to implement siRNA-mediated silencing of expression of a gene of interest is the cDNA sequence of that gene, and commercially available reagents with which to perform the synthesis. Although targeting of siRNAs to any region of an mRNA would be expected to induce degradation of the mRNA and therefore, abolish production of the encoded protein, empirical data suggest that the probability of achieving selective silencing can be increased by targeting the siRNAs to specific regions of the mRNA. Generally following approach is recommended for designing a siRNA. (1) Beginning with the AUG start codon of the target gene transcript, scan downstream for AA dinucleotide sequences; each AA and the 3′ adjacent 19 nucleotides are potential siRNA targets. (2) Compare the sequences of the potential target sequences to sequences in the species-appropriate genome database (www.ncbi.nlm.nih.gov/BLAST/) and eliminate from consideration of any target sequences that are homologous to other coding sequences. (3) Select 3-4 target sequences along the length of the gene for production of siRNAs. Of course it is important for all siRNA experiments to include negative control siRNAs with the same nucleotide composition but a scrambled sequence. Chemical synthesis was the first method used to produce siRNAs, but now they can be produced in any laboratory using in vitro transcription methods. One protocol involves the synthesis of DNA oligonucleotides that include an 8-base sequence complementary to the 5′-end of a T7 promoter primer. Each gene specific oligonucleotide is annealed to the T7 promoter primer, and a fill-in reaction using Klenow fragment produces a double-stranded template for use in an in vitro transcription reaction. The two RNA products of the in vitro transcription reactions are hybridized to each other, treated with DNase (to remove the DNA template) and RNase (to even the ends of the dsRNA), and the RNA is column purified. Another protocol for the production of siRNAs takes advantage of the availability of recombinant human Dicer. Large in vitro transcribed RNA templates are cleaved by Dicer to produce multiple species of 22 base pair siRNAs. An advantage of the latter method is that, because it produces a mixture of different siRNAs directed against the same mRNA target, the probability of obtaining gene silencing is increased.

17.6.2 Construction of Plasmids and Viral Vectors for RNA Interference

There are several reasons why expression plasmids and viral vectors are being used in basic and applied RNAi research. One major reason is that expression vectors allow continuous production of siRNAs in cells and, therefore, sustains depletion of the protein encoded by the targeted mRNA. A second reason is that, particularly with viral vectors, the transfection efficiency of certain types of cells, particularly postmitotic cells, can be greatly increased. A third advantage of viral vectors is that they are typically more effective in obtaining sustained expression (and gene silencing, in the case of RNAi) in vivo. For example, adenoviral vectors have been extensively used to express genes in postmitotic neurons in vivo (Smith and Romero, 1999). Short hairpin RNAs (shRNAs) can be transcribed from RNA polymerase III promoters in cells in culture or in vivo allowing continuous suppression of expression of the targeted mRNA (Paddison et al., 2002). The latter authors proposed the use of this technology in the generation of transgenic mice as an alternative approach to gene knockout mice. Brummelkamp and colleagues (2002) developed a novel vector system for the stable expression of siRNAs in mammalian cells. They used the polymerase-III H1-RNA gene promoter, which produces a small RNA transcript lacking a poly-adenosine tail and has a well defined transcription start and termination signals. The construct also allows cleavage of the transcript at the second uridine after the termination resulting in a transcript that resembles the ends of synthetic siRNAs. They designed a gene-specific insert that specified a 19-nucleotide sequence derived from the target transcript, separated by a short spacer from the reverse complement of the same 19-nucleotide sequence resulting in the production of a 19-base pair stem-loop structure. This vector system was shown to be effective in sustained suppression of target gene expression in several different types of cultured cells. Several laboratories have constructed plasmids that contain DNA templates for the synthesis of siRNAs under the control of the U6 promoter. For example, Sui et al. (2002) inserted DNA fragments that acted as templates for the synthesis of small RNAs under the control of the mouse U6 promoter that directs the synthesis of a Pol III-specific RNA transcript to generate an RNA composed of two identical 21-nucleotide sequence motifs in an inverted orientation, separated by a 6-base pair spacer of nonhomologous sequences. Five thymidines that function as a termination signal for Pol III were added at the 3′-end of the repeat; the resulting RNA is predicted to fold back to form a hairpin dsRNA with a 3′ overhang of several thymidines. Using this plasmid, they were able to demonstrate the efficient inhibition of expression of three different endogenous genes (lamin A/C, CDK-2, and DNA methyltransferase) in cultured human cells. Retroviral delivery systems have been developed based upon several commercially available vectors. For example, a retroviral siRNA vector was developed in which the U6 promoter and anti-target gene hairpin was subcloned into pMSCVpuro at the unique NsiI site just upstream from the 3′ long terminal repeat (Devroe and Silver, 2002). Using this retroviral siRNA delivery system, they demonstrated the efficient and sustained depletion of the NDR kinase and the transcriptional coactivator p75 in cultured cells. Lentiviral systems for shRNA delivery have also been developed. Lentiviruses can infect noncycling and postmitotic cells, and also provide the advantage of not being silenced during development allowing generation of transgenic animals through infection of embryonic stem cells or embryos (Naldini, 1998; Lois et al., 2002; Pfeifer et al., 2002). Using this approach, silencing of green fluorescent protein (GFP) in GFP-positive transgenic mice has been shown after transduction with lentiviruses expressing shRNA directed against the GFP protein (Tiscornia et al., 2003). More recently, Rubinson et al. (2003) used lentivirus-delivered shRNA to induce silencing of CD8 and CD25 in cycling primary T cells and the pro-apoptotic molecule Bim in primary bone marrow-derived dentritic cells. Lentiviral-mediated silencing of CD8 in hematopoietic stem cells was still present after injection of the cells in lethally irradiated congenic mice. Moreover, in vivo silencing for CD8 or p53 was also observed after infection of ES cells or zygotes leading to stable and functional silencing in adult RNAi transgenic mice.

17.6.3 Transfection Methods

Several different transfection methods previously used to introduce oligodeoxynucleotides and DNA plasmids into cells have been used to successfully introduce siRNAs into cells. However, it has become clear there is no single transfection method that can be successfully applied to all cell types under all experimental conditions. It is, therefore, important to optimize transfection conditions so that maximum gene silencing is acheived. The following transfection parameters have been shown to affect transfection and gene silencing efficacy: cell culture conditions, including cell density and medium composition; the type and amount of transfection agent; the quality and amount of siRNA; and the length of time that the cells are exposed to the siRNA. For proliferating cells, a subconfluent cell density is preferable. For post mitotic cells such as neurons, cell densities in the range of 200 to 500 cells per mm^2 of culture surface work well. Because proteins in serum can bind to and/or degrade siRNAs, the transfection should be performed in serum free medium. Differences have been reported in the ability to transfect and silence gene expression between adherent and nonadherent cells. Postmitotic cells such as neurons and muscle cells tend to be more difficult to transfect using liposomes compared to mitotic cells such as stem cells, fibroblasts, and tumour cells. Calcium phosphate-mediated transfection has been used successfully by several laboratories. The most commonly used and effective transfection method for short-term suppression of gene expression using RNAi is to incorporate siRNAs into liposomes. There are an increasing variety of such transfection reagents including Oligofectamine, LipofectAMINE-2000, CellFectin, Effectene, si- PORT-Amine and siPORT-Lipid. Other methods that have proven effective for transfecting siRNAs into cultured cells include electroporation (Calegari et al., 2002; McManus et al., 2002; Randall et al., 2003), microinjection (Calegari et al., 2002; Kim et al., 2002), and hydrodynamic shock (McCaffrey et al., 2002). Similar transfection methods have been used to introduce RNA-expressing plasmids into cultured cells (Iratni et al., 2002; Czauderna et al., 2003).

These studies also demonstrated the therapeutic potential of RNAi. The methods used to transfect cells with viral vectors that produce shRNAs are essentially identical to those used to transfect cells with similar vectors designed to express cDNAs (Abbas-Terki et al., 2002; Barton and Medzhhitov, 2002; Devroe and Silver, 2002; Xia et al., 2002). It should be recognized that as RNAi technology advances it will likely be possible to produce RNAi "knockout" mice (or other mammals) in which the expression of a protein of interest is repressed by the expression of its corresponding RNAi related molecule (shRNA or miRNA, for example). The resulting animal could be seen as an equivalent of its knockout generated by targeted gene disruption but with much more flexibility and efficiency. For example, cell type-specific promoters could be used to effect PTGS only in cells of interest; in many cases this may circumvent embryonic lethality resulting from gene deletion from all cells. It would also be quicker and less costly to produce RNAi transgenic animals compared with conventional knockouts.

17.7 TECHNOLOGICAL APPLICATIONS

17.7.1 Medicine

It may be possible to exploit RNA interference in therapy. Although it is difficult to introduce long dsRNA strands into mammalian cells due to the interferon response, the use of short interfering RNA mimics has been more successful. The first applications to reach clinical trials were in the treatment of macular degeneration and respiratory syncytial virus, developed by Sirna Therapeutics and Alnylam Pharmaceuticals respectively. RNAi has also been shown effective in the reversal of induced liver failure in mouse models.

Other proposed clinical uses centre on antiviral therapies, including the inhibition of viral gene expression in cancerous cells, knockdown of host receptors and coreceptors for HIV, the silencing of hepatitis A and hepatitis B genes, silencing of influenza gene expression, and inhibition of measles viral replication. Potential treatments for neurodegenerative diseases have also been proposed, with particular attention being paid to the polyglutamine diseases such as Huntington's disease. RNA interference is also often seen as a promising way to treat cancer by silencing genes differentially upregulated in tumour cells or genes involved in cell division. A key area of research in the use of RNAi for clinical applications is the development of a safe delivery method, which to date has involved mainly viral vector systems similar to those suggested for gene therapy. Despite the proliferation of promising cell culture studies for RNAi-based drugs, some concern has been raised regarding the safety of RNA interference, especially the potential for "off-target" effects in which a gene with a coincidentally similar sequence to the targeted gene is also repressed. A computational genomics study estimated that the error rate of off-target interactions is about 10%. One major study of liver disease in mice led to high death rates in the experimental animals, suggested by researchers to be the result of "oversaturation" of the dsRNA pathway.

17.7.2 Biotechnology

RNA interference has been used for applications in biotechnology, particularly in the engineering of food plants that produce lower levels of natural plant toxins. Such techniques take advantage of the stable and heritable RNAi phenotype in plant stocks. For example, cotton seeds are rich in dietary protein but naturally contain the toxic terpenoid product gossypol, making them unsuitable for human consumption. RNAi has been used to produce cotton stocks whose seeds contain reduced levels of delta-cadinene synthase, a key enzyme in gossypol production, without affecting the enzyme's production in other parts of the plant, where gossypol is important in preventing damage from plant pests. Similar efforts have been directed toward the reduction of the cyanogenic natural product linamarin in cassava plants.

Although no plant products that use RNAi-based genetic engineering have yet passed the experimental stage, development efforts have successfully reduced the levels of allergens in tomato plants and decreased the precursors of likely carcinogens in tobacco plants. Other plant traits that have been engineered in the laboratory include the production of non-narcotic natural products by the opium poppy, resistance to common plant viruses, and fortification of plants such as tomatoes with dietary antioxidants. Previous commercial products, including the Flavr Savr tomato and two cultivars of ringspot-resistant papaya, were originally developed using antisense technology but likely exploited the RNAi pathway.

17.8 HISTORY AND DISCOVERY

The discovery of RNAi was preceded first by observations of transcriptional inhibition by antisense RNA expressed in transgenic plants and more directly by reports of unexpected outcomes in experiments performed by plant scientists in the USA and The Netherlands in the early 1990s. In an attempt to alter flower colours in petunias, researchers introduced additional copies of a gene encoding chalcone synthase, a key enzyme for flower pigmentation into petunia plants of normally pink or violet flower colour. The overexpressed gene was expected to result in darker flowers, but instead produced less pigmented, fully or partially white flowers, indicating that the activity of chalcone synthase had been substantially decreased; in fact, both the endogenous genes and the transgenes were downregulated in the white flowers. Soon after, a related event termed *quelling* was noted in the fungus *Neurospora crassa*, although it was not immediately recognized as related. Further investigation of the phenomenon in plants indicated that the downregulation was due to post-transcriptional inhibition of gene expression via an increased rate of mRNA degradation. This phenomenon was called co-suppression of gene expression, but the molecular mechanism remained unknown.

Not long after, plant virologists working on improving plant resistance to viral diseases observed a similar unexpected phenomenon. While it was known that plants expressing virus-specific proteins showed enhanced tolerance or resistance to viral infection, it was not expected that plants carrying only short, non-coding regions of viral RNA sequences would show similar levels of protection. Researchers believed that viral RNA produced by transgenes could also inhibit viral replication. The reverse experiment, in which short sequences of plant genes were introduced into viruses, showed that the targeted gene was suppressed in an infected plant. This phenomenon was labelled "virus-induced gene silencing" (VIGS), and the set of such phenomena were collectively called post-transcriptional gene silencing.

After these initial observations in plants, many laboratories around the world searched for the occurrence of this phenomenon in other organisms. Craig C. Mello and Andrew Fire's 1998 *Nature* paper reported a potent gene silencing effect after injecting double stranded RNA into *C. elegans*. In investigating the regulation of muscle protein production, they observed that neither mRNA nor antisense RNA injections had an effect on protein production, but double-stranded RNA successfully silenced the targeted gene. As a result of this work, they coined the term *RNAi*. Fire and Mello's discovery was particularly notable because it represented the first identification of the causative agent of a previously inexplicable phenomenon. Fire and Mello were awarded the Nobel Prize in Physiology or Medicine in 2006 for their work.

17.9 TIPS TO REMEMBER

- RNA interference (RNAi) is the process of mRNA degradation that is induced by double-stranded RNA in a sequence-specific manner. RNAi has been observed in all eukaryotes, from yeast to mammals. The power and utility of RNAi for specifically silencing the expression of any gene for which sequence is available has driven its incredibly rapid adoption as a tool for reverse genetics in eukaryotic systems.
- Short interfering RNAs (siRNAs) were originally identified as intermediates in the RNA interference pathway. siRNAs are 21-25 bp dsRNA with dinucleotide 3′ overhangs that are formed in the cell from longer dsRNA molecules. The fully assembled RNA-induced silencing complex (RISC) contains only one strand of the siRNA, the guide strand. The guide strand is thought to provide target specificity for RISC-mediated cleavage through perfect base pairing with the mRNA target.
- In many cases, endogenous siRNAs originate from repetitive elements within the genome, such as heterochromatic regions at centromeres and telomeres, and are, therefore, known as repeat-associated siRNAs (rasiRNAs). rasiRNAs appear to function, through an RNA-induced transcriptional silencing (RITS) complex, in maintaining the heterochromatic, and hence transcriptionally repressed, state of the region that encodes the rasiRNA.
- MicroRNAs (miRNAs) are a large class of evolutionarily conserved RNAs found in plants and animals. These small RNAs have been shown to play critical roles in developmental timing, hematopoietic cell differentiation, cell death, cell proliferation, and oncogenesis. miRNAs may represent 2-3% of the total number of genes in humans, and estimates of the number of miRNA target binding sites indicate that miRNAs may play a role in regulating as many as 30% of mammalian genes. miRNAs are 19-23 nt single-stranded RNAs, originating from single-stranded precursor transcripts that are characterized by imperfectly base-paired hairpins. miRNAs function in a silencing complex that is similar, if not identical, to RISC.

17.10 VIVA VOCE

Q.1. Explain RNAi.

Q.2. What is the principle of RNAi technology?

Q.3. What is gene silencing?

Q.4. What is RNAi "knockout" and how it is different from gene knockout?

Q.5. What is "Dicer"?

Q.6. What is conditional RNAi?

Q.7. Differentiate between "knockout" and "knockdown".

Q.8. What is RITS and RISC?

Q.9. Differentiate between siRNAs and rasiRNAs.

Q.10. Explain miRNAs.

CHAPTER 18

Qualitative and Quantitative Tests

18.1 CARBOHYDRATES

One of the most important constituents in our food is glucose which we usually obtain in the form of starch from plant sources. In our body glucose is readily utilized or is stored as glycogen. The metabolic processes in our body are mainly centred on glucose, which is a member of a large class of organic compounds called carbohydrates.

These are generally referred to as sugars. Carbohydrates contain C, H and O atoms. Usually, H and O are present in the ratio of 2:1, just as in water; hence the name carbohydrates are in use.

Carbohydrates in general have either an aldehyde group (as in glucose) or a keto group (as in fructose). Those containing an aldehyde group (as in glucose) are called as aldoses and those containing keto groups are called ketoses. They may also be referred to on the basis of the number of carbon atoms contained in them; for example, both glucose and fructose are hexoses as they have six carbon atoms in them.

Ribose and deoxy ribose are pentoses because they have five carbon atoms. Both of them are, however, aldoses. Similarly, arabinose is an aldopentose. They are also trioses, tetroses, heptoses, etc.

Some carbohydrates are formed by the combination of two sugars for instance; the common sugar sucrose contains both glucose and fructose. Therefore, sucrose is referred to as a 'disaccharide',

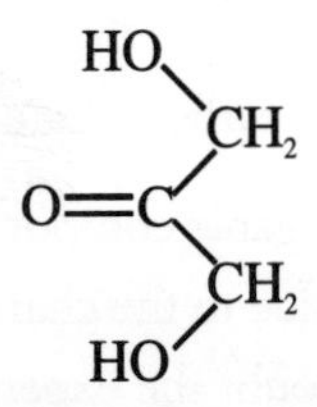

Dihydroxyacetone
(a ketose)

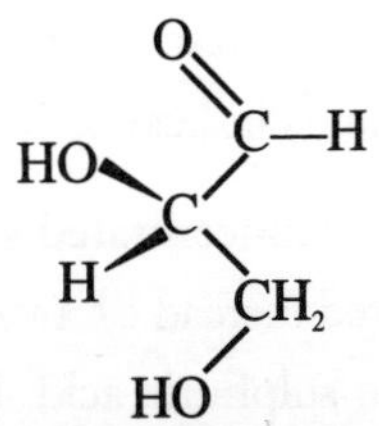

D-Glyceraldehyde
(an aldose)

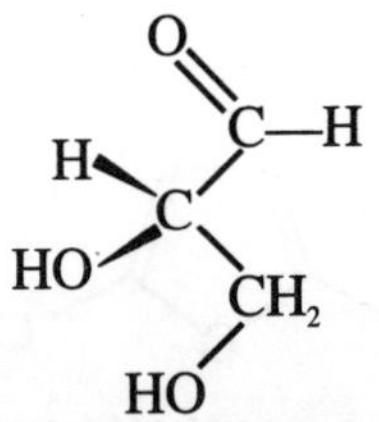

L-Glyceraldehyde
(an aldose)

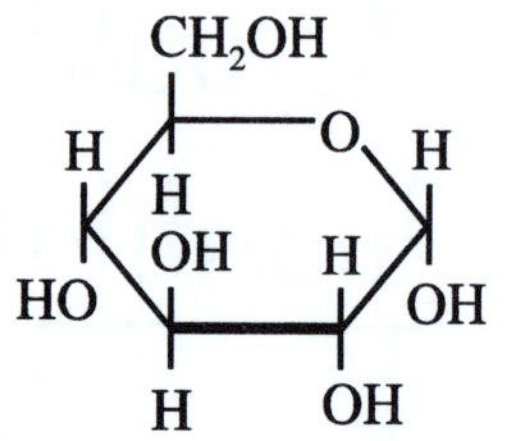

α-D-Glucopyranose

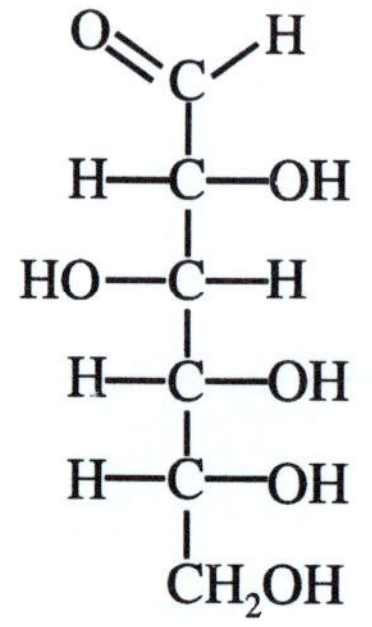

whereas glucose and fructose are monosaccharides. There are many disaccharides like sucrose, e.g., maltose (glucose + glucose), lactose (galactose + glucose) and so on.

18.1.1 Qualitative test for carbohydrates

Most of the tests of the carbohydrates are based on their reducing properties (due to the presence of reducing aldehyde or ketone groups). Fehlings test, benedict's test are the example of this. The unspecific Molisch's test for carbohydrates is one of the examples of some tests which are based on the formation of furfural or furfural derivatives in presence of concentrated acids. Specific complex formation is sometimes used as specific test for carbohydrates. Formation of phenylhydrazone is one such example. For testing polysaccharides, iodine is found to be very useful. These tests will be discussed below:

1. Molisch's Test

Principle

Alcoholic alpha naphthol forms furfural and furfural derivatives, such as hydroxymethylfurfural, by the concentrated sulphuric acid acting on the sugar. This compound forms a reddish-violet coloured ring at the junction of the two liquids. Molisch's reagent is 5% solution of alpha naphthol in alcohol.

Procedure

Add 2 drops of Molisch's reagent to 2 ml of sugar solution in a test tube. Mix thoroughly. Add 2 ml of conc. H_2SO_4 by the side of the test tube slanting the tube. Then errect the test tube slowly. The formation of reddish violet ring at the junction of two liquids indicates the presence of carbohydrates.

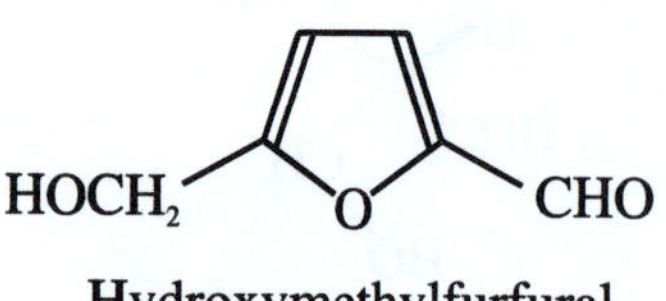

Hydroxymethylfurfural

Concentrated solution of organic compounds may give a red instead of a violet colour due to the charring action of the sulphuric acid. In case of doubt the experiment should be repeated on a more diluted solution of the substance to be tested.

2. IODINE TEST

Principle

The composition of the blue or red or wine red coloured substance is not well defined. This may be an adsorption complex of starch or dextrins or glycogen with iodine rather than a definite compound. Iodine reagent is 0.5 ml of iodine diluted to 5 ml with distilled water. Potassium iodide is added to the reagent solution in order to make the iodine more soluble in water.

Procedure

Add 1 or 2 drops of dilute iodine solution to 2-3 ml of dilute starch or dextrin or glycogen solution. A blue, red and brown colour develops in case of starch, dextrin and glycogen respectively. In case of starch, the blue colour disappears on heating and reappears on cooling. But the red colour and the brown colour in cases of dextrin and glycogen respectively, do not reappear on cooling as in case of starch.

3. Bial's test for pentoses

To 1 ml of sugar solution in a test tube add 3 ml of concentrated HCl and 0.5 ml of Bial's reagent. Heat the tube in a boiling water bath for one minute. Record your observations with different sugars. The Bial's reagent is prepared by dissolving 3 gm of orcinol and 0.1 gm of ferric chloride in 100 ml of ethanol. This is a sensitive test for the detection of pentoses. Heating with strong acid converts the pentose to furfural which then reacts with the coloured compound produced when orcinol and ferric chloride react with each other. A blue green compound is finally formed. This reagent reacts with many sugars but under the condition described above only pentoses yield blue-green colour.

4. Seliwanoff's test

To the sugar (2 ml) add 2 ml of seliwanoff's reagent. A blank without sugar should also be prepared to judge the colour change. Place the tubes in boiling water for exactly 1 min. Note the colour change, if any, and then continue the heating for 5 minutes and periodically observe the colour change. Seliwanoff's reagent is 0.5% resorcinol in conc. HCl diluted 1:1 with water. Ketoses (naturally fructose) give fiery red colour. Aldoses (glucose, etc.) give the test weakly and slowly. If the boiling is prolonged, positive test is obtained with glucose (or maltose) due to its partial conversion to fructose. This test is also given by sucrose which is hydrolysed during the course of the test yielding fructose as one of the products.

5. REDUCTION TESTS

Carbohydrates with free aldehyde or ketone groups have the ability to reduce solutions of various metallic ions. These properties are mentioned below:

A. FEHLING'S TEST

Principle

Carbohydrates with free aldehyde or ketone groups reduce copper sulphate to cuprous oxide forming a yellow or brownish red coloured precipitate. Fehling's reagent is prepared freshly by mixing equal volumes of two stock solutions A and B. Solution A is 6.93 grams of $CuSO_4.5H_2O$ per 100 ml of water and Solution B is 20 grams of KOH and 34.6 grams of sodium potassium tartarate (Rochelle salt) per 100 ml solution.

Procedure

Add a few drops of sugar solution at a time to 5 ml of Fehling's solution and heat the mixture after each addition. The production of yellow or brownish red cuprous oxide precipitate indicates the presence of reducing sugars.

B. Benedict's test

Add 5mL of Bennedict's qualitative reagent to the sugar solution, and place the test tube boiling water bath for 2 minutes. In case of reducing sugars there will be an appearance of red precipitate. Bennedict's qualitative solution is prepared by dissolving 173 gm of sodium citrate and 100 gm sodium carbonate and 100 ml of water, by heating. If there is any turbidity, it should be removed by filtration. Copper sulphate solution (17.3 gm copper sulphate in 100 ml water) is slowly added with constant stirring to the citrate-carbonate solution and the volume is made up to 1 l. This test is based on the modification of Fehling's test by Benedict. The difficulties faced by Fehling's test are, therefore, not faced in case of Bennedict's test. In the presence of even small quantities of reducing sugars the entire body of the solution will be filled with a precipitate which is red. In the case of non-reducing sugar (say sucrose) the solution will remain perfectly clean. This reagent is routinely used and found to be reliable in the examination of urine for pathological amounts of sugars.

C. Barfoed's test

Mix 5 ml of Barfoed's reagent with 1 ml of carbohydrate solution in a test tube and heat in a boiling water bath for 10 min. Appearance of a red precipitate of cupric oxide (Cu_2O) indicates the presence of reducing sugar. Barfoed's reagent is prepared by dissolving 13.3 gm of neutral copper acetate in 200 ml of water and then adding 1.8 ml of glacial acetic acid. Barfoed's test is also copper reduction test but this test differs basically from Fehling's test or Bennedict's test as it is carried out in acidic medium instead of alkaline medium. Under acidic conditions the reduction takes place efficiently. Monosaccharides respond quickly to the test whereas disaccharides respond slowly. When the sugar solution is boiled in contact with the reagent the disaccharide is hydrolysed by acetic acid present in the reagent and the positive test is obtained. Chloride interferes with this assay as it causes the formation of a green precipitate the urine cannot be tested by this method as it contains chloride.

Confirmatory tests

I. Phenylhydrazine test

Take about 300 mg of phenylhydrazine mixture (discussed below), to it add a few drops of glacial acetic acid and then 5 ml of sugar solution. Shake well and heat in a boiling water bath for 30 to 45 minutes. Take the tube out of the water bath and allow it to cool slowly. Yellow crystals of osazone will appear. Examine the crystals under the microscope and describe the nature of crystals.

Phenylhydrazine mixture is prepared by mixing equal weights of phenylhydrazine hydrochloride and anhydrous sodium acetate. The mixing is to be done thoroughly in a mortar. Glucose and Fructose give identical osazones called glucosazones and fructosazones because excepting the first two carbons (which are used in formation of osazone) the remaining four carbon atoms have same configuration in both of them.

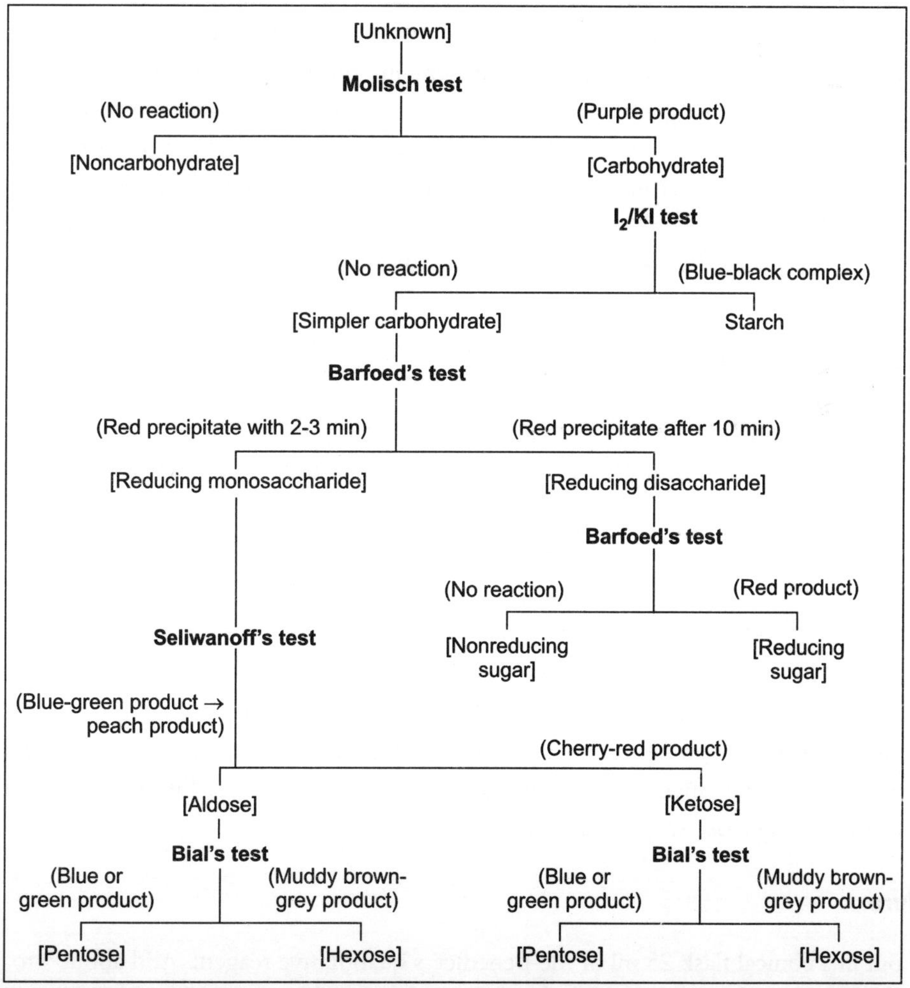

Fig. 18.1: Flow chart for classifying unknown carbohydrate

Sucrose as such does not form any osazone because it has no reducing group available for reaction with phenylhydrazine. On hydrolysis, however, it gives rise to osazone. Osazones of dissacharides are soluble in hot water. Therefore, in their cases the osazones do not precipitate during heating. They appear only after cooling.

18.1.2 Quantitative tests of carbohydrates

1. Estimation of glucose by Benedict's method

During qualitative analysis of sugars we have already learnt that glucose reduces copper sulphate in Benedicts reagent under alkaline conditions and a red precipitate is formed. This qualitative method has been exploited for its use in quantitative analysis.

The Benedicts quantitative reagent contains the following ingredients, copper sulphate, sodium carbonate, sodium or potassium citrate, potassium thiocyanate and potassium ferrocyanide. Of these, copper sulphate has to be very accurately measured as the amount of copper sulphate reduced will correspond to the amount of glucose present in solution. Due to presence of potassium thiocyanate in Benedict's reagent a white precipitate of cuprous thiocyanate instead of red precipitate of cuprous oxide will be formed when copper sulphate is reduced. As the precipitate formed is white it is very easy to determine the end point. Blue tint of Benedict's reagent disappears completely at this point. The small amount of potassium ferrocyanide added helps to prevent the oxidation of cuprous oxide. Sodium or potassium citrate added does not allow the formation of copper carbonate. The alkaline condition is produced by sodium carbonate which is a mild alkali in comparison with NaOH and is, therefore, less destructive for the sugar. The Bennedict's reagent prepared as follows is stable for long periods of time.

To prepare quantitative Benedict's reagent 18.0 gm of crystalline copper sulphate is dissolved in 100 ml of water (solution A). Further, 100 gm of sodium carbonate, 200 gm of anhydrous sodium citrtate and 125 gm of potassium thiocyanate are dissolved in 800 ml of water with heating (solution B). If solution B is not clear it should be filtered. Solution A is added slowly to solution B with stirring. Then 5 ml of potassium ferrocyanide solution is added and the volume is finally made up to 1 litre after cooling.

The reaction of $CuSO_4$ with glucose is quite complicated and a number of molecules of $CuSO_4$ are reduced by one molecule of glucose. Therefore, it is not possible to write the stoichiometric equation for reaction between $CuSO_4$ and glucose. But it has been found that 25 ml of the above mentioned quantitative reagent corresponds to 50 mg glucose. Determination of the unknown amount of glucose will be based on this.

Procedure

Pipette out in a conical flask 25 ml of the Benedict' s quantitative reagent. Add about 5 to 10 gm of Na_2CO_3 and a few porcelain chips to the flask to prevent bumping. Heat the contents of conical flask to boiling and then run in the glucose solution from a burette at first rapidly and then slowly until the blue colour becomes fade. Allow it to boil for 2-3 minutes more and add glucose solution drop by drop till the solution becomes colourless. Note down the volume of the glucose solution used and calculate the percentage of glucose in solution as described below. Sometimes the solution in the flask becomes too much concentrated due to evaporation of water. To avoid it more water may be added.

Suppose 20 ml of the glucose solution is required to titrate 25 ml of Benedict's quantitative reagent. As 25 ml of the Benedict's quantitative regent is equivalent to 50 mg of glucose, hence 20 ml of the solution contains 50 mg of glucose. Therefore, 100 ml of the glucose contains 50×100/ 20 =250 mg of glucose and the strength of the solution 250 mg per cent.

Estimation of lactose by Benedict's quantitative regent

Principle is same as for glucose, only difference being 25 ml of Benedict's quantitative regent is equivalent to 67 mg of lactose.

Even sucrose after acid hydrolysis can be estimated by this method.

2. Glucose oxidase method for estimation of glucose

In this method, the aldehyde group of β-D-Glucose is oxidized by glucose oxidase to give gluconic acid and hydrogen peroxide.

$$\beta\text{-D-Glucose} + H_2O + O_2 \rightarrow \text{gluconic acid} + H_2O_2$$

The hydrogen peroxide may be broken down to water and oxygen by a peroxidase and if an oxygen acceptor is present, it will convert to a coloured compound which can be measured. The reagent usually used is oxidation product of phenol condensed with 4-aminophenazone to give a coloured product as in determination of alkaline phosphatase.

18.2 LIPIDS

A large number of heterogenous compounds are referred to as lipids including fats, steroids, waxes, and related compounds, which are related more by their physical than their chemical properties. They have the common property of being (1) relatively insoluble in water and (2) soluble in polar solvents such as ether and chloroform.

Fatty acids are aliphatic carboxylic acids. If the aliphatic chain contains no double bond then it is called saturated and if it contains one or more double bond it is called unsaturated. Most naturally occurring unsaturated fatty acids have cis- double bonds. Some of the most common fatty acids are palmitic acid and stearic acid. Palmitic has 16 carbon atoms and stearic has 18 carbon atoms. As it is clear from the formulae, both are saturated fatty acids. Some fatty acids like oleic acid may be unsaturated. Naturally occurring animal fats consist largely of mixed glyceride of oleic, palmitic and stearic acids. They are usually mixture of individual fats. Fats have more saturated fatty acids whereas oils have more of unsaturated ones.

Lipids are simple, complex or derived. Simple lipids are esters of fatty acids with various alcohols, e.g., fats (esters of fatty acids with glycerol) and waxes (esters of fatty acids with higher molecular weight of monohydric alcohols). Complex lipids are esters of fatty acids containing groups in addition to an alcohol and a fatty acid, e.g., phospholipids or glycolipids etc. Derived lipids include fatty acids, glycerol, steroids, other alcohols, fatty aldehydes, and ketone bodies, lipid soluble vitamins, and hormones.

Phospholipids yield in addition to alcohol and fatty acids, phosphate and a nitrogenous base like choline, ethanolamine, etc. Lecithins and cephalins are representatives of the phospholipids. Similarly glycolipids contain carbohydrates, and sulpholipids contain sulphate. Lipoproteins are combinations of lipids with proteins.

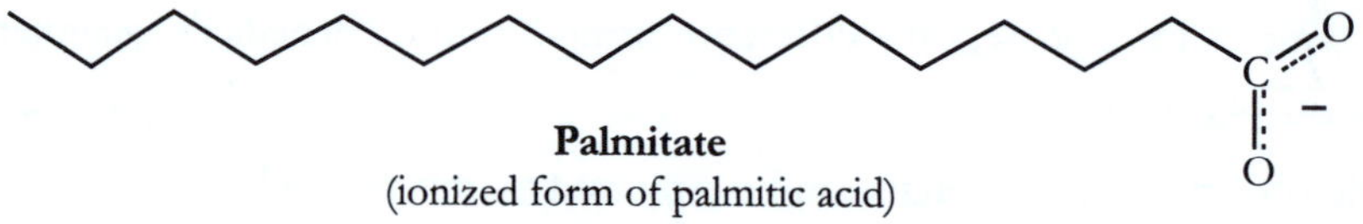

Palmitate
(ionized form of palmitic acid)

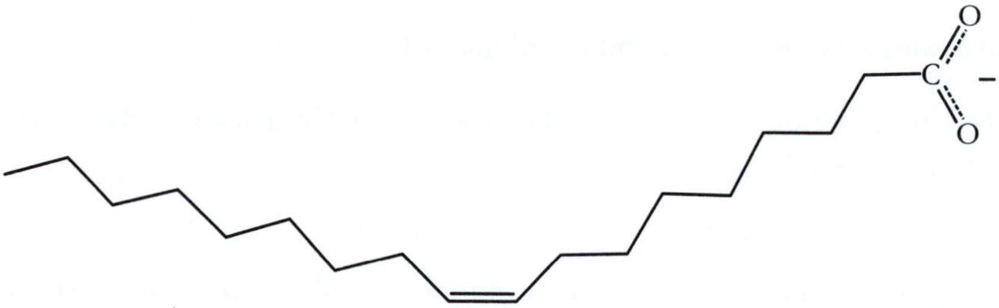

Oleate
(ionized form of oleic acid)

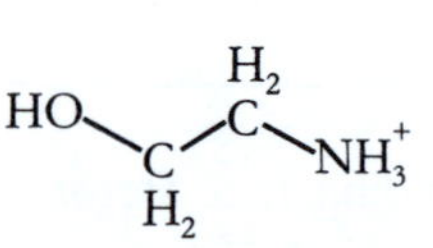

Ethanolamine

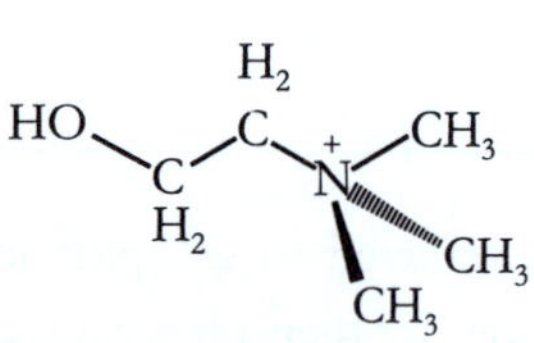

Choline

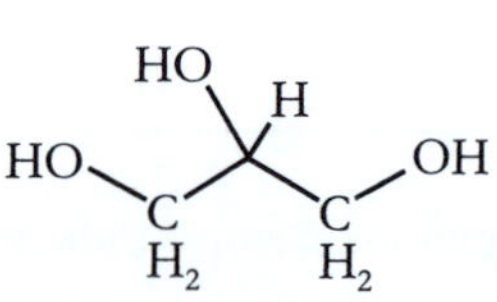

Glycerol

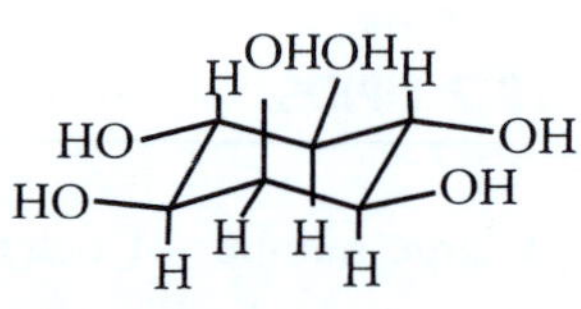

Inositol

Now we will consider some qualitative and quantitative tests for lipids.

18.2.1 Qualitative tests

I. Physical test

1. Grease spot test

Take a small amount of oil on a piece of paper, a greasy spot penetrating the paper will be formed. This happens because lipid does not wet paper unlike water.

2. Test for free fatty acids

Take a few drops of phenophthalein solution in a test tube and add to it one or two drops of very dilute alkali solution, just sufficient to give the solution a pink colour. Now add a few drops of the oil and shake. The colour will disappear as the alkali is neutralized by the free fatty acids present in the oil.

3. Emulsification

Oil or liquid fat becomes finely divided and is dispersed in water when shaken with water to form emulsification. Emulsification is permanent and complete in the presence of emulsifying agent. The important emulsifying agents are bile salts, proteins, soaps, mono- and diglycerides. Emulsification is important in the processes of fat digestion in the intestine. Emulsifying agents lower surface tension of the liquid.

Take 2 clean and dry test tubes, in one test tube added 2 ml water and in other 2ml dilute bile salt solution. Now to each tube added 2 drops of mustard oil and shaken vigorously for about one minute. Allow the tubes to stands for two minutes and note that the water, oil is broken in small pieces and floats on the surface; where as in the bile salt solution, the oil can be seen in minute droplets suspended in the liquid (permanent emulsification).

4. Saponification test

Esters can be hydrolysed by alkali to yield the parent alcohol and salt. When the fatty acid possesses a long chain the salt formed is a soap which we commonly use. This process is called saponification. Oils and fats usually contain long chain fatty acids and are, therefore, the starting materials for the preparation of soap. Take 1 ml of the oil in a test tube and add an equal amount of alcoholic KOH solution, mix them thoroughly and keep the mixture during the course of warming and shake up gently with a little distilled water. Appearance of some oil drops will indicate the incomplete saponification. After complete saponification no oil drops will appear.

5. Tests for unsaturation of fatty acids

Unsaturated fatty acids like oleic acid can react with halogens like bromine and iodine due to presence of double bonds as shown below.

$$CH_3(CH_2)_7CH = CH(CH_2)_7COOH + Br_2 \rightarrow CH_3(CH_2)_7CHBr–CHBr(CH_2)_7COOH$$

The amount of Br_2 or I_2 taken up will indicate the amount of unsaturation present in a particular acid. Approximate idea about the unsaturation in a different oils and fats can be obtained by the following test.

Set up four clean and dry test tubes each containing 5 ml of CCl_4. To the first, add one drop of shark liver oil, to the second, one drop of coconut oil, to the third, a drop of vegetable ghee and add nothing to the fourth tube. Now test for the unsaturation of the added oil by adding bromine water drop by drop to each tube followed by shaking. Record the number of drops required to obtain a permanent yellowish red colour in each tube and infer the relative unsaturation in the three samples used. It may be mentioned here, vegetable ghee is prepared by hydrogenating vegetable oil. Hydrogenation means saturation of unsaturated fatty acid by hydrogen.

6. Isolation of free fatty acids from soap

Take a few ml of 20% H_2SO_4 in a test tube and gradually add 5 ml of some soap solution. The fatty acids will separate out in a distinct layer due to the hydrolysis of the soap.

$$RCOONa + H_2O \rightarrow RCOOH + NaOH$$

Cool the solution which will become hot and skim off the surface layer and wash it several times with water till free from H_2SO_4. Then dissolve it in some water and add alkaline phenophthaelin solution and shake. The pink colour will be discharged indicating the presence of free fatty acids.

Calcium soap formation

To a small amount of the soap solution in a test tube add $CaCl_2$ solution. A white precipitate will be formed. The white precipitate is due to insoluble calcium salt of fatty acid. This is referred to as calcium soap.

Lead soap formation

To a small amount of the soap solution in a test tube add lead acetate solution, a white precipitate will appear. The white ppt is due to insoluble lead salt of fatty acids. This is referred to as lead soap.

7. Tests for Glycerol

I. Acrolein test

Take pure glycerol in a dry test tube, add to it a few crystals of potassium hydrogen sulphate. Warm gently to mix and then heat strongly. A very pungent odour of acrolein is produced. Acrolein is formed due to removal of water from glycerol by potassium hydrogen sulphate.

II. Dichromate test

Take in a dry tast tube 3 or 4 ml of glycerol solution, to it add a few drops of 5% potassium dichromate solution and 5 ml of conc. HNO_3, mix well and note that the brown colour is changed to blue. This test is given by the substances containing primary and secondary alcohol groups. The chromic ions oxidize the glycerol and in this process they are reduced to chromous ions which give the blue colour. This test is also given by reducing sugars, so before confirming glycerol be sure that the reducing sugars are not present.

18.2.2 Quantitative tests

1. Determination of iodine number

The iodine number of a fat is the amount in gm of iodine taken up by 100 gm of fat. Not only iodine but also equivalent amounts of other halogens will add at double bonds; so bromine is often used instead of iodine because it is more reactive. The halogenating reagent used in this method is pyridine sulphate dibromide. This reagent can be prepared by adding carefully 8.1 ml pyridine in 20 ml glacial acetic acid and making the volume up to 1 litre with glacial acetic acid.

Weigh the bottle containing sample of oil plus a medicine dropper and then transfer about 0.1 to 0.3 gm of oil to a flask. Reweigh the bottle containing oil and dropper to find out the exact quantity of the sample transferred. Add 10 ml of chloroform and then 25 ml of the pyridine sulphate dibromide reagent. Shake thoroughly; allow standing for 5 minutes and then determining the residual bromine. To do this, add 10 ml of 10% KI and titrate the equivalent amount of iodine liberated by the residual bromine with the help of 0.1 (N) $Na_2S_2O_3$ (sodium thiosulphate). The titration can be done by adding sodium thiosulphate solution through a burette to the flask. When the colour of the solution in flask becomes light yellow add 1 ml of starch solution. It will become blue. Slowly add the thiosulphate solution again till it becomes colourless. Note the total volume of thiosulphate used.

The total amount of bromine originally added is found by titrating 25 ml of the pyridine sulphate dibromide reagent with thiosulphate after adding KI as in the previous case. The amount of bromine taken up by the fat sample can be determined by the difference between the two titres and then the iodine number can be calculated.

Suppose with a sample of 0.2 gm oil the data obtained are as follows:

0.1 (N) $Na_2S_2O_3$ used for titration of blank = 47.0 ml

0.1 (N) $Na_2S_2O_3$ used for titration of sample = 27.0 ml

0.1 (N) $Na_2S_2O_3$ equivalent to iodine absorbed by the sample = 20.0 ml

As 1 ml 0.1 (N) $Na_2S_2O_3$ = 1.0 ml of 0.1 (N) Bromine = 1 ml of 0.1 (N) Iodine

Hence, 20 ml of 0.1 (N) $Na_2S_2O_3$ = 20 ml of 0.1 (N) Iodine = $20 \times 12.7/1000$ gm Iodine = 0.254 gm Iodine.

Thus 0.2 gm of oil can take up 0.254 gm of iodine.

Therefore, iodine number of oil used = 127.

18.2.3 Qualitative test of cholesterol

Cholesterol is a lipid with a structure quite different from that of phospholipids. It is a steroid, built from four linked hydrocarbon rings. A hydrocarbon tail is linked to the steroid at one end, and a hydroxyl group is attached at the other end. In membranes, the molecule is oriented parallel to the fatty acid chains of the phospholipids, and the hydroxyl group interacts with the nearby phospholipid head groups. Cholesterol is absent from prokaryotes but is found to varying degrees in virtually all animal membranes. It constitutes almost 25% of the membrane lipids in certain nerve cells but is essentially absent from some intracellular membranes.

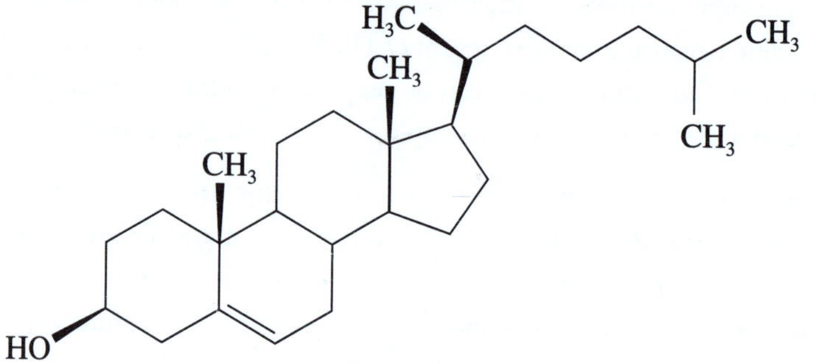

Cholesterol

It is found in bile and a common constituent of gall stones.

The main test for cholesterol is known as Liberman-Burchard test. This is carried in the following way. In a dry test tube take a small amount of solution of cholesterol in chloroform. Add 1 ml of acetic anhydride and 1 drop of conc. H_2SO_4. Mix and observe that a purple colour is formed which soon changes to green. It may take 15-30 min for full development and it is advisable to put the tube in dark during this time.

18.2.4 Quantitative estimation of cholesterol

The above mentioned qualitative test has been developed quantitatively for the estimation of cholesterol. Follow the following protocol for the purpose.

Table18.1: All volumes for test tubes 1 to 5 and B are in ml

	B	1	2	3	4	5
Glacial acetic acid	2.0	1.7	1.4	1.1	0.8	0.2
Cholesterol solution (1 mg/ml)	–	0.3	0.6	0.9	1.2	1.8
Acetic anhydride sulphuric acid reagent	4.0	4.0	4.0	4.0	4.0	4.0

Shake the tubes well and keep them at room temperature for 30 minutes. Blue colour will develop in all the tubes except blank tube. Measure the absorbencies at 625 mμ against the blank tube and plot these against the amount of cholesterol.

Note: Acetic anhydride-sulphuric acid reagent

This reagent has to be freshly prepared before use. Acetic anhydride (20 ml) is taken in a glass stoppered flask which is then chilled in ice water. When cold, add 1 ml of conc. H_2SO_4 to it drop by drop. The contents are mixed and cooled during the addition. After completion of the addition the flask is stoppered and shaken vigorously for a few minutes. The solution has to be kept cold in ice and should be used within an hour.

Enzymatic methods

Assays have been developed in which cholesterol oxidase obtained from the bacterium Nocardia erythropolis is used to convert cholesterol into cholest-4-en-3-one with the formation of Hydrogen peroxide. The cholest-4en-3-one formed has been measured by reading at 240 nm after extracting into isopropanol. Alternatively, hydrogen peroxide has been quantified by formation of chelate complex with quadrivalent titanium and xylenol orange.

18.2.5 Other tests for cholesterol

1. Salkowski's Test (H_2SO_4 test)

Dissolve cholesterol in 2 ml of chloroform in dry test tube. Add equal amount of con. H_2SO_4. Shake gently. The upper layer turns red and the sulphuric acid layer shows a yellow colour with a green fluorescence.

2. Formaldehyde-H_2SO_4 test

Add 2 ml of formaldehyde-sulphuric acid solution (1 part of 40% formaldehyde to 50 parts of the acid) to 2 ml of chloroform solution in a dry test tube. The cherry colour is developed in the chloroform. Pour off the chloroform in another test tube and add 2-3 drops of acid anhydride. The blue colour develops.

This tets is said to be more delicate than Salkowski's test.

18.3 TESTS FOR AMINO ACIDS AND PROTEINS

We divide the food we consume into three main classes: carbohydrates, the body's most readily availble energy source; lipids, the body's principal energy reserve; and proteins, the body's source of energy for growth and cellular maintenance. Proteins also make up the second largest portion of cells, after water. They are large polymeric compounds that cells synthesize from various building blocks called amino acids. The general structure for an amino acid is shown in the following figure:

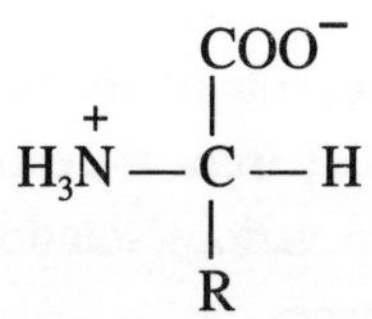

Note that all amino acids contain carboxylic acid groups (—COOH), amino groups (—NH_2), and substituent or replaceable side chains (—R). Twenty different amino acids, which differ only in the structures of their side chains, are used by human cells to build proteins. The side chain structure determines the class of the amino acid: nonpolar, neutral, basic, or acidic. Human cells can synthesize most of the amino acids they need to build proteins. However, about 8 amino acids called essential amino acids cannot be synthesized by human cells and must be obtained from food. Amino acids incorporated into proteins are covalently linked by peptide bonds. Peptide bonds are amide bonds formed between the carboxylic acid group of one amino acid and the amino group of a second amino acid. Equation 1 below shows a peptide linkage formation between two amino acids:

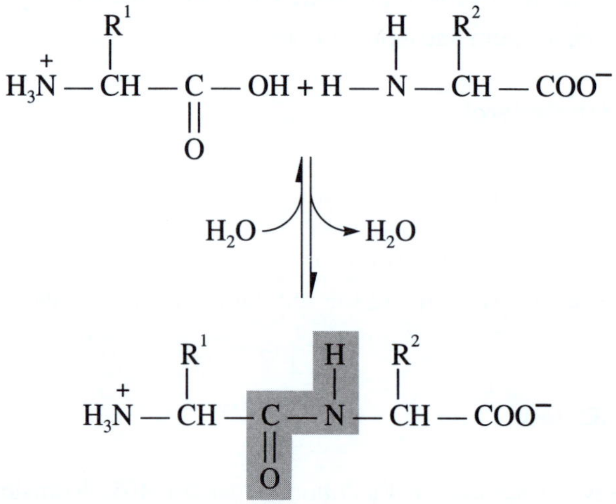

(Eq. 1)

The peptide bond is indicated. Note that, because every amino acid contains at least one amino group and one carboxylic acid group, it is possible for a peptide bond to form in two different ways. For example, with glycine and valine, it is also possible for the peptide bond to form between the carboxylic acid group of valine and the amino group of glycine, producing valylglycine. Proteins are composed of hundreds of amino acids linked by peptide bonds, forming a peptide chain. We define the direction in which the amino acids link by referring to the two ends of the chain as the N-terminus and the C-terminus. The N-terminus is the terminal amino acid in the chain that contains the only amino group not part of a peptide bond. The C-terminus is the other terminal amino acid in the chain, containing the only carboxylic acid group not part of a peptide bond. Note that the N-terminus and the C-terminus are not determined by the side chains. The number of constituent amino acids and the order, in which they are linked starting from the N-terminus, are referred to as the protein's primary structure.

I. Amino Acid and Protein Solubility

The physical properties of amino acids and proteins are mainly a result of their structure, both in the solid state and in various solutions. In this part of the experiment you will investigate the solubility of selected amino acids and proteins in various solutions. Using your data you will compare amino acid and protein structural characteristics.

Solubility as a Function of Solution pH

The presence of amino and carboxylic acid groups enables amino acids to accept protons from and donate protons to aqueous solution, and, therefore, to act as acids and bases. Because proteins contain both acidic and basic side-chains, they too can donate or accept protons. A molecule that functions simultaneously as an acid and a base is known as an amphoteric molecule. In neutral aqueous solutions, amino acids act as amphoteric molecules. For example, an amino acid with a neutral side chain contains two charges: one positive, due to the protonation of the amino group, and one negative, due to the dissociation of the carboxylic acid proton. This double ionic form of an amino acid is the zwitterionic form. Following figure shows an amino acid in the zwitterionic form.

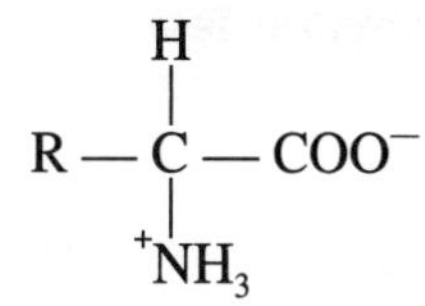

Zwitterion

Amino acids in zwitterionic form have many physical properties that are also associated with ionic salts. For example, zwitterionic amino acids are colourless, nonvolatile, crystalline solids with melting points above 200°C, usually melting with decomposition. They are soluble in water but not in nonpolar organic solvents such as cyclohexane. Compared to organic amines and carboxylic acids of similar molecular weight, amino acids have much lower melting points and are highly soluble in polar organic solvents, but only slightly soluble in water. The amino and carboxylic acid groups of constituent amino acids, as well as the nature of various side-chains, allow proteins to possess some of these same properties. However, there are many other factors that must be considered when discussing protein solubility.

The solubility of amino acids and proteins is largely dependent on the solution pH. The structural changes in an amino acid or protein that take place at different pH values alter the relative solubility of the molecule. In acidic solutions, both amino and carboxylic groups are protonated. In basic solutions, both groups are unprotonated. Following figure shows an amino acid in acidic, neutral, and basic solutions.

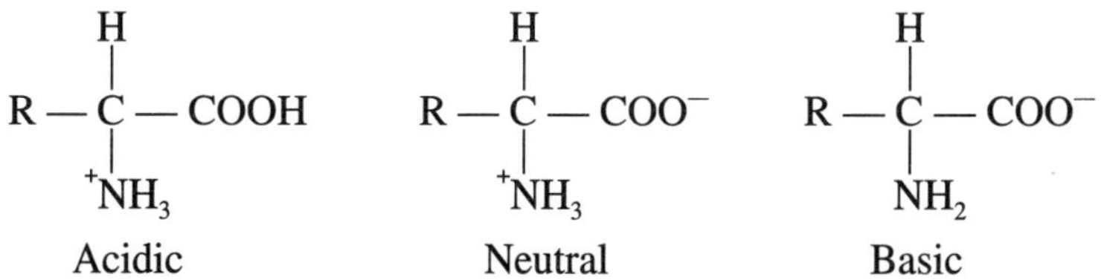

Acidic Neutral Basic

The pH value at which the concentrations of anionic and cationic groups are equal is the isoelectric point for that amino acid or protein. Amino acids and proteins are least soluble at their isoelectric points. Most of the proteins found in human tissues and fluids have isoelectric points below pH 7.0 (below human body pH) and, therefore, exist mostly in their anionic forms.

II. Chemical Reactions of Amino Acid and Protein Functional Groups

Certain functional groups in amino acids and proteins can react to produce characteristically coloured products. The colour intensity of the product formed by a particular group varies among proteins in proportion to the number of reacting functional, or free groups present and their accessibility to the reagent.

Now we will discuss various colour-producing reagents (dyes) to qualitatively detect the presence of certain functional groups in amino acids and proteins.

Ninhydrin Test

Amino acids contain a free amino group and a free carboxylic acid group that react together with ninhydrin to produce a coloured product. When an amino group is attached to the first, or alpha, carbon on the amino acid's carbon chain, the amino group's nitrogen atom is part of a blue-purple product, as shown in Equation 2. Proteins also contain free amino groups on the alpha carbon and can react with ninhydrin to produce a blue-purple product. Amino acids that have secondary amino group attachments also react with ninhydrin. However, when the amino group is secondary, the condensation product is yellow. For example, the amino acid proline, which contains a secondary amino group, reacts with ninhydrin, as shown in Equation 3. Blue-purple and yellow reaction products positively identify free amino groups on amino acids and proteins.

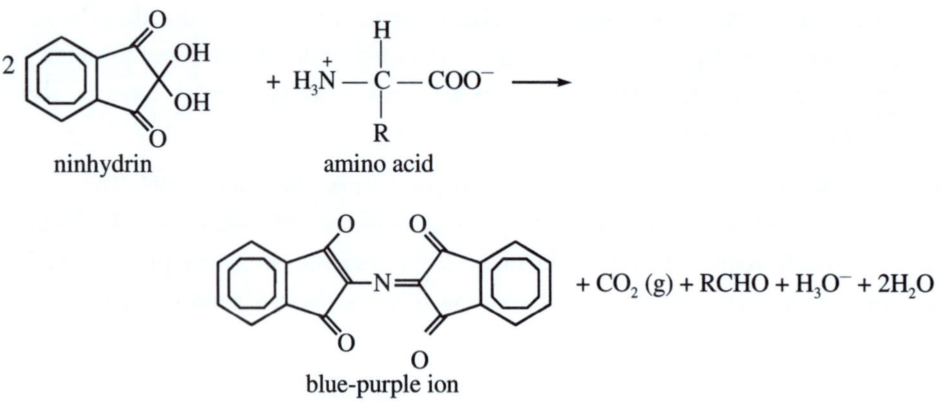

(Eq. 2)

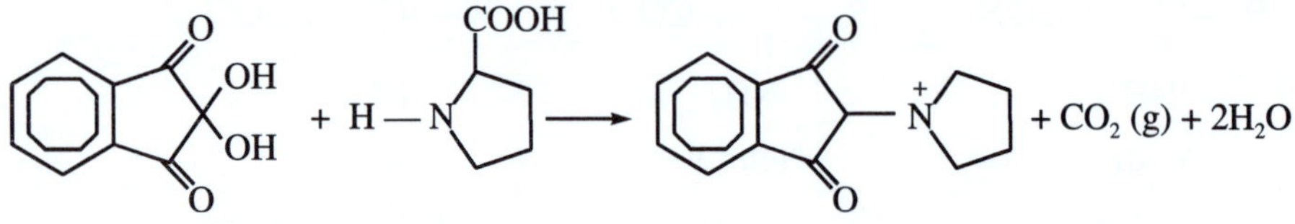

(Eq. 3

ninhydrin proline yellow product

Biuret Test

The biuret test for proteins positively identifies the presence of proteins in solution with a deep violet colour. Biuret, $H_2NCONHCONH_2$, reacts with copper (II) ions in a basic solution to form a deep violet complex. The peptide linkages in proteins resemble those in biuret and also form deep violet complexes with basic copper (II) ions in solution. The general or biuret complex formed between the protein linkages and the copper (II) ion of the biuret test is shown in following figure.

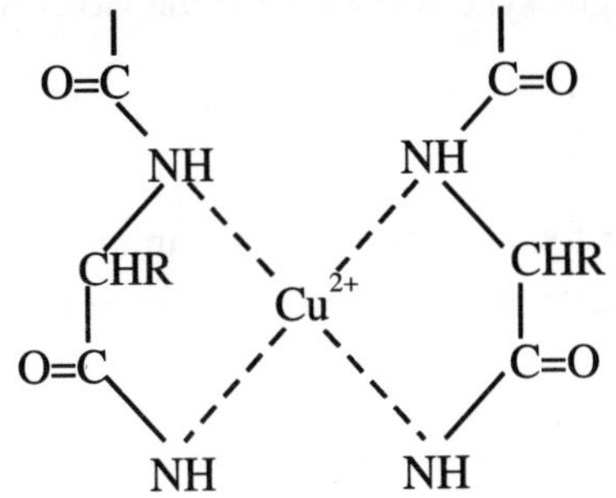

Protein-copper(II) ion complex, also called the biuret complex

Xanthoproteic Test

Some amino acids contain aromatic groups that are derivatives of benzene. These aromatic groups can undergo reactions that are characteristic of benzene and benzene derivatives. One such reaction is the nitration of a benzene ring with nitric acid. The amino acids tyrosine and tryptophan contain activated benzene rings and readily undergo nitration. The amino acid phenylalanine also contains a benzene ring, but the ring is not activated and, therefore, does not readily undergo nitration. This nitration reaction, when used to identify the presence of an activated benzene ring, is commonly known as the xanthoproteic test, because the product is yellow. Xanthoproteic comes from the Greek word xanthos, which means yellow. The intensity of the yellow colour deepens when the reaction occurs in basic solution. This reaction is one of the reactions that occur if you spill a concentrated solution of nitric acid onto your skin. The proteins in skin contain tyrosine and tryptophan, which become nitrated and turn yellow.

Millon's Test

Millon's test is a test specific for tyrosine, the only amino acid containing a phenol group, a hydroxyl group attached to a benzene ring. In Millon's test, the phenol group of tyrosine is first nitrated by nitric acid in the test solution. Then the nitrated tyrosine complexes mercury (I) and mercury (II) ions in the solution to form a red precipitate or a red solution, both positive results. Proteins that contain tyrosine will, therefore, yield a positive result. However, some proteins containing tyrosine initially forms a white precipitate that turns red when heated, while others form a red solution immediately. Both results are considered positive. Note that any compound with a phenol group will yield a positive test, so one should be certain that the sample that is to be tested does not contain any phenols other than those present in tyrosine.

Hopkins-Cole Test

The Hopkins-Cole test is specific for tryptophan, the only amino acid containing an indole group. The indole ring reacts with glyoxylic acid in the presence of a strong acid to form a violet cyclic product. The Hopkins-Cole reagent only reacts with proteins containing tryptophan. The protein solution is hydrolyzed by the concentrated sulphuric acid at the solution interface. Once the tryptophan is free, it reacts with the glyoxylic acid to form the violet product.

Nitroprusside Test

The nitroprusside test is specific for cysteine, the only amino acid containing a sulfhydryl group (—SH). The group reacts with nitroprusside in alkaline solution to yield a red complex.

18.4 TIPS TO REMEMBER

Chemical alert for carbohydrate tests

- Molisch reagent—toxic and corrosive
- Iodine-potassium iodide reagent—highly toxic, corrosive, and irritant
- Barfoed's reagent—corrosive and irritant
- Bial's reagent—toxic and corrosive
- Benedict's reagent—toxic and irritant
- Seliwanoff's reagent—toxic and corrosive
- Concentrated sulphuric acid—toxic and corrosive

- If with Molisch test there is no colour change at the interface between the two layers, the test result is negative. Do not continue the tests for the samples that test negative for the Molisch test.

- In Barfoed's test if a red precipitate before 5 min have elapsed; indicates a positive result for a reducing monosaccharide. If a red precipitate appears at about 10 min, the sample is a reducing disaccharide.

Chemical alert for lipid tests

- acetic anhydride—corrosive and lachrymator
- ammonium molybdate—ammonia solution-toxic,
- 5% bromine—dichloromethane solution-highly
- cyclohexane—flammable and irritant
- dichloromethane—toxic and irritant
- glycerol—irritant
- 6(M) hydrochloric acid—highly toxic and corrosive

- methanol—highly toxic and flammable
- 6(M) nitric acid—highly toxic, corrosive, and oxidant
- 6(M) sodium hydroxide—toxic and corrosive
- concentrated sulphuric acid—highly toxic, corrosive, and oxidant
- toluene—flammable and toxic

- Because of the sensitivity of the ammonium molybdate test in detecting the presence of phosphate, the soap or detergent solution used for washing glassware must not contain phosphate.

Chemical alert for protein tests

- concentrated nitric acid—toxic, corrosive, and strong oxidant
- concentrated sulphuric acid—toxic and corrosive
- 0.1(M) copper(II) sulphate—toxic, skin and respiratory irritant
- 3(M) hydrochloric acid—toxic and corrosive
- Hopkins-Cole reagent contains glyoxylic acid—toxic
- Millon's reagent contains mercury(II) nitrate in concentrated nitric acid—highly toxic, corrosive, and strong oxidant
- 0.5% ninhydrin-ethanol reagent—toxic and irritant
- 3(M) sodium hydroxide—toxic and corrosive
- 2% sodium nitroprusside—highly toxic
- In Hopkins-Cole tests if a purple ring does not form at the interface of the two layers in each test tube, gently tap the side of the test tube once to slightly mix the two layers. If a purple ring still does not form, consider it a negative result.

18.5 VIVA VOCE

Q.1. Show the chemical reaction that takes place when Benedict's reagent (Cu^{2+}) reacts with glucose (use open chain Fisher projection).

Q.2. Do you expect fructose to give a positive reaction with Barfoed's reagent? Explain why or why not.

Q.3. Indicate which of the test carbohydrates (glucose, fructose, xylose, lactose, sucrose, starch) would give furfural when heated with hydrochloric acid.

Q.4. Which of the test carbohydrates would give hydroxymethylfurfural when heated with hydrochloric acid?

Q.5. Describe the reaction that takes place when starch is hydrolyzed by an enzyme.

Q.6. What is the product of starch hydrolysis? What is the name of the enzyme in saliva that would hydrolyze starch?

Q.7. Explain what happened when you added lecithin to the mixture of oil and water. Identify the structural features of lecithin molecules responsible for this result.

Q.8. What is saponification number?

Q.9. What is iodine number?

Q.10. What is the normal value of VLDL and LDL in body?

Q.11. What are the diseases associated with higher level of cholesterol?

Q.12. What are free fatty acids?

Q.13. How can you quantitate the amount of cholesterol in any sample?

Q.14. How can you distinguish between reducing and non-reducing sugars?

Q.15. What is the identifying test for hard water and soft water?

Q.16. What is the normal sugar level in blood and what should be its level in various pathological states?

Q.17. What is β-oxidation?

Q.18. What do you understand by saturated and unsaturated fatty acids?

Q.19. Explain how pH of a solution affects solubility?

Q.20. Define an amino acid.

Q.21. If amino acid is detected what other tests should be done for it?

Q.22. Why biuret test is done and what is the violet colour produced during it?

Q.23. What is the principle of sulphur test for sulphur containing amino acid?

Q.24. Name the test by which presence of phosphorus can be detected in amino acid.

Q.25. Does protein get precipitate? If yes, then what is the principle for this precipitation?

Q.26. If you spill nitric acid on your skin you will observe yellow colour, why this yellow colour is formed and what is the name of this reaction?

Q.27. What is the test for the detection of phenol containing amino acids?

Q.28. What is the class to which most of the naturally occurring amino acid in body belongs?

Q.29. What are essential amino acids?

Q.30. Name two most abundant amino acid in human body.

List of abbreviation

AMP	adenosine 5′-monophosphate
ADP	adenosine 5′-diphosphate
ATP	adenosine 5′-triphosphate
bp	base-pairs
BDD	Backdown Detergent Disinfectant
cAMP	cyclic AMP
CCD	charge couple device
CHAPS	3-[(3-chloroamidopropyl)dimethylamino]-1-propanesulphonic acid
c.p.m.	counts per minute
CTP	cytidine triphosphate
DDT	2, 2-bis-(*p*-chlorophenyl)-1, 1, 1-trichloroethane
DMSO	dimethylsulphoxide
DNA	deoxyribonucleic acid
e^-	electron
EDTA	ethylenediaminotetra-acetate
FITC	Fluorescein isothiocyanate
FAD	flavin adenine dinucleotide (oxidised)
$FADH_2$	flavin adenine dinucleotide (reduced)
FMN	flavin mononucleotide (oxidised)
$FMNH_2$	flavin mononucleotide (reduced)
GTP	guanosine triphosphate
GFP	green fluorescent protein
HAT	hypoxanthine, aminopterin, thymidine medium
Hepes	4 (2-hydroxyethyl)-1-piperazine-ethanesulphonic acid
IPTG	isopropyl-β-thiogalactoside
IPR	intellectual property rights
kb	kilobase-pairs

kDA	kilo Dalton
Mr	relative molecular mass
min	minute
NAD^+	nicotinamide adenine dinucleotide (oxidised)
NADH	nicotinamide adenine dinucleotide (reduced)
$NADP^+$	nicotinamide adenine dinucleotide phosphate (oxidised)
NADPH	nicotinamide adenine dinucleotide phosphate (reduced)
NMR	Nuclear magnetic resonance
OCS	open canalicular system
PEG	Polyethylene glycol
PE	phycoerythrin
Pipes	1, 4-piperazinebis(ethanesulphonic acid)
Pi	inorganic phosphate
PMT	photomultiplier tube
p.p.m.	parts per million
p.p.b.	parts per billion
PPi	inorganic pyrophosphate
PFGE	pulsed-field gel electrophoresis
PVDF	polyvinyledene difluoride
RNA	ribonucleic acid
ROS	reactive oxygen species
r.p.m.	revolutions per minute
SDS	sodium dodecyl sulphate
SOP	standard operation procedures
Tris	2-amino-2-hydroxymethylpropane-1, 3-diol
TRITC	tetramethylrhodamine isothiocyanate
TEMED	N, N, N′, N′-tetramethylethylene diamine
UV	ultraviolet
2-D	two dimensional
3-D	three dimensional